THE PAPER MUSEUM OF CASSIANO DAL POZZO

SERIES B ~ PART ONE

CITRUS FRUIT

THE PAPER MUSEUM OF CASSIANO DAL POZZO

A CATALOGUE RAISONNÉ

Drawings and Prints
in the Royal Library at Windsor Castle,
the British Museum, the Institut de France
and other Collections

General Editors
Francis Haskell and Jennifer Montagu

Assistant General Editor
Henrietta McBurney

Series Editors
Series A ~ Antiquities and Architecture
Amanda Claridge
Series B ~ Natural History
David Freedberg

Patrons and Sponsors
Accademia dei Lincei
Académie des Inscriptions et Belles-Lettres
British Academy
Getty Grant Program
Monument Trust
Ing. C. Olivetti & C
Royal Collection Trust

SERIES B ~ NATURAL HISTORY

PART ONE

CITRUS FRUIT

by David Freedberg
and Enrico Baldini

with contributions by
Giovanni Continella, Eugenio Tribulato
and Eileen Kinghan

HARVEY MILLER PUBLISHERS

HARVEY MILLER PUBLISHERS
Knightsbridge House, 197 Knightsbridge, London SW7 1RB
An Imprint of G+B Arts International

British Library Cataloguing in Publication Data
A catalogue record for this book is available
from the British Library

ISBN 1872501885

This work is published with the assistance of
The Getty Grant Program

The production of this volume has been
supported by Hugh and Catherine Stevenson

Monochrome and colour origination by Neue Schwitter AG, Basel, Switzerland
Printed by Clifford Press Ltd., Coventry
Manufactured in Great Britain

CONTENTS

THE CATALOGUE RAISONNÉ

SERIES A
ANTIQUITIES & ARCHITECTURE
Series Editor: Amanda Claridge

I
Ancient Mosaics and Wallpaintings

II
Early Christian and Medieval Antiquities
Two Volumes

III
Sarcophagi and other Reliefs
Three Volumes

IV
Statues and Busts

V
Classical and later Metalwork, Glass and Terracotta
Three Volumes

VI
Classical Manuscript Illustrations

VII
Inscriptions

VIII
Tombs of Illustrious Italians at Rome

IX
Roman Topography and Architecture
Three Volumes

X
Renaissance and Later Architecture

SERIES B
NATURAL HISTORY
Series Editor: David Freedberg

I
Citrus Fruit

II
Fungi: 'Icones Fungorum'
Four Volumes

III
Geological Collections
Two Volumes

IV
Birds
Two Volumes

V
Animals, Fish and other Fauna

VI
The 'Erbario Miniato' and other Flora
Two Volumes

VII
Flora: 'Syntaxis Plantaria'
Five Volumes

VIII
The Mexican Herbal

AUTHORS' ACKNOWLEDGEMENTS

The authors would like to acknowledge the kind assistance of the following institutions:
Istituto Sperimentale di Agrumicoltura, Acireale; Dipartimento di Colture Arboree, Università di Bologna; Biblioteca Universitaria di Bologna; Istituto di Coltivazioni Arboree, Università di Catania; Rare Book Library, Columbia University; Biblioteca Nazionale Centrale, Florence; Dipartimento di Botanica, Università di Firenze; Sovrintendenza per i Beni Architettonici e Ambientali, Florence; Archivio Gallesio, Genoa; Consiglio Nazionale delle Ricerche, Rome; Rare Book Collection of Dumbarton Oaks, Washington D.C.; Royal Library, Windsor.
The authors are particularly grateful to Giovanni Continella and Eugenio Tribulato from the University of Catania for their taxonomic advice. Thanks are also extended to: Luca Corelli Grappadelli, Bologna; Frederick S. Davies, Florida; Fabio Garbari, Pisa; Eileen Kinghan and Henrietta McBurney, Windsor; Anna Nicolò, Rome; Jane Roberts, Windsor; Francesco Solinas, Rome; Lucia Tongiorgi Tomasi, Pisa.

David Freedberg, New York
Enrico Baldini, Bologna

March 1997

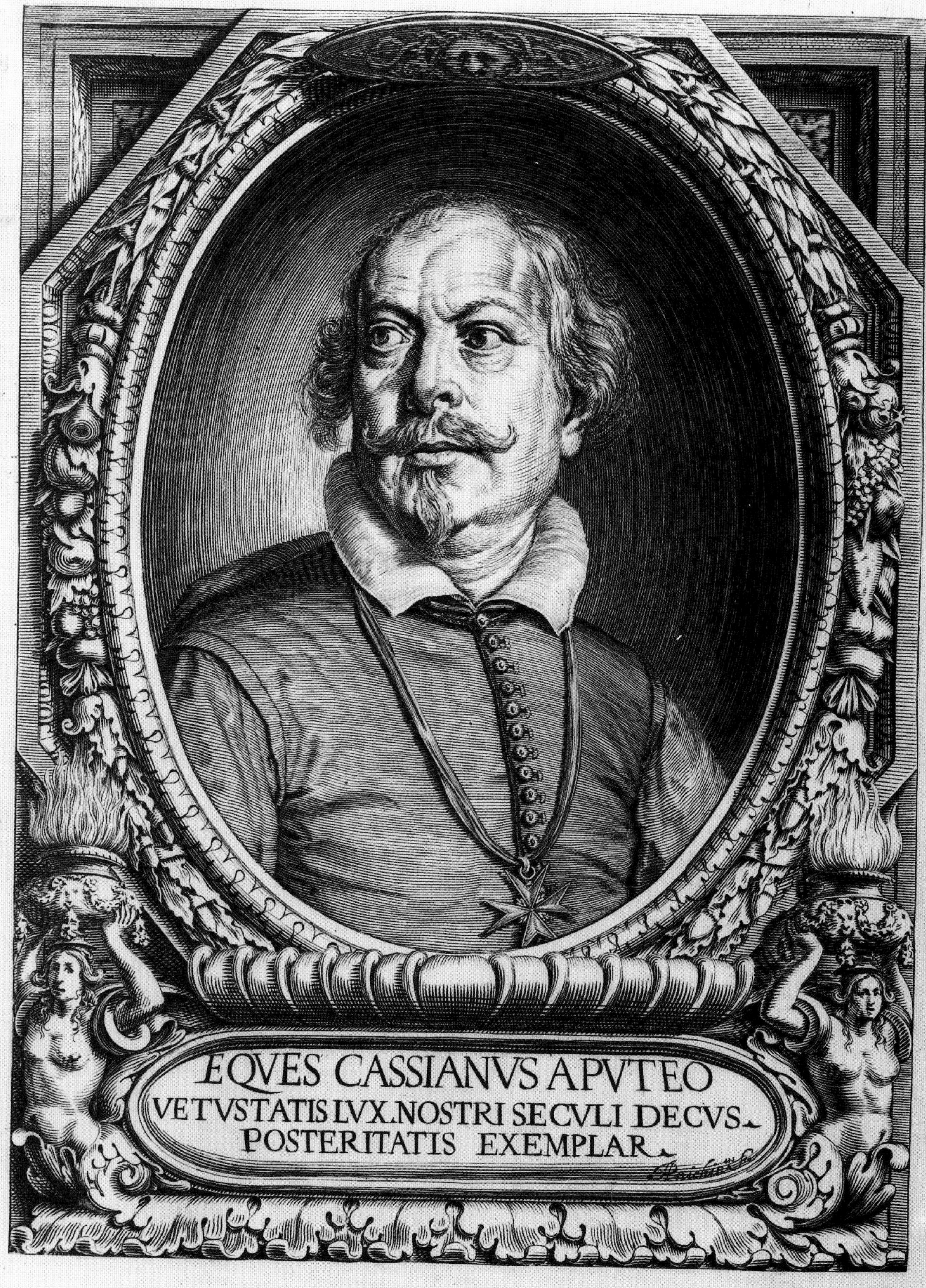

Fig. 1. Pietro Anichini, *Portrait of Cassiano dal Pozzo*, engraving, frontispiece to Dati's funerary oration, 1664

GENERAL INTRODUCTION TO THE PAPER MUSEUM OF CASSIANO DAL POZZO

by Francis Haskell and Henrietta McBurney

CASSIANO DAL POZZO is celebrated today as one of the most important art patrons in seventeenth-century Italy [Fig. 1]. He commissioned more than forty paintings by Nicolas Poussin and owned many others by some of the finest artists of the day. He was also a friend of Galileo and was closely concerned with developments in the political and intellectual life of his time. However, the principal scholars, antiquaries, scientists and collectors in seventeenth-century Europe admired Cassiano above all for his learning in the fields of antiquities and natural history[1] and for the extraordinarily ambitious project which grew out of these interests—a project he described as his *Museo Cartaceo* or 'Paper Museum'. This 'museum' was to consist of drawings and prints of not only as many of the material remains of antiquity as were then known, but also of geological specimens and of living plants and animals from all over the world, including the Americas.[2] Although Cassiano wrote little, if anything, for publication under his own name, his Paper Museum was accessible to artists, scientists and antiquaries, and it is possible that at one stage he hoped to publish at least parts of it.[3] This, however, was not to be, and it was only in 1719, with the appearance of the first five volumes of Montfaucon's *L'Antiquité Expliquée et Représentée en Figures*, that a project of the kind that Cassiano had probably envisaged was realized.[4] Unlike antiquaries of this later period, however, Cassiano had been as deeply interested in natural history as in the ancient world. The volumes of this series will therefore provide a unique indication of the universal interests of one of the most cultivated minds of his day. The whole is definitely more important than the parts, but certain groups and indeed individual drawings are also of great significance for historians of art, archaeology and the sciences.

Cassiano dal Pozzo was born in Turin in February 1588,[5] into an ancient family which had long held prominent positions in the court of the dukes of Savoy. In 1598 Cassiano left Piedmont with his father and spent a short period in Bologna[6] before being put under the charge of his father's cousin, Carlo Antonio dal Pozzo (1547–1607), who was Archbishop of Pisa and an intimate advisor to Ferdinand I, Grand Duke of Tuscany (*reg.* 1587–1609). It was in Pisa that Cassiano received his university education (he graduated in civil and ecclesiastical law in 1607), and it was his

Tuscan connections that were to play a fundamental role in the remainder of his career. It was in Tuscany too that his interests in the worlds of art and science were nurtured—partly through his cousin's connections at the Medici court, where Cassiano would have come into contact with artists working for Ferdinand I, and partly in the intellectual circles centred in the University of Pisa and its Botanic Gardens, where, for example, he met Galileo Galilei (1564–1642). Through his cousin, Cassiano was admitted into the local nobility [Fig. 2]—he was elected to the Order of the Knights of St Stephen from a *commenda* the Archbishop had founded in the order—and in 1608 he was appointed a judge in Siena, a post he held until August 1611. He then, it seems, decided to give up his legal career in order to pursue his intellectual interests. He left Tuscany that year, and in 1612 arrived in Rome, the city that became his home for the rest of his life.

Even before he moved to Rome, Cassiano had been in contact with and worked for the representative there of the Grand Duke of Tuscany, Cardinal Francesco Maria del Monte (1549–1626), an exceptionally influential and cultivated patron of young talent of every kind. It was probably under del Monte's auspices that he began to take part in the aristocratic and intellectual life of the city, though preferment was slow. In 1622 he was elected to the first scientific society in Europe, the *Accademia dei Lincei*; his contact with, and work for, its founder Prince Federico Cesi (1585–1630) and other members of the society were to have a formative influence on Cassiano's own approach to the study of the natural world. In 1623, Cardinal Maffeo Barberini, a major figure in the cultural circles in which Cassiano had been moving, was elected Pope Urban VIII (1623–44), and shortly thereafter Cassiano was appointed to the Household of Maffeo's nephew, Cardinal Francesco Barberini (1597–1679).[7] This was the turning point of his life. Although Cassiano was never to be wealthy by the standards of his patrons, this and other official positions provided him with influence and a comfortable income, and gave him opportunity and time to immerse himself in the cultural pursuits that he shared with Francesco Barberini in particular.

In 1625 Cassiano accompanied Cardinal Barberini on a diplomatic mission to Paris and in the following year to Madrid. His careful accounts of the people he met and the places he visited provide illuminating information about his artistic and scientific interests. His enthusiasm for the paintings by Leonardo that he saw at Fontainebleau, and for the drawings of Mexican plants and animals at the Escorial, were to inspire later projects.[8] It was in France that he developed his friendship with Nicolas-Claude Fabri de Peiresc (1580–1637) [Fig. 3], a lawyer renowned throughout Europe for his learning in antiquarian and scientific fields. This friendship was conducted mainly by letter; it was to put the seal on Cassiano's strongly francophile sympathies.

On his return to Rome, Cassiano and his younger brother Carlo Antonio (1604–89) moved into a small *palazzo* in the Via dei Chiavari, where they began to build up a collection of paintings, books, medals and drawings. Cassiano's patronage of

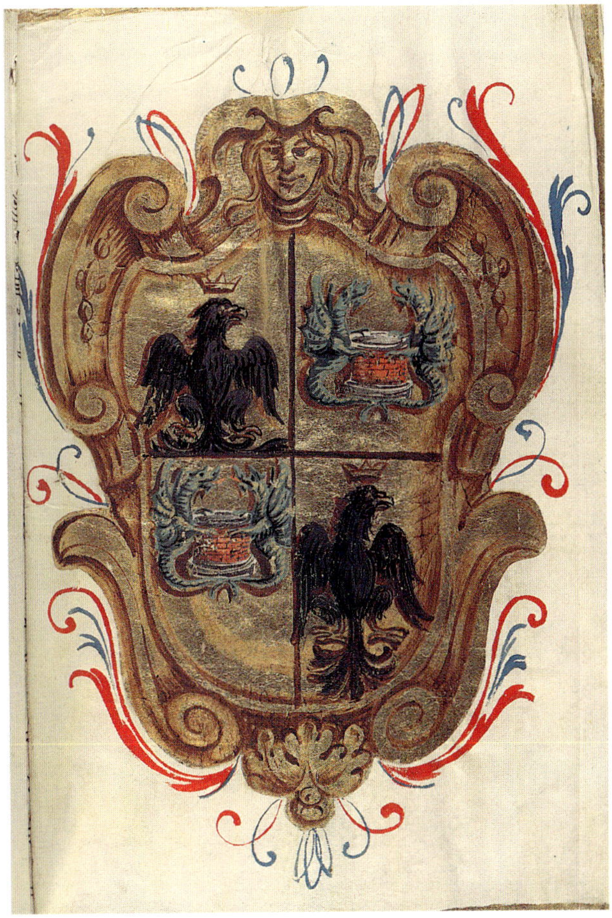

Fig. 2. The dal Pozzo coat of arms, *Ordine dei Cavalieri di Santo Stefano, Provanze di Nobilità*, S 733, no. 9. Pisa, Archivio di Stato

Poussin began in 1626 and culminated in his commissioning of the *Seven Sacraments* in the second half of the 1630s, the series being completed in 1642. The brothers' collection of Poussin's paintings was eventually to occupy the main room of their residence — designated the *Stanza de' Sagramenti*.[9] Carlo Antonio was sixteen years younger than his brother [Fig. 4]. He married Teodora Costa in 1627 by whom he had thirteen children, pursued a successful political career, and was at the same time as passionate a collector as the bachelor Cassiano.[10]

Cassiano's membership of the *Lincei* was of great significance. The scientific investigations of this academy, founded in 1603 by Federico Cesi [Fig. 5], did not always meet with approval in the ostensibly conformist society of seventeenth-century Rome. Despite his position at Court, Cassiano remained loyal to his fellow *Linceo*, Galileo. The question of Cassiano's religious orthodoxy has aroused much controversy in recent years. He was certainly in close touch with French *libertins*, who were contemptuous of prevailing Catholic practices, and also, for example, with Tommaso Campanella (1568–1639) who, although tolerated by the Barberini, held opinions that can only be described as heretical.[11] Less problematic is the influence on Cassiano of Cesi's insistence on the necessity of direct observation and experiment for the study of the natural world. After Cesi's early death in 1630,

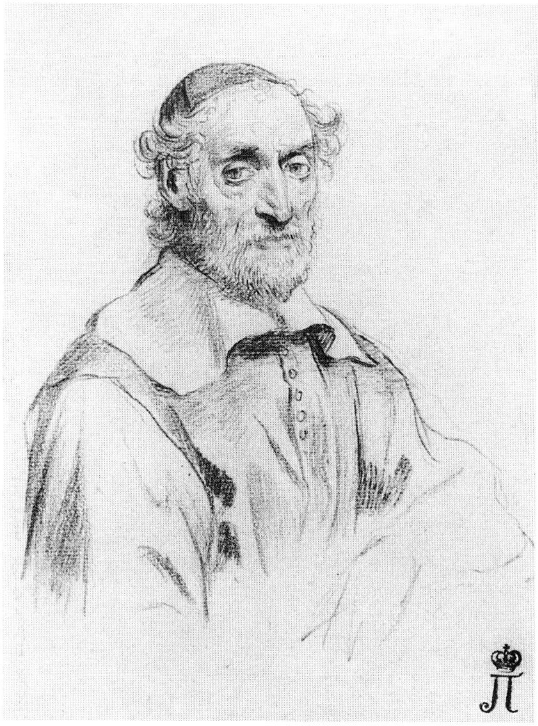

Fig. 3. Claude Mellan,
Portrait of Nicolas-Claude Fabri de Peiresc.
St Petersburg, Hermitage Museum

Cassiano purchased both his scientific instruments and his drawings and manuscripts. The drawings of natural phenomena which Cassiano continued to commission from a number of artists are strikingly accurate by the standards of the day, and set out to record every aspect of the natural world, not exclusively the rare and bizarre items characteristic of fashionable cabinets [Fig. 6].

With the death of Pope Urban VIII in 1644 and the temporary exile of his nephews, Cassiano lost his official positions and the income that went with them. However, some of his posts were restored to him on Francesco Barberini's return to Rome in 1648 and seven years later, with the election of his long-standing friend Fabio Chigi as Pope Alexander VII (1655–67), he regained briefly his favoured position in papal circles before his death in 1657.

Carlo Antonio inherited and continued to add extensively to his brother's collections. No inventory was drawn up at Cassiano's death, so there is no record of the number of volumes of drawings and prints that he had assembled during his lifetime, nor of their contents. Contemporary accounts are limited to references to the drawings after the Antique.[12] There is also the funerary oration published in 1664 by Carlo Dati (1619–76), fellow member of the Accademia della Crusca, which includes a synopsis summarizing the contents of 23 volumes of drawings of Roman antiquities [Fig. 7].[13] However, in a letter two years later, Carlo Antonio wrote that:

> the collection left [by my brother] does not only contain the 23 volumes of drawings after the Antique described in his oration by the good Carlo Dati, but there are still other volumes, and I am continuing to add still more when I find artists talented at drawing the remains of antiquity.[14]

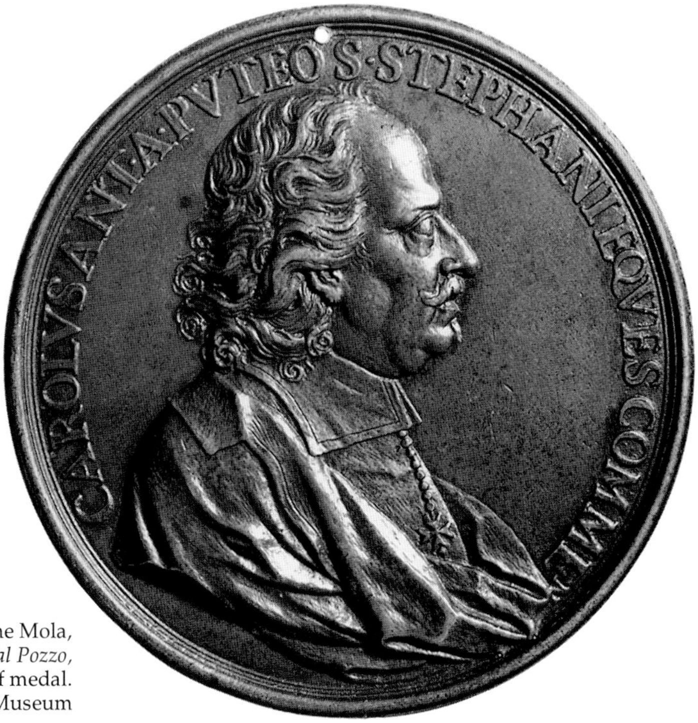

Fig. 4. Gasparo Morone Mola, *Portrait of Carlo Antonio dal Pozzo*, obverse of medal. Cambridge, Fitzwilliam Museum

Further description of the contents of five volumes of drawings of antiquities is given by Filippo Baldinucci (1624–92) who was shown the dal Pozzo collection by Carlo Antonio some time after Cassiano's death.[15]

On the death of Carlo Antonio in 1689, the Paper Museum together with the rest of the art collection and library passed to his son Gabriele dal Pozzo (1642–95). At this stage an inventory was made of the paintings, but unfortunately it provides very little information about the drawings.[16] Neither Gabriele, nor his son Cosimo Antonio dal Pozzo (1684–1740), to whom the collection passed at Gabriele's death, appears to have inherited the scholarly interests of Carlo Antonio and Cassiano, and it is unlikely that either of them added to the Paper Museum.[17] The inventory made on Gabriele's death is substantially the same as the earlier one and throws very little light on the drawings collection.[18]

It was during Cosimo Antonio's custodianship that the family collections first began to be dispersed, mainly to pay off the large debts that he had incurred. The library together with the volumes of the Paper Museum were the first items to go: in 1703 they were sold for a sum of 4,000 scudi to the Pope Clement XI Albani (1700–21), from whom, in 1714, they passed to his nephew, Cardinal Alessandro Albani (1692–1779) for the same sum. The Paper Museum thus joined the important collection of drawings and prints belonging to the Albani family. The library (including the printed books and manuscripts) remained with the Albani family until the French invasion and occupation of Rome in 1798 when it was requisitioned by the French and partly broken up. A substantial part was later lost in a shipwreck off the coast of Gibraltar in 1863.[19] The paintings, however, remained in the possession of the dal Pozzo family until the early part of the eighteenth century,

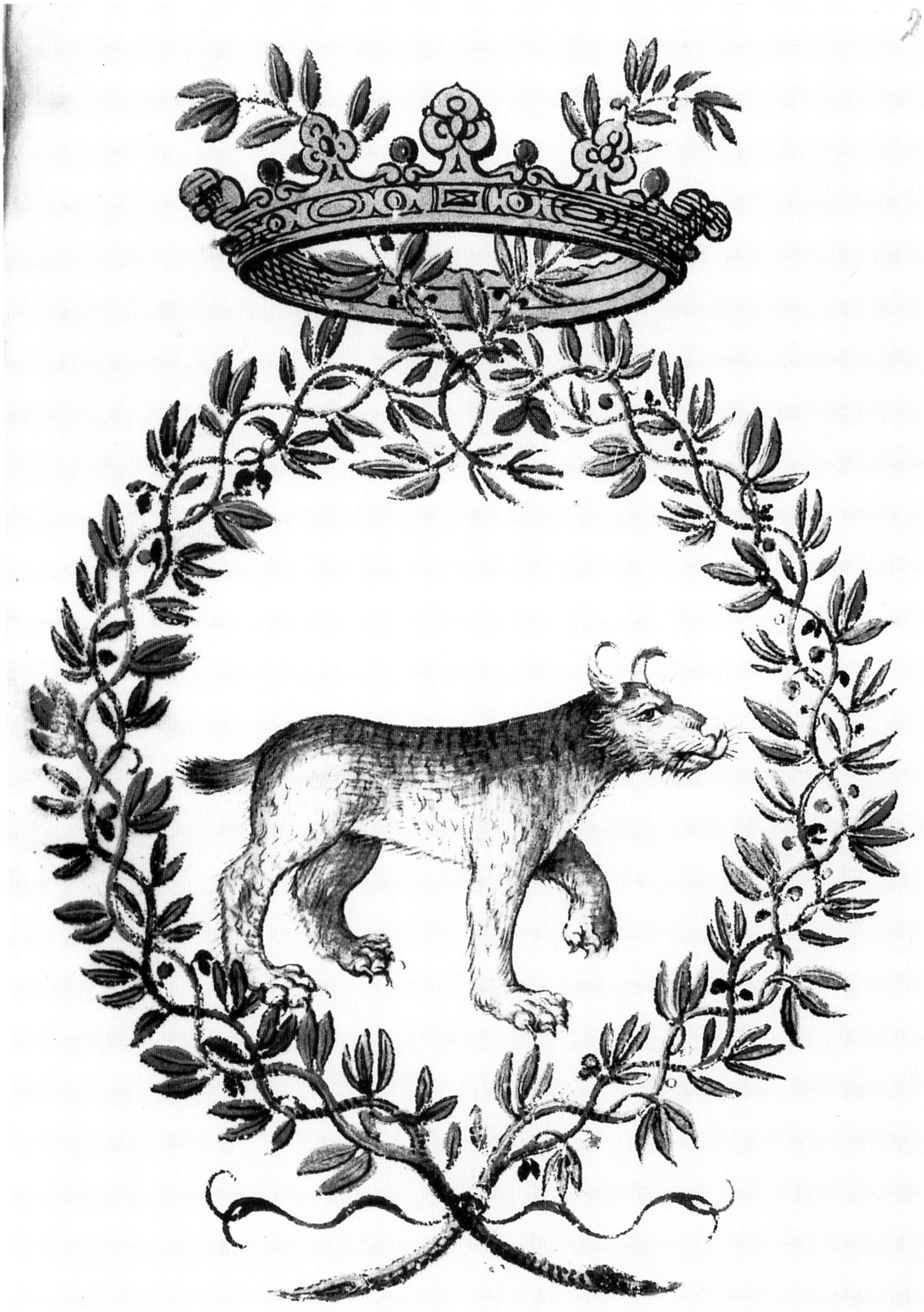

Fig. 5. Emblem of the Accademia dei Lincei. Rome, Biblioteca dell' Accademia Nazionale dei Lincei, Archivio Dal Pozzo

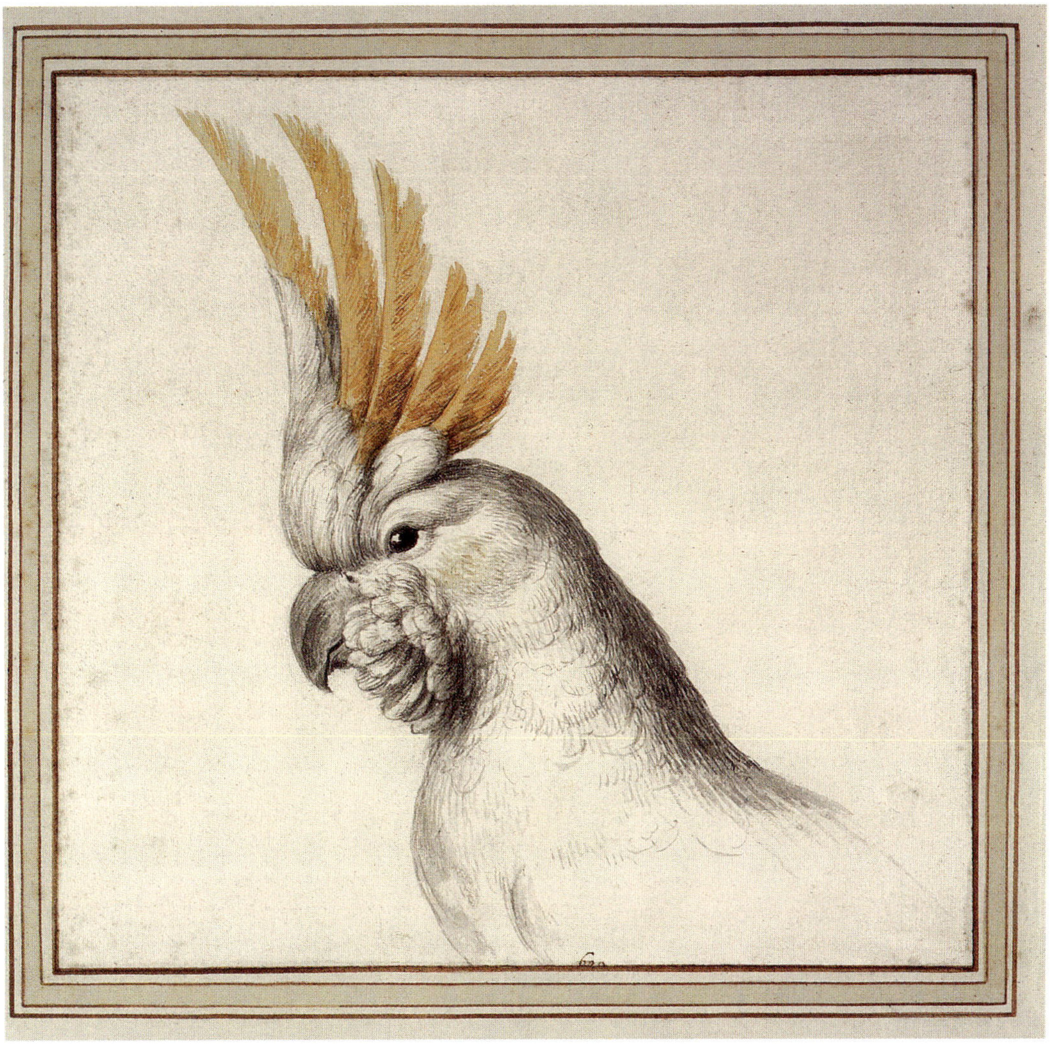

Fig. 6. Vincenzo Leonardi, *Citron-crested Cockatoo*. England, Private Collection

and were then gradually dispersed: Poussin's *Seven Sacraments* came to Britain, where five of them still remain, in the collection of the Duke of Rutland.[20]

The Paper Museum was not kept intact during its period of Albani ownership, although the precise extent of the changes it underwent is difficult to calculate since no inventory has yet been found listing the contents of Clement XI's purchase.[21] The parts of the collection that seem to have been of most interest to the Albani, notably the architectural drawings, were substantially reordered and amalgamated into other sequences of drawings in their collection. It also seems likely that Johann Winckelmann (1717–68), who became personal secretary and librarian to Cardinal Albani in 1743, was responsible for reordering the contents of some of the antiquities volumes.[22]

By the mid eighteenth century, Cardinal Albani's collection was renowned among the *cognoscenti* for being the most extensive and finest of its kind. It was

Synopsis, atque Ordo Antiquitatum Romanarum Illustriss. & Eruditiss. V. Equitis CASSIANI A PVTEO studio, ac impensis xxiii voluminibus digestarum.

- RES
  - DIVINÆ
    - DII
      - Patrij vel peregrini, seu vt Varro vocabat, Certi vel incerti
      - Maiores, Medioxumi, Minores. Siue vt Cicero
      - Cæleftes, Indigetes, & Genij. Vt Lares, Fauni, Satyri, Nymphæ, Flumina.
    - Fabulosæ Deorum actiones. Virtutes, & Vrbes Dearum habitu consecratæ.
    - Templa & aræ, earumq; formæ, & dedicatio: item obelisci, donaria, vota, & ornamenta.
    - *Sacrificia, & ritus.*
      - Publici, victimæ, pompæ, ludi sacri eorumq; apparatus
      - Priuati, nuptiarum, funerum, consecrationes, & monumenta.
    - *Sacrorum ministri.*
      - Pontifices, Flamines, augures, haruspices, vestales, popæ.
    - Instrumenta sacrorum. Litui, acerræ, simpuli, vasa varia.
  - HVMANÆ
    - PACIS
      - PVBLICÆ
        - SERIÆ
          - Magistratus, eorumq; vestitus, insignia, ornamenta Lictores, Fasces, Sellæ, &c.
          - Iudicia, Tribunalia, Subsellia
          - Manumissiones
          - Pondera, & mensuræ.
        - LVDRICÆ
          - *Theatrales, seu Scenicæ.*
            - Theatra, Scenæ
            - Apparatus scenicus, Oscilla, Mimi.
            - Instrumenta musica, tibiæ.
          - Amphitheatrales, gladiatoriæ & venationes
          - Circenses, seu Curules. Currus, aurigæ, circi, metæ.
          - Largitiones, & munera.
      - PRIVATÆ
        - Vestes variæ variorum, & insignia
        - Partes ædium, & varia supellex
        - Hortensia, & rustica
        - Opificia, & artes
        - Exercitia, & ludi priuati
        - Balnea
        - Accubitus, & triclinium
        - Serui, & ministeria.
    - BELLI
      - Castra, eorumq; partes
      - Personæ. Duces, eorumq; habitus, insignia. Tribuni, signiferi, eorumq; aquilæ milites priuati.
      - Classis, naues, earumq; genera, & partes: item Classiarij, & remiges
      - Arma, Tela, Scuta, Machinæ, Fundæ, Glandes.
      - *Actiones Militares*
        - Commeatus
        - Decursiones, & ludi castrenses
        - Allocutiones
        - Munitiones, oppugnationes.
        - Deditiones, & captiui:
        - Victoria, Triumphi, Trophæa, Coronæ, Columnæ, Arcus, eorumq; ornamenta.

Fig. 7. Synoptic Table of the Antiquities section of the *Paper Museum*. Dati 1664

housed in the Cardinal's library in his residence in Rome, and the latter, like the dal Pozzo residence more than a century earlier, was visited by artists, scholars and travellers. Among visitors from England who knew the Albani collection were the architect Robert Adam (1728–92), who studied it during his stay in Rome in 1755–7; John Stuart, 3rd Earl of Bute (1713–92), friend and adviser to the Prince of Wales, later George III (*reg.* 1760–1820); and Richard Dalton (*c.* 1715–1791), the Prince of Wales's librarian, who in 1758 was in Italy looking for works of art to add to the collections of both Bute and the future king. On 10 October 1758, Dalton wrote laconically from Bologna to Lord Bute:

> I've met with a number of very fine drawings in different collections here... One fine collection of Drawings belonging to an Old Curate must let alone, for the old fool asks such an exorbitant price that there is no making even an offer... A little time, as he is very infirm, must bring him to the grave, which if it happens in my time, shall watch the opportunity when his heirs will and must sell them.[23]

Lord Bute seems to have been in agreement that the time was not yet ripe to negotiate with Cardinal Albani. However, Robert Adam revived the idea of the purchase independently, possibly discussing it directly with George III during his audience with the king on his appointment as Royal Architect in November 1761. The king was enthusiastic, and Robert's younger brother James Adam (1732–94), in Rome at the time, took on the role of agent in the purchase negotiations. In order to assure himself and his brother's patron of the quality of the collection, James went to inspect it together with the artists Jacques-Louis Clérisseau (1722–1820) and Antonio Zucchi (1726–95) and was able to report to his brother on 8 May 1762:

> ...we were all three of the opinion that the reputation of the collection was really well founded and that its extent was immense, containing nearly, betwixt drawings and prints, 200 volumes in folio. I am far from saying, however, that these volumes are all interesting. This is not the case. In such a vast collection much rubbish must be expected. The mosaics, paintings and bas-reliefs of the primitive church or the first stages of Christianity will entertain you little, though there is some curiosity even in this subject. But ... there is a collection of drawings after antiquities, bas-reliefs, altars, tripods, cenarie and urns that I believe you will own to be most valuable.[24]

The negotiations lasted from January to May 1762; Cardinal Albani was reluctant to part with the collection, but was persuaded eventually through the intercessions of his mistress and the need of a marriage dowry for their daughter. The price paid was 14,000 *scudi* (the then equivalent of £3,500).[25]

The collection was removed to the Adams' residence in Rome, before being taken to Livorno in June 1762. Its passage to England was, however, delayed for almost a year. George III was also in the process of negotiating the purchase of a large part

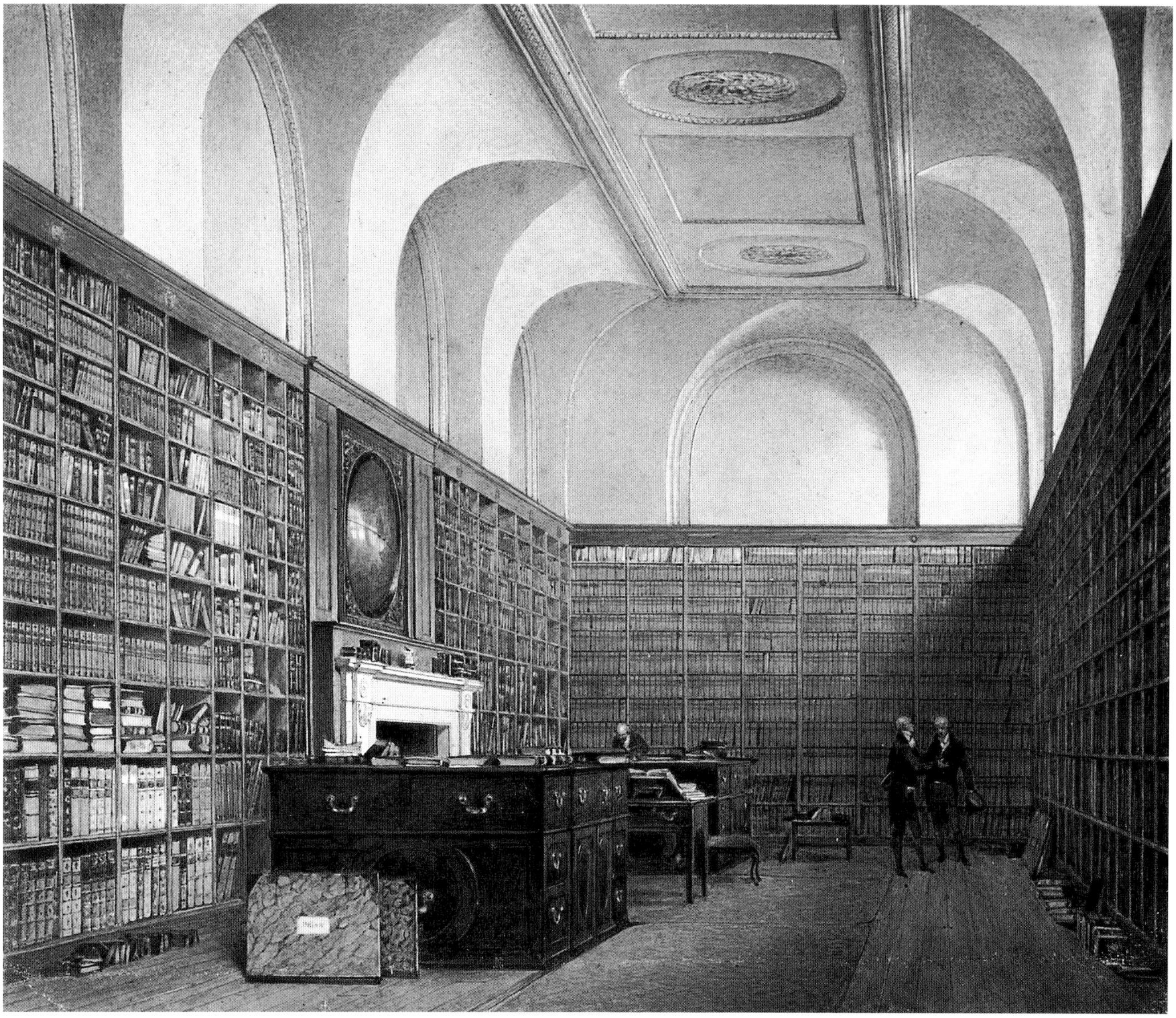

Fig. 8. James P. Stephanoff, *Interior of King George III's library at Buckingham House*, 1817. Windsor, Royal Library

Fig. 9. A volume of the *Paper Museum* open to show the 17th-century arrangement of drawings. Windsor, Royal Library

Fig. 10. Volumes of the *Paper Museum* in George III bindings. Windsor, Royal Library

Fig. 11. Volumes of the *Paper Museum* in 17th-century vellum bindings. Windsor, Royal Library

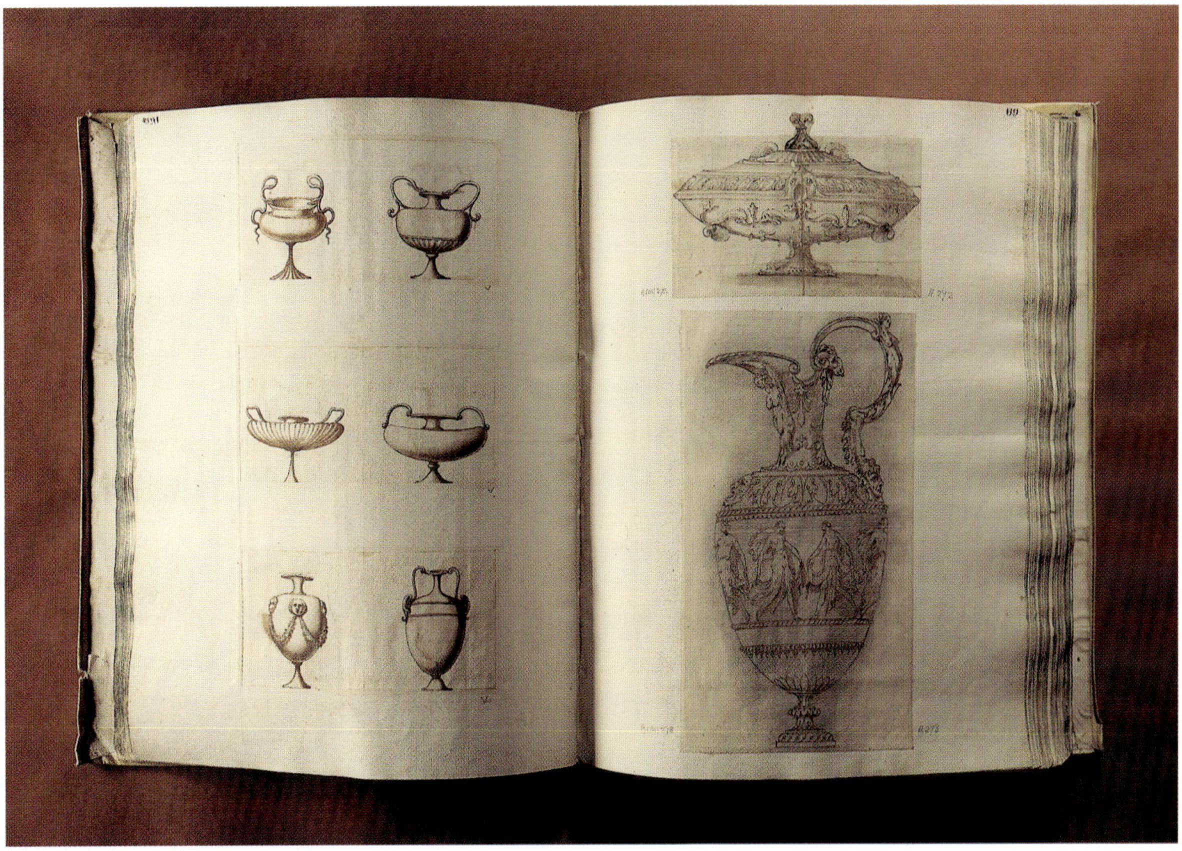

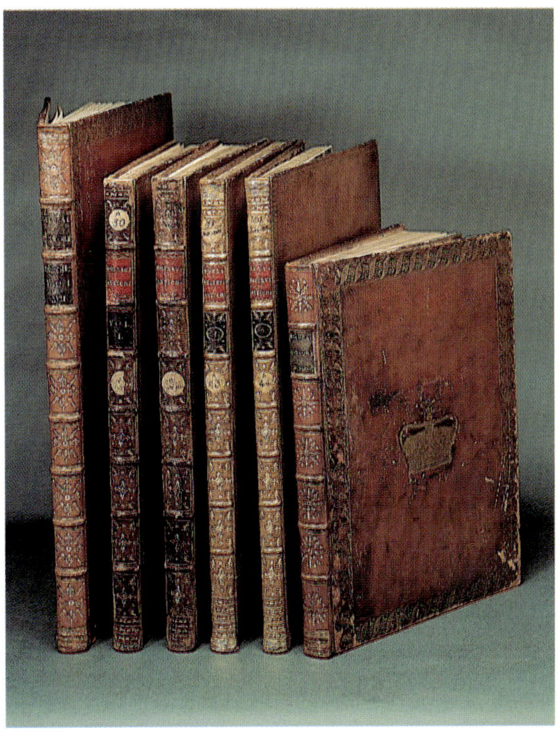

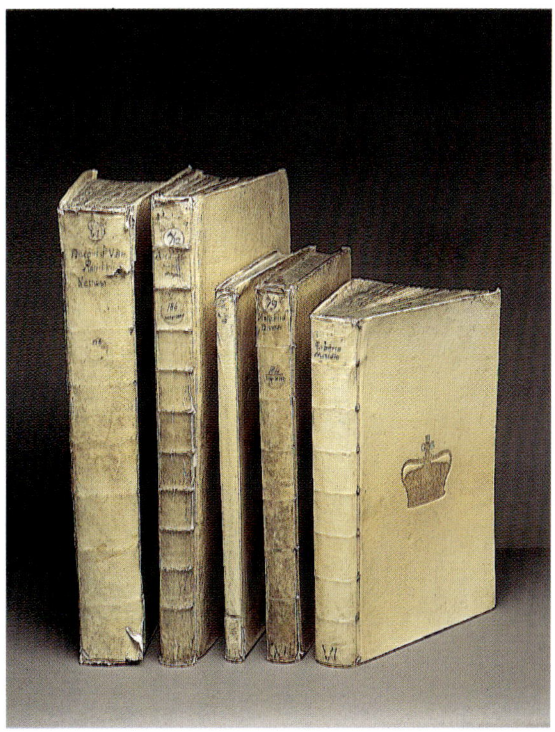

of the collection of Consul Joseph Smith (*c.* 1674–1770) in Venice, a transaction that was completed in January 1763; a few months later the two collections were shipped together from Livorno, arriving in London in July 1763.

The Albani and Smith purchases were added to the royal collections being assembled in Buckingham House, George III's newly acquired London residence [Fig. 8]. Unfortunately no inventory seems to have been made of the contents of the packing cases when they arrived in London. Thus, although it is known that the Albani collection contained 'betwixt drawings and prints, 200 volumes in folio', it is unclear how many of these were from the dal Pozzo collection. The inventory of the drawings in the Royal Collection at this period—known as *Inventory A*—dates from the end of the eighteenth century, by which time substantial sections of the Paper Museum had been split up and integrated into other parts of the Royal Collection. However, all the dal Pozzo volumes of drawings now in the Royal Library at Windsor Castle—notably nineteen volumes of drawings after the Antique, three of architectural drawings and eight of natural history subjects—as well as the volumes dispersed from the Royal Library early this century (for which see Fig. 12 and p. 31), can be identified in *Inventory A*. Although some dal Pozzo volumes were kept intact, the contents of others were reordered and rebound under George III's librarian, Richard Dalton, in the newly-established royal bindery in Buckingham House.[26] The majority of the dal Pozzo albums were rebound in large folio leather bindings, on the front and back boards of which the arms of George III were blocked in gold [Figs 10–11].[27] In some cases the drawings were lifted from their original inlay sheets, laid down onto eighteenth-century mounting sheets and given decorative wash borders [Fig. 9].[28] An unfortunate aspect of the remounting of many of these drawings was the fact that often they were severely trimmed and the numbers and inscriptions, and possibly also signatures on them, were cut away. The volumes of prints, even more than those of drawings, suffered in the process of the reorganization of George III's collection in the eighteenth century, a process that continued in the nineteenth and early twentieth centuries. Only a few albums of prints appear to have remained intact;[29] the contents of others were amalgamated into George III's military and topographical collections and others were incorporated into nineteenth-century sequences of engraved portraits;[30] the evidence that large numbers of prints once belonged to the dal Pozzo collection was removed almost totally in the early twentieth-century reorganization and mounting of much of the Royal Collection of prints.

Aside from the vicissitudes of the Paper Museum within the Royal Collection, brief mention must also be made of the fate of those parts of it that left the main corpus at intervals during the eighteenth and early nineteenth centuries. Four volumes of architectural drawings were apparently reserved by the Adam brothers for themselves while the collection was housed temporarily in their residence in Rome in 1762. These volumes were removed by them from the main body of the collection after its arrival in London, were bought by John Soane (1753–1837) at the

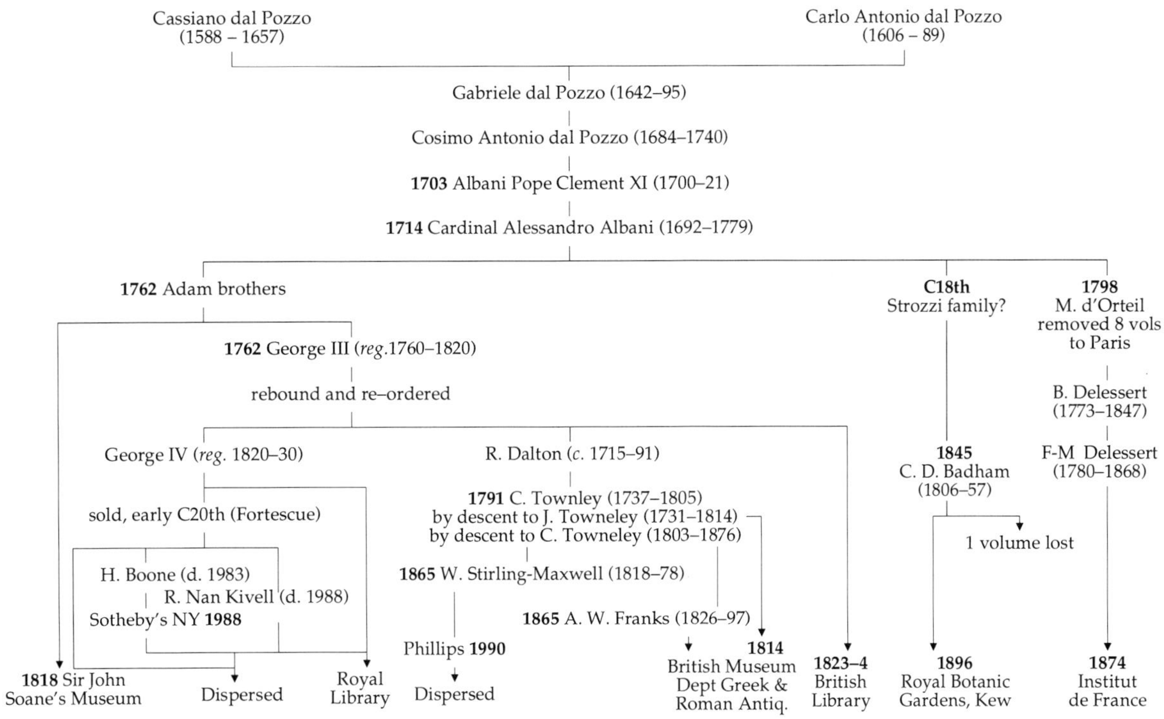

Fig. 12. Table showing the dispersal of dal Pozzo drawings

Adams' sale in 1818, and can now be found in the Soane Museum, London.[31] Several hundred drawings of antiquarian and architectural subjects appear to have been separated from the volumes by Richard Dalton, presumably at the time of the reordering and rebinding of the drawings from the late 1760s onwards. These drawings, kept by Dalton in his own collection, were auctioned after his death in 1791, and the majority found their way into the possession of the collector and antiquary Charles Townley (1737–1805). The bulk stayed with the family until they were sold by Charles's kinsman Charles Towneley (1803–76) in 1865. Some of these, purchased by the collector Augustus Wollaston Franks (1826–97), eventually passed to the British Museum, where, known as the 'Franks' drawings, they are now housed in the Department of Greek and Roman Antiquities. Another group, which the earlier Charles Townley had separated out from the main collection as being of particular interest, came to the British Museum in 1814 together with other drawings and small antiquities from Townley's collection; the dal Pozzo provenance of this group was only identified in 1987.[32] Other drawings which were bought by Sir William Stirling-Maxwell[33] at the Towneley sale in 1865 were sold by the family in 1990 and are now dispersed.[34]

In 1823, three years after George III's death, most of his library was given to the nation by his son George IV (*reg.* 1820–30) and was transferred to the British Museum. As well as his vast collection of printed books, the gift included his collection of topographical drawings and prints, into which many of the volumes of dal Pozzo prints had been amalgamated; several other dal Pozzo print albums and one volume of drawings, presumably mistaken for printed books, were also transferred to the British Museum, where they were rediscovered in 1989 in the King's Library section of the British Library.[35] But the great majority of the volumes of drawings and prints was retained in the Royal Collection; they have been housed since 1834 in the Royal Library created by William IV (*reg.* 1830–37) at Windsor Castle.

Just after the First World War, however, a notable part of the dal Pozzo collection was dispersed from the Royal Library. Fifteen volumes covering natural history subjects—all the zoological and most of the botanical and ornithological material—and one volume of drawings of antiquities, the contents of which were apparently considered 'unimportant' or 'uninteresting', were sold by the then Royal Librarian, Sir John Fortescue (1859–1933), in order to set up a fund from which to finance the upkeep of the Royal Library. No papers have been found relating to the sale, but it appears that just before the Second World War much of the collection was in the possession of a London dealer, Jacob Mendelson (d. 1970). Mendelson, it seems, was responsible for breaking up the volumes and selling the drawings as loose sheets from his two premises in Tottenham Court Road and the King's Road; and it was while the drawings were being stored in his King's Road shop that many of them were badly damaged and water-stained as a result of wartime bombing. Private collectors and dealers bought individual sheets or groups of them from Mendelson, and the drawings became scattered in this country and abroad. An important group of 107 drawings was purchased by an American collector, Herbert Boone (d. 1983), who hung many of them in his country house near Baltimore. In 1988, his art collection, which he had bequeathed to the Johns Hopkins University, was auctioned in New York, and again—despite attempts to prevent their dispersal—the drawings were scattered. Another large group was bought from Mendelson by the dealer Rex Nan Kivell (d. 1978), one of the partners of the Redfern Gallery in London. Some of these drawings he gave away to friends; a few—those he considered to be the finest—he hung in his villa in Morocco; and the rest (numbering well over a hundred) he kept at the Redfern Gallery. Other drawings passed, and continue to pass, through the London sale rooms, where they have not infrequently been catalogued as eighteenth-century; others have been identified with dealers on the Continent.

The process of breaking up the Paper Museum came to an end in the middle of the twentieth century. In the late 1950s, Anthony Blunt (1907–83), then Surveyor of The Queen's Pictures and adviser on the Royal Collection of drawings, began to investigate the whereabouts of the 'lost' drawings. He realized the importance of securing at least a photographic record of them, if any attempt at reconstructing the Paper Museum were to be made possible. Blunt succeeded in locating several owners whose drawings he photographed; and in 1968 he mounted a small exhibition in the Courtauld Institute Galleries of some of Rex Nan Kivell's drawings in order to publicize further the search for the missing drawings. In 1976 Nan Kivell presented the greater part of his Cassiano collection to the Royal Library; his remaining few drawings joined this generous gift at his death two years later. Since then, other drawings have been either returned to or repurchased by the Royal Library. These and the drawings that have been located in other private and public collections currently number over 200, a total that, when added to a group known only from photographs made by Blunt, amounts to about a quarter of the *c.* 1,000 drawings that constituted the contents of the volumes dispersed from the Royal Library earlier this century. Drawings continue to be discovered and it is hoped that others in private ownership will gradually come to light.

The decision to produce the present catalogue has its own history. The Paper Museum remained in relative obscurity after it left Italy in the mid eighteenth century until the late nineteenth century, when scholars began once more to consult Cassiano's drawings. It is to two German archaeologists, Friedrich Matz (1843–74), one of the founders of the great published *corpus* of Roman sarcophagus reliefs, and Adolph Michaelis (1835–1910) that credit is due for the 'modern' revival of interest in the Paper Museum. Their work, and that of the British scholar, Thomas Ashby (1874–1931), provided the inspiration for the study of Cornelius Vermeule, who in the 1960s, in two major articles, catalogued the drawings of sarcophagi and reliefs in the Royal Library and British Museum.[36] In the same period Sheila Somers Rinehart embarked on a comprehensive study of the Paper Museum;[37] and in the 1970s Jean Goldman attempted a general artistic assessment of the drawings.[38] The decision finally to produce a fully illustrated Catalogue Raisonné of the entire Paper Museum was made in the mid 1980s, when work on Cassiano's *Carteggio* brought international scholars to Windsor to inspect the visual counterparts of the documentary evidence for the Paper Museum. However, it soon became clear that it would be impossible to adopt for this body of drawings the same principles as those used in the volumes devoted to the drawings of Holbein, Guercino, Canaletto and other great masters belonging to Her Majesty The Queen. Despite Cassiano's well-informed love of art and the fact that he had access to distinguished artists, including Pietro Testa, Pietro da Cortona and others of real stature, it is clear beyond any doubt that it was the documentary and not the artistic aspect of the enterprise that was important to him. The drawings in the present publication, therefore, are catalogued not by artist, but by subject matter. Viewed from this perspective the drawings readily divide into two categories: those concerned with Antiquities and Architecture (which we have called Series A) and those dealing with Natural History (Series B). The former group encompasses *c.*4,200 drawings and the latter some 2,700.[39] The aim of this Catalogue Raisonné is to reconstruct in some measure both the intellectual and artistic achievements once embodied in Cassiano's Paper Museum.

It is entirely fitting that the present publication should appear under the patronage of some of the most prestigious scholarly European academies: namely Cassiano's own *Accademia dei Lincei*, the *Académie des Inscriptions et Belles-Lettres* and the British Academy. Without the support of these three institutions, the generous sponsorship and encouragement from Olivetti over the last ten years, and significant funding from the Getty Grant Program, this project would never have come to fruition.

Francis Haskell and Henrietta McBurney would like to acknowledge the kind help of the following: Arabella Cifani, Amanda Claridge, Brian Cook, David Freedberg, Antony Griffiths, Ingo Herklotz, Paddee Howard-Wagner, Ian Jenkins, Franco Monetti, Jane Roberts, Francesco Solinas, Donatella Sparti and Simon Towneley.

NOTES

1. He is described, for instance, by Giovanni Faber in Hernandez 1651, as 'very learned in Literature and Humanities, especially Philosophy, Chemistry and Antiquity' (p. 470).
2. In his letter of 15 November 1654 to Reinhold Dehnig (Dehnio), Cassiano speaks of his 'Paper Museum', which he says is divided into many volumes, in the context of his commissioning drawings after the Antique (Rome, BAL, Archivio dal Pozzo, MS XII, c. 75r-v). Although there is no mention of the natural history drawings in this letter, it is clear that these were collected, commissioned and treated in a similar way to those of antiquities, as were also the prints.
3. An intention to publish 'books of drawings copied from various ancient marbles, using Denis Petau's *Preface to the works of Julian the Apostate*, Paris 1630' is found in Cassiano's *Memorie* (Lumbroso 1874, p. 199).
4. For Montfaucon, see Montfaucon/Galliano 1987.
5. We are grateful to Arabella Cifani and Franco Monetti for providing us with information from the baptismal records in the parish register of SS. Stefano and Gregorio, Turin (to be published by them in 'Poussin nelle collezioni piemontesi del XVII, XVIII e XIX secolo' in the *Poussin Actes* 1996). These records show that Cassiano was baptized on 22 February 1588, and was therefore probably born a day or two earlier.
6. It is probable that the purpose of this sojourn was educational; Lumbroso merely says that 'Cassiano left Piedmont...; he went first to Bologna, and then to Pisa to [the house of his cousin] the Archbishop of that city' (Lumbroso 1874, p. 134).
7. He was appointed *gentiluomo ordinario* in 1623; a member of the *famiglia intima* in 1627; and finally *primo maestro di camera* (head of the Cardinal's Household) in 1633. He is also frequently cited in the documents as *coppiere* to Cardinal Barberini.
8. His proposed publication of Leonardo's *Trattato della pittura*, with illustrations by Poussin; and the facsimile of the Aztec herbal (the so-called 'Codex Badianus'). In

Spain he would also have seen the originals of Nardi Antonio Recchi's drawings of Mexican plants and animals which were later to be published in Faber 1628.

9. Described in the room inventory made on Carlo Antonio's death in 1689, see n. 16 below; see also Sparti 1992, pp. 104–5, 211; figs 53–7.

10. It is clear that Carlo Antonio contributed substantially to the Paper Museum. His additions to the section devoted to Christian antiquities are discussed in the present volume, pp. 37–8ff., 62. For his collecting activities in general see Sparti 1990a.

11. We are grateful to Antony Griffiths for bringing to our attention the existence of a copy of Campanella 1631, with an autograph dedication to Cassiano: '*D. Cassiano Puteo ph[ilosoph]o, Benefactori Autor donat*' (in Jonathan A. Hill, *Catalogue 78*, New York 1993, No. 36, pp. 23–5).

12. See, for example, Cassiano's letter to Dehnio (n. 2 above).

13. '*Synopsis, atque Ordo Antiquitatum Romanarum Illustriss. & Eruditiss. V. Equitis Cassiani a Puteo studio, ac impensis xxiii voluminibus digestarum*', in Dati 1664. For a discussion of the synopsis in relation to the drawings contained in the *Bassi Rilievi* volumes, see Herklotz 1992a, pp. 101–4.

14. Letter of 27 February 1666 to Angelico Aprosio; see Lumbroso 1874, pp. 167–8.

15. Baldinucci (1845) V, p. 313. For discussions of Baldinucci's description see Blunt/Cooke 1960, p. 114; *Cassiano* cat. 1993, p. 20; and p. 39 n. 4 of the present volume.

16. Rome, Archivio di Stato, 30 Not. Cap., Ufficio 25, Vol. 419, 12 Sett. 1689, fols 124–126v and 151–152v; 11 Ott. 1689, fols 219–258, transcribed by Sparti 1990b, pp. 556–61.

17. It was on Gabriele dal Pozzo's death that his widow, Anna Benzoni, applied a seal combining the dal Pozzo and Benzoni arms to the first page of many of the volumes; see Ashby 1913, pp. 185–6 and Solinas 1989c, pp. 52–3.

18. Rome, Archivio di Stato, 30 Not. Cap., Ufficio 6, Vol. 210, 5–7 Mar. 1695, fols 253–286, transcribed by Sparti 1990b, pp. 561–9.

19. See Alessandrini 1978, pp. 41–3 for the fate of the library and the manuscripts. Other volumes from the library, including 39 volumes of the Archivio Dal Pozzo, which in the 19th century had been housed in the archives in Turin, were deposited in the Accademia dei Lincei in 1973: see Nicolò 1991, p.V. Further volumes of manuscripts ended up in the École de Médecine, Montpellier (see Alessandrini 1978) and Naples, Biblioteca Nazionale (MS V.E.10).

20. Of the remaining two, *Penance* was destroyed in a fire at Belvoir Castle in 1816, and *Baptism* was sold in *c.* 1939 and is now in the National Gallery of Art, Washington.

21. The fact that the Paper Museum was later rearranged by George III's librarian further complicates the issue.

22. See Wrede 1989, p. 390 and Jenkins 1992, pp. 62–4, who also drew the authors' attention to the following comment by Winckelmann in his *Description des pierres gravées du Baron Stosch*, Florence 1760, p. iv: '*La somptueuse collection de desseins d'après l'Antique de S.E. Monseigneur le Card. Alexandre Albani m'a été faite d'une grande utilité*'. Several instances of Winckelmann 'using' the Paper Museum—in the form of his removing drawings from it for engraving in his own works—have been noted, and will be discussed in future volumes of this series by Helen Whitehouse and Amanda Claridge.

23. Mount Stuart Archive, Rothesay, Isle of Bute.

24. Fleming 1958, p. 167.

25. For a full account of the purchase negotiations, see Fleming 1958. A small section of the Paper Museum remained with the Albani family until the French invasion of Italy in 1798, when eight volumes were removed to Paris and found their way to the Institut de France (MS 968–70 and MS 974–78); see Alessandrini 1978, pp. 39–41. Three further volumes of mycological drawings also remained with the Albani family, from whom they appear to have passed to the Strozzi family in the 18th century. They were purchased by C.D. Badham (1806–57) in 1845, and in 1896 two of these volumes were acquired by the Library of the Royal Botanic Gardens at Kew. The whereabouts of the third volume is unknown.

26. Richard Dalton was made Royal Librarian in 1760; from 1774 he was titled Keeper of the King's Medals and Drawings; and from 1778 Surveyor of the King's Pictures, the title he retained until his death in 1791. By *c.* 1770 the bindery occupied a number of rooms on the ground floor under the Octagon Library, built 1766–7 (Westminster Library, Archives and Local Studies Section, Box 39/29/1–2).

27. While some of the vellum bindings date from dal Pozzo ownership, others may well be 18th-century versions dating from the Albani ownership of the volumes. It is possible, of course, that some groups of drawings arrived in England unbound in portfolios.

28. A large number of the natural history drawings, and some of the architectural drawings were treated in this way.

29. For a discussion of these albums, see Griffiths 1989.

30. For a discussion of these sections of Cassiano's print collection, see *Cassiano* cat. 1993, pp. 247–8, 256–7, nos 156, 159–160 and forthcoming article by Henrietta McBurney in *Print Quarterly*.

31. A catalogue of these volumes, by Lynda Fairbairn, is to be published as part of the *Catalogue of Italian Renaissance Drawings at the Sir John Soane's Museum*, Oxford (Azimuth), 1997–8.

32. Now known as the 'New' drawings, this group of 87 sheets (bearing 110 drawings) was rediscovered by Ian Jenkins in the Department of Greek and Roman Antiquities, see Jenkins 1987, 1989a and 1989b. Jenkins has since discovered further drawings in the British Museum (one of a gem, amongst the Townley drawings, Department of Greek & Roman Antiquities; and one of a bas-relief in the *Museum Secretum*, Department of Medieval and Later Antiquities); these will be published in later volumes of this catalogue.

33. For whose biography and publications, see Stirling-Maxwell 1891, pp. xix-xxxii.

34. Phillips, London, 12 December 1990, lots 219–374. The 225 folios had been arranged (and some 18th-century drawings added) and bound in two volumes described as 'Sculpture' and 'Architecture' on their title-pages.

35. See Griffiths 1989. A further album of prints was identified by Griffiths in September 1995 (BL 136.g.10, entitled on the spine: *Charles III/ Duke of Lorraine/ Cérémonies/ Funèbres*). It is likely that other volumes will be discovered over time.

36. Vermeule 1960 and 1966; see also his articles 1955, 1956, 1958.

37. Rinehart's plan to publish a catalogue was never realized.

38. Goldman 1978.

39. Of these almost 7,000 drawings, just over 3,000 are in the Royal Collection.

CHRONOLOGY

Events in the dal Pozzo household	Year	Contemporary events	Publications
Cassiano dal Pozzo born	1588		
Moves to Pisa	1598		
Made Knight of the Order of S. Stefano	1599		
	1603	Cesi founds *Accademia dei Lincei*	
Birth of Carlo Antonio dal Pozzo	1606		
Cassiano graduates from Pisa University; Carlo Antonio dal Pozzo, Archbisop of Pisa, dies	1607		
Cassiano appointed judge in Siena	1608		
	1610	Galileo moves to Pisa	
Leaves Tuscany	1611	Galileo becomes a member of the *Accademia dei Lincei*	
Arrives in Rome	1612		
	1614	Cesi records Palestrina mosaic	
	1616	Galileo censured	
Cassiano's father (Antonio dal Pozzo) dies	1619		
Carlo Antonio arrives in Rome	1620		
Cassiano elected to the *Accademia dei Lincei*	1622		Olina, *Uccelliera*
Appointed to the Household of Francesco Barberini	1623	Maffeo Barberini elected Pope Urban VIII	Galileo, *Il Saggiatore*
	1624	Poussin arrives in Rome	
Travels to France: meets Peiresc	1625		
Travels to Spain; elected to the *Accademia della Crusca*	1626		
Carlo Antonio marries Teodora Costa; brothers move to Via dei Chiavari	1626/7		
	1628		Faber, *Animalia Messicana*
	1630	Cesi dies	Petau, *Ad Juliani opera* Stelluti, *Persio Tradotto*
	1631		Campanella, *Atheismus*
	1632		Galileo, *Dialogo*
Cassiano buys Cesi's collection	1633	Galileo is condemned	Ferrari, *De Florum Cultura*
	1634-5		Bosio, *Roma Sotterranea*
	1637	Peiresc dies	
Poussin begins *Seven Sacraments* for Cassiano	1638		Stelluti, Trattato del legno fossile minerale
Carlo Antonio nominated *Capitano di Fanteria*	1642		
	1644	Urban VIII dies; Giovanni Battista Pamphilj elected Pope Innocent X	
	1645	Francesco Barberini in exile in Paris	
	1646		Ferrari, *Hesperides*
	1648	Francesco Barberini's return to Rome	
Cassiano's last commission from Poussin: *Pyramus and Thisbe*	1651/2		Hernandez et al, *Rerum medicarum Novae Hispaniae...*
	1655	Fabio Chigi elected Pope Alexander VII	
Carlo Antonio serves as *Conservatore* of Rome	1656-78		

CHRONOLOGY *(continued)*

Events in the dal Pozzo household	Year	Contemporary events	Publications
Cassiano dies; Poussin engaged in designing his memorial. Carlo Antonio takes over the knighthood in the Order of S. Stefano	1657		
	1664		Bellori, *Nota delli Musei* Dati, *Delle lodi del Commendatore...*
	1670-94	Bellori held post of 'papal antiquary'	
Carlo Antonio serves as *Console Capitolino*	1684		
	1687		Mabillon, *Iter Italicum*
Carlo Antonio dies	1689		
	1690-99		Ciampini, *Vetera monimenta*
Gabriele dal Pozzo dies	1695		
	1700	Giovanni Albani elected Pope Clement XI	
Paper Museum sold to Pope Clement XI Albani by Cosimo Antonio dal Pozzo	1703		
Paper Museum passes to Cardinal Alessandro Albani	1714		
	1721	Michelangelo dei Conti elected Pope Innocent XIII	
Cosimo Antonio dal Pozzo dies	1740		
	1743	Winckelmann becomes librarian to Cardinal Albani	
Paper Museum bought by George III	1762		

LOCATIONS OF DAL POZZO MATERIAL

Series A: Antiquities and Architecture

I. ROYAL LIBRARY	folios	RL inventory	Ashpitel	Heaton-Smith	Michaelis	Robert
Bassi Relievi Antichi 1	95	8145-8255	A40	155	I	I
Bassi Relievi Antichi 2	66	8256-8322	A41	156	II	II
Bassi Relievi Antichi 3	70	8323-8400	A42	157	III	III
Bassi Relievi Antichi 4	79	8401-8483	A43	158	IV	IV
Bassi Relievi Antichi 5	69	8484-8551	A44	159	V	V
Bassi Relievi Antichi 6	69	8552-8628	A45	160	VI	VI
Bassi Relievi Antichi 7	69	8629-8701	A46	161	VII	VII
Bassi Relievi Antichi 8	79	8702-8783	A47	162	VIII	VIII
Buste e Statue Antichi 9	77	8784-8860	A48	163	IX	-
Bassi Relievi Antichi	120	7994-8113	A52	153	X	-
Bassi Relievi Antichi	15	8114-8144*	A53	154	IXa	IXa
Mosaici Antichi I	96	8928-9035	A50	165	XI	-
Mosaici Antichi II	103	9036-9221	A51	166	XII	-
Antichità Diverse	108	10189-10297	A9	184	XVIII	XVIII
Architec[tura] Civil[e]	150	10354-10504	A12	186	XV	-
Ancient Roman Architecture	50	10780-10846	A17	190		-
Buildings and Architectural Ornaments	12	10741-45, 10750-56	A16	189		-
Disegni di varie Antichità, Nettuno	220	11120-11521	A31	196	XIII	-
Tombs of Illustrious Italians at Rome	243	11715-11957	-	201	-	-
Terence	23	26645-26667	-	-	XXI	-
Il Vergilio nel Vaticano I	52	26668-26719	-	-	XXII	-
Il Vergilio nel Vaticano II	53	26720-26773	-	-	XXIII	-
Ancient Mosaics & Paintings (loose folios)	34	19201-19234	pf. 5	-	-	-
Architectural drawings (loose folios)	32	19243-19287	-	-	pf. 5	-
Bas–reliefs (loose folios)	2	11880-11881	-	-	-	

* dispersed, 3 in 'Franks' I (see British Museum), others unlocated

II. BRITISH MUSEUM	folios	Drawing no.	Michaelis	Robert
'Franks' I	172	1-202	I	-
'Franks' II	134	203-523	II	-
'New' drawings	87	-	-	-

III. BRITISH LIBRARY	folios	Shelf mark
Plans of Fortified Places	28	118.e.18

IV. SIR JOHN SOANE'S MUSEUM	folios	Vol. no.	Ashby 1904 nos
Montano I: Architectu del Montani	138	125	-
Montano II: Montan 1	97	124	-
Montano III: Tempj del Montani	98	123	-
'Codex Coner'	165	115	1-165

V. DISPERSED	folios	Phillips 1990 lot nos.
Stirling-Maxwell I - Sculpture	124	219–298
Stirling-Maxwell II - Architecture	101	299–374

RL inventory = Royal Library inventory number
Robert = number given in Robert *ASR*
Heaton-Smith = number given by Evelyn Heaton-Smith
(worked in Royal Library between 1905 and 1916, under Sir John Fortescue)
Michaelis = number given in Michaelis 1882
Ashpitel = number given in Ashpitel 1862

LOCATIONS OF DAL POZZO MATERIAL
Series B: Natural History

I. ROYAL LIBRARY		Folios	RL inventory no.
Natural History of Birds	I	44	dispersed
" " "	II	49	dispersed
" " "	III	54	dispersed
" " "	IV	44	dispersed
Natural History: Fossils	V	45	25480-25535
Natural History: Fishes	VI	61	dispersed
Natural History: Fruits	VII	53	dispersed
Natural History: Plants	VIII	74	dispersed
" " "	IX	67	dispersed
" " "	X	56	dispersed
" " "	XI	57	dispersed
" " "	XII	53	dispersed
" " "	XIII	58	dispersed
Natural History: Fossils	XIV	62	25536-25591
" " "	XV	48	25592-25639
" " "	XVI	57	25640-25696
" " "	XVII	41	25697-25734
Natural History: Quadrupeds		25	dispersed
Natural History: Corals		16	dispersed
Con[cae] & Plant[ae]		29	dispersed
Natural History of Birds		92	27599-27690
Erbario Miniato		214	27691-27901
Erbar[io] Minia[to]/ 'Libellus dei Medicinalibus Indorum herbis...'		41	27902-27938

II. KEW: Library, Royal Botanic Gdns	Folios	File no.
Icones Fungorum Ineditorum I	122	QG 3884
Icones Fungorum Ineditorum II	123	QG 3884

III. PARIS: Institut de France	Folios	MS nos
Fungorum genera et species I	216	968
Fungorum genera et species II	206	969
Fungorum genera et species III	177	970
Plantae et flores I	129	974
Plantae et flores II	108	975
Plantae et flores III	193	976
Plantae et flores IV	193	977
Plantae et flores V	178	978

IV. DISPERSED

Located, as of 1996 :	
Courtauld Institute Galleries	Witt colln cat. nos: 3218, 3745, 3865
Fitzwilliam Museum	Accession nos: PD.42-1956, PD.43-1956
Victoria & Albert Museum	Accession nos: Circ.13-1925, E.731-735-1949, E.2776-2777-1962
Rijksmuseum, Amsterdam	Inv. no. 1966: 87
Private collections	
Herbert Boone	Sotheby's NY 1988 lot nos. 125-182. 99 drawings
Other	56 drawings
Unlocated:	
Witt photos	6 drawings
Courtauld Inst. microfilm	16 drawings
Ex-sale room	4 drawings

INVENTORY A
References to volumes containing dal Pozzo material

Antiquities and Architecture

[p. 3] Terence, illustrated with Figures. 23 D°

[p. 3] Il Vergilio nel Vaticano 2 Vols. V[ol] 1, 52 pages. V 2, 54 D°

[p. 4] Ancient Roman Architecture 50 pages. Fol

[p. 4] Buildings & Architectural Ornaments. 43 [*in pencil* 42] D° Fol.

[p. 4] Antichita Diverse. 173 Drawings. Fol.

[p. 4] Mosaici Antichi. 2V. Fol. VI–96 pages. V2–102 D°

[p. 4] Bassi Relievi Antichi. 9 Vols Fol. V 1–95 pages V 2–67 D° V 3–70 D° V 4–79 D° V 5–67. V 6–70. V 7–69. V 8–78. V 9–120 D°

[p. 4] Busti & Statue Antichi. 76 pages

[p. 4] Disegni di varie Antichita. 220 pages. Fol.

[p. 4] Architectu: Civile & Milit: (containing a medley of Plans in different parts of Italy & elsewhere & entire Buildings) 147 pages

[p. 4] Tombs of illustrious Italians at Rome etc. MS: Fol. 243 D°

[pp. 166-7] Six Portfolios of Architecture
No. 5
1. 18 large coloured Drawings of the Mosaic Pavement at Palestrina, with 2 Copies of the same engraved in 1721. The Abbé Barthelmy says that the *Commandeur Pozzo fit copier en grand la Mosaique, mais par parties detachées, conservant dans les desseins les couleurs des differentes objets qu'elle representent, les Dessins etoient retracées en dix-huit morceaux, dont Suares, Evêque de Vaisons donna une description succincte dans son histoire imprimée a Rome en 1655.*
2. 2 Prints of separate parts of the same Pavements
3. 7 coloured Drawings of Mosaic
4. 43 Drawings of ancient Buildings in Rome [*with a later annotation in pencil* These are partly prepared for binding]

Natural History

[p. 3] Drawings of Birds. 4 Vols Fol. V1, 44 pages. V2, 49 D° V3, 54 D° V4, 44 D°

[p. 3] D°...of Fruits. 1 Vol Fol. 53 pages

[p. 3] D°...of Fishes. 1 Vol Fol. 61 D°

[p. 3] D°...of Plants. 6 Vols Fol. V1–74. V2–67. V3–56. V4–57 [*in pencil* 40 only]. V5–53. V6–58.

[p. 3] D°...of Fossils. 5 Vols Fol. V1–45. V2–62 [*in pencil* including] 12 of Index. V3–48. V4–57. V5–38 & 3 of Index.

[p. 3] Sixteen Drawings of Plants. Fol.

[p. 3] Twenty five Plates of Drawings in Water colours of Quadrupeds. Fol.

[p. 4] Conc: & Plant: Fol. 16 Shells & 13 Plants

[p. 4] Drawings of Birds. Fol. 92. Drawings

[p. 4] Erbario Miniato. 209 Drawings, with an Index

[p. 4] Libellus de Medicinalibus Indorum herbis a Martino, Collegii Sanctae Crucis. compostus: cum 37 figuris coloratis. MS.

[p. 143] A Volume containing twenty-five pages of Drawings of Animals

[p. 143] A Volume (sixteen pages) containing drawings of Corals?

[p. 143] A Volume (92 pages) cont: Drawings of Birds. Small Fol. let.d 'Natural History of Birds'

[N.B. on page 141 of Inventory A is a list of 17 of the above volumes labelled 'Natural History' with a running sequence of numbers assigned to them (for this sequence see 'Locations of dal Pozzo material', I. Royal Library, p. 30)]

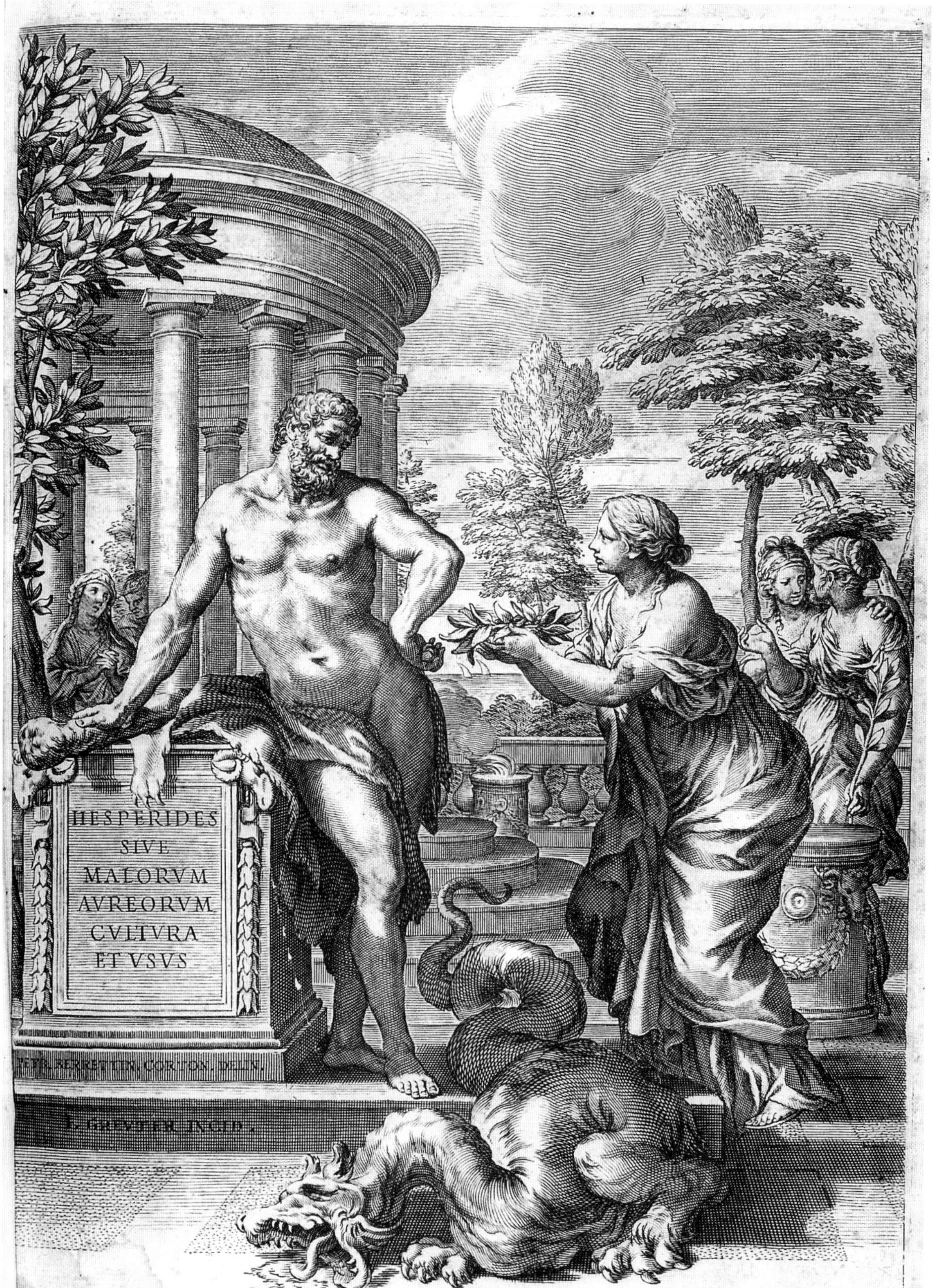

Fig. 13. J.F. Greuter after P. da Cortona, *Hercules as Guardian of the Garden of the Hesperides receiving a wreath from the Nymphs of the Hesperides*, engraving, Ferrari 1646, Frontispiece

SERIES B: NATURAL HISTORY

David Freedberg

THE IMPORTANCE of Cassiano's drawings of natural history cannot be overestimated. Largely commissioned by a group of Galileo's closest friends and supporters, they provide the visual evidence for some of the most crucial developments in the modern study of palaeontology, mycology, botany and zoology. Amongst them may be found some of the earliest illustrations ever produced with the aid of a microscope. Together they constitute one of the most ambitious attempts ever to classify the natural world through the use of visual description. Ultimately that attempt would fail, as the very men who commissioned these drawings came to realize themselves. But before they came to the conclusion that visual description in and of itself was not entirely adequate to revealing the mysteries of nature — and was even of limited use in analysing them — they provided the world with one of the most beautiful and comprehensive records of how the natural world looked at any period in history.

Made at the same time as Galileo's pioneering investigations in astronomy, mathematics and physics, the drawings and the documents associated with them have remained unknown to the vast majority of historians of science, and have been omitted from almost every treatment of the scientific revolution of the seventeenth century; and while there were certainly precedents for encyclopedic collections of drawings of nature — commissioned by men like Conrad Gesner and Ulisse Aldrovandi — none can match the present one in terms of size, comprehensiveness, or accuracy. No one ever before had paid quite such meticulous attention to local flora, fauna, or geology, or to the ever-increasing variety of species that were daily being reported and imported from the Americas. And none had been even remotely as rigorous in excluding the fabulous and the fantastic specimens that until then had so bedevilled the study of nature. For while there is no shortage of anomalous specimens in Cassiano's collection of drawings, each one of them had actually been seen and studied by one or other of his friends. Every specimen, in other words, actually occurred in nature, unlike the thousands of fantastic monsters, prodigies and wonders that populate the pages and sheets of Gesner, Aldrovandi, and their like. This was no longer a world in which the chief motor for explanation was either historical or based on the sheer force of similitude (generally anthropomorphic); it was one in which nothing could be taken as given unless it had been recorded and analysed by direct observation. It is not surprising that the first systematic use of both the telescope and the microscope should have emerged from precisely the group of people who were responsible for the drawings commissioned in the present series, and who themselves went into the field to make detailed notes on everything they saw.

But who were these people? Above all, they were the young men who had gathered round Federico Cesi, Prince of Acquasparta (1585–1630), ever since he founded the Accademia dei Lincei (or Academy of Lynxes) in 1603, at the age of eighteen. First came Cesi's lifelong friend and follower, Francesco Stelluti (1577–1653), then the obsessive Dutchman and ever-travelling Johannes Heck (1576–1618?), and the more obscure Anastasio de Filiis (1577–1608). Seven years later, in 1610, they invited the 75 year-old Giovanni Battista della Porta (1535?–1615) to join them, perhaps the most famous Italian student of nature, science and magic in the age before Galileo. In the following year the Lynxes invited none other than Galileo (1564–1642) himself to join them, and he willingly accepted. This was the year in which Galileo made his famous demonstration of the capabilities of his telescope on the Janiculum in Rome, and when he was received in triumph and celebration by the Jesuits there. Later on in the same year, three talented Germans, Johannes Schreck (1576–1630), Johannes Faber (1574–1629), and Theophilus Müller (1576–1618) were invited to join the group. The distinguished botanist from Naples, Fabio Colonna (1566–1640), became a member in 1612, but it would be six more years before those key figures in the active core of the group, Vincenzo Cesarini (1595–1624) and Giovanni Ciampoli (1590–1643) were added to their number. By then Heck had gone mad and died, and Galileo had encountered the first serious opposition to his Copernican ideas. In 1622 Francesco Barberini, whose uncle would soon be elected Pope, joined them, and in the same year Cassiano himself was elected a Lynx. These, basically, were the chief players in the story, and they will feature time and again in the volumes in this series. They were the ones who went into the field, incessantly collecting botanical, zoological, and geological specimens, who supported Galileo and edited the *Saggiatore* in the years in which he was discouraged by the threat of excommunication, and who put together the bulk of the drawings in the present corpus. And it was they who realized the prime importance of making visual records (or commissioning such records) of everything they saw and encountered.

Cassiano, of course, became a prime mover in all this. It was he who commissioned the beautiful drawings of birds and citrus fruit in the corpus and he who had the facsimile made of an old Aztec herbal (the *Codex Badianus*) when he visited the Escorial in 1626. But — most importantly of all in the present context — it was he who bought the contents of Cesi's great library when he died in 1630, thus preserving for posterity the visual heritage of the Lincean researches. The library contained not only books and instruments, but also the drawings of natural specimens which Cesi had commissioned from the earliest days of the academy onwards. But since we know that Cassiano continued to add to this inheritance, and that he commissioned natural history drawings himself in order to aid his own researches, as well as to make them available to others, the following questions arise: which of the surviving drawings come from Cesi's collection, and which were added to Cassiano's Paper Museum after Cesi's death? And who made them? These are difficult questions to answer, although there is a considerable amount of internal and external evidence to help us along the way.

At the time of writing, the natural history drawings that survive from Cassiano's Museum number almost 3,000. They are chiefly to be found in the Royal Library, Windsor Castle and in the Institut de France, Paris, but also in a number of other public and private collections. Since this is material that has generally not been known, and since it only represents a portion of what we know Cesi and Cassiano to have commissioned and collected, it is very likely that more will come to light, even as the volumes in this Catalogue Raisonné appear.

The surviving drawings are preserved in a variety of bindings and as loose sheets. The numbers may be broken down as follows: in Windsor there are a total of 253 folios of drawings of fossil and mineral specimens bound into the five volumes entitled *Natural History of Fossils*. The so-called *Erbario Miniato* has 211 folios of drawings of plants and mushrooms, while that entitled *Libellus de medicinalibus Indorum herbis* has 41 folios of Mexican plants with their Aztec names. In the *Erbario Miniato* there is sometimes more than one drawing to a page; in the Aztec herbal, always. The volume entitled *Natural History of Birds* has 92 folios of ornithological drawings. The volumes described as *Plantae et flores*, numbered I–V (MSS 974–978 in the Institut de France), contain over 800 numbered folios of botanical specimens and the occasional drawing of insects and other animals (in MS 978). Although entitled *Plantae et flores* on their spines, these are presumably the volumes of the so-called *Syntaxis Plantaria* often referred to in the Lincean documents. The volumes *Fungorum genera et species*, numbered I–III (MSS 968–970 in the Institut de France) have 599 folios of fungi, while those entitled *Icones Fungorum* in the Library of the Royal Botanic Gardens, Kew, contain a further 245 folios of the same subject, mostly copies of the drawings in Paris. Some 460 loose sheets of animals, fish, crustacea, corals and sponges, citrus fruit, plants, mushrooms and birds have been located so far. Many of these must once have been bound into now dismembered volumes. For example, the original numbering of the drawings in the *Erbario Miniato* runs from 499 to 705, and the numbers on all of the loose drawings of plants so far located fall between 1 and 498. The spine of the *Erbario* carries the number VI, thus suggesting that it was part of a series of earlier volumes, presumably at least one of which was of plants, and the others of related natural history subjects. Even from these figures the enormous scale and ambition of the corpus is immediately apparent.

The Catalogue Raisonné will broadly follow the chief divisions already present in this material, along these lines: (i) citrus fruit; (ii) fossils and other mineralogical material; (iii) fungi; (iv) birds; (v) animals, fish, and other fauna; (vi) the *Erbario*

Miniato and other flora; (vii) the five volumes of the *Syntaxis Plantaria* in Paris, and (viii) the Mexican Herbal (see p. 6). Some subdivisions will be inevitable, according to the quantity of drawings concerned and the internal logic of whatever may be discovered about their original or early organization.

The *Erbario Miniato* and Mexican Herbal in Windsor are in vellum bindings which almost certainly date from the time they were in the Albani collection; both have green wash fore-edges. The drawings in the *Erbario Miniato* are still on their original folios; those in the Aztec herbal are all inlaid. The volumes entitled *Natural History of Fossils* and *Natural History of Birds* are in calf bindings specially made some time after the drawings they contain entered the Royal Collection in 1763; here the drawings are laid down. The volumes in the Institut de France are in eighteenth-century boards with leather spines, perhaps also made at the time they belonged to the Albani collection. On the basis of internal evidence it would seem that the drawings in the Albani volumes reflect their earlier seventeenth-century ordering much more closely than the sometimes haphazard ordering in the volumes re-bound in the Royal Library in the late eighteenth century.

Many of the drawings carry inscriptions and numbers. These are in a variety of hands, and refer to many different things — the name of a specimen, the date of its discovery or its reproduction, its possible occurrence in other sources, and so on. A large number of the fossil and fungus drawings and several of the botanical drawings contain inscriptions in Cesi's own hand. Especially in the case of the fossil drawings, these are often erased or cut off; fortunately Cassiano had had them copied by a scribe who reproduced these now truncated inscriptions elsewhere on the same sheet. This same scribal hand appears in the extensive indices now bound into the flower and fungus manuscripts in the Institut de France and in the plant and fossil volumes in Windsor. To what extent these indices reflect the original ordering of the folios they refer to (whether in Cassiano's or in Cesi's collection) is not clear.

A similar uncertainty applies to the surviving indications of numbering. These are mainly in lower case arabic numerals, but some are roman, and others take the form of lettering, occasionally capitals. Almost all are in brown ink in a small, often identical seventeenth-century hand. If there is any area that needs further research it is this; but given the lacunose state of our knowledge about the original ordering and binding of the majority of these drawings no conclusions may yet be reached. In any case the numbers rarely correspond exactly (and sometimes they do not correspond at all) to any order in which they may still remain today, and they do not correlate in any helpful way with the order of reproductions in the printed books associated with them. The extent to which the numbers reflect their organization (or proposed organization) in existing treatises, such as those on birds and citrus fruit (and perhaps others yet to be discovered), remains uncertain. In the case of the Antiquities drawings in Series A of this Catalogue Raisonné, one set of these numbers refers to an inventory of a group of drawings kept and acquired by Cassiano's brother, Carlo Antonio dal Pozzo, but there is little evidence at present to suggest that Carlo Antonio went on expanding the collection of natural history drawings, even though one or two of these may indeed date from the period after Cassiano's death.

One of Cassiano's favourite methods of mounting was by inlay. A window a few millimetres smaller than the drawing sheet was cut out of the mount sheet, and the drawing pasted on it, with an overlap of about 5 mm. The overlap was burnished at the back so that the sheet could lie completely flat, and the drawing was often framed within double lines drawn on the mount sheet. But most of the drawings in the natural history series are not thus inlaid, and very many have the typical wash borders added at the time of their entry into the Royal Collection, when they may have been cut down further. It is possible, however, that many of Cesi's drawings were already cut down when they were in the collection of Cassiano, his brother, or his heirs. All that can be said is that much valuable evidence has been cut off or erased with the passage of time, and that we must look for other evidence for the original ordering and provenance of the drawings.

Sometimes the situation is very clear. For example, on the basis of their inscriptions, the occasional date, and a wide variety of external evidence (all of which will be discussed in the appropriate volumes in this series), it can be established that the drawings in the Institut de France were commissioned by Cesi himself. The drawings of fungi are the *Icones Fungorum* which he and his friends repeatedly discussed from the very beginning of their association onwards. The botanical drawings both in the so-called *Syntaxis Plantaria* in Paris and in the Windsor *Erbario Miniato* also carry Cesian inscriptions and may be related in many ways to the Lincean researches in the field. In the *Erbario Miniato* there are constant cross-references to discussions and illustrations of the corresponding specimen in 'my Matthioli', presumably Cesi's own copy of an edition of the great sixteenth-century botanist. The simple but proud inscriptions referring to the use of the microscope chiefly in the Paris volumes may be paralleled in Cesi's letters to his Lincean friends and in the occasional passage in their great compendium of South American flora and fauna, the *Rerum Medicarum Thesaurus Novae Hispaniae, sive Plantarum Animalium Mineralium Mexicanorum Historia* which was published in preliminary form in 1628 but in its complete form only in 1651. We shall return to this work, the so-called *Tesoro Messicano*, later.

Then there are the drawings which relate to other, chiefly Lincean publications. The drawings now bound into volumes XIV–XVII of the *Natural History of Fossils*, for example, were also commissioned by Cesi himself: a few were reproduced as etchings in the small *Trattato del Legno Fossile Minerale* published by Stelluti in 1637, and the group as a whole is repeatedly referred to there, as well as in a variety of Cesi's and Stelluti's correspondence and other papers still preserved in the Vatican Library, in the library of the Accademia dei Lincei, and in that of the École de Médecine in Montpellier.

If the fossil, fungus, and most of the flower drawings have a clearly Cesian provenance, many others were verifiably commissioned by Cassiano himself. As already noted, the volume in Windsor known as the *Libellus de medicinalibus Indorum herbis* is the facsimile which Cassiano had made of the Aztec herbal, the *Codex Badianus*, which had so fascinated him when he visited the Escorial in 1626. That had been made over seventy years earlier (in 1552) by a converted Nahuatl Indian for the son of Philip II's viceroy, Antonio de Mendoza. It probably inspired Philip to send his personal physician, Francisco Hernandez, to Central America to explore its botany and zoology. Hernandez, subsequently known as the *Protomedicus* of the Indies, spent over six years there, from 1571 to 1578, gathering all kinds of information and many drawings about these matters. He brought this material back with him and deposited it in the Escorial, where it was seen by Cassiano in 1626.

Several further groups of drawings, now largely disbound from the volumes which once contained them, actually exceed most of the Cesian ones in terms of finish and refinement, and can only have been commissioned by Cassiano. Firstly, there are the drawings which he had made in preparation for his own entry presentation to the Accademia, the *Uccelliera* of 1622. These are by the same meticulous, skilful and attractive hand as the other bird drawings ordered by Cassiano in connection with the small unpublished ornithological treatises he wrote himself and which may still be found amongst his papers. Indeed ornithology seems to have been his particular area of interest and competence. Then come the excellent drawings of citrus fruit, documented to have been commissioned and collected by Cassiano, and made in preparation for the illustrations of a book that eventually appeared in 1646, Giovanni Battista Ferrari's *Hesperides sive de malorum aureorum cultura*. Of all the drawings that came from Cassiano's Museum, this is by far the largest group ever to have been published.

In their almost complete suppression of any acknowledgement of his role in their production, both the *Uccelliera* and the *Hesperides* are typical of the natural history works published with the help of Cassiano. For although we know from his papers that he played a key role in their preparation and in the preliminary research for them, they were actually written by friends of his, in these instances the lawyer Giovanni Pietro Olina and the Jesuit professor of Hebrew and Rhetoric, Giovanni Battista Ferrari respectively; but Cassiano's name is scarcely mentioned in their pages. Both books name Vincenzo Leonardi as the author of the drawings on which their illustrations were based. We thus have the name of at least one artist in the corpus. Furthermore, one of the bird drawings turns out to have his signature on its verso, and it is reasonably clear that it is by the same hand as the best of the citrus drawings, documented as being by him. On stylistic grounds alone, there is much to be said for the view that many other of the finest animal drawings in the corpus which are not thus documented are also by Leonardi (for further discussion see p. 58). Unfortunately his is the only name we have discovered so far which can definitely be associated with any drawing or group of drawings in the present series.

Certainly a number of other possible candidates may be excluded. While we know that Heck, Stelluti and Colonna all made drawings of natural history and other specimens, their hands are not to be found amongst the drawings from Cassiano's Paper Museum. In the course of his travels, Heck made a large number of rather crude drawings of what he saw, but these are preserved in his manuscripts in Rome, Florence, Montpellier, and elsewhere, and are very different from anything in this series. So too is the occasional surviving drawing attributable to Stelluti, while for Colonna we have only the occasional reference to drawings which he made. It is certain that the other Lincei, especially Faber, Schreck, and Cesarini, made use of drawings in the course of their manifold and intense researches, but we have almost no evidence (except rarely in the *Tesoro Messicano*) of what they may have commissioned or even drawn themselves. Whatever the case, however, there can be absolutely no doubt of the importance these men all attached to the graphic recording of everything they saw.

But why, more specifically, were these drawings made? It was not just to have a visual record and proof of their omnivorous researches, or to make clear the errors and shortcomings of the ancient authorities, from Aristotle to Theophrastus, Pliny to Dioscorides. It was to help them in the complex taxonomical tasks in which they were constantly engaged. How much better to have a clear drawing than to describe something in ambiguous words, or to have a multiplicity of different names for what might, after all, be the same thing. With the aid of drawings, people might at least attempt to ensure that one name was given to one thing, not one name to many things or many names to one. If they could not send a specimen around to their correspondents, or ask for one to be sent, they could at least circulate a drawing with a name attached to it.

The other reason for the production of these drawings was that they should serve as a means of assisting in the distinction between species, of ensuring that a particular specimen was not assigned the wrong species, or that it did not partake of more than one. Nothing obsessed Cesi and the early Lynxes more than phenomena which seemed to stand on the borderlines between classes, such as the fossil woods which are catalogued in the fossils volume, or what they called species of ambiguous or doubtful nature. Indeed, if we turn to Cesi's vast attempt to classify all of the plant world, the *Tabulae Phytosophicae* appended to the *Tesoro Messicano* from 1628 on, or to his even more ambitious attempt to put all of nature in order, the great outline plan still preserved in Naples and entitled the *Speculum Rationis* (a kind of draft for his planned *Theatrum totius naturae*), we find one overriding concern. This is the problem of how to deal with species that were of a middle nature, as Cesi put it, intermediate between one species and another, or apparently on the borderline between two species, or partaking of the character of two or more species.

The problem of anomaly was evidently closely allied to this; and so in the *Tabulae*

Phytosophicae we find a whole section on plants that were of a *mezza* or *mezzana natura*, and in the *Speculum Rationis*, a main division headed *De mediis naturis in universo*. Here would be included species of a doubtful or ambiguous nature, *de ancipiti natura, seu de ancipiti specie, seu ambiguis*. And the next division was that entitled *De Plantis imperfectis, Quo in genere toto plantarum insigni aliquo defectu vel nota peculiari monstruosa mirabiliaque existunt* ... One of Cesi's great breakthroughs was his clear-headed realization that the problem of the ambiguous or borderline or intermediate case would only be resolved by close attention to the reproductive systems of plants. In this respect he was one of Linnaeus's most important predecessors.

Very swiftly then, Cesi and his friends realized that in order to discover the truth in such matters, they would have to penetrate the mysteries of reproduction, and to this end they studied ferns and fungi of every description, anatomized the marine turtle and the bee, and began to make use of the microscope. Indeed, there are several puzzling drawings of eggs and foetuses, and the earliest illustrations ever made of botanical specimens with the aid of the microscope appear in the volumes of plants in the Institut de France where the newly-perfected instrument (a rudimentary example had been used previously by Giovanni Battista della Porta) was used to scrutinize the sori and sporophytes on the underside of ferns. It would be several years before the first *printed* illustrations made with the aid of the microscope appeared. These were the anatomy of the bee prepared by Cesi, Stelluti and Colonna in honour of the papal jubilee in 1625, and the enlargements of the seed-pod and seeds of the 'Chinese Rose' (*Hibiscus mutabilis* L.) in Giovanni Battista Ferrari's *De Florum Cultura* of 1633.

While it is clear that most of the drawings in this corpus were made for the purposes of research and taxonomy, it is just as clear that many were intended for eventual publication. This emerges above all in the case of the citrus drawings, very many of which were published in Ferrari's *Hesperides* (for which, as we have already noted, Cassiano himself did a considerable portion of the preliminary research, in addition to organizing much of the necessary funding), and in the few fossil drawings selected for reproduction in Stelluti's strangely minimal and evasive work on fossil woods of 1637, the *Trattato del Legno Fossile*. That work *had* to be minimal and evasive because the subject of the origins of fossil woods was evidently controversial, particularly in the years following Galileo's trial. A few of the beautiful series of ornithological drawings were used in the *Uccelliera* of 1622, and others (as we may judge from the excellent little treatises written by Cassiano himself and preserved in the *carteggio* now in the Library of the Accademia dei Lincei in Rome) may well have been intended for further, never-realized publications in this domain. The same probably applies to the animal and flower drawings too. Certainly Cesi himself must have intended to publish at least part of his great series of mushroom drawings, his *Icones Fungorum*, but this project, like so many others of his, would never see the light of day.

The one project to which this does not apply was the *Rerum Medicarum Thesaurus*

Novae Hispaniae, sive Plantarum Animalium Mineralium Mexicanorum Historia. This was the great compendium of material on American plants and animals that first appeared in incomplete form in 1628 but was eventually published in full as late as 1649/51, long after Cesi had died and the Academy was effectively defunct. Moreover, its many illustrations bear no direct relation to the drawings in the present corpus. But its story is an instructive and telling one. Cesi had been preparing this work ever since he had heard of the Neapolitan doctor Nardo Antonio Recchi's summary of the massive collection of notes and illustrations gathered in Mexico by Hernandez between 1571 and 1577 and neglected ever since. Already in 1610 Cesi had proposed to publish Recchi's manuscript (which by then he had bought from Recchi's heirs), accompanied by a commentary by his fellow Lynxes. When Galileo made his epochal visit to Rome in 1611 he paid a special visit to Cesi's palace in the Via Maschera d'Oro to look at pictures of over 500 Mexican plants (presumably those in Recchi's manuscripts), and when asked his opinion confessed that neither he nor anyone else could identify them correctly.

By the end of the decade Cesi had already had prepared the set of over 800 woodcuts for the proposed work, several of which he then annotated himself; and finally, by 1628, a first edition, funded by Cardinal Francesco Barberini, appeared. This contained Schreck's edition of the botanical material, Faber's great and expansive commentary on Mexican animals, a set of important additions and annotations by Fabio Colonna, and the first thirteen tables of Cesi's own vast attempt to classify all of the botanical world, the *Tabulae Phytosophicae*. Then Cesi died, and had it not been for the devoted and patient efforts of Cassiano and Stelluti, none of this material would ever have appeared. But for years they kept the project alive, bringing it up to date, writing hither and thither for more information, and always seeking new sponsors for a revised and expanded version. Thus it was that the definitive edition of the *Tesoro Messicano* only appeared in 1651 (though several copies carry title-pages dated 1648 and 1649), long after the death of Cesi and almost all his original *confrères*. This time the book was not subsidised by the now disgraced Barberini, but rather by the Spanish Ambassador Alfonso de las Torres, who had been involved in the project by Cassiano. It contained the parts already printed in 1628, but with new front and end matter, as well as seven of Cesi's *Tabulae*, which the latter had apparently left incomplete but which were now edited and completed by Stelluti. It also included Hernandez's *Liber Unicus* on Mexican animals (including birds, fishes, and insects) and minerals, which Cassiano also had had copied in Madrid at the time of his visit there in 1626.

But the woodcuts in the *Tesoro Messicano* are positively crude and rudimentary and lacking in detail compared to the bulk of the surviving drawings. Presumably they were based on the now lost illustrations Hernandez made and had made in South America, or on Recchi's copies of them. But at the same time as Cesi realized the importance of publishing Recchi's material, as evidence of a world unknown to the ancient authorities, he also knew that it was possible to have much better illustrations made, with closer indications of colour, texture, and detail, both inner and outer, than had ever come from the Americas. And he and his fellow Lynxes made it their business to find draughtsmen — and then engravers — who could convey with as great as possible precision and detail both the things they had examined with their own eyes and the remarkable specimens seen by reliable eyes abroad. And of all the Lincei it seems to have been first Cesi and then Cassiano who went furthest in this direction.

This is why the drawings preserved in the Royal Library, the Institut de France and in many private and public collections are so important. They, more than anything else, provide the visual evidence for the Lincean researches into natural history. Aside from Ferrari's *Hesperides*, also published long after Cesi's death, and written by someone who was never a Lynx himself, the published material offers only the palest shadow of the intensity and quality of their researches in this domain. The illustrations in works such as the *Trattato del Legno Fossile* and the *Tesoro Messicano* have nothing like the precision, refinement, and analytic intensity of the drawings. All of these testify time and time again to the fact that they were based on direct observation of the specimens they show, and on a kind of observational persistence that the printed illustrations hardly ever suggest. The drawings show the same specimen from one side and from another, in sections and in segments; they focus on specific parts of plants or animals or minerals, from the snout of a porcupine to the seed pods of plants, from the sulphuric florets and pyritic growths on fossil woods to the interiors of shells and the details of leaves and grasses, from the legs and beaks of flamingoes to their wings and feathers. They carry inscriptions about colours or actually show them; sometimes they record the date they were made, often the location. Repeatedly they testify to the desire on the part of the Lincei to look ever more closely into the interior of things that until then had only been described from the outside. This was done either with the aid of the microscope or simply by picking apart and delicately dismantling the specimens which they had found or which had been sent to them by friends and colleagues from all over the scientific world of the day.

Together these drawings represent the high water mark of the belief in visual reproduction as a means of analysing and classifying the creations of nature, indeed of everything in the visible world. There had, as we have noted, been ambitious attempts to reproduce much of nature before, and there would, of course, be more; but it is hard to think of any other collection that can match the drawings from Cassiano's Paper Museum in terms of quantity, range, scope, and in so many cases, descriptive and artistic refinement. At the very height of the Lynxes' engagement with drawing as a tool of natural history research, however, their own fellow and friend Galileo laid to rest the idea that the subjective qualities of things were capable of being recorded with the same objectivity as number, mass, volume, weight, and size. And so the Lynxes themselves began to lose faith in the possibilities of pictorial reproduction as an instrument of what we would now simply call scientific research. Cesi would bury himself in his efforts to clarify the natural world in its entirety not by drawing but in tabular form; and the basis of this effort, he realized, had to be geometrical and schematic, not densely descriptive and pictorial. But neither he nor any of his fellows ever felt the same hostility to picturemaking that Linnaeus, for example, was to feel; and whatever scepticism they may have had about its possibilities lies silent beside the observational intensity, descriptive beauty, and analytic closeness of the drawings that will finally be reproduced fully in the volumes of this corpus.

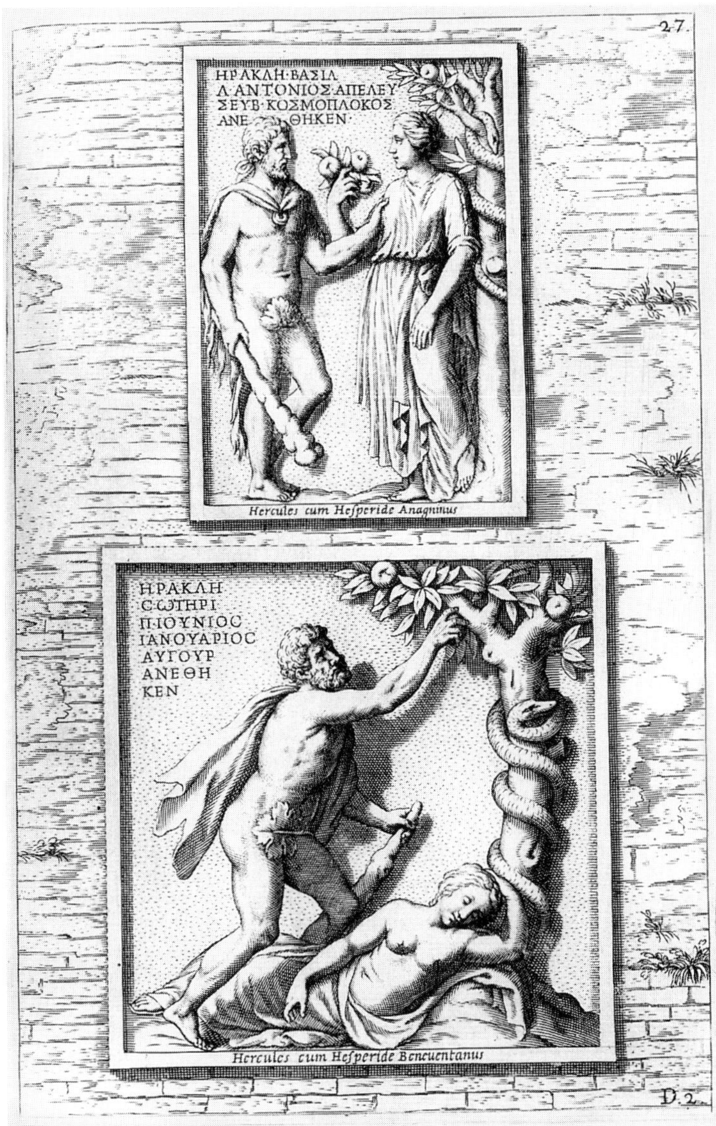

Fig. 14. C. Bloemaert (?), *Antique Reliefs from Anagni and Beneventum showing Hercules in the Garden of the Hesperides*, engraving, Ferrari 1646, p. 27

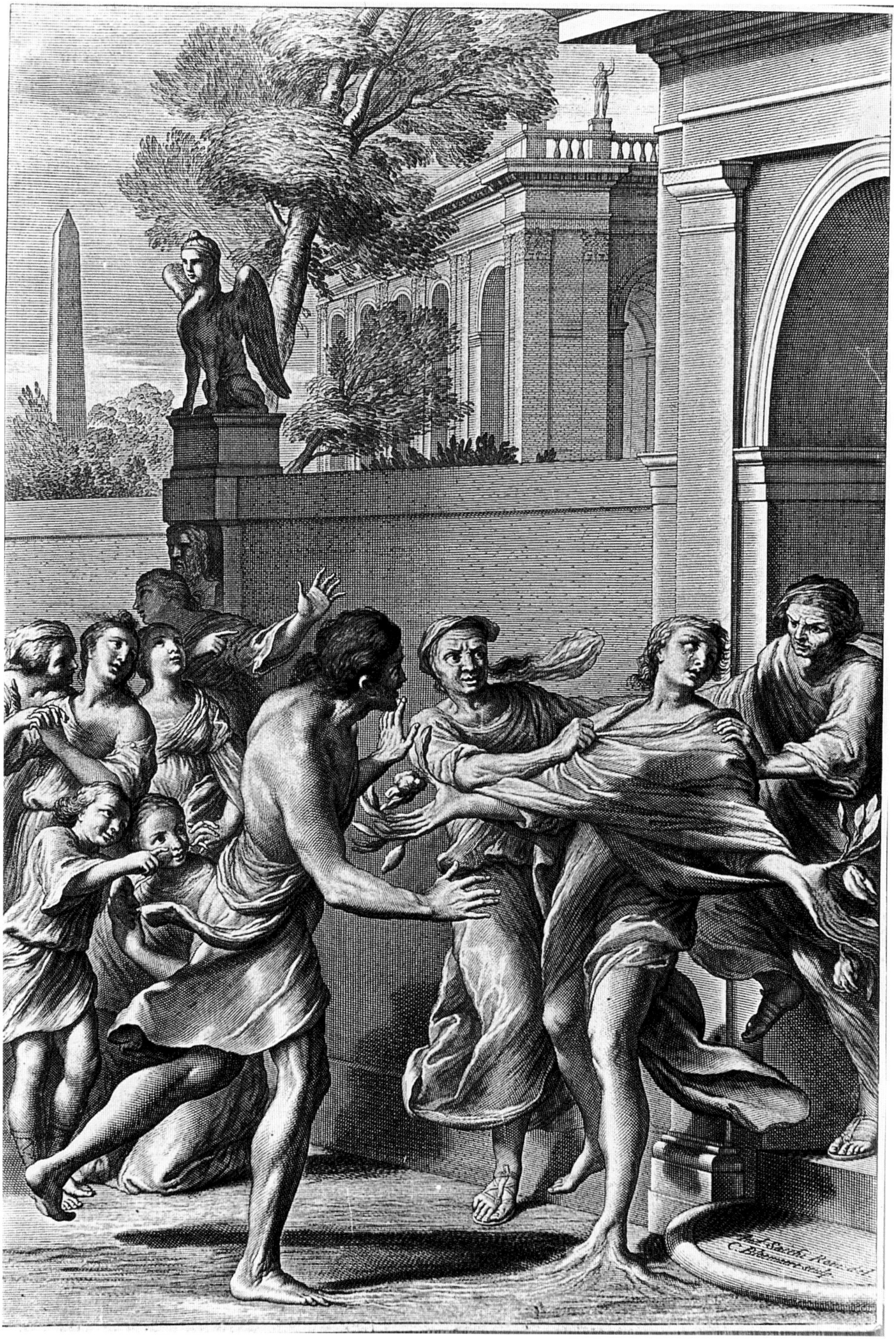

Fig. 15. C. Bloemaert after A. Sacchi, *The Transformation of Harmonillus*, (see Fig. 28, p. 75)

CASSIANO, FERRARI AND THEIR DRAWINGS OF CITRUS FRUIT

David Freedberg

JUST BEFORE CHRISTMAS 1652, Cassiano dal Pozzo wrote from Rome to his friend Francesco Nardi in Florence. As usual, there was much to discuss. For many years they had been corresponding about the wonders of the gardens of Nardi's employer, Cardinal Gian Carlo de' Medici. There was hardly a horticultural problem they did not touch upon; but above all they exchanged views about hybrid fruits, especially citrus. On this occasion Cassiano acknowledged receipt of a particularly remarkable branch Nardi had sent him. It came from a tree they had been discussing for many months. This strange tree produced a combination of both oranges and citrons; and, as so often, Cassiano had commissioned a drawing of yet another wonder of the Medicean gardens. In his letter, he wrote to Nardi that the drawing would add even further to the preciousness of 'the book I have containing the coloured figures of all the citrus fruits, which Father Ferrari used in his book on the Hesperides, and which were copied from drawings which I had made for that work'.[1] But who was Ferrari, what was the book on the Hesperides, and why should Cassiano have collected coloured drawings of citrus fruits? The story begins at least twenty years earlier.

On 30 September 1631, the famous French antiquarian and correspondent of everybody who was anybody in the *République des lettres*, Nicolas Claude Fabri de Peiresc, wrote to Cassiano from his country house at Boisgency [Belgentier]. He thanked him for three manuscript botanical treatises Cassiano had sent him about the cultivation of flowers and citrus fruit, and then gave way to the expression of what can only be described as mild panic. This seems to have been provoked by the shortest of the treatises, innocuously entitled 'Brief Instructions on Vases and Espaliers for [the citrus species then known as] *Cedrati* and *Lumie*',[2] written by Filippo Magalotti, the friend, disciple, and defender of Galileo:

> I have no difficulty with the flowers and other plants, but in the case of the citrus fruit, I really cannot grasp either the differences between the fruits or between their names.

This was the point of the letter, the nub of the matter. But first Peiresc noted that 'it would be worth the expense to have them all reproduced in the Jesuit Father's book, which the Cardinal is having printed'; and then went on to explain the real difficulty:

> I have some vases containing them, and a number of other plants in espaliers — a dozen oranges, and about fifteen or twenty different lemons or *limoncelli*. But we don't understand the difference between the *cedrato broncone* and the common one, between the Apple of Paradise and that of Adam, between the large citron from Salò and the others, between the large sweet lemon and the Spanish sweet lemon; nor do we have any record of names such as the *spongino*, the *rosso ricciuto*, the *Monna Laura*, of the *Calcedoni*, of those of San Remo and Oneglia, in the lemon group, just as for the oranges we do not have the furrowed one, the one striped with white and green, the curly-leafed and star-shaped one, and we lack the *melangolo*, though we may have some of these fruits under other names or *cognomi*. As for the *lumia of San Domenico*, we certainly have never seen it; but I would like to see the drawing of flower and fruit, and even better the plant itself, if it could be found for a reasonable price. But what you write to me about the various graftings which are now being made in Florence with citrus fruits, of so many different kinds of flowers and fruit, is worthy to be marvelled at, and deserves to be studied, if permitted, and to be practised with all diligence and delicacy.[3]

Poor Peiresc! He could not sort out the differences between the many different kinds of citrus fruit he knew, and he was even less able to associate them with groups which Cassiano may have known. He was seriously preoccupied with a problem that one might not have thought would cause much brow-rubbing to the famous student of Roman antiquities, troubadour poetry, and so much else. This was the problem of the nomenclature and taxonomy of citrus fruit, and to Peiresc, as we may judge from this letter, it seemed pressing and urgent. His chief concern was with the matter of 'difference' — something that may strike us as a perfectly straightforward concept, but was in fact a specific technical term, used in its old Aristotelian sense, for the taxonomic distinctions between species.

Typically associated with the taxonomic problems were the nomenclatural ones, but in this case Peiresc also showed his characteristic concern with vernacular nomenclature. His attention to the popular names of different kinds of citrus fruit now seems particularly striking in the light of the extreme paucity of English equivalents. But the Italian nomenclature in those days (and to some extent still now), as will be apparent from the catalogue, was exceptionally rich and colourful. Furthermore Peiresc, like Cassiano, was fascinated by the problems of grafting. And while it may be tempting to identify the 'Jesuit Father's book' with the book on the *Hesperides* mentioned by Cassiano in his letter to Nardi (which would deal with all such subjects at much greater length), the reference was undoubtedly to an earlier book by the Jesuit Father, namely his *De Florum Cultura* published in Rome in 1633.[4] It was produced with the financial assistance of Cardinal Francesco Barberini, favourite nephew of Urban VIII and well-known patron of the arts and sciences. But it would not, in the end, contain any of the illustrations of citrus fruit so much desired by Peiresc (it only illustrated flowers). For this one had to wait until the much-delayed publication of the *Hesperides* in 1646;[5] by then Peiresc was dead, and the Barberini pope as well.

The De Florum Cultura

The *De Florum Cultura* was the first of two masterpieces by Giovanni Battista Ferrari, born of prosperous parents in Siena in 1583.[6] In 1602 the young Ferrari went to Rome to join the Jesuit novitiate there.[7] His was a strange career, unusual even for the multifaceted members of the then vigorous but still youthful Order. He began as Professor of Hebrew and Rhetoric at the Jesuit College in Rome, went on to become horticultural adviser to the papal family, and faded into oblivion in his native Siena, where he died in 1655. His first book was a Syriac Lexicon;[8] his last a treatise on Sienese saints.[9] Between them he published a remarkable series of *Orations*, specimens of his skills in the old discipline of Rhetoric. Published everywhere from Lyons to London and Vilna, the *Orations* served as a model for good Latin and enjoyed a remarkable popularity throughout the century.[10] But most important of all were the two books on flowers and on citrus fruit.

The *De Florum Cultura* was an epoch-making work in the history of gardens, horticulture, and botanical illustration. Its focus was on the cultivation, culture, use, and decorative qualities of flowers, especially the exotic species then so enthusiastically being imported into Rome. It was not especially concerned with the medicinal properties of plants, as had traditionally been the case with illustrated botanical works. Above all it was a horticultural book, with much advice on the planting and planning of gardens.[11] In at least four respects the *De Florum Cultura* broke new ground: first, in its concentration on decorative and ornamental flowers and their suitability for the new gardens of Rome; second, in its profound concern with the names of plants, and establishing them with certainty; third, in its botanical illustrations; and fourth, in its allegorical plates designed by four of the chief Roman artists of the time, but following the literary inventions of Ferrari himself.

The concentration on the decorative and horticultural certainly offers some of the chief delights of the book; but these, clear and apparent enough, need no further discussion here. There are other aspects of Ferrari's work, however, that do merit comment at this stage. To a large extent the preoccupation with nomenclature arose from two rather different factors: on the one hand Ferrari's philological training, and on the other the practical problems he encountered daily as a gardener and horticulturist. After all, he had been placed in charge of a brand new garden — that of the freshly-built Barberini Palace on the Quirinal — where some of the newest and rarest species then being imported from Asia, Africa, and America (generically regarded as 'the Indies') were being cultivated. At a time of nomenclatural confusion it was important to get the names right, and to establish the correlations between the old Latin names (sometimes there were several for the same plant) and

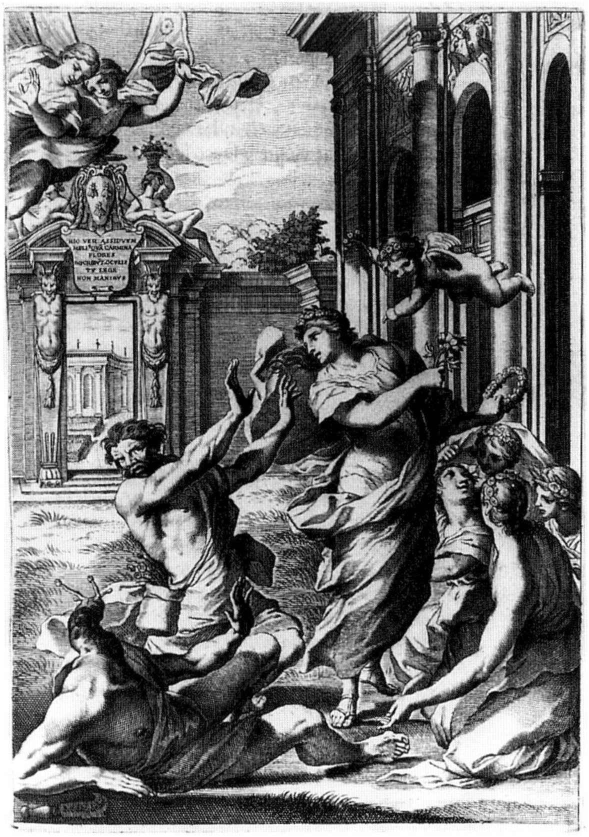

Fig. 16. J.F. Greuter after A. Sacchi, *The Transformation of Limax and Bruchus*, engraving, Ferrari 1633, p. 55

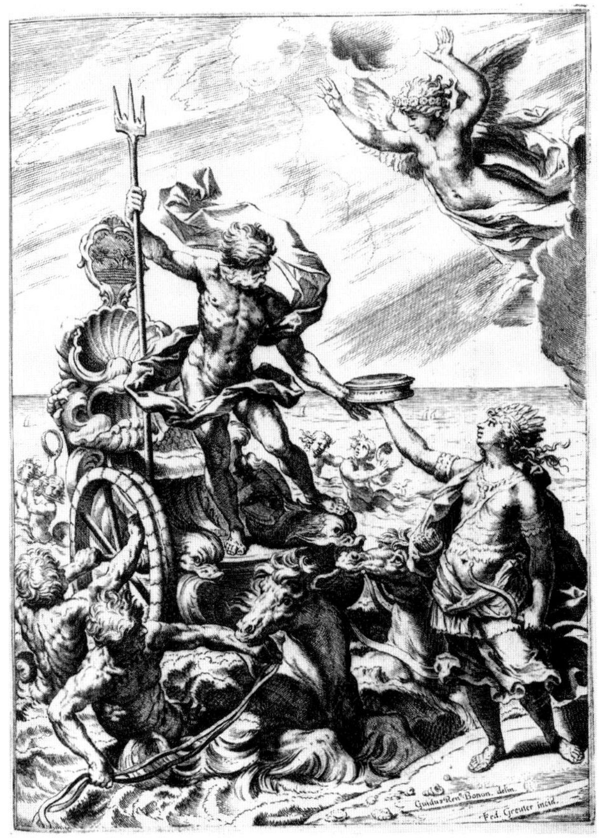

Fig. 17. J.F. Greuter after G. Reni, *The Indies handing seeds to Neptune to transport to the Barberini Gardens in Rome*, engraving, Ferrari 1633, at p. 376

the vernacular ones (often there were many for the same plant). In communicating about plants one had to be sure that one was talking about the same species. Naming went hand in hand with the desire to classify and order. The old guides and authorities — Aristotle, Dioscorides, Theophrastus, Pliny — were no longer adequate. It had become increasingly obvious that their coverage of the plant world was insufficient, not only because of the rapidly growing number of local plants which were being discovered and could ever less frequently be forced into the Aristotelian and Dioscoridean taxonomies, but above all because of the ever-increasing number of plants that were being imported into Europe and which the ancients could never have dreamt of. New names had to be devised to fit new systems of ordering.

The taxonomic pressures were clear, and Ferrari, realizing the urgent need for taxonomic clarification, made a start with his drive, throughout the *De Florum Cultura*, towards nomenclatural resolution. He adduced all the names of the specimens he discussed, considered their etymologies, their linguistic and vernacular variations, and sought to arrive at firm conclusions. No one previously had been quite as persistent in this domain, and few had made it their business to provide such solid nomenclatural coverage.

But if the concentration on names and naming was a notable and comparatively novel feature of Ferrari's book, its visual component was even more significant. On the one hand there were the botanical plates, almost all engraved by Cornelis Bloemaert (1603–?84), a Dutch printmaker working in Rome, which were executed with an unprecedented degree of care and precision. They display a prettily efficient economy of line that manages not just to be clear and decorative, but also to convey the essential differences between species — sometimes very closely related ones. The clarity and meticulousness of the botanical plates were essential to the success of Ferrari's finely-tuned nomenclatural drive, less efficient and consistent by far, it is true, than that of either Bauhin or Linnaeus, but conscientious, deliberate and certainly clearer than many other taxonomic attempts in this field previously.

Then there were the allegorical illustrations, by Pietro da Cortona (who provided five designs), Andrea Sacchi, Guido Reni, and Lanfranco.[12] All were made to illustrate stories which Ferrari made up himself, to lighten, as it were, the serious scientific fare of the book, to mix the *utilis*, as the ancients would have had it, with the *dulcis*. Ferrari told long and windy tales about the presence of the gods on the horticultural scene, or about such events as the transformation of the thieving gardeners Limax and Bruchus into a snail and a caterpillar respectively. This particular scene, for example, was illustrated by Sacchi [Fig. 16], while Guido Reni designed the plate showing the personified Indies handing over seeds to Neptune to transport to the gardens of the Barberini in Rome [Fig. 17]. Pietro da Cortona designed the frontispiece as well as several plates showing the freshly-built Palazzo Barberini in the background, such as the delightful one of the dance of Vertumnus in the presence not just of Flora, Nature and Art but also of the hibiscus tree, or the 'Chinese Rose', as Ferrari called it.[13] He proudly noted that he had been the first to grow this exotic flower in Rome.[14]

Ferrari's tales are notable not just for their rhetorical flair (for that they have, despite their long-windedness) but above all for their independence from any earlier literary or iconographic source. They are wholly his own; and by some happy set of circumstances he managed to have them illustrated by four of the leading young artists of his time. Of course, he would not have been able to pay for these commissions himself, and so it is not surprising to discover that it was through the agency of Cassiano dal Pozzo that the book as a whole was subsidized by the papal nephew and patron of all the arts and sciences, Francesco Barberini. Not only did the young Cardinal pay for the costs of production and the commissions from the artists, he also paid for the engraving of the plates.[15] The printmakers concerned were Cornelis Bloemaert for the large majority of the botanical plates, Johann Friedrich Greuter for most of the allegorical plates, and then a range of other figures, such as Claude Mellan, who engraved one allegorical plate, Camillo Cungio, and Anna Maria Vaiani, who did at least one of the illustrations of bouquets, and one or two other figures whose identities have yet to be established with certainty.

For the sake of completeness it should be mentioned that the *De Florum Cultura* also contains a series of charming and inventive designs for garden plans and parterres, as well as a number of illustrations of horticultural implements, flower bouquets, and a variety of ingenious vases and boxes for transporting flowers and flower arrangements. Nor ought it to be forgotten that the book also contains the first printed botanical illustration ever made on the basis of a microscopic investigation, namely the plate showing the seeds and seed-pods of Ferrari's beloved 'Chinese Rose'.[16]

For the rest of the decade, Ferrari continued to teach Rhetoric and some Hebrew at the Collegio Romano, as well as advising on the Barberini gardens. The *De Florum Cultura* turned out to be a fine success, with a charming translation into Italian by Ludovico Aureli appearing in 1638 and negotiations for a new edition soon under way (it was finally published in Amsterdam in 1646).[17] From his correspondence with Cassiano dal Pozzo we know that almost immediately after the appearance of his flower book, Ferrari began to project a large book on fruit, a *Pomona*, which in the course of the later 1630s seems to have turned into a plan for a book on citrus fruit alone. This was a project that occupied the bulk of his attention — and much of Cassiano's — for many years, and they seem to have had constant difficulty in getting adequate funding for the planned work. Once again, we may turn to the correspondence between Cassiano and Peiresc for more light on the genesis of the book on the apples of the Hesperides.

The Compilation of the Hesperides

In a postscript to a letter to Peiresc dated 22 December 1635, Cassiano writes that 'the treatise on citrus fruit is beginning to be put together by the same Jesuit Father who did the book on flowers; and so I am enclosing herewith a list of instructions for you to use when you gather what information you can over there about the citrus fruit of Provence'.[18] This list of instructions happily survives on a small slip of paper in another manuscript in the Bibliothèque Nationale, Paris. Under the heading of *Capi d'Informatione*, Cassiano enumerates all the different categories of information Ferrari required about 'oranges, lemons, and citrons, as well as the species in which these are divided that are to be found in Provence and its surroundings'. He asked for:

> 1. The names of each kind and the origins of each name. 2. The shape of the tree, the leaf, the flower, and the fruit. 3. How it is sown. 4. What site it requires. 5. What kind of earth. 6. What manure. 7. How to protect it against cold and heat. 8. How and how much it has to be pruned. 9. To what diseases it is exposed and how to cure them. 10. How hard. 11. How they are grown in vases or in the ground. 12. Other pleasures and uses of this kind of plant ...[19]

These are exactly the same *Capi d'Informatione* which Cassiano requested from a large number of correspondents throughout Italy, from North to South, especially in the mid 1630s, as we learn from the extraordinary compilation of notes bound together under the title of *Notizie diverse del sig. Abb. Cav. Cassiano dal Pozzo; originale spettante a' Agrumi et Historie d'essi; stampata in Roma del P. Gio. Batta. Ferrari della Compagnia di Giesù sotto Titolo d'Hesperides* that still survives in the Archivio Dal Pozzo in the Accademia Nazionale dei Lincei — and which also happens to contain the very covering letter with which Ferrari sent Cassiano the outline of all the *Capi d'Informatione* he needed.[20]

Now writing from Aix, Peiresc acknowledged receipt of Cassiano's letter of 22 December (and the accompanying request for information) on 31 January, 1636. His response seems exhaustive in its detail. He wrote:

> In order to satisfy the note about citrus fruit along the lines suggested by Father Ferrari, I should really be at Boysgency … I think it's necessary to see his description of the trees that occur in the Italian gardens, so that if we have any species here that don't occur there I can make an exact description, and send you a living shoot. As for the others, I will simply add the different names, and the different ways in which they are cultivated according to our usages. I will also send the different denominations. These cause great confusion, since here we do not know how to distinguish between the species of lemons and citrons and *lumie*. We would be very happy to learn the real differences, as in your usage, which are perhaps not the same as the usages in Genoa, Florence, Naples, Sicily, and Spain. Here citrons are called *pomcyres* (as if they were Syriac or Assyrian apples [*pomi*]) and are very large; the lemons are smaller, but taste better. There is also another species called *limes*, or *une lime*. They are feminine, smooth, and have a very unpleasing taste, but look both pretty and bizarre. We have three or four species of the citrons called *pomcyres*, some smooth, others ridged, others ordinary, others *bigarrati*, that is horned, and cut, as it were, into various fingers. The Genoese called them fingered citrons [*cedri dalle die*]. I have only seen one kind of lime. In the lemon family, here called *limons*, there are both large and small ones, of the kind called *pomcyrats* (rather than *cedrati*) which imitate the forms of the *pomcyres* and which have a much better taste. There are also the *bigarrati*, that is, horned, and some in the likeness of a pear, which the Genoese call a *lanza spada*. These can be sweet, large, small, and medium-sized. We have many kinds of oranges, but the best are the *bigarrati* or horned ones, and finally we have had those which have a white flower but are almost the colour of wine on the outside, as in the case of citrons. And there is another which has leaves, bark, and peel veined with yellow and green, that has not yet been seen. Also in the genus of oranges are the Adam's apples. They are very large and good to eat, sometimes even with the skin. Then there are certain small oranges, called chinese; one eats the peel together with the fruit, and even the flower eaten with sugar is very tasty. These are the principal species. Oranges and

> lemons *dalle die*, or horned, have also been seen. They have several rinds, one on top of the other, but we have not been able to keep the trees or the grafts that were derived from them. I understand that one finds lemons with three rinds in Candia, with the juice in the centre, but I have not been able to find one …
>
> For some time several people have been so involved in these citrus pleasures here in Provence that they have had specimens sent from afar; if they could not get them from Nice they had them sent from Genoa, and especially from San Remo.[21]

In this letter the obsession with taxonomy and vernacular names emerges even more strongly than before. Here again one hears the voice of the philologist, not so much the classical philologist, but the vernacular one. It is the man who was deeply interested in troubadour poetry, and the native languages of Provence, who speaks, rather than the learned antiquarian. Also worth noting is the emphasis on (and pleasure in) teratological forms such as the horned and digitated specimens which feature so prominently amongst Cassiano's citrus drawings. And then there is the close descriptive detail: to us this detail may sometimes seem trivial, but it is altogether typical of the attentiveness to surface and to what would come to be called the secondary qualities of things — namely those which were dependent on objective, sensual perception — which defined the observational practices of the time, at least in the field of natural history. In letters such as these, natural history had not yet become scientific in the modern sense of the word, despite the emphasis on direct observation. But direct observation, as we shall also see, would not be sufficient for Ferrari.

On March 7 Cassiano replied to Peiresc. His letter casts further light on their joint citrus investigations and on the long and complicated history of the *Hesperides*.

> I have communicated what you have written about the Provençal plants to Father Ferrari, who is contemplating his new labour on citrus fruit, and I shall see to it that you have the information you desire, since there are so many species — and those all poorly divided … The species there called *pomcyres* probably correspond to our *ponzini*, which are very similar to citrons, but superior to them, since they have a very delicate flavour (the citron is really no use unless it is turned into jam or candied). We will send you a note of those species which we have here and which we need … Father Ferrari intends to divide all this information into kinds, that is into Citrons, Lemons, and Oranges, since all citrus fruit can effectively be reduced into these three species, and he will begin with the fable of the Golden Apples of the Gardens of the Hesperides. I have given him a brief account of those which have been seen both in Genoa and Florence, which come to about fifty species. Among the plants that have been noted is one which has a very sweet-scented flower with many leaves, that seems like a double rose. It came from the garden of

> the Prince of Caserta, today Gaetano, Duke of Sermoneta. The citron from Candia is a very noble plant, and we are hoping that it will bear fruit in spring. It was sent from Candia by my friend Monsignor Stella, Bishop of Vicenza. The *ponzini* of Sorrento are so highly regarded that they are given as presents to His Holiness. The *cedrati* of Florence are of equally extraordinary renown. One finds double examples of these at Lucca. Both are very delicate and suitable for use at the table. Here in Rome we have a lemon from the Mattei garden which is superior to all others, as a result of which it is called "without equal" [*nullapari*]. Ferrari's discourse will be a most interesting one, because he discusses very many things regarding citrus fruits, both from the point of view of taste and their medical use ... [22]

These are hardly systematic reflections. Their randomness, alas, is all too typical of Cassiano's scientific and natural history notes. But the whole exchange between him and Peiresc on the subject of citrus fruit gives one some idea of the extraordinary intensity with which the subject could then be discussed. Cassiano's own letter also offers some insight into his role as a crucial intermediary in the dissemination of natural history information, and, more specifically, into his close collaboration with Ferrari that would eventually result in the publication of the *Hesperides*.

The nature of this collaboration can be specified quite precisely. As in the case of the ornithological book that served as Cassiano's presentation piece upon his inscription as a member of the Accademia dei Lincei in 1622, Cassiano acted as a kind of impresario, gathering information from a large variety of his correspondents all over Italy, and then handing this information over to someone else to edit, compose, and write up in the form of a book. The book for the Accademia was called the *Uccelliera* or 'The Aviary: a discourse on the nature and properties of different birds, in particular of song-birds, and how to catch, recognize, breed, and keep them'. But it was not written by Cassiano. It was written, according to its title-page, by a lawyer friend of his, Giovanni Pietro Olina, and published by Andrea Fei in Rome in 1622.[23] It deals with everything from the taxonomy of birds to birdsong, hunting, and recipes for cooking. Although his name is barely mentioned in the *Uccelliera*, Cassiano was clearly the *genius libri*: he supplied both the information on which it was based (many of Cassiano's own ornithological notes, analyses and small treatises still survive), and — most importantly of all — he commissioned the drawings from which its illustrations would be engraved by Francesco Villamena and Antonio Tempesta.[24] Much the same thing would happen in the case of the *Hesperides*, though in this case Ferrari was obviously much more of an expert on citrus fruit himself than Olina was ever likely to have been in the field of birds. Neither book could have appeared without Cassiano's active assistance, both scientific and financial. He was both the moving spirit of each book, and a kind of *éminence grise* behind them. His name hardly appears in either of them, two or three times in each case. It is altogether typical of Cassiano's personality and attitude that unless one had the manuscript evidence of his work in both spheres, one would never guess from the books themselves quite how large a role he played in their genesis.

The compilation *Notizie diverse del sig. Abb. Cav. Cassiano dal Pozzo; originale spettante a' Agrumi et Historie d'essi; stampata in Roma del P. Gio. Batta. Ferrari della Compagnia di Giesù sotto il Titolo d'Hesperides con il disegno della Veduta della Riviera di Salò* consists of 131 folios containing the information about citrus fruit which Cassiano had gathered from his correspondents all over Italy.[25] It includes his own requests for information, generally made on behalf of Ferrari, as well as his correspondents' often very lengthy and detailed replies. Drafts and cancelled drafts, outlines, chapter-headings and many lists of names, often but not always written in the neat hand of Cassiano's favourite scribe, mingle with reports from aristocrats, ecclesiastics, and gardeners from north to south. There are reports of the arrival of ships containing citrus fruit from overseas.[26] Instructions for planting are interspersed with recipes for candied fruit and pot-pourri.[27] The *disegno della Veduta della Riviera di Salò* may well be the design which Poussin provided for eventual illustration in the *Hesperides*, showing the Nymphs of the Hesperides bringing citrons to the shores of Lake Garda [Fig. 20]; but still loosely inserted at the end of the manuscript is another drawing measuring 288 x 429 mm, which, according to an inscription on its verso, shows a view of Nervi (a particularly favoured location near Genoa for the growing of oranges and lemons), drawn by the Flemish artist Cornelis de Wael [Fig. 18]. The drawing was commissioned, as it transpires from a note on the back of the drawing itself, by one of Cassiano's most assiduous and careful correspondents on the matter, Cattaneo Cataneo of Genoa, and was perhaps intended for use in the *Hesperides*.[28] At the end of the manuscript is another drawing (220 x 325 mm), done in chalk and now much worn, of a group of citrus fruit, almost certainly of a selection from Reggio Calabria and keyed to numbered indications on various pages in the manuscript itself [Fig. 19].[29]

At the very beginning of the *Notizie diverse* is the detailed contract for the *Hesperides* signed by Cassiano, Ferrari, and the publisher Herman Scheus. This is followed by Ferrari's own rider to the contract and lengthy descriptions of how he wanted the title-page, and some of the allegorical and archaeological illustrations, to look.[30] Then, forming the large bulk of the manuscript, come the reports on citrus fruit and various pieces of evidence, chiefly letters and drafts, for the discussions that evidently took place between Ferrari and Cassiano on the subject of citrus fruit. Some of this material bears directly on the drawings in the present catalogue, and supplements the information provided in the printed version of the *Hesperides*. That the manuscript was rebound some time after its initial compilation by Cassiano is clear from the several discrepancies between the page references in Cassiano's own index and the actual page-numbering within the manuscript as presently constituted.[31]

The earliest notice in the manuscript dates from 1626, describing oranges grown in the famous Ricciardi gardens in Via Gualfonda (currently Via Valfonda) in

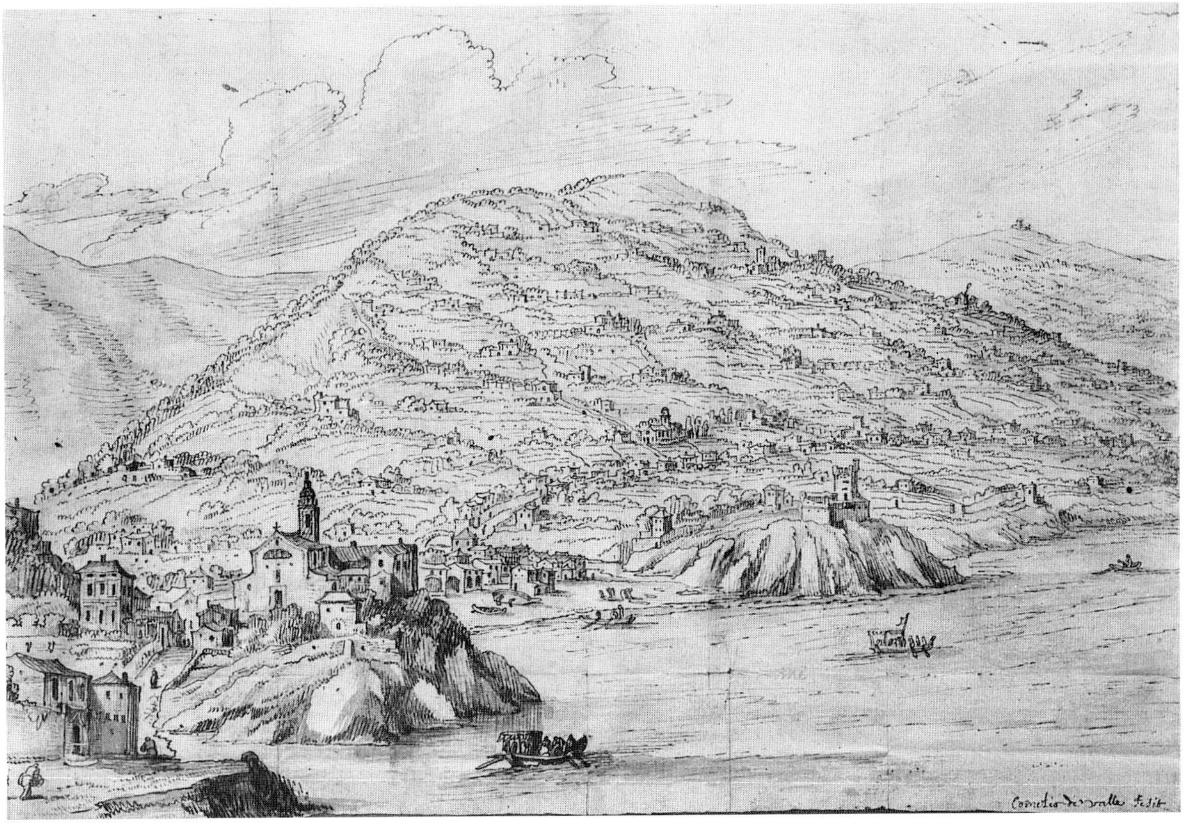

Fig. 18. C. de Wael, *View of Nervi*, pen and ink, Archivio Dal Pozzo MS 39, loosely bound in at end

Fig. 19. Seventeenth-century Italian, *Arrangement of Citrus Fruit*, black chalk, Archivio Dal Pozzo MS 39, loosely bound in at end

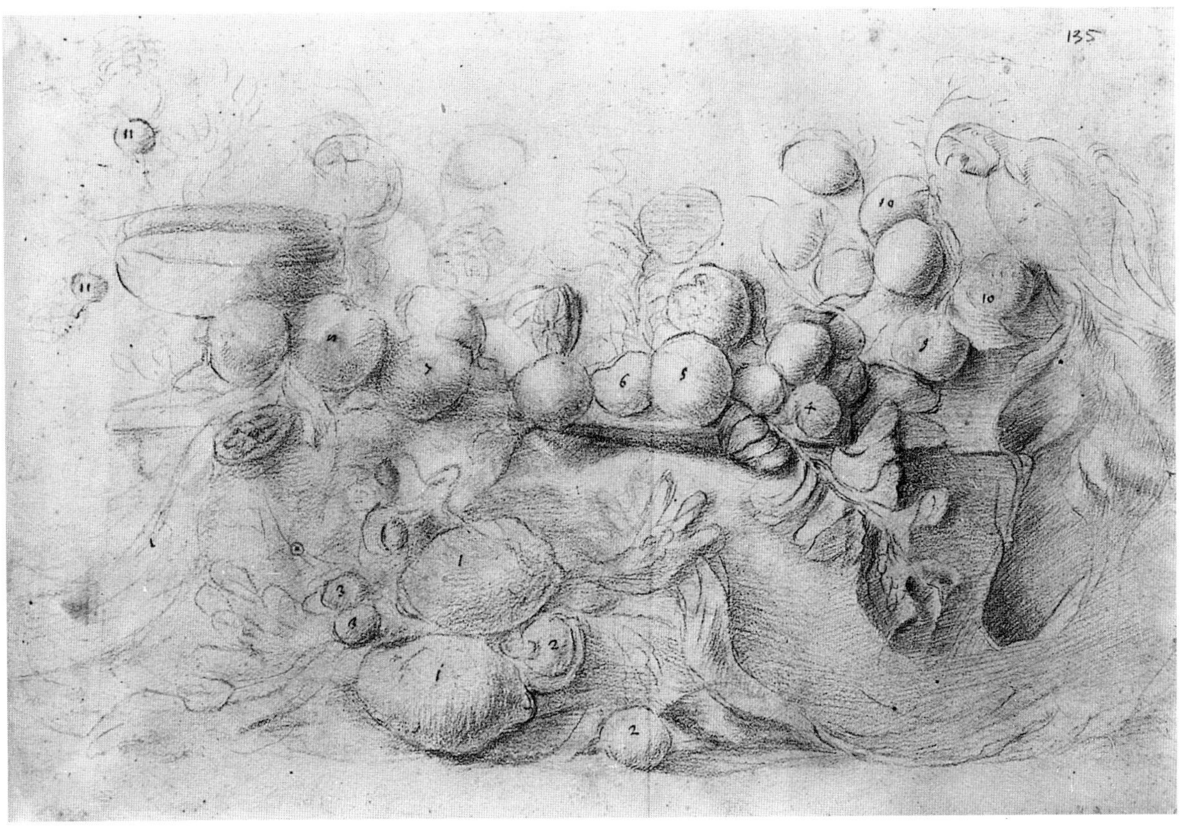

Florence. It was provided to Cassiano by the Ricciardi gardener, Colombini, at the time of the return, via Florence, of the legation headed by Francesco Barberini to Paris and Madrid, a legation of which Cassiano was a member.[32] That Cassiano continued to collect citrus material after the publication of the *Hesperides* in 1646 is clear not only from one or two later notices in this manuscript, but also from his correspondence of the 1640s and 1650s with Nardi, also preserved in the Archivio Dal Pozzo in Rome.[33] The bulk of the material, however, dates from the one or two years after 22 December 1635, when Ferrari sent Cassiano his *Nota de' Capi che si desiderano sapere per havere compita informatione di tutti gli Agrumi*.[34] It was this *Nota* which Cassiano then had copied and sent to Peiresc in Provence and to correspondents all over Italy, ranging from Alessandro Gaetano in Sicily[35] and the Provincial of the Dominican Order in Naples to Cattaneo Cataneo in Genoa (his most voluminous source),[36] Scipione Barone in Tropaea,[37] and Silvestro Morosini, Bishop of Treviso, who sent information about the citrus fruit around Lake Garda.[38] The Capuchin fathers of Naples sent an account of the citrus fruit of Sorrento,[39] while other first hand descriptions came from Nervi, from San Remo, from Toscanella,[40] and from the two Reggios. Observations about oranges freshly arrived from abroad come from Massa, Pietrasanta and Livorno, often by the most circuitous routes,[41] and there are detailed observations about some of Italy's most venerable orange trees, such as those that still grow in the gardens of Santa Sabina in Rome.[42] The Flemish gardener Adriaen Mynsicht provided information about gardens and garden practice in Flanders,[43] just as another Fleming, Willem de Blasere, did for Ferrari himself.[44]

There is much on the folklore of citrus fruit, including their use in popular games, and Cassiano seems to take a particular delight in enumerating culinary and household uses. In his own correspondence one often finds references to gifts of delicacies and wine, either received or sent, and here there are several mouth-watering pages on citrus jams, candied fruits, and juices. Wherever possible actual specimens were exchanged, and it is perhaps significant that when Ferrari himself received a specimen, he would send it to Cassiano to be drawn and for a detailed description. Throughout the manuscript, the problems and implications of grafting are a paramount concern.[45] Particular attention is paid to hybrids and teratological specimens, often referred to in the manuscript as species *che scherzano* ('that joke'), as well as the whole group of specimens deformed in such a way that they appeared to be digitated.[46]

For several years, then, the collaboration between Cassiano and Ferrari was very intense indeed, with Ferrari devising the plan of research and the programme for his forthcoming book, and sending specimens to Cassiano for examination and description. Cassiano, for his part, passed on the notes and descriptions he received or had made himself, for Ferrari to digest and reduce to suitable form for inclusion in his book. But what is perhaps most pertinent in the present context is the fact that several correspondents refer to drawings of citrus fruit which they were sending to Cassiano along with their responses to his request for information, while on several occasions Cassiano mentions drawings which he himself had had made. When Ferrari receives a specimen directly from Cardinal Antonio Barberini, he writes to Cassiano begging him to have coloured drawings made as soon as possible, both of the whole and the half fruit.[47] The more one studies this material, the clearer it becomes that coloured drawings formed an essential part in the exchange of information about citrus fruit, and played a key role in the fundamental problems of identification, naming, and classification.

The Drawings in the Paper Museum

Anyone who takes more than even a cursory glance at the reproductions in the present catalogue will immediately be struck by a general qualitative discrepancy between the drawings. To some extent this is not at all surprising, since the quality of the drawings in most other sections of Cassiano's Paper Museum also varies widely. But in this case it seems possible to associate *some* (but certainly not all) of the poorer drawings with those sent to Cassiano by his correspondents (and then passed on to Ferrari), while the more finished drawings, clearly by a single hand, are presumably those commissioned by Cassiano himself, and then used as the basis for the engravings of citrus fruit in the *Hesperides*. The name of the artist who made them can be established with an unusual degree of certainty — unusual, at least, for the Paper Museum, where attributions are often hazardous in the extreme. But in this instance the evidence is compelling.

On page 69 of the *Hesperides* Ferrari tells us who made the drawings for the engravings of citrus fruit. Ferrari rhapsodizes:

> O Vincenzo, you double nature with your art, since you produce real fruit by what you paint within this volume; indeed, you bring about a new and real miracle, since the same things are born on these pages as in the soil. Thus in this light volume a large *pomarium* may easily be carried around. And is it not another miracle, that the fruits which are enjoyed here need not be torn from their branches?[48] They are solely for the pleasure of the eyes. It is forbidden to pluck them with the hand, to eat them with the mouth; and so these marvellous fruit may be preserved uncorrupted for eternity. As long as they endure, your fruitful talents will never be contested.[49]

And in the margin the rest of his name is given: *Vincentius Leonardus* it simply says, Vincenzo Leonardi. Despite Ferrari's rapturous tribute, we do not, at present, know anything else about this otherwise forgotten artist, except that his name also occurs in the *Uccelliera*[50] and on the verso of the drawing for a bird preserved in the Royal Library.[51] Like many of the best of the citrus drawings, this is done in a subtle and meticulous watercolour technique, finished with the application of gum arabic, just as in the case of those which were made for reproduction in the *Hesperides*. In the following catalogue, therefore, Leonardi will be credited with the authorship of a reasonably coherent group of drawings that are clearly by a single hand and are of a superior quality (in terms of draughtsmanship, attention to detail, preparation of the paper, and finish), to the others. The drawings that may be attributed securely to Leonardi are almost all characterized by the use of heavy bodycolour, gum arabic (in the highlights and in the dark shadows cast by the fruits), brush stippling in the fruits themselves, and black chalk underdrawing. Most are on cream prepared paper.

Leonardi generally began each drawing with a rough sketch of the specimen in black chalk, and then established its basic form in coloured wash. At this stage he would have added shadows, stalks, and leaves where necessary. Next he applied successive layers of bodycolour, with white highlighting and gum heightening. In order to give a sense of the texture of each fruit, he delicately stippled its surface in watercolour, and sometimes incised — or even scraped into — the bodycolour itself. At some stage or another in these complex processes of graphic development he often deepened the shadows by the further addition of black watercolour. The results are exceptional for the delicacy of surface colour and detail, and the sense of volume which almost all the securely attributable drawings convey to the viewer.

But there is also another large group of drawings which at first sight seem to be more coarsely executed and lacking in the precision of detail and finesse of handling of those by Leonardi himself. In these drawings the use of the brush seems much broader and the wash much thinner. Often the white of the paper still shows through, and the leaves have a characteristic coppery-green colour. Surface detail is generally treated in a more cursory, even casual, way. Little if any gum arabic may be discerned in most of the drawings in this group. But on closer inspection it becomes clear that these drawings are probably not by another hand, but are more likely to be drawings by Leonardi himself, which for one reason or another he left unfinished or decided not to develop further. Thus drawings such as **15**, **28**, **48** and **63** all seem to show a progression from wash to finished drawing. **15**, for example, shows Leonardi's typical black chalk underdrawing and a basic form indicated in wash, with some layers of bodycolour added — but not quite sufficient to prevent the paper itself from appearing through. **48**, on the other hand, shows all the typical characteristics of Leonardi, but the leaves are still unfinished and the shadow, rather than being sharply defined, is still in a comparatively light grey wash. Other drawings such as **14**, **66** and **107** seem to be at a still earlier stage of development. In the catalogue this group of drawings will therefore be attributed to Leonardi with a query.

Two drawings — **97** and **98** — are on vellum, and are at least as good as any by Leonardi in terms of refinement of handling. But unlike the fruits he shows, the specimens represented by this artist are shown in a delicately oblique perspectival setting, in two cases even with the indication of a shelf or step on which they seem to rest. This suggestion of a setting in depth is unlike any of Leonardi's drawings. The stippling in these drawings on vellum is very fine and minute indeed, and bears a marked resemblance to the fruit and still-life drawings by Giovanna Garzoni, such as those shown in the watercolours on vellum of *The Old Man of*

Artiminio, and the three still lifes with cherries, an open pomegranate, and melons, all in the Palazzo Pitti.[52]

Four drawings (**103**, **104**, **117** and **118**) have graphite underdrawing and are distinguished by a combination of yellow and orange watercolour in the fruits not found in the other drawings. For these drawings no attribution may be suggested at the moment, although they seem clearly to be by an artist familiar with Leonardi's work, who may even have been trying to emulate him (or even to copy existing drawings).

Finally there are a few drawings (e.g. **37** and **71**) which cannot be classified within any of the above groups and whose authorship must therefore be left an open question.

Since in the *Hesperides* there are eighty plates showing citrus fruits (and many more images, as almost all the plates show the fruits both whole and in sections, sometimes from a variety of viewpoints), and since only fifty-five drawing sheets, some bearing more than one drawing, may be associated with thirty-eight of these plates, one must assume that several are lost. On the other hand, there are also a number of surviving drawings which cannot be associated with any illustration in the *Hesperides*.

Some may have been the drawings referred to in the reports of Cassiano's correspondents, others may have been made earlier and come from Cesi's own vast holdings of botanical drawings, and others yet could well have been made after the printing of the *Hesperides*. One or two may even be copies of one or another illustration in the *Hesperides*, or other drawings. What is clear, however, both from the *Notizie diverse* as well as from Cassiano's correspondence with figures such as Francesco Nardi, is that he continued to have citrus drawings both made and sent to him for many years after 1646. Even of those made before then, some may have been intended solely as instruments of discussion and research, rather than for eventual publication. But a more likely explanation is either that Ferrari had nothing to say about them, or that it would have been too expensive — and too unwieldy — to reproduce them all. The book itself was large enough, and Ferrari may well have had some trouble in finding funds to publish it, even with the aid of Cassiano. That this was so is suggested by the difficulty Ferrari had in finding a suitable dedicatee.

By the early 1640s, when Ferrari was bringing the book to a close, Francesco Barberini seems to have been much less of a presence, and a concerted effort was made to obtain the patronage of Louis XIII of France, as is clear from several letters which Cassiano wrote on Ferrari's behalf to the French painter Poussin at the time of the latter's sojourn in Paris in 1641–2. But the King had more local projects to sponsor, and even if he had been favourably inclined — which is unlikely — Poussin himself left Paris before anything could be done. The King died in 1643, the Pope in 1644, and soon after Cardinal Francesco himself left Rome in disgrace. In the end the book was dedicated to Ferrari's home town of Siena, which probably offered a small subvention (a letter of thanks to the *Balia* survives in the Archivio di Stato in Siena,[53] and a specially bound dedicatory copy in the Biblioteca Comunale there).[54] But it seems that Cassiano himself supplied the drawings for most of the illustrations and probably paid for at least some of them.[55] Who exactly covered the rest of the expenses for the book has not yet been discovered, though it may be that the publisher himself took it on as a prestigious commercial venture, a venture supported by one of the greatest scholarly entrepreneurs of the time.

The Hesperides

The *Hesperides sive de Malorum Aureorum Cultura* was the largest and most sumptuous book on citrus fruit ever yet published. It started a vogue for beautifully illustrated books on the subject that would be continued with Jan Commelin's *Nederlantze Hesperides* of 1676[56] and J. C. Volkamer's even more spectacular *Nürnbergische Hesperides* of 1708–14,[57] and which culminated in the Abbé du Breuil's 1872 edition of Risso and Poiteau's 1818 *Histoire naturelle des orangers*.[58] Prior to Ferrari, of course, many natural historians and poets had written about citrus fruit, precisely because of their aesthetic, gastronomic and mythological appeal and interest. Celio Calcagnino's *Commentaria de Citro* and Giovanni Pontano's two books *De hortis Hesperidum* were both cited repeatedly by Ferrari, and he certainly knew both the illustrations and the discussions of citrus fruits in the books by the great sixteenth-century botanists, notably Matthioli,[59] Clusius,[60] and Dodonaeus and Lobelius too[61]. But Ferrari conceived a work of a different scale and order than any of these.

The *Hesperides* combined literature, art, mythology, etymology, ethnography and botany in a way that had never been done before. On the surface it may seem to be very similar in approach to the works of the great pioneer of encyclopedic natural history illustration, Ulisse Aldrovandi (1522–1605).[62] Like Aldrovandi's mostly posthumous books, which were based on a very much larger corpus of visual material than ever assembled before, Ferrari's treatise appears at first sight to be a promiscuous jumble of every imaginable piece of information about citrus fruit. It seems to be deliberately encyclopedic, in the rather random tradition that began with Pliny the Elder and was still reflected in Aldrovandi. Suffused with classical learning, it deals with every possible aspect of the mythological and archaeological resonance of citrus fruit. It tells the ancient tales involving oranges, lemons, and citrus fruit, beginning with the story of Hercules's guardianship of the garden of the Hesperides and his fight with the dragon Ladon. It reproduces and discusses ancient reliefs and coins of Hercules and the Hesperides [Figs. 14, 31]. But at the same time Ferrari was also interested in modern usage and lore, and so he recounts matters such as the use of citrus fruit in popular games — notably the famous citrus fights of Reggio Calabria and Treviso — and discusses the origins of such games.[63] Particularly attentive to what can only be called the ethnography of citrus fruit, he provides valuable information, for example, about the Jewish use of *ethrogim* at the Feast of Tabernacles, how the Jews of various parts of Europe both import and export them and how they take special care to select them.[64] He is interested not only in complex Latin and Greek etymological issues, but also in the origins of citrus terms in other languages, from Portuguese to Dutch, from Chinese to Arabic and Hebrew. At several points he even cites the brilliant but very deluded sixteenth-century etymologist Goropius Becanus, who believed that all languages derived from Flemish.[65]

The *Hesperides*, however, was not simply intended to be a parade of antiquarian, philological knowledge. Some of the most engaging passages in the book are on the culinary uses of citrus fruit. In even greater abundance than in Cassiano's compilation, here may be found details of citrus candies, from candied peel to nougat, accounts of how pastilles impressed with letters are made,[66] and delightful recipes for the new Turkish invention of lemon sherbet. One was provided by that famous seventeenth-century traveller in the Orient, Pietro della Valle, the other by Cassiano himself[67] (one of the only two occasions in the book when Cassiano's name is mentioned).[68] Above all — and predictably enough — Ferrari refers to the medicinal uses of citrus fruit. Here, for example, are remedies against cholera, apoplexy, gallstones, hair-loss, impetigo, dysentery, headaches, intestinal worms, excessive menstrual bleeding, haemorrhoids, and so on and so forth — even bad breath. He outlines a variety of antidotes against diseases and pests that afflict citrus fruit themselves. He records the climatic factors that affect the well-being of citrus plants. As in the case of the *De Florum Cultura* he assesses the relevance of the phases of the moon for sowing and planting. And of course he supplies details of every possible aspect of the cultivation of citruses, from irrigation and manuring to pruning, grafting, potting, and propagation, and of their decorative possibilities.

Yet Ferrari's *Hesperides* differs significantly from anything by Aldrovandi or by any of the writers who still formed part of what Michel Foucault would have called the Renaissance paradigm, where 'history was the inextricable and completely unitary fabric of all that was visible of things, and of the signs that had been discovered or lodged within them.'[69] In writers such as Gesner and Aldrovandi, the miscellaneous accounts of the legends and stories of particular plants and animals were mixed in with 'exact descriptions, reported quotations, fables without commentary, remarks dealing indifferently with a plant's physiology, its use in heraldry, its habitat, its mythological values,' its medicinal or magical uses, the foods it provided, what the ancients recorded and what travellers said about it.[70] Mythology, history, and crude judgements about resemblance all feature in Aldrovandi's haphazard, non-systematic and thoroughly random and inconsistent taxonomic scheme.

In almost every respect, beginning with the organization of his book, Ferrari was more systematic than any of his predecessors. His concentration on establishing the right and secure names for individual fruits had no precedent (and indeed this is where Ferrari's etymological skills and attentiveness to vernacular nomenclature emerged most clearly). The effort at providing a firm classification for this family of fruit was vastly more rigorous than ever before. Citrus classification, as Enrico Baldini acknowledges in his essay in this volume, still remains in a comparatively confused state, but Ferrari took the fundamental step of dividing the entire family of citrus fruits into the three basic divisions of oranges, lemons, and citrons (the mandarin and the grapefruit had not yet appeared in Europe). He also approached the problem of hybrid and monstrous forms in a much less fanciful way than any of his predecessors. Although the *Hesperides* shows an interest in anomalous and teratological forms that one might hardly have expected from the author of the *De Florum Cultura*, Ferrari renounced all explanations driven by the idea of similitude, which provide so much of the motor for Renaissance efforts at classification and botanical explanation. There was little concern with the zoomorphic and anthropomorphic dimensions of monstrous specimens, as there was so consistently in the work of Aldrovandi. Monstrosity was here integrated into natural explanation, rather than excluded from it.

But at the same time, Ferrari did something that was distinctively his own. He engaged poetical narrative and what he himself called 'poetical explanation' in the service of natural history. Instead of simply recounting the antique fables around his subject (which of course he does too), he made up his own stories to supplement the deficiencies in his own botanical and agricultural knowledge. Right at the outset he declared that he would devise his own tales, believing that pleasant fables might help in the understanding of more serious matters.[71] They were pressed into service precisely in order to fill out what he himself called the 'philosophical' explanations for aberration, anomaly, and monstrosity. Finally, and most importantly of all, Ferrari used visual illustration in a way and of a quality that had never been done or seen before.

The *Hesperides* is divided into four books. The first deals expertly with the archaeological, numismatic, mythological and etymological background to citrus lore, while the remaining three books are named after each of the Hesperidean nymphs — Aegle, Arethusa, and Hesperthusa — and are devoted to the citrons, lemons and oranges respectively. By making firm decisions as to where to place hybrid specimens and monstrosities (the term itself was used with unprecedented thrift in the *Hesperides*), Ferrari resolved some of the old taxonomic confusions at a stroke. He was not deflected by phenomena such as the hybrid citrated lemons and citrated oranges, or the so-called Adam's Apples and Apples of Paradise or the fruit popularly called *ponzini*, *giarette*, and any number of fruit which carried similar names in the popular terminology of particular regions. Although he made every possible effort to assign accurate and reasonably revealing names to both teratological and hybrid forms (such as citrated lemons and oranges), he understood the importance of retaining the primary groups. But at the same time no one in the history of botany had made even remotely comparable an effort to distinguish the huge variety of citrus species, or to illustrate it in all its wondrous diversity.

If the taxonomic drive in the *De Florum Cultura* is strong, it is almost overwhelming in the *Hesperides*.[72] The book is drenched, as Foucault would have put it, in the desire to name. And in order to do so, Ferrari observed with a care and meticulousness that few other botanists had yet displayed — let alone an amateur and a Professor of Hebrew from the Collegio Romano. Page after page of the book reveals his efforts to examine specimens at first hand; he constantly refers to gardens and places he had visited himself (and his knowledge of Roman — and, apparently, Neapolitan — gardens seems to have been exceptionally wide, and intimate too); and he painstakingly recounts the firsthand reports of friends who supplemented his own knowledge. Over and over again he provides evidence of the importance he attached to observation and close description. When he cannot find a particular fruit in Latium, he has a branch specially sent to him from Reggio Calabria, 'so that having seen it with my eyes, I could faithfully describe it';[73] and this kind of remark appears and reappears with great regularity throughout the work. Just as in the case of the *De Florum Cultura*, then, the emphasis on direct observation places him firmly in the context of the so-called scientific revolution around Galileo; and he thus joins the ranks of the new breed of practical botanists, instead of simply being an amateur escaped from the confines of his pedagogical duties as Professor of Hebrew and Rhetoric.

The accounts of individual fruits are marked by an astonishing attention to the texture and colour of every part of the particular specimen. Ferrari tastes each part of the fruit — from rind to segments; and, in anticipation of that essential ingredient of what Foucault called the new *episteme* of 'classical' knowledge, he counts segments and seeds. Although in his taxonomic distinctions he relied heavily on what Galileo and Locke would have called the secondary qualities of things, there is, nevertheless, a clear sense of the importance of number (and, for that matter, of weight).[74] But for the most part, the descriptions now seem superfluously close and purely external, and one has no sense of the possibility of the essential distinguishing features of a species. The accounts of individual specimens proceed as if the dense accumulation of descriptive detail somehow amount to the basis for taxonomic differentiation. This material, however, is always varied by some interesting aspect of the fruit concerned, with culinary, etymological, cultural or ethnographic observations; and so the chapters devoted to specific fruits are never tedious reading.

In each of the three books of the *Hesperides* (devoted respectively to citrons, lemons, and oranges), the taxonomically and descriptively oriented chapters are preceded by a general chapter on the particular family as a whole, and then followed by a more or less systematic procession of chapters devoted to the practical aspects of cultivation and use, just as required by the *Capi d'Informatione* sent out by Cassiano to his correspondents all over Italy and abroad. Their topics range from sowing and propagation to irrigation, manuring, pruning, from protection against insect and ant-infestation to protection against heat and cold, from efficient forms of picking and conservation to the splendidly rich chapters on the *utilitates* of citrus fruit, whether popular or 'occult', whether local (*nostrates*) or exotic (*apud esteros*).

But it is the illustration of the *Hesperides* that is of most concern in the context of the present catalogue. Indeed, as Ferrari repeatedly makes clear, he was deeply committed to the effectiveness of visual illustration. However close his verbal description of individual specimens might be, they had to be illustrated in order to ensure accurate communication, in a way that words never could. And then, as he himself notes, there were the fables he made up to adorn the serious matter of the book. These would be supplemented by the 'artifice of painting', as he put it, so that however rude his narrative efforts might have seemed, the eyes would at least be satisfied by the work of outstanding artists.[75] When it came to Romanelli's illustration of just such a fable, for example, he observed that 'what is a beautiful spectacle in the telling must now be looked at hard'. 'Look now', he exhorts his readers, 'at the adult glory of the art of Francesco Romanelli'[Fig. 29].[76]

Whereas the *De Florum Cultura* contained twenty plates of different flowers, the *Hesperides* restricted itself just to citrus fruits, and had no fewer than eighty botanical plates representing the flowers, whole fruits, and sections of a hitherto unsuspected variety of oranges, lemons, and citrons. And whereas the poetic narratives in the *De Florum Cultura* were rather disparate, with only an occasional attempt to link one with the other, and were illustrated by four artists, the stories in the *Hesperides* had two strong narrative threads running through them, and were illustrated by no fewer than seven artists, who were among the leading artists of their day: Cortona, naturally enough, and then Albani, Sacchi, Poussin, Romanelli, Lanfranco, and Domenichino. The engraving of the plates was divided between Johann Friedrich Greuter and Cornelis Bloemaert. The designs by Albani, Poussin and Lanfranco showed the arrival of the apples of the Hesperides in different parts of Italy — in Rome, Lake Garda, and Naples, respectively; as in the case of the *De Florum Cultura*, Cortona did the title-page [Fig. 13] while Sacchi, Romanelli and Domenichino illustrated a series of long tales all revolving around three transformations: that of the youth Harmonillus — who had been transported to Apollo's singing school, or *Ephebeum* — into a citron tree bearing multiform digitated fruits (see **72** and **81**), that of his mother Tirsenia into a lemon tree bearing the kinds of 'pregnant' fruits described in **91** and **92** of the present catalogue, and that of his sister Leonilla into an *Aurantium distortum* very like **98-100**. If one looks carefully at the plates by these three artists one discovers a number of fruits of exactly the same teratological variety as those represented in the botanical plates. The plate by Sacchi [Figs. 15, 28] shows a *Malum Citreum multiforme* and a *Limon a Rio* (**72** and **81**), that by Romanelli [Fig. 29] a 'pregnant' hybrid lemon (**91** and **92**), and that by Domenichino [Fig. 30] a pregnant hybrid lemon again and an *Aurantium distortum* (see **98-100**).

As is clear from the repetition of the double lemon in the illustrations after Romanelli and Domenichino, these artists, strangely enough, seem to have made their drawings in ignorance of each other (for all their close adherence to Ferrari's text). So while the narrative thread from one episode to another in Ferrari is clear and consistent enough, the same cannot be said of the plates. Accompanying each of the plates were not only the long and flowery texts of these tales, but also long

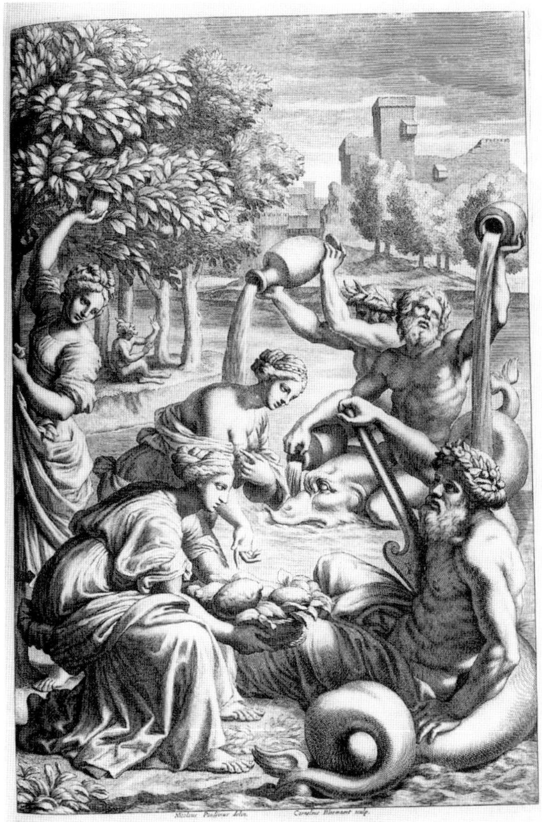

Fig. 20. C. Bloemaert after N. Poussin, *The Nymphs of the Hesperides presenting Citrons to the Deities of Lake Garda*, engraving, Ferrari 1646, p. 97

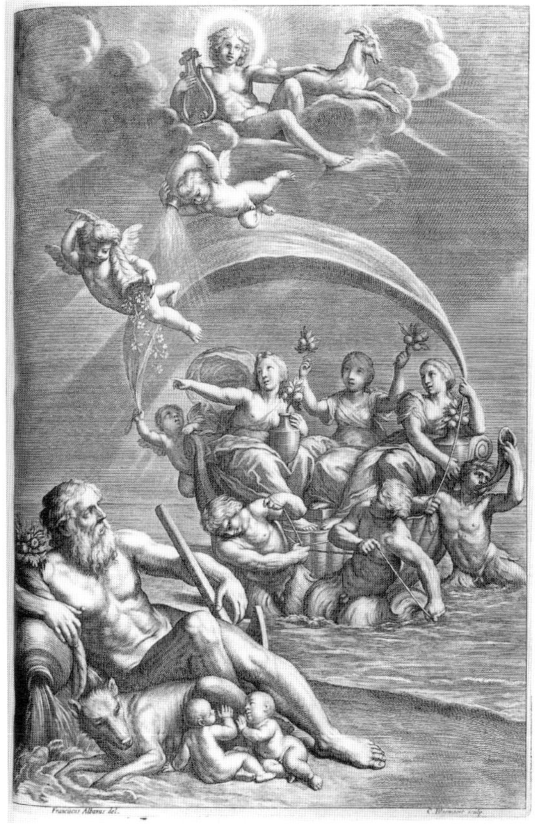

Fig. 21. C. Bloemaert after F. Albani, *The Nymphs of the Hesperides bringing Citrus Fruit down the Tiber to Rome*, engraving, Ferrari 1646, p. 51

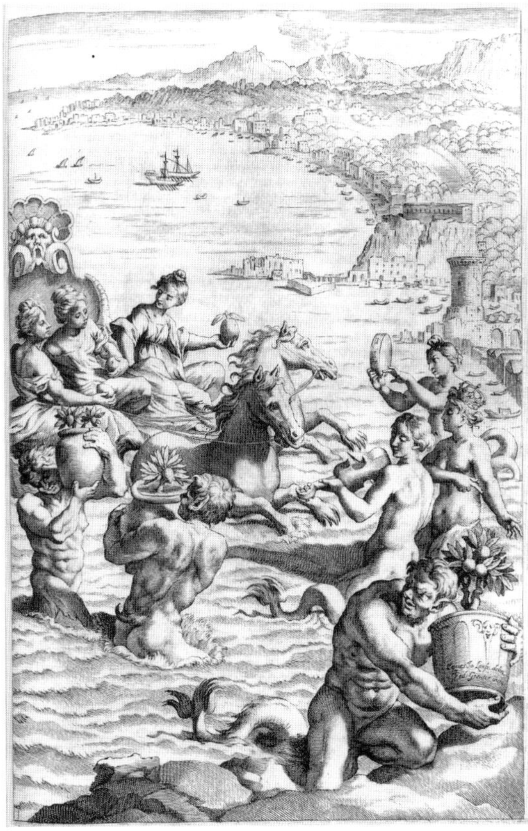

Fig. 22. J.F. Greuter after G. Lanfranco, *The Arrival of Citrus Fruit in Naples*, engraving, Ferrari 1646, p. 447

tributes to each of the artists concerned, which have still not all entered the standard literature on them — perhaps because of the difficulties of Ferrari's ornate Latin.

As in the case of the *De Florum Cultura*, there were other kinds of illustrations too. But whereas the subsidiary illustrations — if they can be called that — in the flower book were of garden implements, bouquets, vases and parterres, those in the *Hesperides* are much grander. They show the orangeries and other citrus-related structures of the Aldobrandini, the Pii, Cardinal Lante, and of the Farnese in Parma [Fig. 23], and — in the course of the first book of *Hesperides*, devoted to the mythology, lore, and archaeology of Hercules and the Golden Apples — a number of impressive plates of ancient statues and coins [Figs. 14, 31]. These were engraved by Cornelis Bloemaert after designs by Pietro Paolo Ubaldini and François Perrier, while the architectural plates were engraved by Camillo Cungio and Claude Goyrand after Filippo Gagliardi and Girolamo Rainaldi. A more impressive array of artists and printmakers could hardly have been assembled in Rome at the time.

Yet it is the citrus plates, naturally enough, that are of most relevance to the present catalogue. As Ferrari noted so enthusiastically early on in the *Hesperides*, the plates were made after especially lifelike and accomplished drawings by Vincenzo Leonardi. The etchings (and in some cases engravings, and in a few cases a combination of both) were by Cornelis Bloemaert, though a few plates were signed by Dominique Barrière. Once again, the prints have a kind of decisive clarity of line and attentiveness to detail that are unprecedented in the history of botanical illustration; and no single family of plants or animals had yet been so extensively illustrated. As in the case of most of the botanical illustrations in the *De Florum Cultura*, each citrus plate in the *Hesperides* has a fluttering scroll that decoratively bears the name of the particular specimen, or set of specimens, represented. But for the rest the plates are marked by a neatness and precision that seem to go beyond even that of the flowers of thirteen years earlier. And there is a clear interest in anomaly, monstrosity, and texture in general that is almost entirely absent from the earlier book. The plates themselves show an almost obsessive attentiveness to every variety of texture, rugosity, corniculation and digitation — all phenomena which particularly interested Ferrari. In almost every case, the sections of each kind of fruit are shown, either on the same plate or on succeeding ones, generally only cross-sections, but sometimes longitudinal sections as well. The disposition of images on each plate, as can be judged from these pages, varied according to the size of the whole fruit: if it was especially large, it was shown on its own; if it was small, it was shown along with other views or one or more sections. Particular care seems to have been taken to show the number of segments — and even of seeds — with each cross-section, and on one or two occasions a few seeds alone were shown (see **73**).

Very little in the history of botanical illustration prepares one for the sumptuousness and refinement of Vincenzo Leonardi's drawings of citrus fruit. They are executed with much descriptive precision and with such an extraordinary command of colouristic nuance, that they do indeed seem to be 'doubles of nature', as

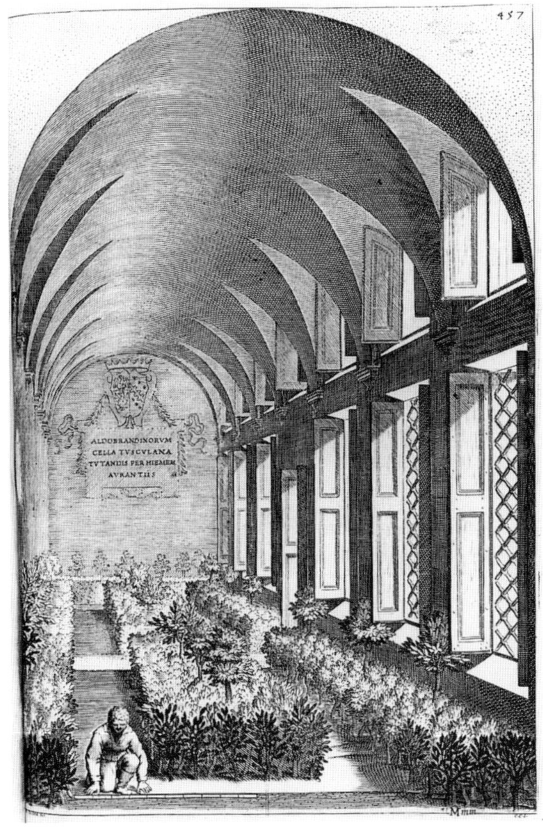

Fig. 23. C. Cungio after F. Gagliardi, *The Aldobrandini Orangery*, engraving, Ferrari 1646, p. 457

Ferrari put it in his tribute to Leonardi.[77] It is true that artists like Jacopo Ligozzi and Joris Hoefnagel, for example, had previously made naturalistic drawings of arguably even greater skill and refinement, but such drawings were not made, by and large, to serve as the basis for printed illustrations.

But why, one might ask, did Leonardi's drawings need to be in colour at all? The possibility (which may be the first to occur to one) that they were intended to serve as the basis for the colouring of the prints in copies of the *Hesperides* itself may be immediately discounted. Very few coloured copies survive. It is unlikely that this was one of those books intended to be coloured from the outset. On the other hand, Ferrari's colouristic notes in the text of the *Hesperides* (just as Cassiano's, characteristically, in the *Notizie diverse*) are so detailed that one can only regard these drawings as the graphic equivalents of such verbal attentiveness. While Ferrari made a point of trying to see as many of the specimens that he described as possible, it could well be that some of his descriptions were made on the basis of drawings such as those by Leonardi, and that Leonardi was thus expressly instructed to be as attentive as possible to colour. At any rate, such drawings would then have served as substitutes for direct observation, and there can be no doubt whatsoever that they were made directly from the specimens themselves.

Although all the drawings published in the present volume may be assumed — as far as can be ascertained — to have come from Cassiano's Paper Museum (very many have the standard wash bordered mounts onto which they were pasted following their acquisition by George III in 1762, and many show the typical hand with which the drawings that entered Cassiano's collection or were commissioned by him were numbered or annotated), the existence of other citrus drawings that may be related to these in one way or another ought briefly to be recorded here. There are, for example, eleven pages with drawings of citrus fruit (five in brown ink alone and six in watercolour), in the great volumes of Cesi's *Syntaxis Plantaria* preserved in the Institut de France in Paris[78], while — at a further remove — one should probably also note that there are twelve good watercolour drawings of citrus fruit preserved among Aldrovandi's *Iconographia Plantarum* manuscript in the University Library in Bologna [Fig. 32]. The strange black chalk drawing of an arrangement of citrus fruit that survives at the back of Cassiano's *Notizie diverse* should also be mentioned here [Fig. 19]. Later on in the century, of course, Bartolommeo Bimbi would make paintings of citrus fruit that offer even more exhaustive visual inventories of citrus (and other) fruits,[79] while the manuscripts of Pierantonio Micheli [Fig. 24] and Giorgio Gallesio contain coloured drawings of oranges, citrons and lemons — normal specimens, hybrids, and teratological specimens as well — that seem to take up the tradition begun by Leonardi's drawings.[80] While it is not clear whether either Micheli or Gallesio knew any of the latter at first hand, by then the vogue for citrus books and for citrus drawings (and paintings) had been well established. While Commelin's *Nederlandtze Hesperides* and Volkamer's *Nürnbergische Hesperides* are both wholly indebted to Ferrari's *Hesperides*, at least in terms of their representations of citrus fruit, the luxurious 1872 edition by the Abbé du Breuil of Risso and Poiteau's 1818 treatise is unimaginable except at the fertile confluence of the tradition of printed illustration and painted representation established by the joint efforts of Cassiano and Ferrari in their conceptualization, preparation and publication of the *Hesperides* of 1646 [see Fig. 25].

The Monstrous and the Poetic

Few of the subsequent collections of citrus drawings, however, have quite the same concentration on digitated, 'pregnant' and other teratological forms.[81] This concentration emerges not so much from the *Hesperides* itself, as from the group of surviving drawings. The present catalogue has been divided largely according to modern taxonomic principles, and it will escape no one that by far the largest groups are the hybrids and the teratological fruits. This reflects both Ferrari's and Cassiano's strong interest in such forms, as well as the fact that such drawings may well have aroused the special interest of subsequent collectors. Ferrari and Cassiano were interested in hybrid forms not just because of the problems of propagation that they posed, or because they were often (as is clear from Cassiano's and Nardi's correspondence) fine instances of the wonders or games of nature (the *mirabilia naturae* or the *lusus naturae*) which continued to fascinate students of natural history. They were interested in them precisely because such forms acutely posed the problem of the borderline or intermediate case. How was one to classify

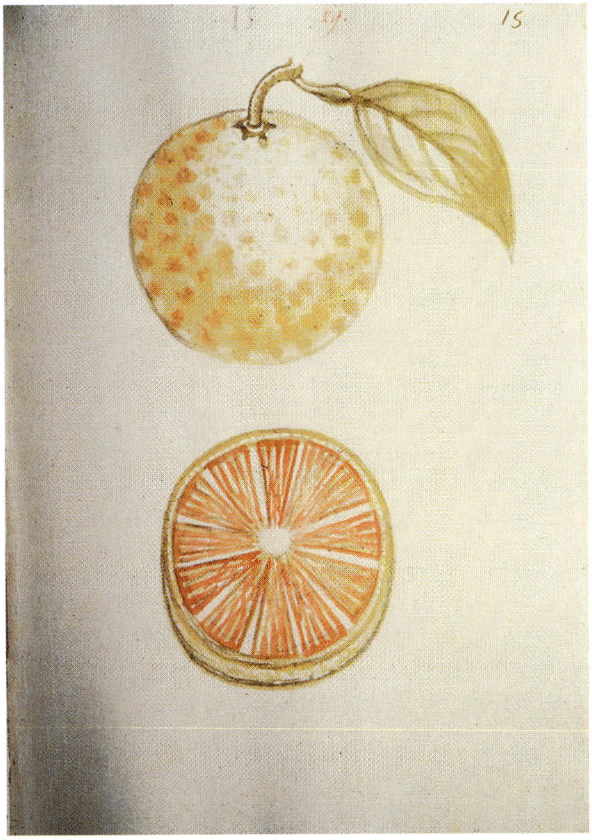

Fig 24. Eighteenth-century Italian, *Aurantium hierocunticum* (blood orange), Micheli *Enumeratio*, p. 39

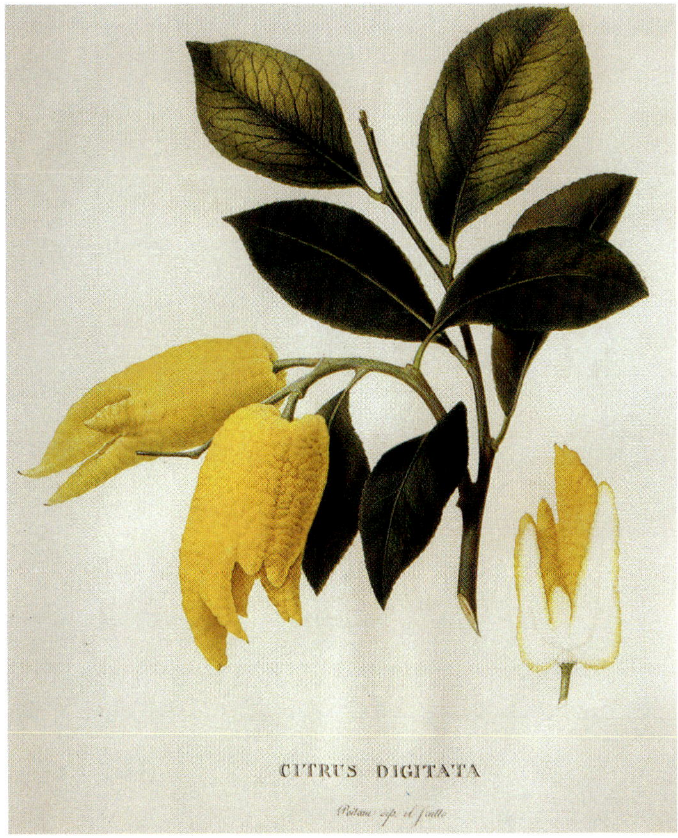

Fig. 25. A. Poiteau, *Limone multiforme*, watercolour and bodycolour on vellum, loose unnumbered folio, Gallesio Archives, Genoa

the citrated lemon or orange, for example? Were these forms really citrons, or hybrid oranges and lemons? And how was one to establish the criteria for secure classification in cases where the specimen seemed to show the essential characteristics of more than one group. Hybrid forms, and even more so monstrous ones — ones which were not just *lusus naturae*, but which seemed to go beyond the normal rules — were both of intense aesthetic and scientific interest and as such demanded explanation of one kind or another.

'Abortus & monstra in animantibus plerunque horremus, amamus in pomis', 'Generally we shudder at deformation and monstrosity in humans and animals, but we love them in fruit', Ferrari began his chapter on the *Aurantium distortum*.[82] If there is any theme that may be said to dominate the *Hesperides*, it is this, just as it seems to dominate Cassiano's collection of citrus drawings. Of the *Limon scabiosus* Ferrari observed, with a nice sense of Baroque paradox: *'Scabies & lepra pomo aureo nomen ac decus faciunt. Nimirum in auro etiam foeditas ac deformitas placet'*.[83] 'For indeed, even in gold hideousness and deformity may be pleasing', he says, almost as if he were commenting on forms such as the *Aurantium callosum multiforme* so beautifully drawn by Leonardi (**90**).

But the two men were not just interested in the aesthetic aspects of strangely shaped fruits, nor did they simply seek to find visual analogies — generally anthro-

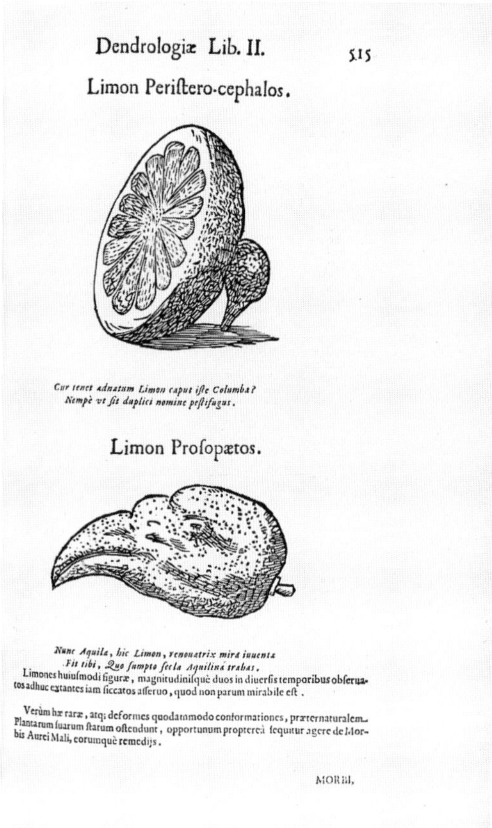
Dendrologiæ Lib. II. 515

Limon Periſtero-cephalos.

Cur tenet adnatum Limon caput iſte Columbæ?
Nempè vt ſit duplici nomine peſtifugus.

Limon Proſopætos.

Nunc Aquila, hic Limon, renouatrix mira iuuentæ
Fit tibi, Quo ſumpto ſecla Aquilina trahas.

Limones huiuſmodi figuræ, magnitudiniſquè duos in diuerſis temporibus obſeruatos adhuc extantes iam ſiccatos aſſeruo, quod non parum mirabile eſt.

Verùm hæ raræ, atq; deformes quodammodo conformationes, præternaturalem Plantarum ſuarum ſtatum oſtendunt, opportunum propterea ſequitur agere de Morbis Aurei Mali, eorumquè remedijs.

MORBI,

Fig. 26. *Limon peristero-cephalos* and *Limon Prosopaetos*, woodcut, Montalbani 1668, p. 515

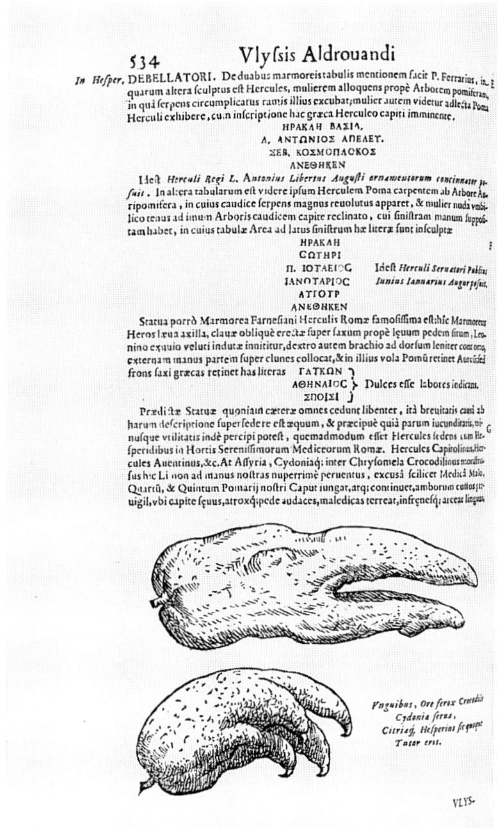
534 Vlyſsis Aldrouandi

In Heſper. DEBELLATORI. De duabus marmoreis tabulis mentionem facit P. Ferrarius, in quarum altera ſculptus eſt Hercules, mulierem alloquens propè Arborem pomiferam, in quâ ſerpens circumplicatus ramis illius excubat; mulier autem videtur adlecta Poma Herculi exhibere, cum inſcriptione hac græca Herculeo capiti imminente.

ΗΡΑΚΛΗ ΒΑΣΙΛ.
Λ. ΑΝΤΩΝΙΟΣ ΑΠΕΛΕΥ.
ΣΕΒ. ΚΟΣΜΟΠΛΟΚΟΣ
ΑΝΕΘΗΚΕΝ

Id eſt *Herculi Regi L. Antonius Libertus Auguſti ornamentorum concinnator poſuit.* In altera tabularum eſt videre ipſum Herculem Poma carpentem ab Arbore Auripomifera, in cuius caudice ſerpens magnus reuolutus apparet, & mulier nudâ vmbilico tenus ad imum Arboris caudicem capite reclinato, cui ſiniſtram manum ſuppoſitam habet, in cuius tabulæ Area ad latus ſiniſtrum hæ literæ ſunt inſculptæ

ΗΡΑΚΛΗ
CΩΤΗΡΙ
Π. ΙΟΥΝΙΟC — Id eſt *Herculi Seruatori Publius*
ΙΑΝΟΥΑΡΙΟC — *Iunius Iannarius Augur poſuit,*
ΑΥΓΟΥΡ
ΑΝΕΘΗΚΕΝ

Statua porrò Marmorea Farneſiani Herculis Romæ famoſiſſima eſt; hîc Marmoreus Heros læua axilla, clauæ obliquè erectæ ſuper ſaxum propè læuum pedem ſitum, Leonino exuuio veluti indutæ innititur, dextro autem brachio ad dorſum leniter contorto, externam manus partem ſuper clunes collocat, & in illius vola Pomū retinet Aureū. frons ſaxi græcas retinet has literas ΓΛΥΚΩΝ ΑΘΗΝΑΙΟC ΕΠΟΙΕΙ } Dulces eſſe labores indicans.

Prædictæ Statuæ quoniam cæteræ omnes cedunt libenter, itâ breuitatis causâ ab harum deſcriptione ſuperſedere eſt æquum, & præcipuè quià parum iucunditatis, minuſque vtilitatis indè percipi poteſt, quemadmodum eſſet Hercules ſedens cum Heſperidibus in Hortis Sereniſſimorum Mediceorum Romæ. Hercules Capitolinus, Hercules Auentinus, &c. At Aſſyria, Cydoniaq; inter Chryſomela Crocodilinus monſtroſus hic Li non ad manus noſtras nuperrimè peruentus, excusâ ſcilicet Medicâ Malo, Quartū, & Quintum Pomarij noſtri Caput iungat, atq; contineat, amborum cuſtos peruigil, vbi capite ſæuus, atroxq; pede audaces, maledicas terreat, infrenesq; arceat linguas.

Vnguibus, Ore ferox Crocodile
Cydonia ſeruas,
Citriaq, Heſperias ſic quoque
Tutor eris.

VLYS-

Fig. 27. Clawed lemon, *Unguibus, Ore ferox Crocodile*, woodcut, Montalbani 1668, p. 534

pomorphizing — for deformed or anomalous specimens, in the usual manner of the day. Certainly Ferrari had much to say on the phenomenon of digitation, and the surviving corpus probably has more recognizably digitated forms than almost any other anomaly; but when one turns to someone like Aldrovandi and his editors, one finds a yet stronger drive to what Foucault would have described as explanation in terms of resemblance and similitude. Thus, in addition to a few digitated forms of the kind later to be found in Ferrari, such as the plain *Limon digitatus* and the *Limon chirodes*, Aldrovandi has forms such as the lemon with a face of an eagle, the *Limon Prosopaetos* [Fig. 26], or the lemon with a head attached to it, the *Limon peristero-cephalos*, also called, Aldrovandi's editor notes, *pestifugus*, because it put birds and other pests to flight. There is a fiercely clawed specimen whose ferocity is clearly exaggerated in the interest of visual spectacle, *Unguibus, Ore ferox crocodile* [Fig. 27].[84]

There is little of this sort of forced visual analogy in Ferrari; indeed, one of the most striking aspects of his interest in anomaly in general is his almost total renunciation of the usual terms of monstrosity. He rarely refers to the anomalies and deformations he discusses as *monstra* or *monstrosa*, because in the end his aim is to insert such specimens into his overall taxonomic scheme. He has to find a place for such forms in nature, rather than to extract them from it; they must be seen as part of the general fabric of things, rather than as inexplicable aberrations; and — as we have seen — when his attempts at explanation in terms of what we now call 'biology' fail, he turns to what he himself calls 'poetry', and what by any reckoning still counts as poetic narrative, written in his distinctive, rhetorically rooted, poetic diction.

In the compilation of notes on citrus fruits prepared by Cassiano to aid Ferrari, the *Notizie diverse*, one finds an increasing interest in what are described, throughout the manuscript, as citrus forms *che scherzano* ('that joke'). At one point there is a reference to a *Cedro mostruoso a dita o che scherza* that came from Reggio Calabria, thus assembling in one complex taxonomic description the most important available descriptive terms for the digitated group.[85] Indeed, it is clear that for Cassiano such forms were actually a distinctive class of their own, a whole group whose members could be generalized as belonging to the generic *che scherza* or *che scherzano* type: *'questi cedri che scherzano [sono] i digitati di Reggio di Calavria massime questi di forma lunga o divisi come in rami sono dentro tutti bianchi senz'agro...'*, 'These joking citrons are the digitated kind from Reggio di Calabria, especially those that are long in shape or divided into branches, completely white inside and without juice', he observed on fol. 105.[86] *Le cedri sono due sorti, cedri ordinarj e cedri della dita...'*, he notes on fol. 18. In his letter to Cassiano of 21 May 1652, referring to a citrated orange, Nardi said it belonged to a group *'che dicono bene spesso suole la Natura scherzare'*.[87]

But for Ferrari (as indeed for Cassiano too), the matter of joking forms had to be taken quite seriously. So too had the apparently related phenomena of double fruit (that is, specimens which seemed to be pregnant with others), and the whole problem of hybrids and fruits that seemed to cross the borderline between species. Occasionally Cassiano and his correspondents referred to double and to pregnant fruits, and often to types such as the *Cedrato arancio*, or the *Melangolo cedrato*, the *Cedrangolo* or *Limoni cedronati* and so on and so forth.[88] But the *Hesperides* is full of such phenomena. There is hardly a chapter in which one or another anomalous or borderline form does not appear. Very dutifully Ferrari records the normal, 'vulgar', or ordinary types, but it is clear that it is the other forms, the forms that writers such as Aldrovandi would insistently have called 'monstrous', that most occupy his interest and attention. Indeed, along with the commitment to illustration in visual form, the drive to taxonomy and nomenclature, and the rhetorical diction, this is precisely the commitment that forms one of the most distinctive features of Ferrari's work; and it emerges with equal clarity from the drawings in the present catalogue.

According to Ferrari, 'Nature plays no more licentiously in any other fruit'[89] than in the multiform or digitated citron (**72**). Such citrons take on a whole variety of wonderful shapes: in addition to the similarities with fingers and toes, they may look like the beaks of birds or the crests of poultry. Some are horned like deer, others like cattle.[90] It would be futile, announces Ferrari, to enumerate all such forms. But how do such shapes come about? Ferrari suggests two possible explanations. Could it be that since the force of the seeds is so weak when they come to join together to form the different parts of the fruit, they cannot form the accustomed round shape, but rather abort into different monstrous appearances? Or is it rather because at the moment of generation the same force gives rise to a number of different foetuses, and since the generative force is thus dissipated, some parts of the fruit are left defective — even though these might be elegant and graceful?[91] Ferrari leaves the matter open, and takes, as we shall see, another approach.

Then there was the problem of 'pregnant' fruits, those cases where one fruit seemed to be included within another. Many instances appear in the present catalogue as well, such as **89-95**. Ferrari suggests that such phenomena may have to do with the special fertility of the soil in which they grow, where the parts of the seeds are exuberantly diverse, with the result that the generative force causes the multiplication of forms and the consequent folding of one within the other. Once again Ferrari speaks of a playful and lascivious nature at work.[92] This idea of a playful nature is always present; but it does not suffice as an explanation, as little as do the weak attempts at explanation in terms of the moment of vegetable conception.

For Ferrari it was the orange group that contained the greatest number of deformed, anomalous, and hybrid types (although it will be clear from the present catalogue that many of these were in fact lemons or hybrids). In addition to the predictable double oranges, there were also, for example, the *Aurantium foemina sive foetiferum* (see **95**), the *Aurantium hermaphroditum, sive corniculatum*, the *Aurantium callosum multiforme* and a large and interesting variety of what Ferrari would have called *Aurantia distorta* (see **96-98**). In each case Ferrari offered what we would call a scientific explanation (however deficient or reductive his 'science' may seem to be) and what he called a 'philosophical' one. According to Ferrari, the origin of the *Aurantium foemina sive foetiferum* had much in common with certain kinds of 'pregnant' lemons, or lemons containing one whole fruit within another. This early form of the modern navel orange is accounted for in very similar terms to the more complex type of pregnant citrated lemon which Ferrari simply called the *Limon citratus alterum includens*[93] (see **89-90**). As always, he sought his explanation in terms of the processes of reproduction. At stake were what appeared to be the aborted foetuses (the 'navel'), the result — once again — of generative weakness (*invalido nisu*). In such cases the fruits could only be female or androgynous (*'Ita & hoc & proxime exhibendum, genere neutrale, a se ipso degenerat, vel in foeminam, vel in androgynum'*).[94]

This minimal account of the phenomenon was then supplemented by an equally minimal and unsatisfying explanation of why the 'foetus' appeared to protrude from the fruit, divided in such a way that it looked like a small flower or navel. It had something to do, Ferrari claimed, with the subtlety of the seminal material, or some other congenital quality.[95] In the case of the rather spectacular 'horned' or hermaphroditic orange, a marginal note refers to the *Androgyni pomi causa*; but this 'cause' turns out to be nothing more than a statement about how a small horn shape forces its way out through the thinner part of the skin (hence *hermaphroditum*).[96]

Although the *Aurantium callosum multiforme* (see **95**) ought to be counted, according to Ferrari, amongst the choice number of horticultural miracles, and although it was yet another instance of the *lusus naturae*, he could come up with no adequate explanation at all. In the case of the *Aurantium distortum* — of which there are several related types in the corpus of drawings published here (see **96-98**) — he simply said that the basic cause of these aborted fruits, was the same as in the case of the digitated lemons. The fruits are aborted and so are the flowers; and these fruits often have the appearance of other living things, forced into small, compact, mendacious forms, which can then grow together to make larger fruits. The drive to similitude reappears in this suggestion of illusory mendacity, but only briefly. As in the case of almost all the other instances of deformed and anomalous fruits, these are phenomena that seem to defy Ferrari's capacity for 'philosophical' explanation. Perhaps one ought not to expect more (even some of the modern accounts remain equivocal); but sometimes he does not even seem to try. Obviously he realized that his attempts at scientific explanation were little more than shots in the dark. *'Et hactenus quidem philosophicas inter coniecturas caecutisse sit satis'*, he abruptly observed at the end of his conjectural account of the origins of the *Malum citreum digitatum seu multiforme* (see **72**).[97]

So what does Ferrari do? He turns to a form of explanation which he finds more comfortable, namely a poetic one. He says exactly this: 'I have a philosophical and a poetic cause of this admirable pregnancy in fruits, to bring forward to anyone who wants to know the reason for them', he says of the *Limon citratus alterum includens* (**89**).[98] And of what do these poetic explanations consist? We already have some idea, but the thread must now be tied together.

The 'poetry' of the *Hesperides* is to be taken in its broad sense. The book as a whole ought certainly not to be classified as a *'poème assez pédant sur la botanique, où plus exactement sur les maladies des plantes'*, as Anthony Blunt once put it.[99] The *Hesperides* is neither a 'poem', nor even a botanical work, strictly speaking; and to call it a poem on the diseases of plants is wholly to misunderstand the interest in anomalous and teratological specimens that so characterize both this book and much of seventeenth-century botany in general. It is, however, a work in which the importance of poetic diction is paramount. No work of botany in the sixteenth or seventeenth century had anything like the same amount of sophisticated literary and rhetorical care expended upon it. All *too* sophisticated and clever at times: the accumulation of cloying adjectives, the incessant use of diminutives, assonance, and alliteration,[100] as well as the abundance of metaphors and other tropes, sometimes make the book a little too rich to digest. But this may be so simply because these aspects of Ferrari's style are hardly novel (except, perhaps, in their very quantity and charming overstatedness). What *is* novel, however, is his use of poetic invention to account for the digitated, pregnant, and 'distorted' citrus fruits. The narrative thread that forms the basis of his poetic explanations of such fruits, as well as of the illustrations by Sacchi, Romanelli, and Domenichino, runs through the book with an originality that has few parallels in seventeenth-century literature or art. The stories he made up are not just adaptations of ancient mythology, or of some tale from Ovid's *Metamorphoses*. Instead, as we have seen, they are wholly new inventions by Ferrari, and are set out in elaborate and often seductive detail.

While the illustrations by Albani, Poussin, and Lanfranco show the arrival of the Nymphs of the Hesperides with their citrus fruit in Rome, Salò, and Naples respectively [Figs. 21, 22], and while Guido Reni's depiction shows the Nymphs planting orange trees in the gardens of Genoa, it is the three illustrations by Sacchi, Romanelli, and Domenichino that are of crucial interest in the present context.

It would be impossible to give an adequate sense of the literary flavour of these stories, since they are so dependent on Latin word-play, on the use of charming diminutives, and on the marriage between rhetorical figures and a kind of slightly overwrought sentimentality. A summary of the narrative which accompanies Sacchi's illustration [Figs. 15, 28] gives some idea of the contents of Ferrari's inventions:

There was once a beautiful young boy called Harmonillus, who as his name implies, had a most lovely voice. Naturally he came to the attention of Apollo, who enrolled him in his Ephebeum in Arcady, where young male singers were trained. The prefects of the school were fierce old ladies. One year a particularly fierce matron was put in charge, with the suitable name of Vafronia Rabiria. Failing in her attempts to lighten up the earnest young singer and make him devote himself to Thalia, the muse of comedy (he, of course, persisted in cultivating Urania), and intensely jealous of the beauty of his voice, she decided to plot against him. In this she was assisted by a nasty female attendant and a traitorous friend of Harmonillus. On the spurious grounds that he would spoil his precious voice by too much practice, she banned him from all singing for a certain period. The boy was profoundly upset — and then fell into the trap. Pretending to help him out of his distress, the traitorous friend, called Musculus Naserna ('nosy little rat' might be a good English equivalent) said to him that he knew of a sequestered place in the woods where he might sing. Having led Harmonillus there, he assured him that no one was within earshot; and the young man poured forth a lovely melody. But barely had he begun when Vafronia Rabiria's wicked assistant, secretly installed nearby, pounced on him, stopped him from singing, and denounced him to the old lady.

While the psychological implications of all this may not be far to seek, what has it to do with oranges and lemons? The point is that at the very moment in which the women pounce on poor Harmonillus, as is clear from the illustration commissioned from Sacchi [Figs. 15, 28],[101] he begins to weep profusely, calling upon his old tutor Cleomedes to come and rescue him. Cleomedes rushes forward, but — too late! Watered by the abundant flow of tears, Harmonillus begins to turn into a citron tree. His feet turn into roots, his arms into branches; but by the greatest miracle of all, his very hands turn (as is clear from Sacchi's illustration of the story)

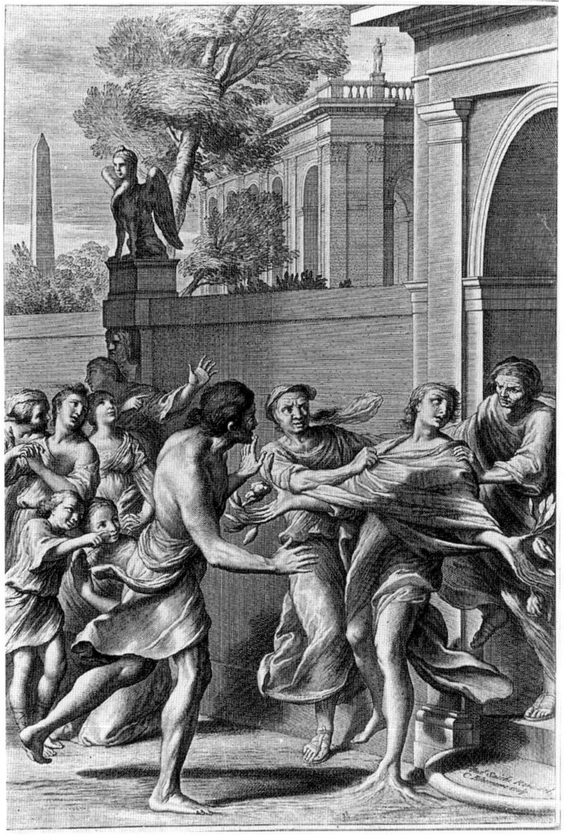

Fig. 28. C. Bloemaert after A. Sacchi,
The Transformation of Harmonillus,
engraving, Ferrari 1646, p. 89.
See also Fig. 15

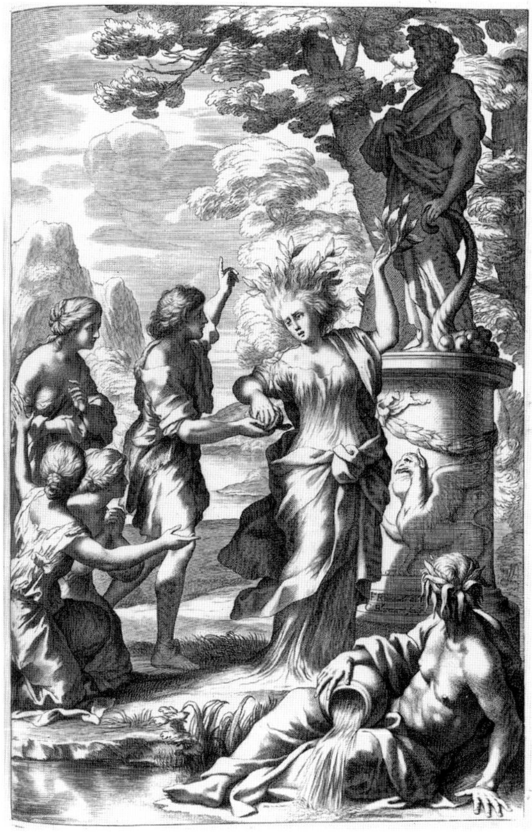

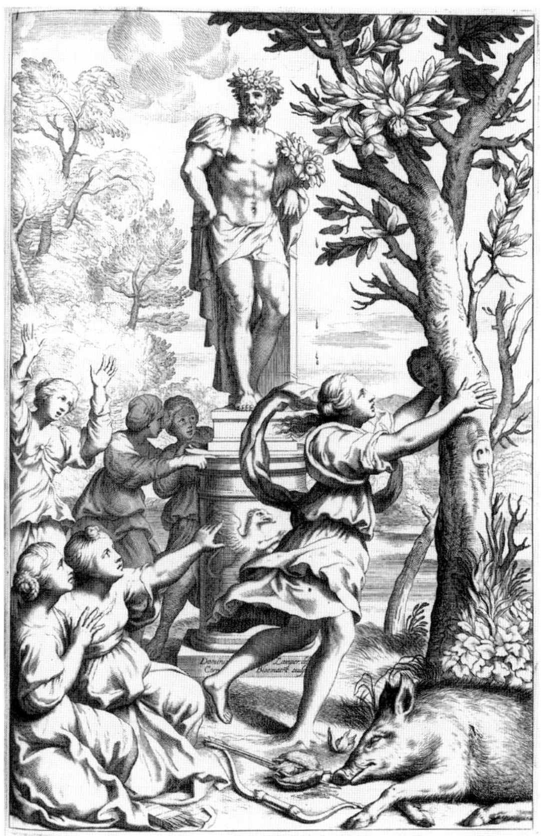

Fig. 29 (above right). C. Bloemaert after G. Romanelli,
The Transformation of Tirsenia,
engraving, Ferrari 1646, p. 277

Fig. 30. C. Bloemaert after Domenico Zampieri,
called Domenichino,
The Transformation of Leonilla, engraving,
Ferrari 1646, at p. 418

into digitated fruit — the one exactly like the *malum citreum multiforme* (see **72**), the other rather closer to the *Limon a rivo seu rio* (see **86** and **87**).[102]

This is a very abbreviated summary of a long and detailed narrative, which concludes with a remarkable (and underutilized) eulogy of Sacchi. The story is fundamental for two later sets of explanations for digitated, pregnant and corniculated fruit. When Ferrari comes to the *Limon a rivo seu rio* itself (see **86** and **87**), he refers back to the Harmonillus story, not at all inappropriately, since Sacchi had also illustrated this fruit in his design. We have already seen how scanty a 'philosophical' explanation he offered for the phenomenon of pregnancy; but he then embarks, swiftly enough, on another poetic account. It takes off from the Harmonillus tale, and is now illustrated by Gianfrancesco Romanelli [Fig. 29], whose design provides the salient features of yet another lengthy and emotional narrative.

As Romanelli's illustration shows, the mother of Harmonillus, Tirsenia, herself turned into a lemon tree when she heard of the sad end of her son. The scene is a grove dedicated to Vertumnus whose statue is adorned with all kinds of flowers and fruits. Tirsenia's servant Ergastus arrives back from Apollo's Ephebeum with the bad news, and presents her with the sweet-smelling fruit from Harmonillus's citron tree. As he hands it over, it is enclosed within a new and paler fruit, the double *Limon citratus* of course (see **91** and **92**), a kind of pregnancy, as Ferrari says, that recalls the son once born within the womb of the mother.[103] But the story contains an apt surprise. Just before his panegyric on the talents of the youthful Romanelli, Ferrari relates (typically enough) how this fruit, generally found in Pietrasanta, was carried to Florence and her neighbourhood, *'ubi rerum omnium praestantia vere perpetuo floret'*, and from there was carried off to grace other *pomaria* and *agrumeti*.[104]

When it comes to the anomalous and teratological oranges, Ferrari begins, with the so-called *Aurantium foetiferum seu foemina* and concludes with the *Aurantium callosum* and the *Aurantium distortum*. For all of these Ferrari offers very perfunctory explanations, but it is only with the *Aurantium distortum* that he once again takes up the thread of the narrative that underlies the illustrations by Sacchi, Romanelli, and Domenichino. Thus Domenichino's plate [Fig. 30] shows how Harmonillus's sister Leonilla, having returned from a boar hunt, rushes to embrace her arboreal mother. Crying *nulla me dies a te divellet o mater*, she too is turned into a citrus tree, (with the fruit of the *mala aurea* naturally matching the colour of her golden hair, and distorted like the sad expressions of pain on the faces of mother and daughter).[105] This is probably the least satisfying of the stories and the least satisfactory of the illustrations. Domenichino does not show the *Aurantium distortum* (which it is supposed to illustrate), but — once again — the fruit of the *Limon citratus*, already illustrated by Romanelli [Figs. 30 and 29 respectively]. Nevertheless, Ferrari's prose immediately swells into a characteristic panegyric on the art of Domenichino. Although the artist had to represent an aborted fruit, *non abortavit ars tua errare nescia*, Ferrari assures his readers.[106] And with this Ferrari neatly combines one of his chief botanical interests with artistic eulogy. His two greatest interests, apart from the art of rhetoric itself, are thus seamlessly epitomized, in the true fashion of a rhetorical trope.

Taxonomy and Representation

Thrice in their great treatise on citrus fruit, Risso and Poiteau launched a fierce attack on the Linnaean system. They alleged that following the reform of Linnaeus, botanists had ceased to occupy themselves with an infinity of useful plants, on the misguided grounds that they were nothing but varietals and monsters, unworthy of the attention of even the most minor botanists. And they berated the great naturalist for producing a system which allowed his followers to confuse the sweet oranges with the *bigarades* under the single name of *Citrus aurantium*.[107]

Risso and Poiteau had a point. Linnaeus had allowed only a very small number of species of citrus fruit — *Citrus trifoliata*, *Citrus medica* (which included both citrons and lemons), and *Citrus aurantium* (from which he later separated the pummelo, the *Citrus decumana*). Almost all Ferrari's species were simply regarded as varietals or cultivars, produced by natural hybridization or grafting — even after the crucial taxonomic improvements introduced by Osbeck between 1757 and 1765.[108] But perhaps more than in the case of any other plant, the Linnaean classification has continued to fray around the edges. While the species of Risso and Poiteau no longer serve as much of a basis for classification, even less, perhaps than Ferrari's, the plain fact is that citrus classification is still in a state of some disorder. Modern research, for example, has suggested that the endogenous growth of specimens like the *Aurantium callosum* (**101** and **102**), for which recent parallels have been found in Sicily, may be the result of casual, synthetic chimeras,[109] and it is now established — though only since the late 1930s — that most of the digitated forms are the result of anomalous flower structure or infestation by the so-called citrus bud mite *Aceria sheldoni*.[110]

But for the rest there is much uncertainty. Despite the limited number of names in the botanical gardens, the terminological vacillation and proliferation continues, from Acireale, the seat of the Istituto di Agrumicoltura, to the shores of the gulf of Florida. Enrico Baldini's introduction to the present volume gives some sense of the variety of possible accounts of the causes of polymorphism. Throughout the citrus world there remains much hesitation not only about the classification of evidently problematic species such as Ferrari's *Aurantium citratum*, but even of ones that one might have thought, after so many years, to be secure, such as the *Pomum Adami* already illustrated by Dodonaeus and Clusius.[111]

Ferrari's contribution was a fundamental one. He isolated a large number of forms and set them decisively into his firm tripartite family. He despatched the problem of borderline cases with remarkable assurance and clarity, as in his descriptions of the *Lumia* and the *Pomum Adami*.[112] As a result, the problem of ambiguity and *aequivoca* would henceforward seem much less urgent and therefore less distracting. His descriptions were always meticulous and sober; and where his explanations were in doubt, he candidly provided alternatives. Where the fanciful forms of species caught his imagination — or could not adequately be explained by science — he devised narratives in which poetry supplemented philosophy (in the form of natural history),[113] and had them illustrated by the finest artists of his day. It would be easy to dismiss the stories for their longwindedness and to say that they are irrelevant to the scientific program of the work. But they are not; they are integral to it. Where what Ferrari called philosophical explanation and what we call scientific explanation could not sufficiently account for a type, and generate an appropriate category for it, poetic narrative was pressed into service. Where the laws of nature could not be bent to embrace its variety, Ferrari told a story to explain it. The poetic imagination, harnessed by the rules of rhetoric, became a heuristic tool to supplement empirical logic and the limited evidence of the eyes.

It is, of course, chiefly through the illustrations of citrus fruit that Ferrari is now remembered, if he is remembered at all.[114] And rightly so. Never before had visual documentation been so extensively pressed into the service of the empirical investigation and documentation of a single family of plants, never indeed, of any family within the whole world of nature. And certainly not in such an accurate, detailed, analytic and beautiful way as in the *Hesperides*. The inspiration for this nexus between visual description and taxonomic ordering may have come from Federico Cesi, the brilliant founder and leader of the Accademia dei Lincei, and Cassiano dal Pozzo; but the quality of art in the service of description was never higher than here. It is both comprehensive and analytic, in its range and in its attention to the detail that distinguishes. The book is a landmark in the field of that most empirical of all sciences, the science of taxonomy. But for the evidence of the extraordinary artistic labours that went into its preparation we have only to turn to the corpus of citrus drawings that survives from the collection of Cassiano and is recorded, if still only partially, in the present catalogue. They do not all correlate with the illustrations in the *Hesperides*; but this makes the collection as a whole even more notable. They give the present-day reader a sense, however glimmering, of the intensity with which the small group of scholars sponsored by Cesi and Cassiano, and working in the immediate orbit of Galileo, pursued their researches into the world of nature.

As is clear from the other groups of drawings also assembled by Cassiano and being published in the Catalogue Raisonné — on fossils, fungi, birds, animals, and geological specimens, *inter alia* — the young members of the Academy of Lynxes and friends and associates of theirs like Ferrari, went into the field and gathered information about new and old species of all things with unparalleled intensity. Realizing that the old authorities, especially Aristotle, were insufficient and incomplete, they discovered unheard-of new things and new phenomena. In order to continue their studies once they returned from the field, and to preserve, disseminate and publish their discoveries they made use of visual illustration in a more systematic and more extensive way than ever before. This commitment to visual description is also to be found, to some extent, in the work of Ulisse Aldrovandi, and in this respect he is their only predecessor. But they were more methodical than Aldrovandi, and the number of species and varieties they recorded was very much greater. One has only to turn to the drawings of oranges, lemons and citrons, and to the marvellously teratological hybrid forms recorded, however fragmentarily, in the present catalogue, to gain an understanding of a faith in pictures that would never again be equalled. The natural history drawings collected by Cassiano provide testimony to a profound belief that visual description would cast light on some of the deepest secrets of nature. Soon natural historians — even Linnaeus himself — would begin to be sceptical about the representation of objects as an aid to the study and classification of nature;[115] but for the time being — not, indeed, for very long — that faith remained unshaken. Who, looking at Cassiano's drawings of citrus fruit, could doubt that?

NOTES

1. *Il libro, che ho delle figure colorite di tutti gl'Agrumi, che il P. Ferrari nel suo libro dell'Hesperidi pose, le quale furno ricopiate dalle quelle [figure deleted] che havevo fatto far'io per quella sua fatica*, Archivio Dal Pozzo MS 32, fols. 220–220v. See also Archivio Dal Pozzo MS 39, fol. 106 for a report on the same fruit.

2. *Breve Istruttione per vasi, e spalliere di cedrati, e lumie*. This is still bound along with the other two treatises sent by Cassiano to Peiresc, namely the *Discorso del modo di governar diversa sorte di fiori* by Tranquillo Romauli, and the *Modo di custodir fiori* by Alessandro Gaetano, as Montpellier, BEM, MS H.502.

3. Montpellier, BEM, MS H.271, fol. 54; for further taxonomic concern about the *spongino*, see Archivio Dal Pozzo MS 39, fol. 53; for the *Limon di Monna Laura*, see Archivio Dal Pozzo MS 39, fol. 78, also Ferrari 1646, p. 219. For *limoncelli* see n. 36 below; for the *lumia* of San Domenico, see **59**; for *ponzini* in general, see **20**, **55**, **81**; for striped and furrowed 'oranges', see, for example, **105–116**.

4. Ferrari 1633; Ferrari 1638. On Ferrari and the *De Florum Cultura*, see Freedberg 1989b and Freedberg 1996b. For extracts from the Barberini accounts recording the payments for the illustration, printing and publication of the *De Florum Cultura*, see Merz, pp. 326–8.

5. Ferrari 1646.

6. On Ferrari, see Freedberg 1989a and Freedberg 1989b, with the earlier bibliographical references. In the earlier reference works Ferrari's birthdate was generally incorrectly given; and I myself erred in placing it in 1582. In fact, his Baptismal Act is preserved in Siena, Archivio di Stato, Biccherna 1137 (Battezati da 1661 al 1586), fol. 324 [2 May, 1583]: *Giovanni Battista figlio dello Excellente Signore doctor Messer Pasquino di Matteo Ferrari e di Madonna Caterina sua moglie fu battezato dal Piovano il dio 2 maggio, fu compare Thomasso di fabio spagni*).

7. Rome, ARSI, Rom. 172 [Ingressus Novitiorum 1594–1630], fol. 59.

8. Ferrari 1622.

9. Although the *Fasti Senenses ab Academia Intronatorum Editi*, as its title suggests, passed as the joint work of the Accademia degli Intronati, and was completed and edited by Sebastiano Conti, it was largely researched and written by Ferrari, as its Introduction makes clear. Its first edition appeared with no place and no date of publication, though it seems most likely to have been published in Siena in 1659; the second, less sumptuous edition was published in Siena by Bonetti in 1669. On this work see De Backer/Sommervogel, II, cols. 1390–1, s.v. Conti, Sébastien; and Freedberg 1989b, p. 40, n. 27.

10. The first edition of Ferrari's *Orationes* appeared in Lyon in 1625 (Rouille heirs) with 25 orations; several were added to the editions that appeared in Rome in 1627 (Corbelletti), 1635 (Facciotti) and in Venice, 1644 (Baglioni), Cologne, 1650 (Cornelius ab Egmont), London 1657 (Roger Daniel), London, 1668 (John Redmayne), Vilna, 1730. See De Backer/Sommervogel III, cols. 676–9, as well as the *Biographie Universelle*, XIII, p. 614 for references to other editions which I have not been able to trace.

11. For more on Ferrari and the *De Florum Cultura*, see especially Freedberg 1989b and Freedberg 1996b, both with further relevant bibliographical references.

12. It is not clear whether Lanfranco was originally commissioned to provide a *Feast of the Gods* or not; but in some copies of the 1638 Italian edition, his design replaces that of Pietro da Cortona in the earlier edition. On the editions, illustrations and various issues of the *De Florum Cultura*, see Petrucci 1985, pp. 140–2.

13. On the relationship between Cortona's designs (especially the *Dance of Vertumnus* and Poussin's *Realm of Flora* in Dresden), see Freedberg 1996b.

14. Ferrari 1633, p. 468.

15. For a very representative selection of the relevant Barberini accounts, see their excellent publication by Merz, pp. 326–8.

16. Ferrari 1633, p. 499. See also Tongiorgi Tomasi/Ferrari, pp. 2–15, as well as Freedberg 1996b.

17. *Flora ovvero Cultura dei fiori, ... e trasportata dalla lingua Latina nell'Italiana da Lodovico Aureli Perugino*, Rome 1638; and *Flora seu De florum cultura Libri IV. Editio nova, accurante Bernhard Rottendorfio*, Amsterdam 1646.

18. Paris, BN, MS FF. 9539, fol. 81v.

19. Paris, BN, MS FF. 9542, fol. 46.

20. Archivio Dal Pozzo MS 39, fol. 110.

21. Montpellier, BEM, MS. H. 271, fol. 185. On the problem of *ponzini*, see especially **55**; but see **20** and **81** as well. For *lumie*, see **59–62**; for the Adam's Apple, **58**, **63**, **71**, and **93**; for pear-shaped fruits, especially **45**; for small Chinese oranges, **34–36**; and, of course, for the horned and/or fingered fruits (*dalle die* in this letter), the whole section on teratological and pregnant specimens in the Catalogue, and further below in this essay. For further references, see also n. 3 above.

22. Paris, BN, MS FF. 9539, fols. 68-68v.

23. Olina 1622; on this work, see McBurney 1989, pp. 37–47, as well as the articles referred to in following note. The drawings associated with the *Uccelliera* will be discussed in volume B.IV of the present series.

24. In addition to the article cited in the previous note, see also McBurney 1992a, especially pp. 355–8 and McBurney 1992b.

25. Archivio Dal Pozzo MS 39.

26. For example, Archivio Dal Pozzo MS 39, fols. 50–52.

27. See especially the splendid lists on Archivio Dal Pozzo MS 39, fols. 20–21, as well as fols. 17–18.

28. *Veduta di Nervi in Riviera di Genova fu mandato da S. Cattaneo Cattaneo per servirsi nel libro degl'Agrumi*; see Archivio Dal Pozzo MS 39, fols. 114–115v for Cattaneo Cataneo's report.

29. Archivio Dal Pozzo MS 39, fols. 67–67v, 88–89v, 103–103v.

30. Archivio Dal Pozzo MS 39, fols. 4–5.

31. For example, the two Genoese reports from Cattaneo Cataneo and Luca Fabiano are referred to in the Index as being on fols. 5 and 8 respectively; but in fact they are bound as fols. 114 and 112 respectively.

32. Archivio Dal Pozzo MS 39, fol. 23.

33. In Archivio Dal Pozzo MS 32. See also note 1 above. Archivio Dal Pozzo MS 39, fol. 89 bis is dated 1689 and refers to a drawing of a Calabrian *limoncello* by Pietro Santi Bartoli; the page is clearly a later insertion.

34. Archivio Dal Pozzo MS 39, fol. 110.

35. The various reports credited to Gaetano are dispersed throughout the early part of Archivio Dal Pozzo MS 39, and they give some idea of the difficulties facing any student of this complicated manuscript. Fols. 28–28v (from Gaetano) are rewritten, in a different hand, on fols. 30–30v; similarly, fols. 25–26v are repeated on fols. 32–34v, but this time incorporating Cassiano's own pencilled additions on the bottom of fol. 26.

On the other hand, fols. 27–27v show the *corrected* form of fol. 29 (that is, again incorporating significant corrections by Cassiano, such as his insistence on changing *limoncelle* — Gaetano's term — to *limone* — Cassiano's). Fol. 31v is in the same neat hand as fols. 28–28v, repeating material on fol. 35; fol. 31, also in this hand, repeats fol. 40, while fols. 53–54 present further material by Gaetano, not repeated elsewhere.

36. Archivio Dal Pozzo MS 39, fols. 114–115v; see fols. 14–18.

37. Archivio Dal Pozzo MS 39, fols. 38–39.

38. Archivio Dal Pozzo MS 39, fol. 24.

39. Archivio Dal Pozzo MS 39, fol. 36; see fol. 108.

40. Archivio Dal Pozzo MS 39, fol. 53, sent to him personally by Alfonso Bonnini.

41. Archivio Dal Pozzo MS 39, fol. 50 (via Livorno from Lisbon).

42. Archivio Dal Pozzo MS 39, fols. 43, 45 (an early report dated April 1629); see Ferrari 1646, pp. 372–3.

43. Archivio Dal Pozzo MS 39, fol. 41.

44. Ferrari 1646, p. 139.

45. See for example Archivio Dal Pozzo MS 39, fols. 14v–15, 41.

46. See notes 87 and 88 below.

47. Letter from Ferrari to Cassiano, 4 June 1639, in Archivio Dal Pozzo MS 4, fol. 350: *Questa mattina ho ricevuto dal giardino dell' Eminentissimo Cardinale Antonio Barberino il limone del agro dolce di Spagna con il suo ramo, foglie, et fiori. Supplico V.S. Illustrissima a favorirmi di farlo quanto prima dipignere intero e smezzato; e poi favorirmi ancora di parte del frutto per poterlo meglio descrivere.*

48. Since *legere* can mean both 'read' and 'pluck', Ferrari's typical pun and its playful rhetorical extension into the next metaphor cannot be translated directly: *Alterum hic miraculum est, ut, quae leguntur poma, non avellantur* ... Ferrari 1646, p. 69.

49. Ferrari 1646, p. 69. Although Leonardi is apostrophized in this passage, his full name is given *in margine*.

50. In the *Avviso a chi legge*, on sig. A4 verso (*Ma havendo doppo fatto rifare tutte le sudette figure a bulino, cavate da disegno fatto assai diligentemente da Vincenzo Leonardi....*); the only plate in which his name appears is the one facing p. 63.

51. Windsor, RL 27628, *Natural History of Birds*, fol. 30. Signed on the verso, top edge of the sheet, *Vin Leonardi F. 1629*. See *Cassiano* cat. 1993, no. 106, p. 177.

52. But for references to citrus drawings on vellum, see also the correspondence with Giovanni Nardi (see note 1 above) in Archivio Dal Pozzo MS 32, fol. 219v, where he refers to *Otho Merceus, che fece ritrasse i pomi dell'altra volta in carta pecora assai diligentemente* (Cassiano dal Pozzo to Giovanni Nardi, 21 December 1632).

53. Siena, Archivio di Stato, Balia 795, No. 20, 6 October 1646.

54. Siena, Biblioteca Comunale B.LXXIII.A.7.

55. This is implied, for example, by the fact that the contract at the beginning of the *Notizie diverse* states that Cassiano had Poussin's drawing of the *Nymphs of the Hesperides bringing citrons to Lake Garda* on p. 97 of the *Hesperides* made for that book, and it stipulates that at least one of the other designs for the allegorical illustrations — Albani's on p. 50 of the *Hesperides* — be returned to him after being engraved; Archivio Dal Pozzo MS 39, fol. 2.

56. Commelin 1676.

57. Volkamer *Nürnberg. Hesp.* I and II.

58. Risso/Poiteau 1872.

59. See, for example, Matthioli 1582, p. 266 (lemon). Most—but by no means all—of the illustrations first appeared in the Czech edition published by Georgen Melantrich at Prague in 1565. There was also a series of editions between 1554 and 1596, chiefly coming out of the house of the Valgrisi in Venice, which had considerably smaller woodcut illustrations. On the 16th-century editions of Matthioli, see Fabiani 1872.

60. Clusius 1601, p. 7.

61. See for example Dodonaeus 1616, pp. 791–2; and Pena/de Lobel 1576, p. 425.

62. The literature on Aldrovandi is rapidly growing; see, *inter alia*, the important articles by Giuseppe Olmi, such as those collected in Olmi 1992 and Olmi 1978. Montalenti 1960, pp. 118–34 is a useful sum-

mary, and has a good bibliographic overview. On the Aldrovandian illustrations in general, as well as the relationship with other botanical illustrations, see the excellent articles by L. Tongiorgi Tomasi and P. Tongiorgi in *L'immagine e Natura*. For the Aldrovandian corpus of fruit illustrations, see Baldini 1990 (p. 66 on citrus fruit).

63. Ferrari 1646, pp. 70–71. See also Freedberg 1992 for more on these and other descriptions of games by Ferrari and Cassiano, as well as on the interest in the ethnography of games in their circle. To this day a fierce orange battle is still fought every year at carnival time in the town of Ivrea, Piedmont.

64. Ferrari 1646, p. 39. See also Archivio Dal Pozzo MS 39, fols. 13v–14.

65. Ferrari 1646, pp. 41, 45, and elsewhere in the extraordinary chapter (I.VIII) entitled *Nominum etymologiae* (pp. 41–6).

66. Ferrari 1646, p. 357.

67. Ferrari 1646, p. 360.

68. The other reference is in connection with the illustration by NicolasPoussin of the arrival of citrons borne by the Nymphs of the Hesperides at Lake Garda, where Ferrari refers to Cassiano's patronage of art, and in particular to his patronage of Poussin (Ferrari 1646, p. 98).

69. Foucault 1973, p. 129.

70. See Foucault 1973, p. 39.

71. See Ferrari 1646, p. 4: *Ad ejusdem silvae umbram & iucundam & salubrem saepius lectorem placet fabulis, rei seriae subservientibus*.

72. For a much fuller account of Ferrari's contribution to citrus taxonomy, see the essay by Enrico Baldini here.

73. *Rhegio Roman cum ramulo arcessivi, ut oculata describerem fide*, Ferrari 1646, p. 239; see p. 271.

74. On this, see Freedberg 1994.

75. Ferrari 1646, p. 4.

76. Ferrari 1646, p. 279.

77. Ferrari 1646, p. 69.

78. Paris, IF, MS 976, fols. 84 bis, 85–85 quinquies, 187A-D.

79. On Bimbi's fruit paintings, see Baldini 1982.

80. On Gallesio, see especially Baldini/Tosi 1994.

81. On all of these see the introductory essay by Enrico Baldini in this volume, with further bibliographic references; but note especially Baldini 1989a as well.

82. Ferrari 1646, p. 413.

83. Ferrari 1646, p. 257.

84. Montalbani 1668, pp. 527, 497, 515 and 534, respectively.

85. Archivio Dal Pozzo MS 39, fol. 103; there is, of course, much more on digitated fruits: see, for example, the extended discussions on fols. 14 and 114.

86. See Archivio Dal Pozzo MS 39, fols. 15, 34 and 54 on the *Melangoli della dita*, the *Cedri che scherzano*, and the *Aranci che scherzano* respectively.

87. Archivio Dal Pozzo MS 32, fol. 197.

88. For example, Archivio Dal Pozzo MS 32, fols. 23 (*Cedrangolo gigante*), 24 (*Limoni cedronati*), 40 (*Melangolo cedrato*) 106 (*Del Cedrato detto Cedrato Arancio*).

89. *Caeterum nullo ludit in pomo natura licentius* ... ('But in no fruit is nature more freely playful ...'), Ferrari 1646, p. 75.

90. Ferrari 1646, p. 75.

91. Ferrari 1646, pp. 75–6.

92. *Igitur cum seminis partes inter se diversae, quas etherogenias vocant, loci fertilate exuberant: cumque foetum informans arboreum vivida vis, quam vulgo spiritum appellant, multiplicata dividitur, uni facile pomo alia ingeneratur, prout artes easdem vario lusu lasciviens natura digesserit*, Ferrari 1646, p. 264.

93. Ferrari 1646, pp. 263–4.

94. Ferrari 1646, p. 403.

95. Ferrari 1646, p. 403.

96. Ferrari 1646, p. 407.

97. Ferrari 1646, p. 67.

98. *Iam vero querenti causam praegnantionis in pomo admirandae philosophicam habeo & poeticam afferre* ('Now to those who ask I must bring forward the poetic and the philosophical cause of the remarkable pregnancy in the fruit ...'), Ferrari 1646, p. 264.

99. Blunt 1964, p. 52.

100. Especially in the 'fables'; a typical pair occurs, for example, with the use of *corculum*, and phrases like *flebilis susurri rumusculum* on p. 86.

101. For the preparatory drawing in the Louvre, Cabinet des Dessins, see now Harris 1977, no. 86, pp. 101–2. See also the excellent discussion in the landmark catalogue by Bean 1959, no. 28, p. 56.

102. Ferrari 1646, pp. 81–8.

103. Ferrari 1646, pp. 273–5.

104. Ferrari 1646, p. 275: *merito Florentiam finitimasque in oras translata est: Ubi rerum omnium praestantia vere perpetuo floret.*

105. Ferrari 1646, pp. 417–18.

106. Ferrari 1646, p. 421.

107. Risso/Poiteau 1872 (first edition, Paris, 1818), pp. 50-51 (... *Depuis la réforme opérée en botanique par Linné, les botanistes ont cessé de s'occuper d'une infinité de plantes et d'arbres utiles ou agréables ... parce que ce naturaliste célèbre prétendait que toutes ces plantes et tous ces arbres n'étaient que des variétés indignes de l'attention du plus petit botaniste* ...). See also p. 13 (*C'est une chose assez remarquable que la légèrté avec laquelle les botanistes décident, du fond de leur cabinet, que telle plante est une variété, et qu'en conséquence elle est indigne de leur attention. On les voit blanchir, la loupe à la main, sur une mousse, sur un lichen de la Nouvelle Hollande ou de la Cochinchine, et ils négligent de savoir à priori quel est le froment que fait le meilleur pain et quel est le raisin qui fait le meilleur vin. A peine trouve-t-on quelques Orangers dans leur répertoire ...), and p. 63 (... On ne conçoit pas comment, à la voix de Linné, la botanique a pu rejeter de son domaine les plantes les plus utiles et les plus agréables, en les qualifiant de monstres et de variétés: on ne conçoit pas mieux comment on peut parvenir aux honneurs et à la célébrité, en décrivant des mousses ou d'inutiles moississures ... Depuis que l'Oranger qui nous occupe a doublé ses profits et nos jouissances en doublant ses fleurs, il est devenu un monstre pour les botanistes, et comme tel, banni de leur répertoire. ...*).

108. For a good summary of the Linnaean classification and the subsequent modifications and refinements, see Russo 1986.

109. Baldini 1991, pp. 13 and 17. See also the discussion, illustration and observations on pp. 86, 87, and 281 below.

110. See **74**, **87–91**, **95** etc., as well as pp. 91 and 281 below. For both these forms see Baldini 1991, and Baldini 1982, p. 41f.

111. Baldini 1982, pp. 17–44, especially pp. 39–40. But see also Baldini 1989 for a more detailed treatment of 'polymorphism'.

112. Both were placed at the end of the section on lemons, just prior to the concluding chapters on the lime; Ferrari, 1646, pp. 309–29.

113. It will be remembered that under the Jesuit *Ratio Studiorum*, which had formed the basis of Ferrari's education, the second year of the three-year Philosophy course was devoted to natural philosophy (*physica*, which included mathematics as well) chiefly in the shape of the Aristotelian treatises *De Anima*, *De Coelo*, and *De Generatione*.

114. For an early appreciation, see Evelyn 1662, p. 78.

115. See Freedberg 1994.

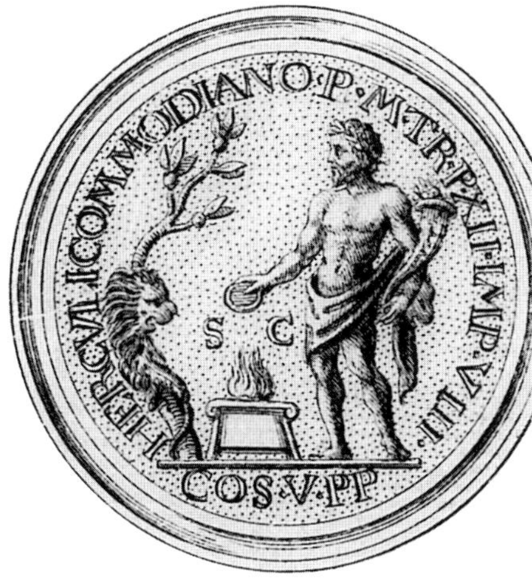

Fig. 31. C. Bloemaert (?), *Ancient coin showing Hercules, from the collection of Francesco Gottifredi*, engraving, Ferrari 1646, p. 23 (detail)

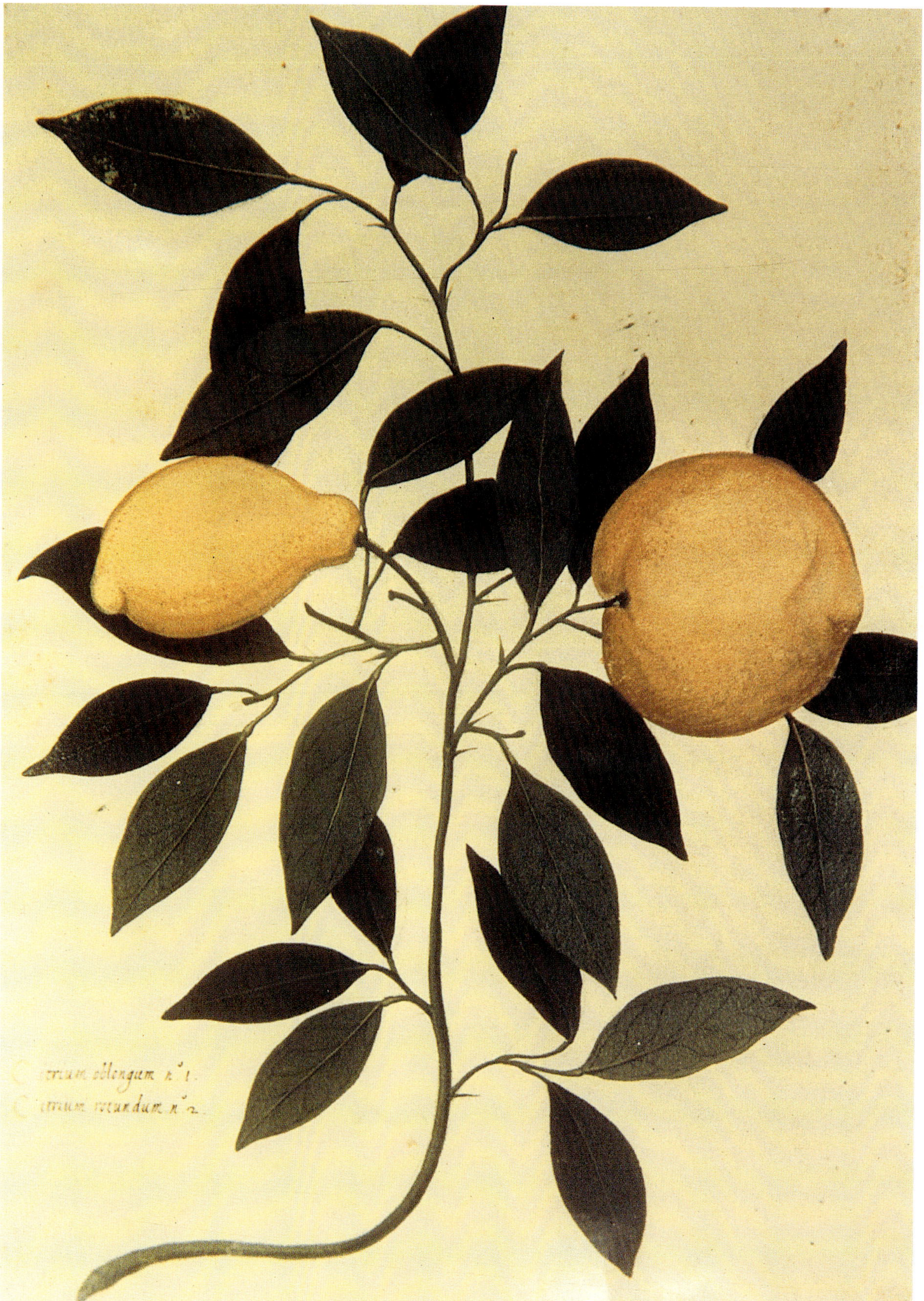

Fig. 32. Citrus heteromorphism, where two kinds of lemon fruits — *Citrium oblongum* and *Citrium rotundum* — hang from the same twig: Aldrovandi I.P., I, fol. 16

THE ROLE OF CASSIANO DAL POZZO'S PAPER MUSEUM IN CITRUS TAXONOMY

Enrico Baldini

CITRUS TAXONOMY has posed vexing questions for centuries, puzzling even modern scientists who have confronted it.[1] The marked nomenclatural confusion which persists in citrus botany and pomology depends primarily on the wide range of genotypic and phenotypic diversity within this genus and on the plethora of terms used by such leading early scholars as Giovanni Battista Ferrari, Francis van Sterbeek, Jan Commelin, Carl Volkamer, Pierantonio Micheli, Giorgio Gallesio, Etienne Michel, Antoine Risso and Antoine Poiteau[2] to distinguish the great many specimens included in their works. Quite apart from the obvious taxonomic differences pertaining to the various citrus species, the startling biodiversity among and within the cultivated species is to be sought in such factors as intergeneric and interspecific hybridization, plant heteromorphism, the development of nucellar embryos, the rejuvenation of senescent species long propagated asexually, the natural occurrence of synthetic and autogenous chimeras (the so-called graft-chimeras and bud mutations) and, finally, in teratological events, long seen as a rich source of intriguing *monstrosa et mirabilia* ('monsters and wonders').

I

In *Citrus*, spontaneous cross-fertilization easily occurs not only within the same but also between different species and genera; normally progenies exhibit a wide range of characteristics which set them apart from one another, yet partially recall those of their parents.[3] Although Ferrari suspected that intermediate genotypes could occur, he was nevertheless resolute in his determination to reduce them drastically to only three distinct botanical categories: citrons, lemons and oranges.[4] Specimens showing morphological characteristics of both citron and lemon or of both lemon and orange, as for example *Malum citreum dulci medulla*, *Limon citratus primae notae laevior* and *scaber*, and *Aurantium citratum*[5] among others, were also ascribed to this restricted taxonomic frame. This quite limited classification influenced later citrologists up to Giorgio Gallesio, who considered, in addition to his four basic citrus species (*citronier, limonier, bigaradier, oranger*), a number of 'hybrids' covering awkward intermediate genotypes like *Cédrat de la Chine*, *Cédrat de Florence*, *Limonier à fruit doux*, *Limon-cédrat*, *Bigaradier à fruit doux*, *Pomme d'Adam* or *Pompoléon*, *Oranger à fruit blanc* and *Oranger turc*.[6]

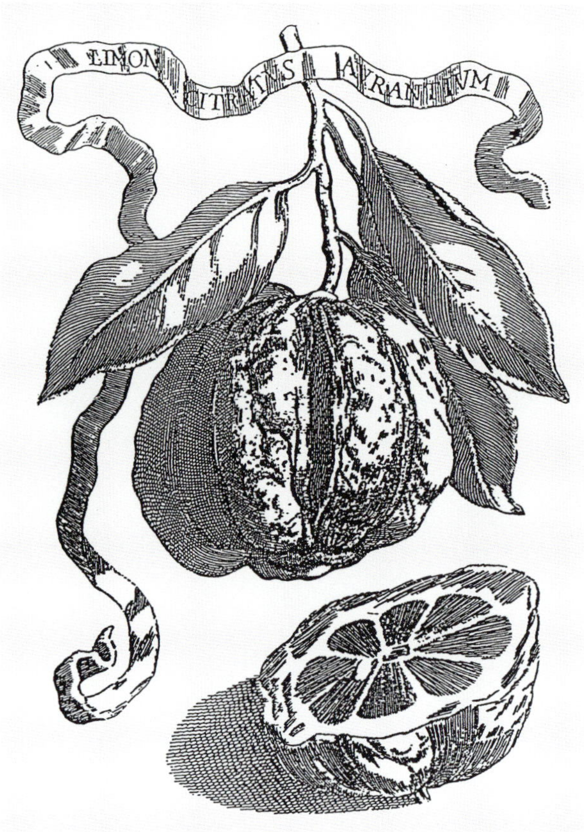

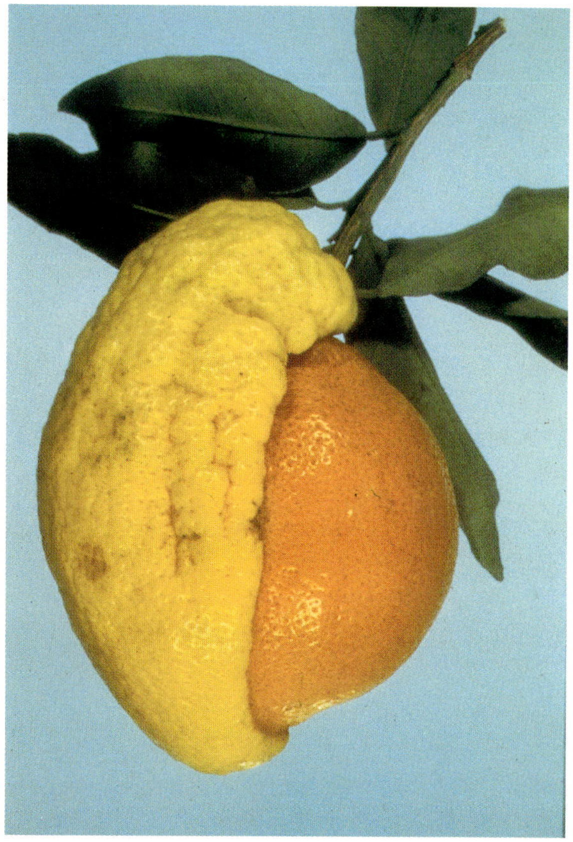

Examples of fruit polymorphism, an epigenetic condition in which differently shaped fruits are produced by the same trees in the course of the year or even simultaneously by a single tree, are well rendered in watercolour in Aldrovandi's *Iconographia Plantarum*[7] [Fig. 32] and similarly in Montalbani's illustration of *Citreum polyschematicum*,[8] each a faithful depiction of a citron twig bearing both an ellipsoid and a globose fruit; similarly the *Limon pusillus Calabriae duplex*, depicted by Ferrari,[9] and the *limone spinoso da premere ordinario*, depicted by Micheli,[10] exhibit both ellipsoid and globose fruits. Yet the most classic example of citrus polymorphism is perhaps offered by the present-day everbearing Sicilian lemon *Femminello*, whose fruits are either ellipsoid or globose depending on their ripening time.[11]

Strains of old cultivars rejuvenated through nucellar embriony[12] can easily be mistaken for distinct genotypes because of their vigorous growth, large leaves and fruits and luxuriant development of spines, the latter generally missing from cultivars that have long been propagated asexually.[13]

Chimeric and teratological fruits are other striking sources of citrus biodiversity. The freakish *Malolimonia citrata-aurantia, vulgo Bizzarria* [Fig. 33] was recorded in 1674 by the Florentine physician Pietro Nati, who found it *c.*1645 on a citron grafted onto a sour orange tree growing in the gardens of the Panciatichi family's villa near Florence.[14] The specimen was in fact a synthetic graft-chimera [Fig. 34].[15] While Nati believed he was the first to discover this, the *Aurantium callosum* (see **101, 102**)

Fig. 33 (facing page, far left).
Synthetic chimera: 'Bizzarria' *Limon citratus aurantium*. Etching by A. Haelweg after B. Franceschini, Nati 1674

Fig. 34 (facing page, left).
Synthetic chimera: citron+orange fruit occasionally found in Sicilian citrus groves

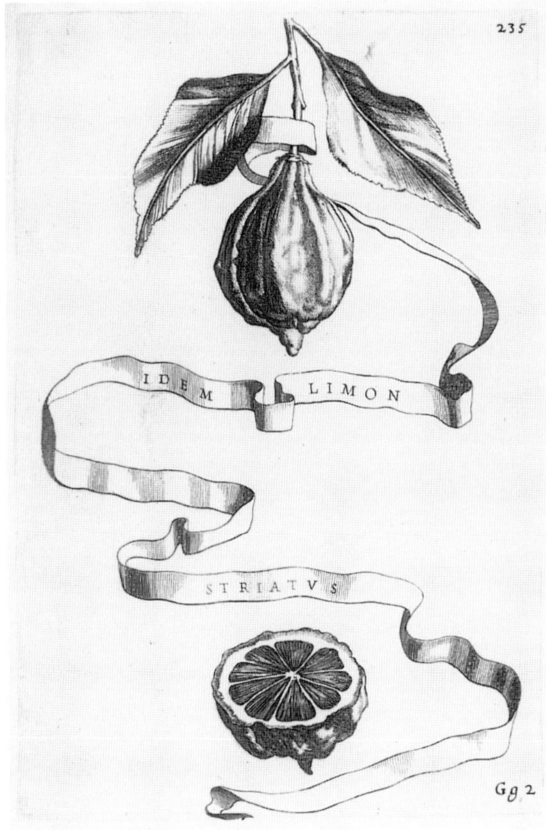

Fig. 35. Autogenous cytochimera, bud mutation: *Limon striatus*, etching and engraving, Ferrari 1646, p. 235

described thirty years earlier on page 411 of Ferrari's *Hesperides* clearly belongs, as well as other dal Pozzo drawings (see **103**, **106**), to this chimerical type, as do the heteromorphic fruits resulting from the budding of genetically different half-buds in the even earlier and, for the time, astonishing method postulated in 1558 by J.B. della Porta.[16]

A remarkably wide-ranging diversity of autogenous cytochimeras, which are produced by naturally occurring bud mutations, has been recorded since the seventeenth century in *Citrus*. These include the furrowed, ribbed or striped fruits (see **100**, **107**) as well as the variegated leaves which appear in Ferrari's *Hesperides* [Fig. 35];[17] in Bartolomeo Bimbi's paintings;[18] Volkamer's *Nürnbergische Hesperides*;[19] Risso and Poiteau's *Histoire Naturelle des Orangers*;[20] Targioni Tozzetti's *Raccolta di fiori, frutti e agrumi ...* ;[21] and, finally, in the unpublished drawing of *Limone Bizzarria*, commissioned in 1826 from the Italian artist Domenico Del Pino by Giorgio Gallesio and now preserved in the Gallesio Archive at Genoa.

Yet it is the three basic types of teratological citrus fruit, 'digitated,' 'pregnant' and 'twin', which since the seventeenth century have attracted the special attention of citrologists because, in the words of Ferrari: 'aborted foetuses and monsters are prized in fruits yet abhorred in living things'.[22] Digitated fruits, so called because they end in irregular, finger-like appendages, are recorded in Aldrovandi's *Iconographia Plantarum*;[23] in Ferrari's *Hesperides* [Figs. 36, 37],[24] in Montalbani's *Ulyssis*

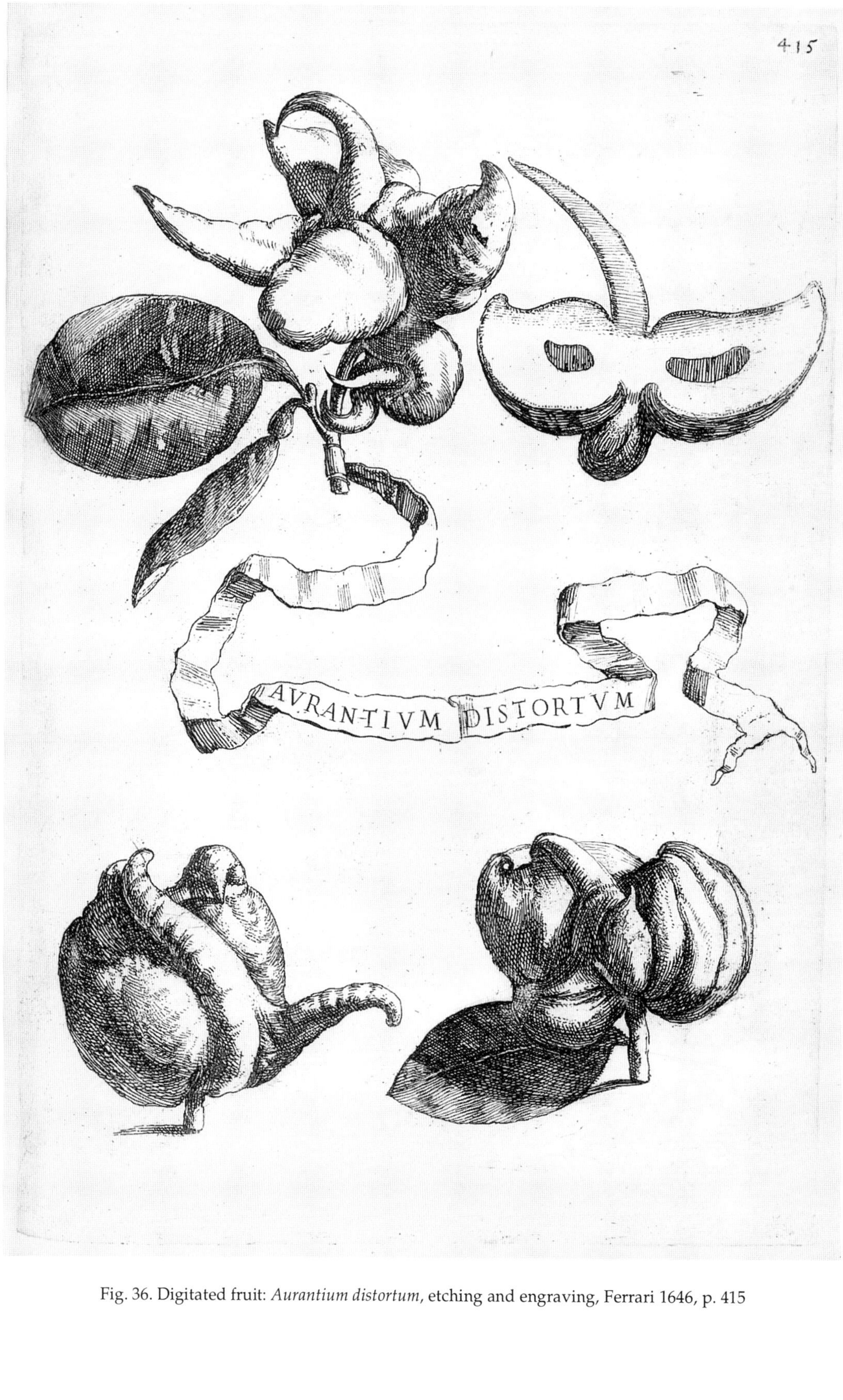

Fig. 36. Digitated fruit: *Aurantium distortum*, etching and engraving, Ferrari 1646, p. 415

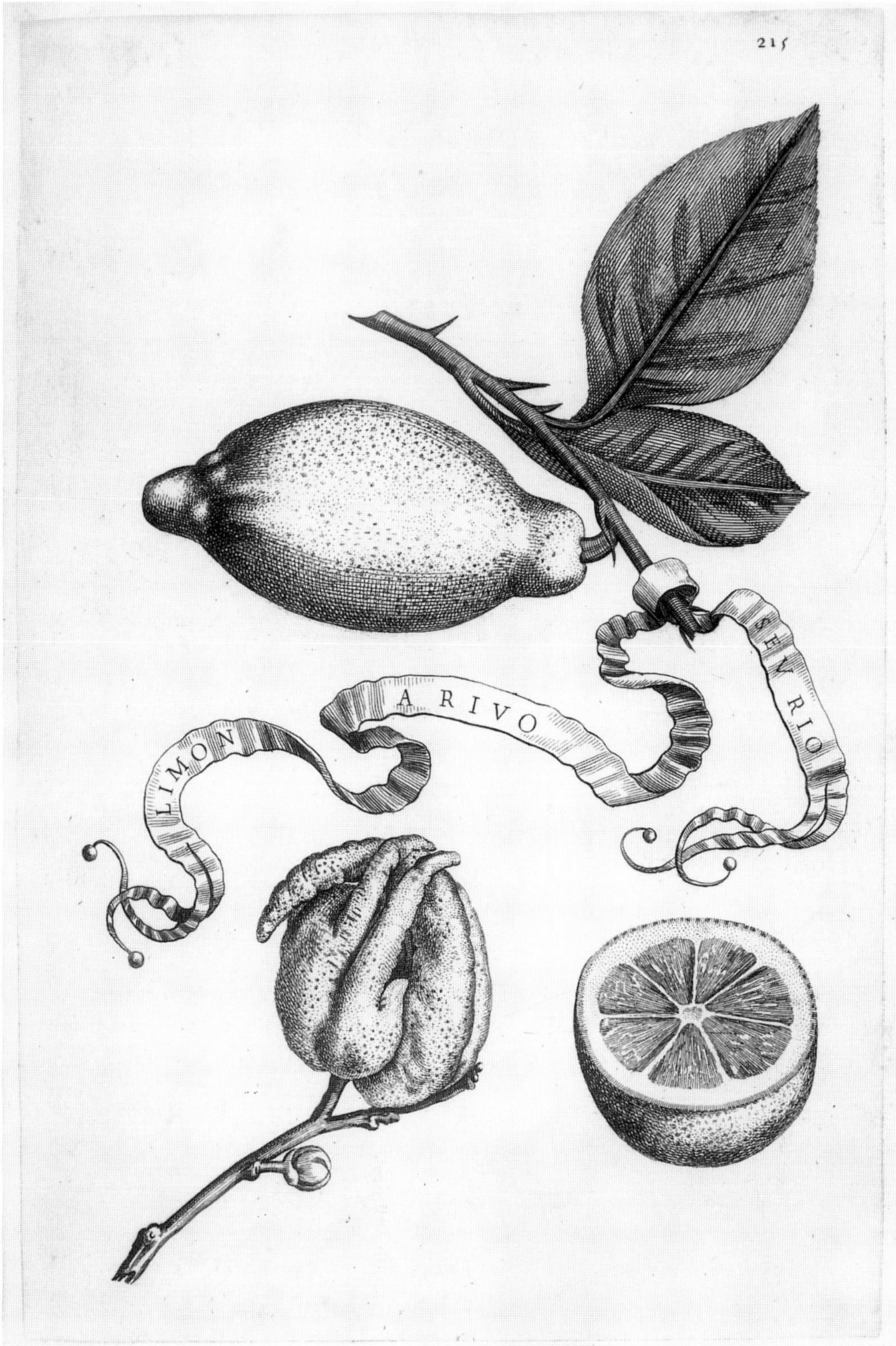

Fig. 37. Digitated fruit: *Limon a rivo seu rio*, etching and engraving, Ferrari 1646, p. 215

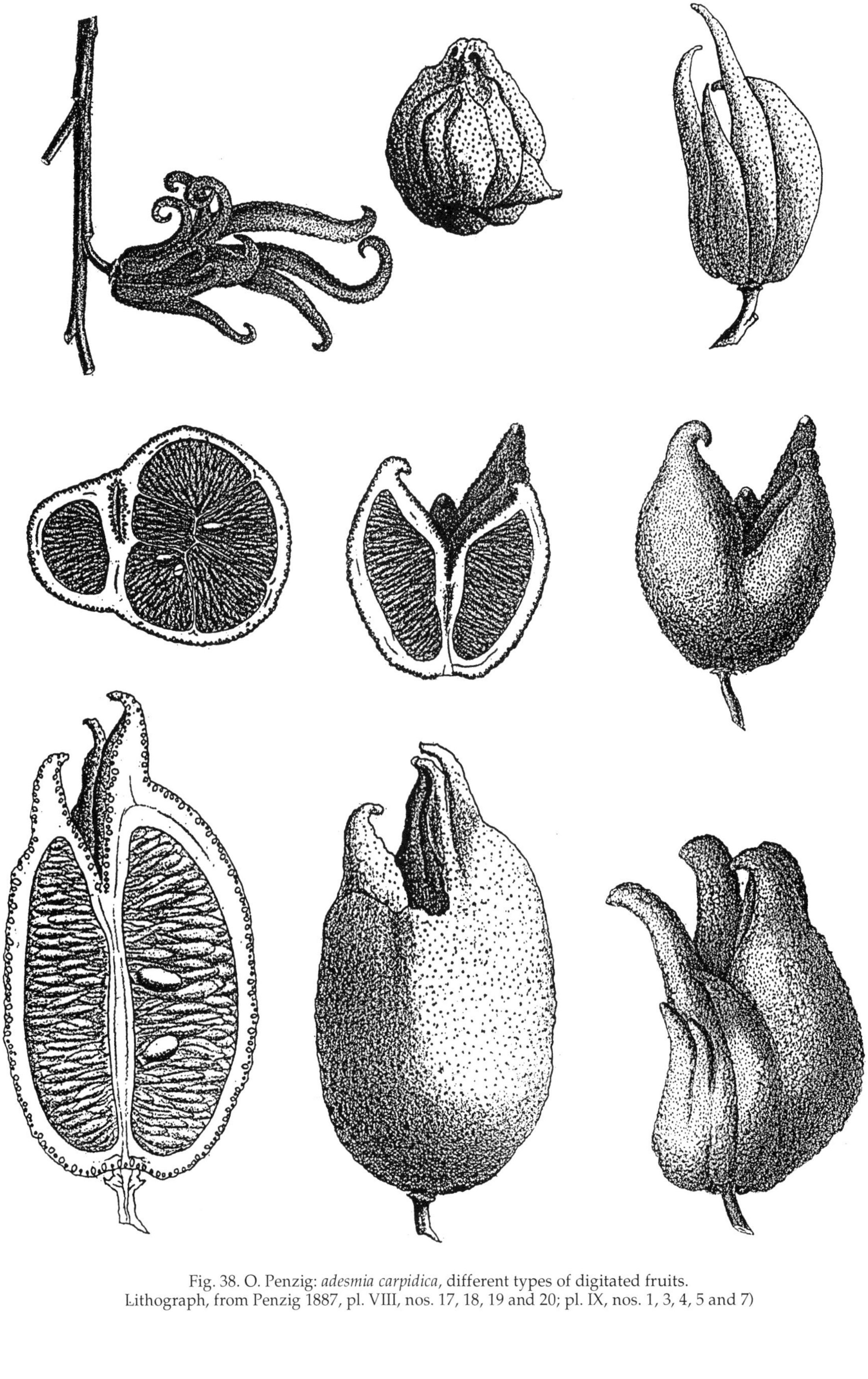

Fig. 38. O. Penzig: *adesmia carpidica*, different types of digitated fruits.
Lithograph, from Penzig 1887, pl. VIII, nos. 17, 18, 19 and 20; pl. IX, nos. 1, 3, 4, 5 and 7)

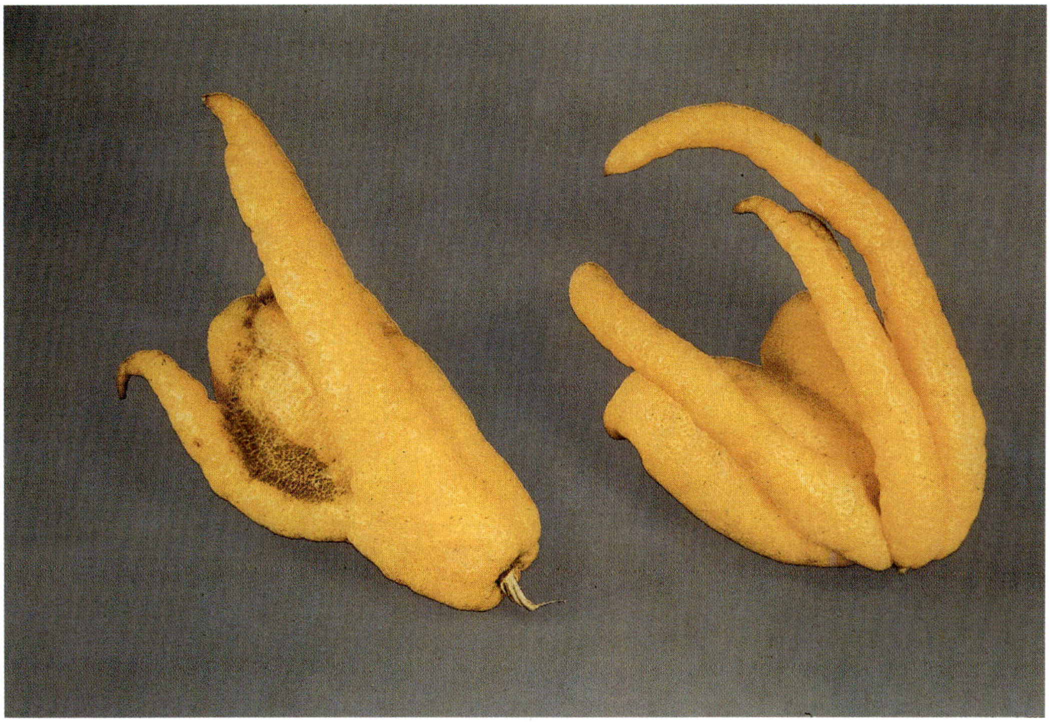

Fig. 39. Digitated lemons resulting from damage to fruit buds caused by the *Aceria sheldoni* mite

Aldrovandi ... ;[25] a Bimbi painting;[26] Volkamer's *Nürnbergische Hesperides*;[27] a manuscript entitled *Miscellanea rerum naturalium* by Luigi Ferdinand Marsili (1658-1730);[28] Risso and Poiteau's *Histoire naturelle des orangers*;[29] and in the unpublished drawing of *Citrus digitata* commissioned by Giorgio Gallesio from the French artist Antoine Poiteau and now preserved in the Gallesio Archive at Genoa.[30] Penzig (1887) applied the scientific term *adesmia carpidica* to the occurrence of digitated fruits [Fig. 38].

Ferrari offers a neat 'philosophical' explanation for the origin of these teratological fruits,[31] ascribing it either to a weakness of the seeds and the consequent inability to form regular fruits, or to a particular type of fertilization that engenders foetuses which turn into graceful finger-like shapes.[32] Needless to say, these conjectures stand in contrast to the current scientific explanation which identifies the most common cause as *Aceria sheldoni* Ewing (formerly *Eriophyes sheldoni*) (see p. 281), a worm-like mite which disrupts fruit development by damaging the buds and flowers of lemon trees, seldom of other species [Fig. 39]. *Aceria* can induce remarkable deformities in shoots and leaves too.[33]

Digitated fruits are also genetically produced by *Citrus medica* var. *sarcodactylis* (Noot.) Swing., a citron native to Japan and China but rare in Europe and North America: these misshapen fruits are prized both for their shape, being commonly known as 'Buddha's hands' (*Fo Shou Kan* in Chinese, *Bushukan* in Japanese), and for their marked fragrance, being widely used to scent rooms and clothes.

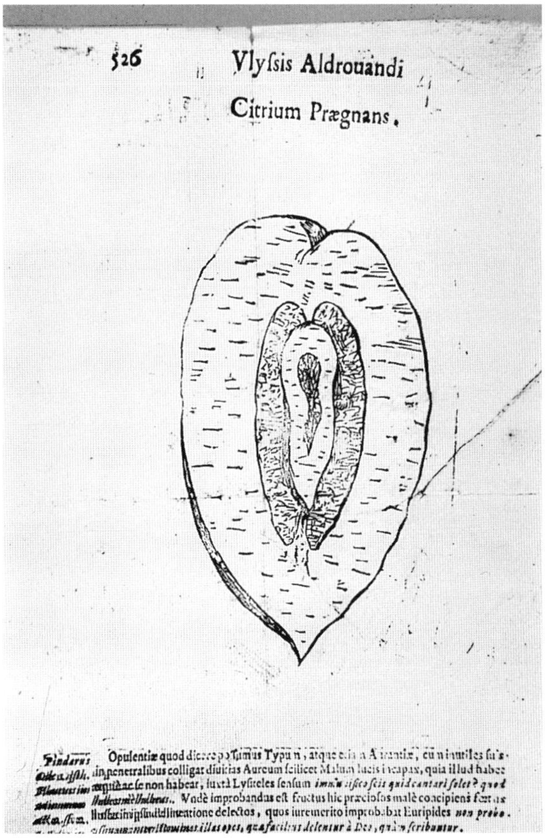
526 Vlyfsis Aldrouandi

Citrium Prægnans.

Fig. 40. *Citrium praegnans*, an unpublished proof for Montalbani 1668. Aldrovandi MS 88, fol. 4.

Fig. 42 (opposite, right).
Twin citrus: *Limon dulci cortice melopeponis*, whole and half fruit, woodcut, Montalbani 1668, p. 494

Fig. 43 (opposite, far right).
Twin lemon, lateral view

Fig. 41. O. Penzig: peeled navel orange and its longitudinal half, showing a small fruitlet at the top of the axis. Lithograph, from Penzig 1887, pl. X, nos. 4 and 5

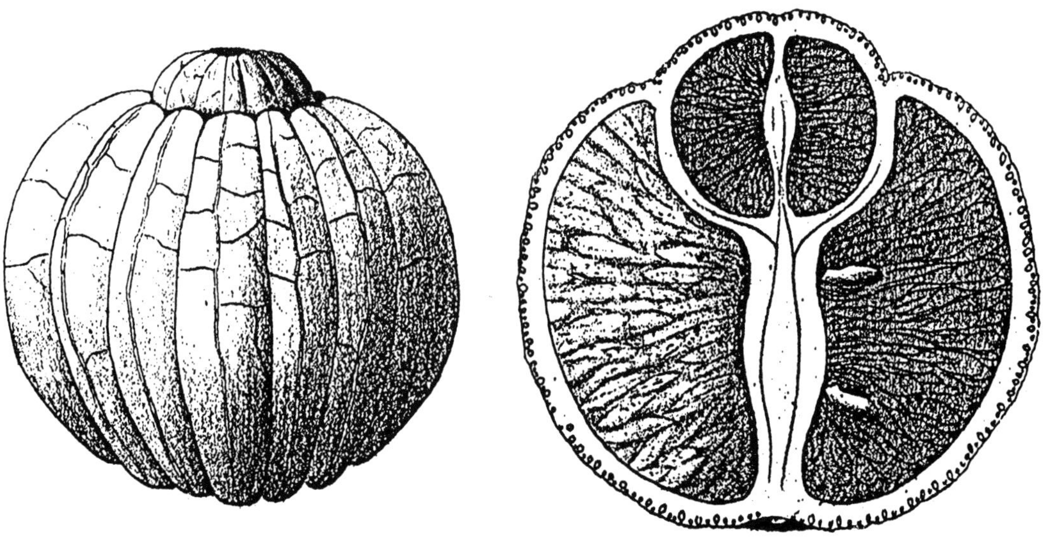

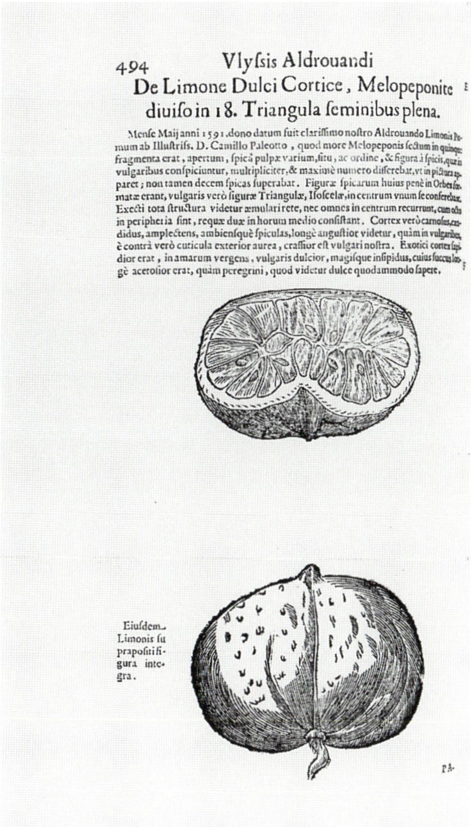
494 Vlyſsis Aldrouandi

De Limone Dulci Cortice, Melopeponite diuiſo in 18. Triangula ſeminibus plena.

Menſe Maij anni 1591. dono datum fuit clariſſimo noſtro Aldrouando Limonis Pomum ab Illuſtriſs. D. Camillo Paleotto, quod more Melopeponis ſectum in quinque fragmenta erat, apertum, ſpicâ pulpæ varium, ſitu, ac ordine, & figura à ſpicis, quæ in vulgaribus conſpiciuntur, multipliciter, & maximè numero differebat, vt in pictura apparet; non tamen decem ſpicas ſuperabat. Figuræ ſpicarum huius penè in Orbes formatæ erant, vulgaris verò figuræ Triangulæ, Iſoſcelæ, in centrum vnum ſe conferebat. Excêti tota ſtructura videtur æmulari rete, nec omnes in centrum recurrunt, cum octo in peripheria ſint, reque duæ in horum medio conſiſtant. Cortex verò carnoſus, candidus, amplectens, ambienſquè ſpiculas, longè anguſtior videtur, quàm in vulgaribus, è contrà verò cuticula exterior aurea, craſſior eſt vulgari noſtra. Exotici cortex ſapidior erat, in amarum vergens, vulgaris dulcior, magiſque inſipidus, cuius ſuccus longè acetoſior erat, quàm peregrini, quod videtur dulce quodammodo ſapere.

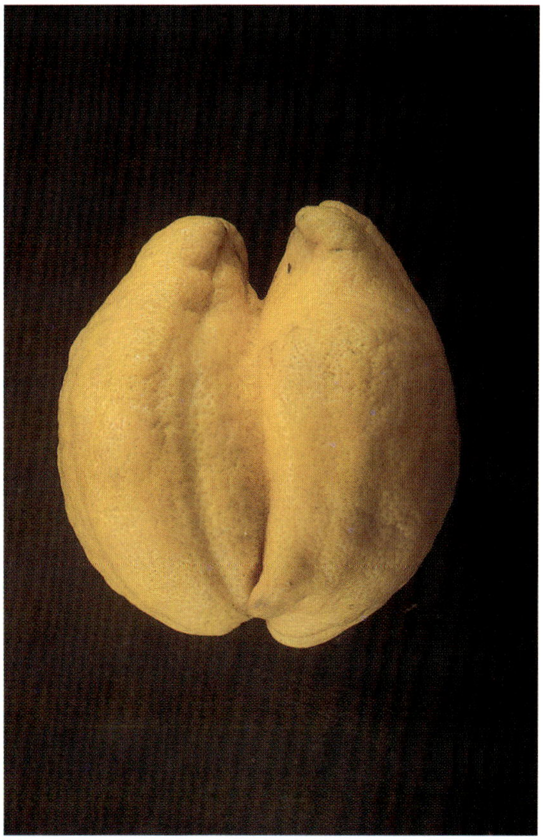

Another teratological phenomenon found in citrus fruits is pregnancy,[34] so called because of the fruitlet that forms inside and can be seen through an apical hole of the larger mother-fruit, called the 'navel' (see p. 283). The fruitlet may remain relatively small and totally enclosed in its mother, as in some oranges [Fig. 41], or develop to such an extent as to protrude beyond the broken stylar scar, which is the case in the teratological fruits. The unpublished proof for page 526 of Montalbani's *Dendrologia*, 1668, [Fig. 40][35] shows in longitudinal section a large citron with a second fruit inside it (Baldini/Tagliaferri 1990). Freakish 'pregnant' fruits are found in some of Cassiano's drawings (see **89-95**) and in Ferrari's *Hesperides*[36]. They are also depicted in some classic citrological treatises of the eighteenth and nineteenth centuries[37] and in Del Pino's unpublished drawing of *Citrus medica foetifera*, commissioned by Giorgio Gallesio in 1828 and now preserved in the Gallesio Archive. The 'poetic and philosophical' explanation Ferrari proffers for the origin of 'pregnant' citrus fruit makes a generic appeal to the relationship between the particular fertility of the ground and exuberant seed production, or to a remarkable generative vigour.[38] Gallesio tentatively suggested that pollination might be involved in this process, though he suspected that foetuses might form even before flower fertilization.[39] Later scientific findings have confirmed this suspicion, showing that navels can arise during the earliest stages of bloom from *A. sheldoni*-induced floral disorders or from genetic factors causing supplementary carpel formation by stamens (see p. 284), or by receptacle extrusions.[40] 'Twin' fruits may occur occasionally in citrus, as well as in apple, peach and other fruits. They are recorded in Cassiano's drawing, **98**, in Montalbani [Figs. 42, 43][41] and Volkamer[42]. Twin citrus fruits can arise from the concretion of stamens and ovary. According to Evreinoff, who thoroughly investigated this phenomenon in peaches, twin fruits can also arise from double-pistillate flowers that form under the influence of peculiar climatic conditions,[43] such as drought during flower induction.

Nearly all the genetic and epigenetic examples of citrus biodiversity discussed so far are portrayed in Cassiano's drawings and in the etchings of Ferrari's *Hesperides*. These pomological 'mirabilia' undoubtedly posed vexing questions when their scientific identification was undertaken.[44] Indeed, in many cases an obstinate uncertainty still persists that the depicted fruits do not represent real cultivars but rather single specimens chosen for their particular size or shape or teratogeny and even gathered from the same tree.

At present, the citrus images in Cassiano dal Pozzo's Paper Museum[45] constitute 118 drawing sheets, which represent 161 specimens: 55 sheets bear preparatory watercolours for 38 of Ferrari's etchings by Cornelis Bloemaert, which, on comparison, show faithful rendering; 63 were not reproduced and represent the unexplored part of the Cassiano-Ferrari citrus germplasm. Yet 42 of Ferrari's 80 citrological etchings do not have counterparts among the known Cassiano drawings, although their preparatory watercolours must have existed.

II

Nearly all of Cassiano's drawings are life-size renderings of whole and half fruits 'scientifically' displayed, in agreement with his personal instructions,[46] in a highly rigorous pomological arrangement that enables one to ascertain fruit shape, peel surface and colour, mesocarp (albedo) thickness and the characteristics of the endocarp (segments vesicles, axis, seeds). The Latin nameplate found in the scrolls of all Ferrari's etchings is missing in all but four of Cassiano's drawings (see **20**, **29**, **31**, **49**). In some cases twigs, leaves and flowers are also depicted, showing the blade form and margin, the incidental presence of spines and the size of the petiolar wings, according to the *Nota de' Capi che si desiderano sapere per havere compita informatione di tutti gli Agrumi* (that is, a sort of indexing system devised by Cassiano for organizing all citrological data; see p. 50).[47] In addition, two atypical drawings show fingered fruits assembled as if in still-life composition (**97-98**), and two others represent bearing branches (**117-118**).

Cassiano's Paper Museum drawings encompass a greater number of species than those found under the three restricted headings Ferrari employed to classify the specimens he had to hand, that is *Malum citreum*, *M. limonium* and *M. aurantium*. Species like *Citrus medica* L., *C. limon* (L.) Burm. f., *C. aurantium* L., *C. sinensis* (L.) Osb. *C. grandis* (L.), Osb., and *C. myrtifolia* Raf.[48] are duly represented in Cassiano's Paper Museum along with a great many hybrids and teratological fruits. However, *Citrus deliciosa* Ten., *C. bergamia* Risso and *C. paradisi* Macf. were unknown to either Cassiano or Ferrari because they make their appearance in Italy at a later epoch:

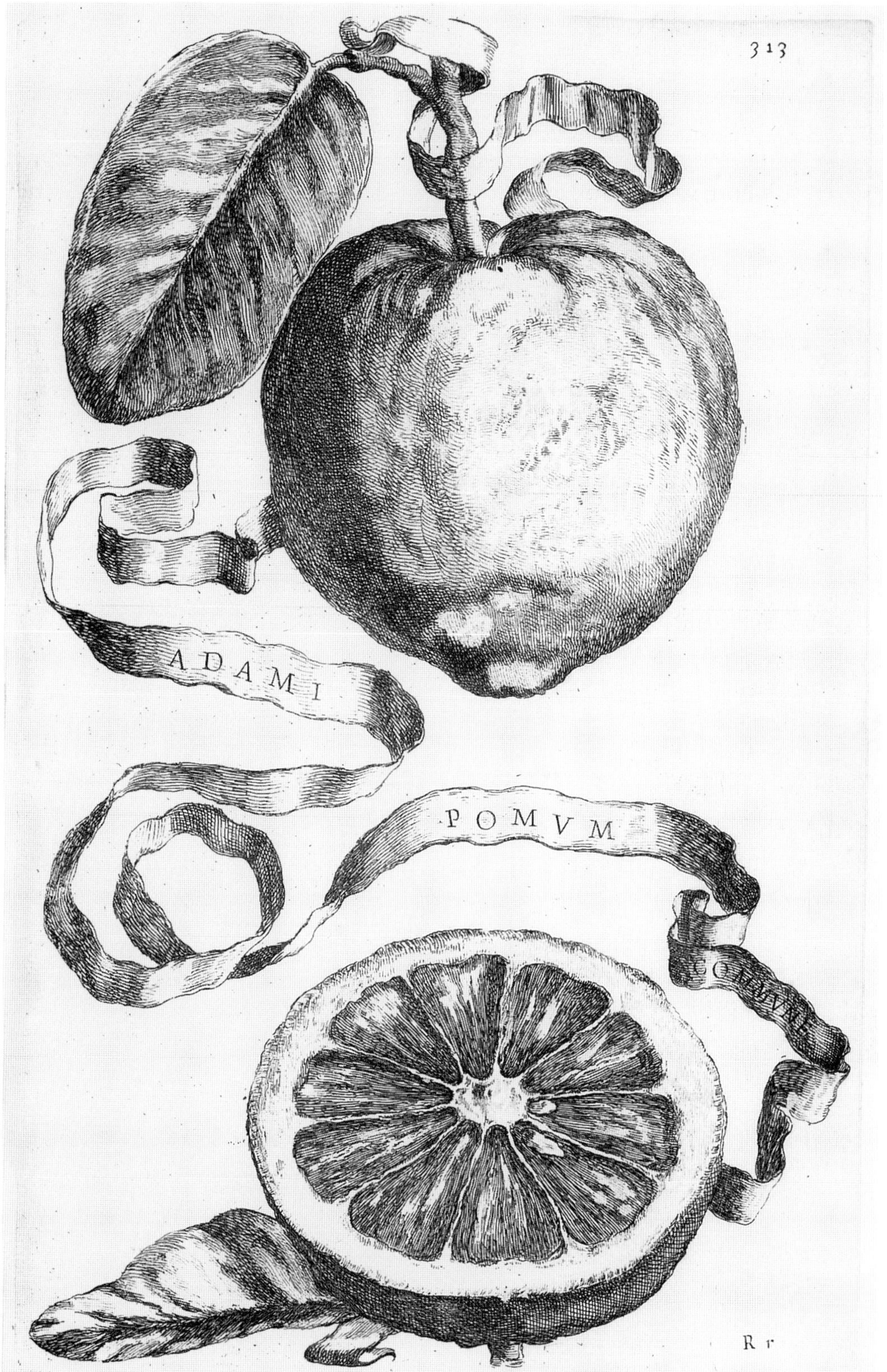

Fig. 44. *Adami pomum commune*, a taxonomically puzzling citrus, so called because some linear scars on its rind resemble a bite mark. Etching and engraving. Ferrari 1646. p. 313

the mandarin orange *C. deliciosa* Ten. dates to the early nineteenth century;[49] the bergamot *C. bergamia* Risso appeared in Tuscany during the time of Grand Duke Cosimo III (1642–1723), portrayed at the end of the seventeenth century by Bimbi in his paintings[50] and also reported by Paolo Bartolomeo Clarici in the early eighteenth century;[51] finally, the grapefruit *C. paradisi* Macf. (a spontaneous hybrid pummelo × sweet orange), first discovered *c.*1750 in the Bahamas, reached Italy's botanical gardens only in the late nineteenth century.

While we have ascribed a number of the ambiguous Cassiano-Ferrari hybrid specimens to the citron-lemon taxonomic category (*C. limonimedica* Lush.) because they share the traits of both species, another curiously ambiguous large fruit, the *Pomum Adami* or 'Adam's apple' [Fig. 44] has been assigned an indeterminate taxonomic position. Narrowly constrained by Ferrari to his lemon rubric,[52] this citrus fruit has been portrayed several times in differing shapes in Aldrovandi's *Iconographia Plantarum* as *Pomum Adami*[53] and *Pomum Ade,*[54] and even confused by some later citrologists with *Citrus decumana.*[55]

Cassiano dal Pozzo and Ferrari lived in an age that accorded citrus fruit a far greater variety of use and enjoyment than is true today. The contemporary citrus industry is decidedly given over to the narrow commercial pursuits of the market-place, whether in fresh produce (lemons, limes, oranges, mandarins, grapefruits) or processed (oranges, grapefruits and citrons), and consequently relies upon a restricted number of selected species and cultivars. The Renaissance, by contrast, saw, in the exuberant biodiversity of citrus fruit, myriad occasions to enjoy and to employ both trees and fruits: citron, orange and lemon plants, dressed as espaliers or potted, were living ornamental components that were fundamental to the 'Italian garden' (see p. 307). Their fruits served in numerous culinary, aesthetic, medicinal and even 'social' capacities: many lemons that were devoid of palatable qualities, for example *Limon sponginus, Limon racemosus* or *Limon Ponzinus chalcedonicus* ('hard as stones'), were prized as ornaments; many others, poor in juice and acidity but endowed with a pleasantly sweet rind, were sliced and eaten, sometimes dressed with salt (eg. *Lumia salis sive Spatafora notha*); sour oranges found service as seasonings, appetizers or powders to be dissolved in wine to prevent worms and the Black Death; citrons, much esteemed as gifts when bizarre or particularly large, were also an ingredient of the *panforte* cake[56] called *torrone;* finally, from the peel of many citrus fruit and the flowers of the sour orange came essential oils for perfumes.[57]

Considering the drastic geographical and genetic modifications which have occurred in Italian citriculture during the last two centuries, the attempt to match the Cassiano-Ferrari drawings with modern-day cultivated varieties of citrus fruit is somewhat hazardous. Nevertheless, some old cultivars (see **16, 34, 45, 46, 48, 52, 71, 100, 103**) seem to have survived in the historical gardens of central and northern Italy, and a few areas in the South, as well as in the fruit tree repositories of several Sicilian scientific institutions.[58] These old cultivars have a specific archeocitrological interest, but also can play an important role in the projects devoted to preserving the citrus germplasm resources.

Needless to say, Ferrari's *Hesperides* was instrumental in the first attempts at citrus classification thanks to its superlative plates, which were skilfully etched by Cornelis Bloemaert from dal Pozzo's drawings. Yet, in our opinion, Cassiano's drawings, when compared to the pictorial complement of Ferrari's compendium, make other remarkable contributions to citrus taxonomy: the great many hitherto unpublished documents of the Paper Museum extend the assortment of specimens to be studied, and the incomparable accuracy of the watercolours render in faithful, 'scientific' images both the botanical details and the chromatic shadings of the leaves, flowers and fruits of the trees in the *Hesperides*.

NOTES

1. Marcovitch 1926; De Noter 1926; Tanaka 1961; Hodgson 1961; Swingle/Reece 1967; Scora 1988; Davies/Albrigo 1994.
2. See Ferrari 1646, Sterbeek 1682, Commelin 1676, Volkamer *Nürnberg. Hesp.*, Micheli *Enumeratio*, Gallesio 1811, Loiseleur-Deslongchamps 1819, Michel 1816, Risso/Poiteau 1818.
3. With the aid of DNA molecular markers the genealogy of taxonomic entities can be determined. The RAPD test enabled Deng *et al.* 1995 to ascertain that citron and lemon are the irrefutable ancestors of the variety described in **45** (*Limone cedrato Piretto*, according to Continella/Tribulato, 1992), while citron and sour orange are the ancestors of *Pomo d'Adamo cedrato* (according to Continella/Tribulato 1992), described in **71**.
4. Freedberg 1992b.
5. Ferrari 1646, pp. 72-3, 263-7, 422-3.
6. Gallesio 1811, pp. 100, 103, 112, 115, 135, 138, 165, 166.
7. Aldrovandi I.P. I, fol. 16.
8. Montalbani 1668, p. 526.
9. Ferrari 1646, p. 211.
10. Micheli *Enumeratio*, pp. 134, 135.
11. See Casella 1935; Baldini 1991.
12. Apomictic adventitious embryos in *Citrus* can form within a single seed from somatic tissues of the nucellus (see p. 283). These embryos give rise to strains at once genetically uniform and virus-free but very vigorous and thorny, just as those arising from sexual embryos.
13. Reuther/Batchelor/Webber II, 1948, pp. 327-8.
14. Penzig 1887, p. 86, and Tanaka 1927, p. 77, suggested that Nati's 'Bizzarria' arose from an adventitious bud spontaneously formed on a citron + sour orange grafted tree with the histological contribution of both scion and rootstock. Penzig also supported the identity of Nati's 'Bizzarria' as Ferrari's *Aurantium callosum* (**101**), whereas Tanaka notes that it should be a periclinal chimera with an epidermal skin of sour orange over a citron core, the sectorial structures resulting from breaches of the orange skin by the underlying citron tissues. Images of 'Bizzarria' are also found in a painting by Bimbi (Bimbi: Castello 594) and in the engravings of Volkamer (*Nürnberg. Hesp.* I, p. 172a), Risso/Poiteau (1818, pl. 52), Targioni Tozzetti (1825, pl. 33), and in an unpublished drawing by Gallesio (GAG).
15. Penzig 1887; Tanaka 1927; Tilney Bassett 1986; Baldini 1991.
16. *...gemmas per medium findes, ut simul ritè coalescant, & cicatricis locus non appareat, sed gemma una videatur* (Della Porta 1558, p. 48). Della Porta's positing the budding of two halved buds to induce them to grow together as one is a curiously striking anticipation of the fascinating contrivances of modern biotechnology and its techniques.
17. *Malum citreum vulgare belluatum*, pp. 65, 70; *Limon striatus vulgatior*, pp. 245, 247; *Limon striatus Amalphitanus*, p. 249; *Aurantium virgatum*, p. 379; *Aurantium striatum*, pp. 397, 401.

18. Bimbi, Villa Med. C.594 (*Arancio scannellato, Limone scannellato tondo*); C.597 (*Limone di Galauria a fette di popone*); C.612 (*Limone scannellato, Arancio virgolaro bello e buono, Arancio a fette di mellone nuova specie, Arancio detto scannellato della China*); C.616 (*Arancio colle foglie gialle, Arancio biccio, Bergamotta scannellata*).

19. *Limon peretto Spadafora incanellato*, I, p. 140b; *Aranzi incanellati*, I, p. 194a; *Aranzo rigato*, I, p. 198; *Limoni Spadafora incanellati*, II, p. 131; *Aranzi stellati*, II, p. 191; *Aranzo con frutto & foglia variegato*, II, p. 193.

20. *Oranger à fruit toruleux*, pl. 18; *Oranger à fruit changeant*, p. 28; *Oranger turc*, pl. 29; *Bigaradier bicolor*, pl. 51; *Bergamottier à fruit toruleux*, pl. 54; *Limonier à fruit cannelé*, pl. 72; *Cedratier à fruit à côtes*, pl. 105-06.

21. *Arancio turco*, p. 36; pl. 35; *Arancio scannellato*, p. 78; pl. 77.

22. *Abortus et monstra in animantibus plerumque horremus, amamus in pomis* (Ferrari 1646, p. 431).

23. Aldrovandi I.P. VIII, fol. 61: *Mala aurantia seu narantia figura melopeponis*.

24. See Ferrari 1646: *Malum citreum digitatum multiforme*, pp. 75, 77; *Limon a rivo seu rio*, pp. 213, 215; *Aurantium corniculatum*, pp. 407, 409; *Aurantium distortum*, pp. 413, 415.

25. See Montalbani 1668: *Malus aurantium alatum*, p. 486; *Aurantia mala segmentata*, p. 487; *Limon digitatus*, p. 513; *Limon chiroides*, p. 514; *Malum aurantium monstrificum*, p. 714.

26. *Arancio cornuto detto dei minimi, Arancio detto strasizzecca nuova specie, Limone del fior doppio bizzarro famoso* (Bimbi, Villa Med. C.612, nos. 13, 18 and 30).

27. See Volkamer *Nürnberg. Hesp.*: *Cedro a dittela*, I, p. 116b,c; Limon zattella, I, p. 145; *Aranzo distorto o monstroso*, I, pp. 206a,b; *Limoni dolci multiforme*, II, pp. 96–7; *Limon monstroso di Molfetta*, II, p. 107; *Bergamotti monstrosi*, II, p. 148.

28. Marsili, c. 136-48, drawings in pen and ink of misshapen citrus fruits.

29. See Risso/Poiteau 1818: *Oranger à fruit cornu*, pl. 12; *Bigaradier à fruit corniculé*, pl. 32; *Limonier à fruit digité*, pl. 9.

30. Baldini 1996a.

31. See Ferrari 1646, pp. 75-6.

32. *An quia semina imbecilla vis, cum, pomi partes coniungere, consuetumque in citrei mali vultum conglobare non possit, in plura iucundéque monstrosas facies abortat? An potiùs quia vis eadem genitali primum impetu multos simul foetus enititur: eosdemque deinde viribus & foecundo humore defecta, medio in conatu relinquit informes, graciles, medulla & semine aut prorsus aut prope destitutos? Et hactenus quidem philosophicas inter coniecturas caecutisse sit satis.* (Ferrari 1646, p. 75f.).

33. Boyce/Korsmeyer 1941.

34. Called *diafisi centrale fruttipara* by Penzig 1887, Tav. X, fig. 4.

35. *Citrium praegnans* in Aldrovandi MS 88, fol. 4r.

36. *Limon citratus alterum includens*, pp. 264, 269-71; *Pomum Adami foetum*, pp. 310, 315; *Aurantium foetiferum*, pp. 403, 405.

37. For example, in Volkamer *Nürnberg. Hesp.*: *Cedro di fiore e sugo doppio* (I, p. 128a); *Aranzo cornuto hermaphrodito* (I, p. 192a); *Pomo d'Adamo foetifero* (II, p. 141); *Bergamotto foetifero di Padova* (II, p. 145); *Bergamotto monstroso* (II, pp. 147-48). Also in Risso/Poiteau 1818: *Bigaradier à fruit fetifère* (p. 78; pl. 33); *Bigaradier riche depouillé* (p. 35; pl. xxxv).

38. Ferrari 1646, p. 264: *Iam verò quaerenti causam praegnationis in pomo admirandae philosophicam habeo & poeticam afferre. Igitur cum seminis partes inter se diversae, quas etherogenias vocant, loci fertilitate exuberant: cumque foetum informans arboreum vivida vis, quam vulgò spiritum appellant, multiplicata dividitur: uni facilè pomo alia ingenerantur, prout partes easdem vario lusu lasciviens natura digesserit.*

39. *Le pollen change la forme et la nature de l'ovaire, et multiplie quelque fois les foetus [...] L' 'Aurantium foetiferum' présente une superfétation, un développement imparfait de plusieurs germes renfermés l'un dans l'autre, ou réunis sous l'enveloppe d'un germe extérieur. Mais ces germes existoient-ils dans cet ovaire, ou bien y ont-ils été formés par le pollen qui l'a fécondé ?* (Gallesio 1811, p. 169).

40. Tenore 1844; Savastano 1884; Penzig 1887; Tillson/Bamford 1938; Leroy 1953; Reuther/Batchelor/Webber.

41. Montalbani 1668: *De limone dulci cortice melopeponite diviso in 18 triangula*, p. 492, and *Limon orchites*, p. 514.

42. Volkamer *Nürnberg. Hesp.* II, pp. 72-4: *Limon di Salerno monstroso*.

43. Evreinoff 1946.

44. Professors Giovanni Continella and Eugenio Tribulato, University of Catania, have made a substantial contribution to the botanical identification of Cassiano's drawings. Yet, despite this scientific consensus, the taxonomic position of some of Cassiano's citrus specimens remains doubtful or indeterminate.

45. Fifty-four of them are preserved today in the Royal Library at Windsor (see Table B *List of Locations*), although a number of dispersed drawings have now been identified (McBurney 1989d and 1993). See also p. 21.

46. Archivio Dal Pozzo MS 39, fol. 110.

47. Archivio Dal Pozzo MS 39, fol. 110.

48. Mediterranean regions had known the citron since the second or third century B.C., the lemon and the sour orange from about A.D. 1000, and the sweet orange possibly from about A.D. 1500, brought by Portuguese or Genoese travellers.

49. Palermo, 1810; Naples, 1816; Florence, 1824.

50. Bimbi, Villa Med. C.612 and C.616. Targioni Tozzetti 1853; Baldini 1982.

51. Clarici 1726.

52. Ferrari 1646, pp. 309-315.

53. Aldrovandi I.P. VIII, fol. 65.

54. Aldrovandi I.P. III, fol. 161, and V, fols. 312, 314.

55. Gallesio 1811; Calabrese 1990.

56. *Panforte* is a traditional Italian cake made with citron and orange candied rinds, spices, honey and almonds.

57. Archivio Dal Pozzo MS 39; Gallo 1578; Cato 1677; Clarici 1726; Baldini/Rossi 1983; Freedberg 1992b.

58. Istituto Sperimentale per l'Agrumicoltura, Acireale; Istituto di Coltivazioni Arboree, Catania; Istituto di Coltivazioni Arboree, Palermo.

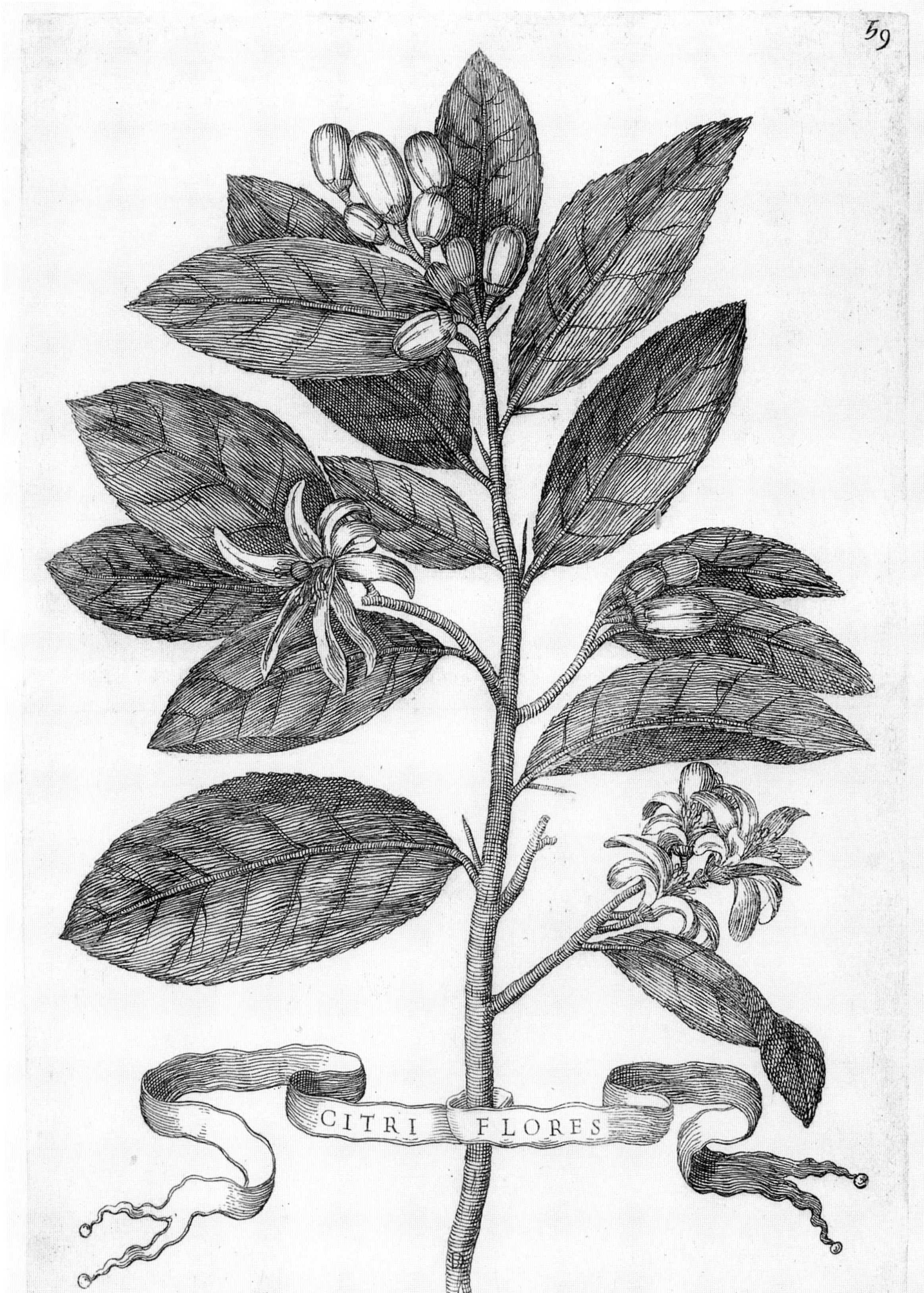

Fig. 45. *Citri Flores*, etching and engraving, Ferrari 1646. p. 59

CATALOGUE

CONTENTS OF CATALOGUE

Subdivided according to broad botanical groups followed by unidentified citrus

CITRONS

(1–17)

***Citrus medica* L.**

Trees small, with irregular habit of growth. Leaves elliptic-ovate or ovate-lanceolate, serrate, bluntly pointed or rounded at tip; petioles wingless. Flowers in clusters with petals pinkish on the outside; stamens 30–40 or more. Fruits large, yellow, ellipsoid or ovoid, normally rough and bumpy, with a very thick albedo and small, whitish segments; juice acid. Seeds long, pinkish and numerous.

LEMONS

(18–30)

***Citrus limon* (L.) Burm. f.**

Trees tall, with irregular habit of growth. Leaves serrate, elliptical, long-ovate or lanceolate, pointed at tip; petioles normally wingless. Flowers in clusters, reddish in the bud; stamens 20–40. Fruits yellow, ovoid or ellipsoid-apiculate, protruding into an apical, normally pointed nipple, with a rather thick rind and normally yellowish segments; juice acid. Seeds ovoid, yellowish, small.

SOUR, SWEET AND MYRTLE-LEAF ORANGES

(31–37)

***Citrus aurantium* L. (sour orange)**

Trees medium-sized, with regular habit of growth and rounded top. Leaves ovate, bluntly pointed at tip; petioles winged. Flowers on clusters, white, large and very fragrant; stamens 20. Fruits subglobose or oblate, orange, with a thick, rather rough, bitter rind and a bitter and acid flesh. Seeds yellow-whitish, small, numerous.

***Citrus sinensis* (L.) Osb. (sweet orange)**

Trees medium-sized, with regular habit of growth and rounded top. Leaves long-ovate or lanceolate, pointed at tip; petioles narrowly winged. Flowers in clusters, white, smaller and less fragrant than those of sour oranges; stamens 20–25. Fruits deep orange, subglobose or oblate, with a thin, taut rind and a juicy orange or red flesh. Juice sweet. Seeds small, yellow-whitish, absent in parthenocarpic cultivars.

***Citrus myrtifolia* Raf. (myrtle-leaf orange)**

Small dwarf bushes. Twigs with very short internodes and generally thornless. Leaves alternate, small, lanceolate, pointed at tip: petioles short and wingless or slightly wing margined. Flowers small, white. Fruits small, orange to deep orange, oblate to spherical, with a pitted rind; flesh bitter, sour and characteristically flavoured. Seeds few or none, small. This species probably arose as a mutation of *Citrus aurantium* L.

PUMMELO

(38–42)

***Citrus grandis* (L.) Osb.**

Trees large and round-topped, with thorny twigs. Leaves very large, oval, with a blunt point at tip; petioles broadly winged and more or less cordate. Flowers large, white; stamens 20–25. Fruits large or very large, yellow, subglobose, oblate-spheroid or subpyriform, with a smooth rind. Flesh yellow. Seeds large, thick and wrinkled.

CITRUS HYBRIDS

(43–71)

In *Citrus*, spontaneous cross-pollination and fertilization easily occurs between different species and even between different genera, producing progenies which show a wide range of variability in their intermediate morphological and physiological characters. Ferrari (1646) suspected that individuals looking like both a citron and a lemon or an orange and a lemon could exist, but Gallesio, in 1812, was the first expressly to call them hybrids.

TERATOLOGICAL CITRUS

(72–104)

Citrus fruit show a wide range of teratological forms which were greatly appreciated in the past, as were all sorts of wonders and monsters. Fingered, foetiferous and twinned fruit were actively sought for their freakish shapes and troubling beauty. Citrus teratology includes a great many synthetic and autogenous chimeras.

UNIDENTIFIED CITRUS

(105–118)

Despite the efforts of citrologists and the checking of Cassiano's drawings against specimens held in germplasm collections, some have remained unidentifiable. Their presence confirms the remarkable nomenclatural difficulty affecting citrus taxonomy, which has made it a vexing question since ancient times.

Scientific nomenclature and authorities:

The botanical name is given in italics, followed by the authority (see abbreviations below). Where there has been a nomenclatural change, the original authority is given in parenthesis, followed by new author.

Abbreviations

Burm. f. = Nicolaas Laurens Burman (1734–93) [f. = filius]
L. = Carl von Linnaeus (1707–78)
Lush. = Alfred Wyndham Lushington (*c.* 1860–1920)
Macf. = James Macfadyen (1798–1850)
Osb. = Pehr Osbeck (1723–1805)
Raf. = Constantin Samuel Rafinesque-Schmaltz (1783–1840)
Risso = Joseph Antoine Risso (1777–1845)
Roem. = Johann Jakob Roemer (1783–1819)
Swing. = Walter Tennyson Swingle (1871–1952)
Ten. = Michele Tenore (1780–1861)
Ewing = H. E. Ewing (1883–1951)
× = crossed with + = synthetic (graft) chimera

NOTES FOR THE READER

The following notes are intended to clarify the meaning of key terms, headings and typographical conventions used in the Catalogue.

CATALOGUE NUMBER: each Catalogue entry has a unique number composed of the SERIES letter (A = Antiquities and Architecture, B = Natural History); the PART number in roman numerals, and the ENTRY number in **bold** type, e.g. B. II. **32**. Cross-references to entries within a Part are indicated by the **bold** entry number only. References to entries in other Parts of the Catalogue Raisonné are to the full Catalogue number. A list of all planned Parts in the Series appears on p. 6.

DRAWING SHEET: size, in millimetres (height followed by width); condition and *Watermark*. For discussion and types of watermarks see p. 277.

ANNOTATION: numbering or lettering, in *italic*. The position of the annotation is given only when it is not in the lower right-hand corner (the most common position).

MOUNT SHEET: size, in millimetres (height followed by width); condition and *Watermark*. Most of the drawings were removed from their 17th-century mount sheets in the late 18th century, when they were laid down onto contemporary mount sheets bearing the watermark of D & C Blauw; these mount sheets very often retain the coloured wash borders applied at that time (see p. 20).

LITERATURE: this pertains to the drawing only. Full details of titles cited in abbreviated form in the Catalogue will be found in the Bibliography.

REFERENCES: information is given regarding the fruit drawn.

ENGRAVINGS: made after the catalogued drawing.

LOCATION UNKNOWN: all such drawings were dispersed from the Royal Library earlier this century (see p.21). A number of them are known only from photographs made by the Courtauld Institute: **7** and **38** were photographed in the 1950s on black and white miniature film; **2-3**, **8-9**, **42**, **45** and **78** were photographed as 35 mm colour slides. A record of the measurements of the drawings in the second category survives on the slide frames; for those in the former category no measurements or other details survive. Thirty-two of the dispersed drawings were bought by the American collector Herbert Boone (d. 1983). Details of all of these were collected at the time of the sale of Boone's collection (Sotheby's, New York 16 September 1988). Where the present owners of the ex-Boone drawings have been identified, the details of the drawings have, in most cases, been rechecked.

For reproductions of all the citrus fruit engravings in Ferrari 1646, see pp. 285–316.

The diagram opposite is a key to the terms used in the Catalogue to describe the drawings and their mounts.

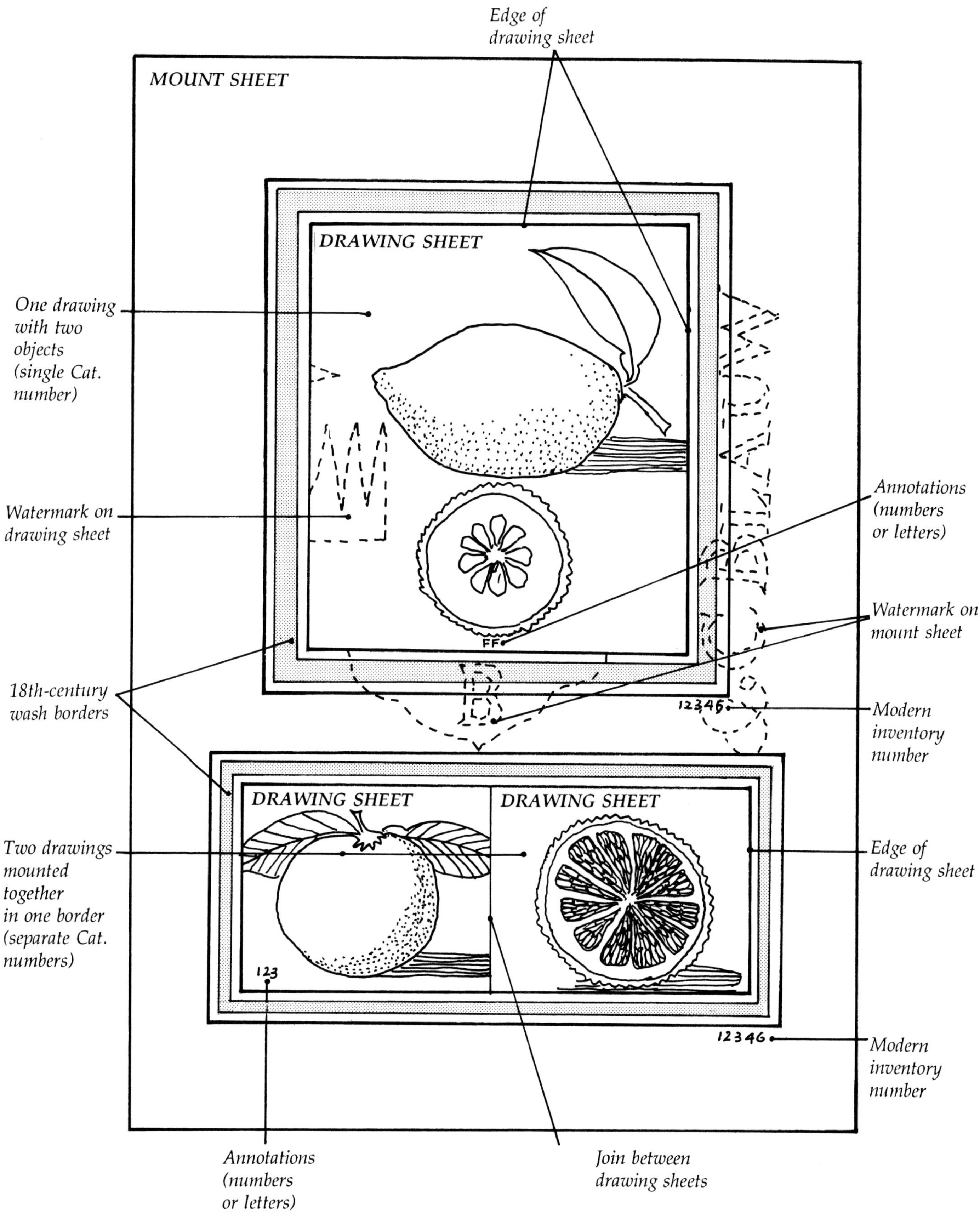
Edge of drawing sheet
MOUNT SHEET
DRAWING SHEET
One drawing with two objects (single Cat. number)
Watermark on drawing sheet
Annotations (numbers or letters)
FF
Watermark on mount sheet
18th-century wash borders
12346
Modern inventory number
DRAWING SHEET
DRAWING SHEET
Two drawings mounted together in one border (separate Cat. numbers)
Edge of drawing sheet
123
12346
Modern inventory number
Annotations (numbers or letters)
Join between drawing sheets

1. Citron, *Citrus medica* L.: whole and half-fruit

VINCENZO LEONARDI (fl. 1621–46)

Windsor, RL 19334

Watercolour and bodycolour, with gum heightening, over black chalk, with incised outlines on half-fruit and on part of whole fruit

379 × 296 mm. Survival of 5 mm overlay of original 17th-century mount at right edge. Stained after present mounting

ANNOTATION: [below half-fruit] *ee*

MOUNT SHEET: 556 × 401 mm. Stained and marked. *Watermark:* D & C Blauw

These are the first illustrations of citrus fruit reproduced in Ferrari's *Hesperides*. They represent the ordinary citron, the *Malum citreum vulgare*. The upper drawing is reproduced on p. 61 of the *Hesperides* (although there it is shown vertically and is lacking one leaf), while the half-fruit below appears on p. 63.

Ferrari begins his chapter on this fruit by noting that, although it was not indigenous to Italy, it had long become 'a citizen of our gardens' (Ferrari 1646, p. 56). In typical fashion, he proceeds to give a brief account of its description in ancient writers such as Theophrastus and in more recent ones such as Cesalpino and Ruel. As usual, he begins with his own very detailed description of both the tree and the flower, before moving on to that of the fruit. No previous writer had ever described citrus fruit with anything like the same attention to form, texture and colour. Wherever he possibly can, Ferrari notes variety, to such a degree that he even lists the different weights which the ordinary citron can attain: usually 6 or 7 *librae* (more than 3 kg) in Calabria, 16 in

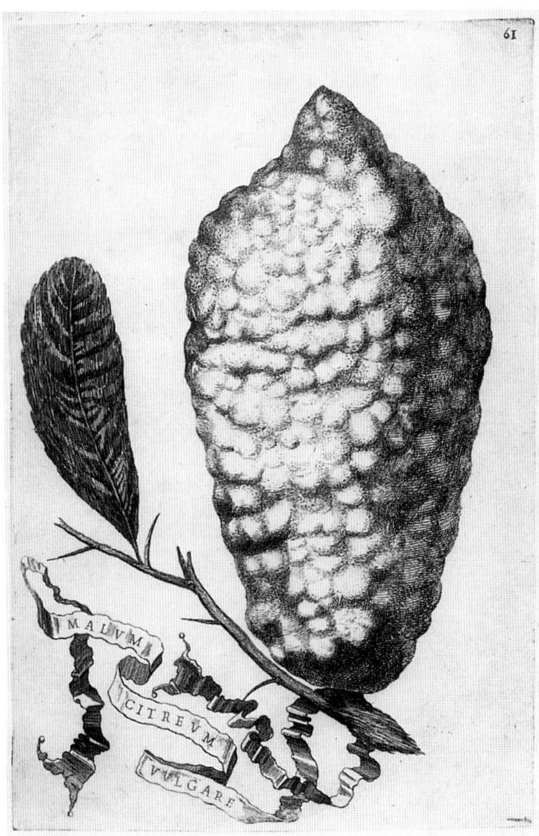

Comp. fig. 1a: Ferrari 1646 p. 61

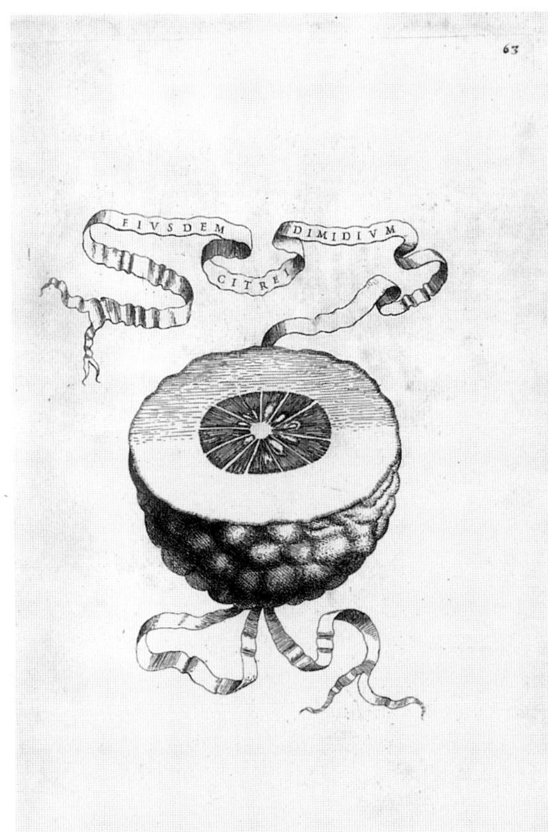

Comp. fig. 1b: Ferrari 1646 p. 63

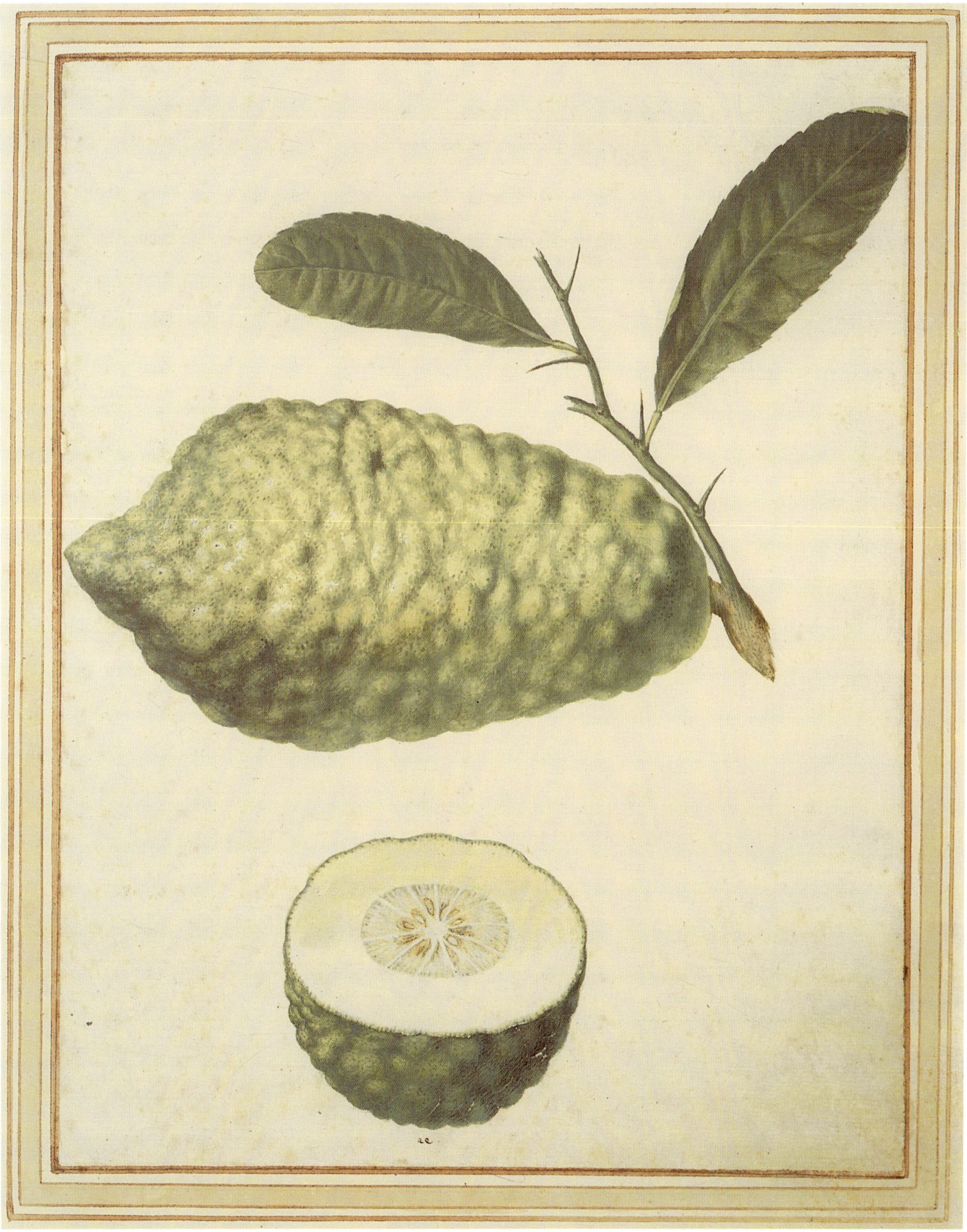

1

Salò, and up to 20 in Rome. Then he notes the colour and texture of immature, 'adolescent', and mature fruits ('golden, wrinkly, and warty'), the colour of their flesh, the quantity of seeds, and the quantity of juice. 'For those who can distinguish still more clearly', he continues, 'the common citron can be divided into three kinds', the first coming from the shores of Salò on Lake Garda, the second from Genoa — in terms of sweetness of taste and smell superior to the others — and then a third much smaller and abundant type, not localized. Finally he refers to two types of citrons, one from Corfu and another from the island of Crete, which the local inhabitants call *Mazzacani.*

Following the pages on which the present drawings are reproduced, Ferrari illustrates two strange forms, the *Malum citreum vulgare belluatum,* with a monstrous ridge-like growth along its length (see **74**), and the *Malum citreum cucurbitinum vulgare,* a fruit that is almost all flesh and has the shape of a small gourd (see **2-4**).

Although the chapter on these fruits is uncharacteristically devoid of anecdote, it is entirely typical of Ferrari's approach in its extraordinary attentiveness to the colouring, shape and feel of every possible part of the fruit and its plant — as careful and as methodical as one could imagine verbal description of this kind to be. No wonder that the drawings supplied by Cassiano for the engravings of citrus fruit had to meet high standards of precision and meticulousness! Immediately following these first reproductions of citrus fruit, Ferrari launches into a panegyric on their hitherto forgotten artist, Vincenzo Leonardi. Then he speaks of the citron hedges of Reggio, and gives a glowing account of the royal 'pomarium' in Valencia, so beautiful that it was rightly called the 'Spanish Elysium'. He also describes the garden of one Cristoforo Cavanilla there, which included a labyrinth of citron trees. Finally Ferrari gives a long excursus on the game or 'war' of citrus still played around Carnival time in Reggio Calabria, and the similar spectacle in Treviso, where, in the citrus battle between young Venetians and Paduans, hard immature (greenish) citrons were used as especially effective missiles. Then Ferrari moves on to his next chapter, on the *Malum citreum dulci medullae.*

The specimen shown here, both whole and in section, is a large, bumpy, nippled, greenish citron. The half-fruit (*eiusdem citrei dimidium*) reveals the remarkable thickness of the rind, which is typical of citrons. There are numerous seeds. The twig is thorny, and the leaves relatively small, narrow, elliptical and serrate, with wingless petioles.

PROVENANCE: dal Pozzo; Albani; George III, by descent to George V; London art market (Mendelson); Rex Nan Kivell; Her Majesty The Queen

LITERATURE: Freedberg 1993, p. 181f., fig. 109

REFERENCES: Ferrari 1646, pp. 56–72; Risso/Poiteau 1818, p. 194, pl. 96; De Noter 1926, p. 145f.

ENGRAVINGS: Ferrari 1646, pp. 61, 63

2. Gourd-shaped citron, *Citrus medica* L.: whole fruit

VINCENZO LEONARDI (fl. 1621–46)

Location unknown

Watercolour and bodycolour over black chalk

412 × 222 mm (joint measurement with **3**. Sheet cut and rejoined). Trimmed through shadow at right

ANNOTATION: *58*

MOUNT SHEET: mounted with **3**, within the same wash border

The drawings catalogued here and in the following entry (**3**) are preparatory drawings for the *Malum citreum cucurbitinum vulgare* in Ferrari's *Hesperides*, p. 67. Curiously enough, Ferrari seems to give no description of this citron in his text. The drawings show the whole fruit and the longitudinal half of two large gourd-shaped citrons. The albedo, under their thin, smooth, yellow peel, is remarkably thick. The seeds are aborted. This particular specimen resembles another of Cassiano's drawings of gourd-shaped citrons (see **4**), as well as to the specimen represented in Bartolomeo Bimbi's painting in the Villa Medicea (C. 612).

The two drawings must originally have been on the same drawing sheet, later separated, and then in the 18th century mounted together after entering the collection of George III. They are known only from a 35 mm colour slide, made by the Courtauld Institute in 1967. The only known measurement is that of the two drawing sheets together.

PROVENANCE: dal Pozzo; Albani; George III, by descent to George V; London art market (Mendelson)

REFERENCES: Bimbi, Villa Med. C. 612; Volkamer *Nürnberg. Hesp.* II, p. 42, pls. 41, 43–4; Micheli *Enumeratio*, p. 104

ENGRAVINGS: Ferrari 1646, p. 67

3. Gourd-shaped citron, *Citrus medica* L.: half-fruit

VINCENZO LEONARDI (fl. 1621–46)

Location unknown

Watercolour and bodycolour, with blackened lead white highlights, over black chalk

412 × 222 mm (joint measurement with **2**. Sheet cut and rejoined)

ANNOTATION: *59*

MOUNT SHEET: mounted with **2**, within the same wash border

See **2**

PROVENANCE: dal Pozzo; Albani; George III, by descent to George V; London art market (Mendelson)

REFERENCES: Bimbi, Villa Med. C. 612; Volkamer *Nürnberg. Hesp.* II, p. 42, pls. 41, 43–4; Micheli *Enumeratio*, p. 104

ENGRAVINGS: Ferrari 1646, p. 67

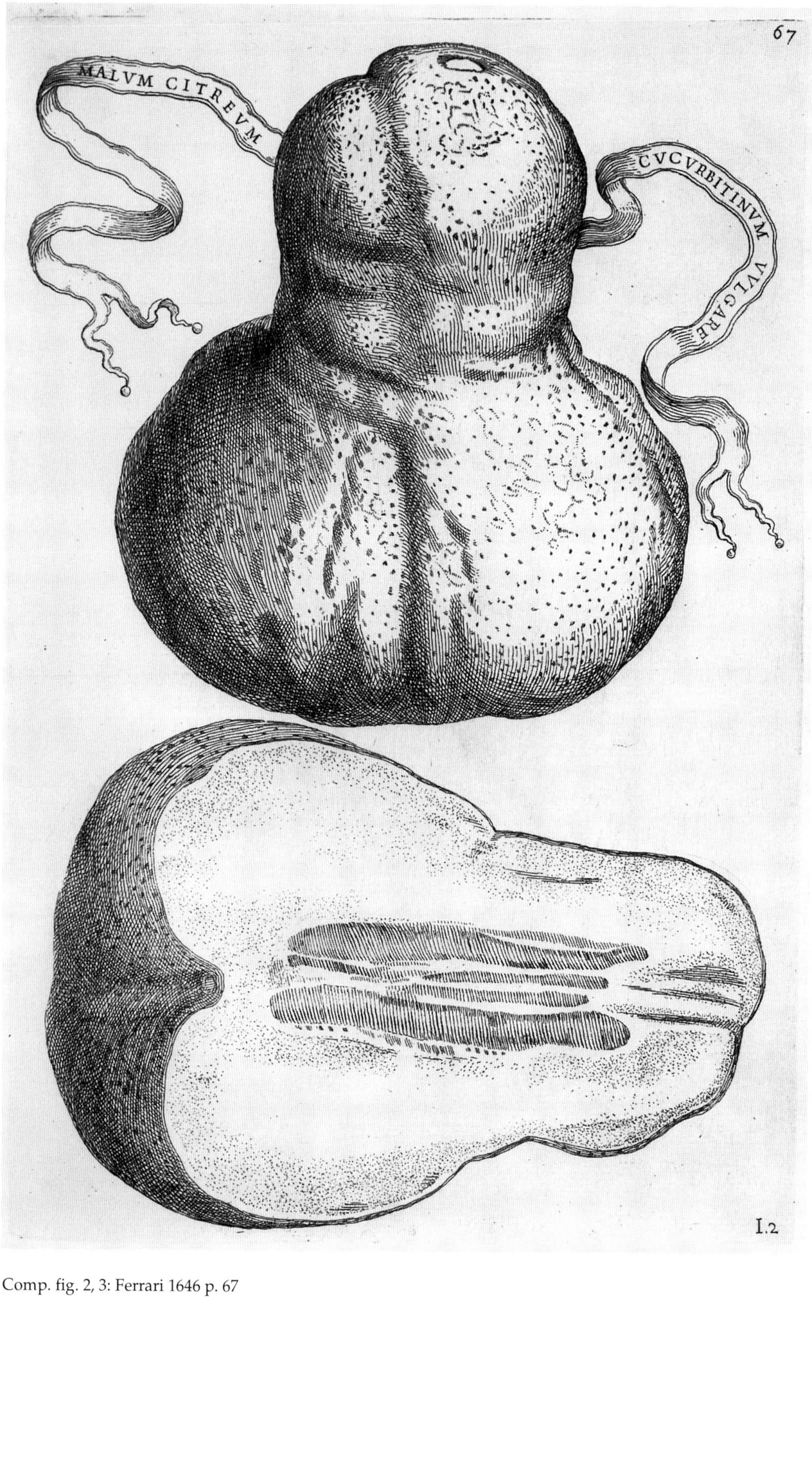

Comp. fig. 2, 3: Ferrari 1646 p. 67

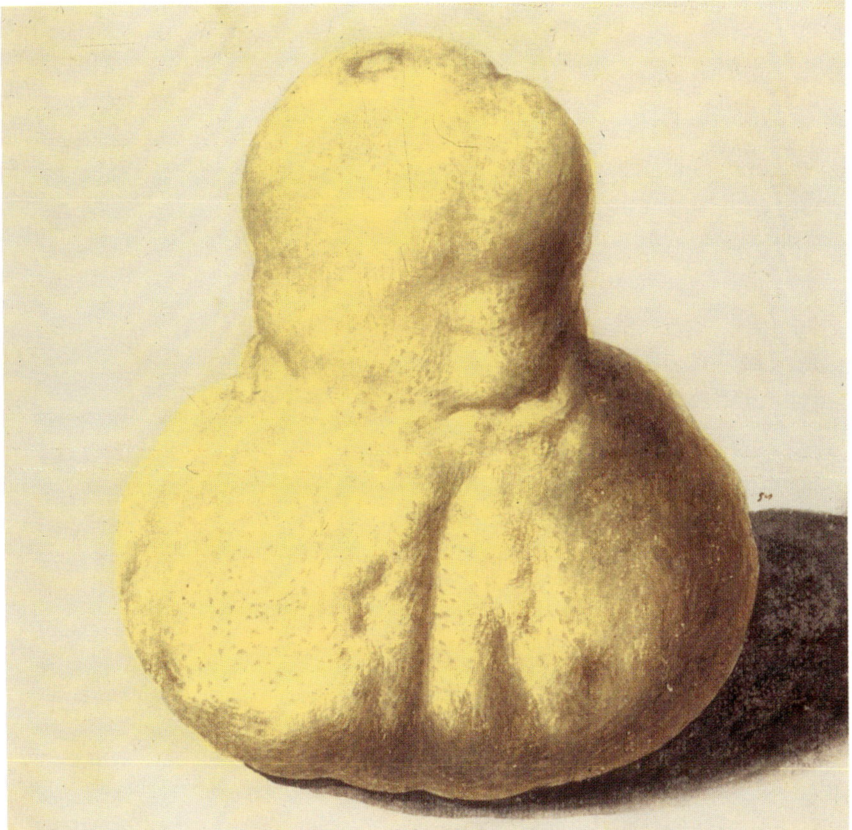

2

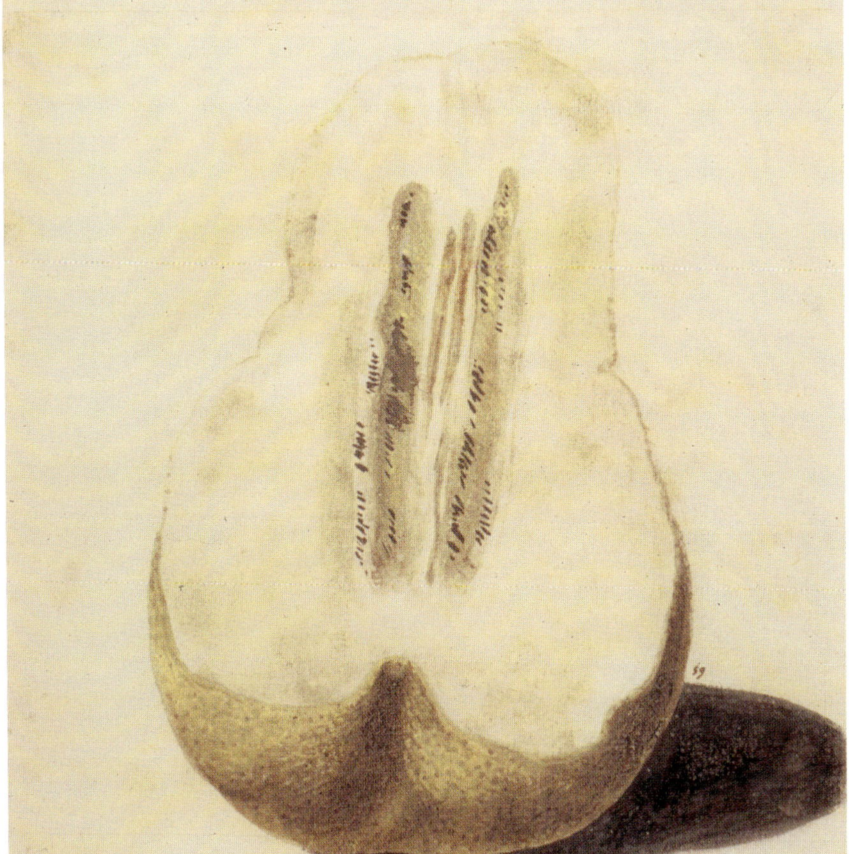

3

4. Gourd-shaped citron, *Citrus medica* L.: whole fruit

VINCENZO LEONARDI (fl. 1621–46)

Windsor, RL 19355

Watercolour and bodycolour, with gum heightening, over black chalk

182 × 160 mm.

ANNOTATION: [lower right] *62*

MOUNT SHEET: 289 × 435 mm. *Watermark*: D & C Blauw (cut). *Annotation*: [verso: pencil, in a 20th-century hand] 79

Although not reproduced in the *Hesperides*, this drawing does look similar to the *Malum citreum cucurbitinum vulgare*, as described in **2**. It represents a gourd-shaped, necked, yellow whole fruit, with shallow ribs towards the apex. The half-fruit is not present amongst the surviving drawings.

PROVENANCE: dal Pozzo; Albani; George III, by descent to George V; London art market (Mendelson); Rex Nan Kivell; Her Majesty The Queen

REFERENCES: Bimbi, Villa Med. C. 612; Volkamer *Nürnberg. Hesp.* II, p. 42, pls. 41, 43–4; Micheli *Enumeratio*, p. 104

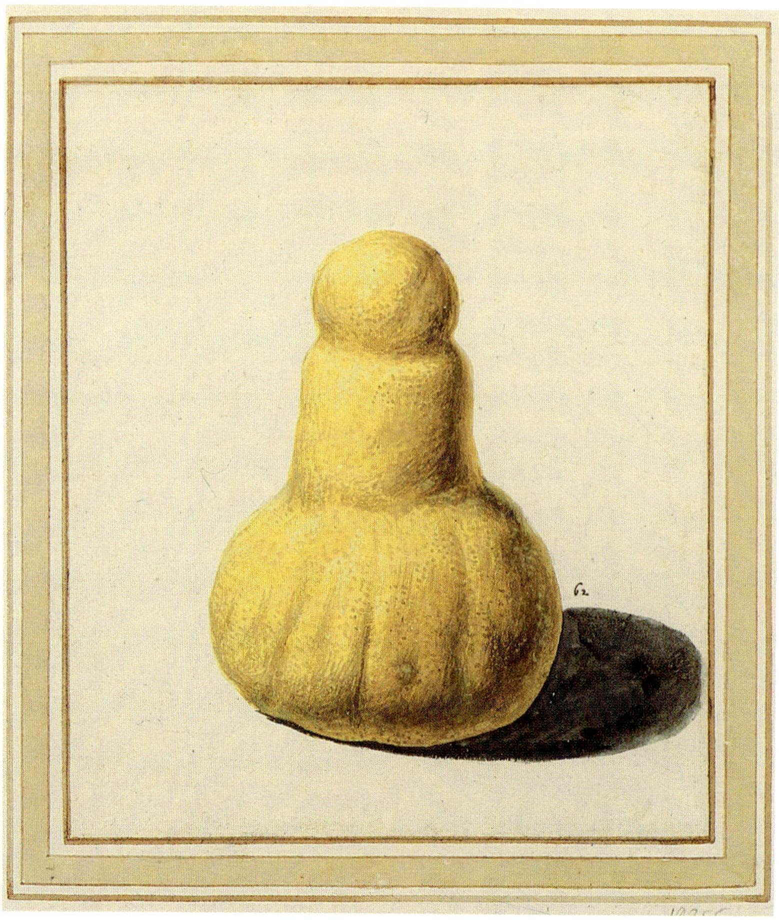

4

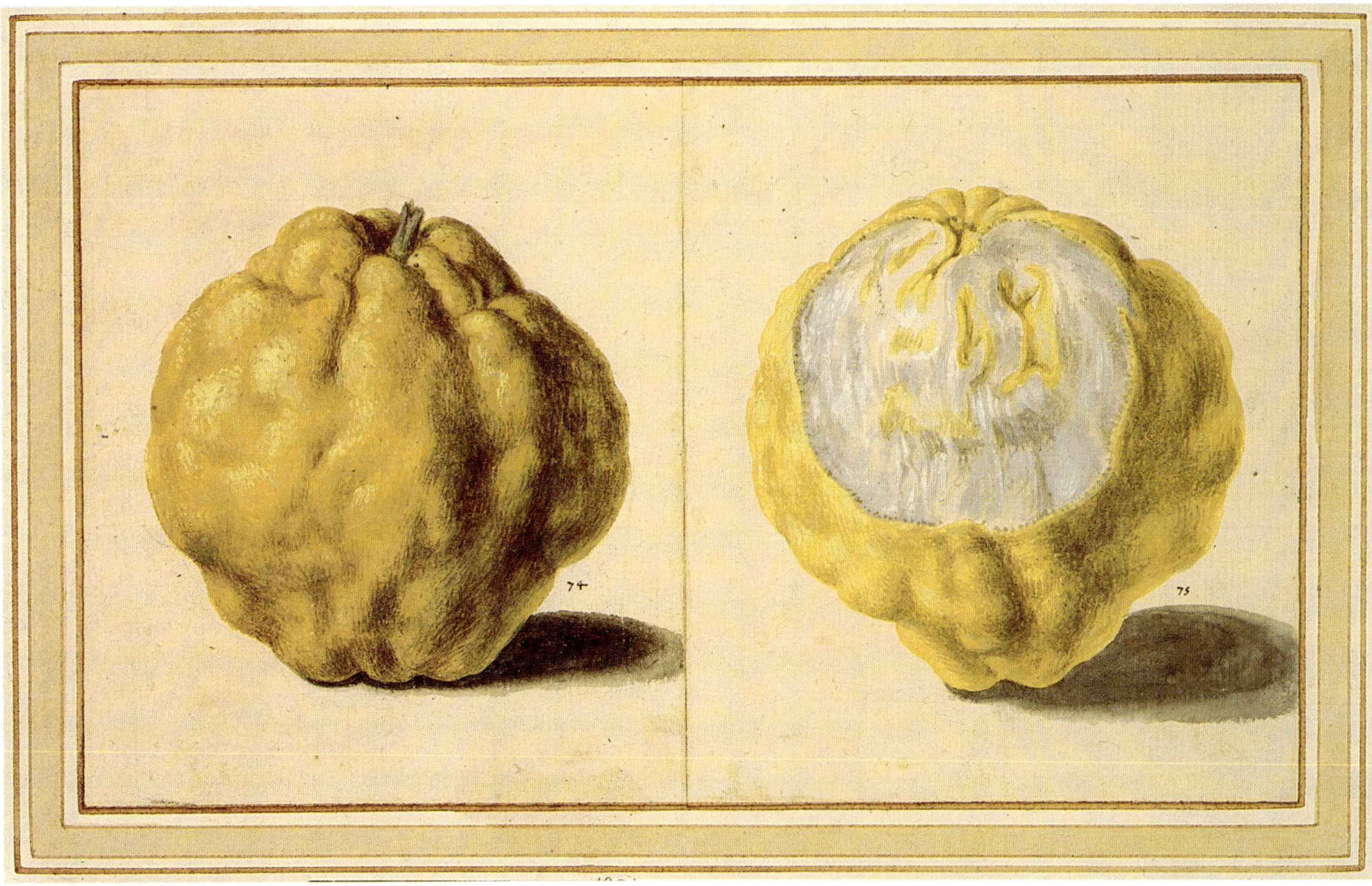

5 6

5. Citron, *Citrus medica* L.: whole fruit

VINCENZO LEONARDI (fl. 1621–46)

Windsor, RL 19360

Watercolour and bodycolour, with gum heightening, over black chalk

158 × 135 mm. Cream prepared paper

ANNOTATION: *74*

MOUNT SHEET: 435 × 586 mm. Stained. Mounted with **6**, within the same wash border. *Watermark*: D & C Blauw

This drawing, which represents an irregularly globose, bumpy citron with a deeply depressed and furrowed base, is not reproduced in the Hesperides. No drawing of the half-fruit survives among the known drawings in the Paper Museum.

Both this drawing and that catalogued in **6** probably once formed part of the same sheet. They were then separated and again mounted as one in the 18th-century rearrangement.

PROVENANCE: dal Pozzo; Albani; George III, by descent to George V; London art market (Mendelson); Rex Nan Kivell; Her Majesty The Queen

6. Citron, *Citrus medica* L.: whole fruit, part peeled

VINCENZO LEONARDI (fl. 1621–46)

Windsor, RL 19361

Watercolour and bodycolour, with gum heightening, over black chalk

158 × 138 mm. Cream prepared paper

ANNOTATION: *75*

MOUNT SHEET: mounted with **5**, within the same wash border

See **5**. The drawing shows what appears to be the other side of the fruit shown in **5**, part of the rind peeled off to show the thick, whitish albedo, typical of citrons.

PROVENANCE: dal Pozzo; Albani; George III, by descent to George V; London art market (Mendelson); Rex Nan Kivell; Her Majesty The Queen

7. Citron, *Citrus medica* L.: whole fruit

VINCENZO LEONARDI (fl. 1621–46)

Location unknown

Watercolour and bodycolour, with gum heightening, over ?black chalk

Measurements unknown

ANNOTATION: *66*

Not represented in the *Hesperides*, this drawing shows a typical, large, ellipsoid citron with a deeply depressed base, a prominent apical nipple and a bumpy, coarsely pitted, yellow rind. It looks like the large fruit (*Spongino di libbre 4, once 7* on a silver dish) depicted by Bartolomeo Bimbi in his painting *c.* 1685–95, now in the Botanical Museum, University of Florence (Bimbi, DBF 36). No drawing of the half-fruit survives among the known drawings in the Paper Museum.

The drawing is known only from an image on 35 mm miniature film made by the Courtauld Institute in the late 1950s.

PROVENANCE: dal Pozzo; Albani; George III, by descent to George V; London art market (Mendelson)

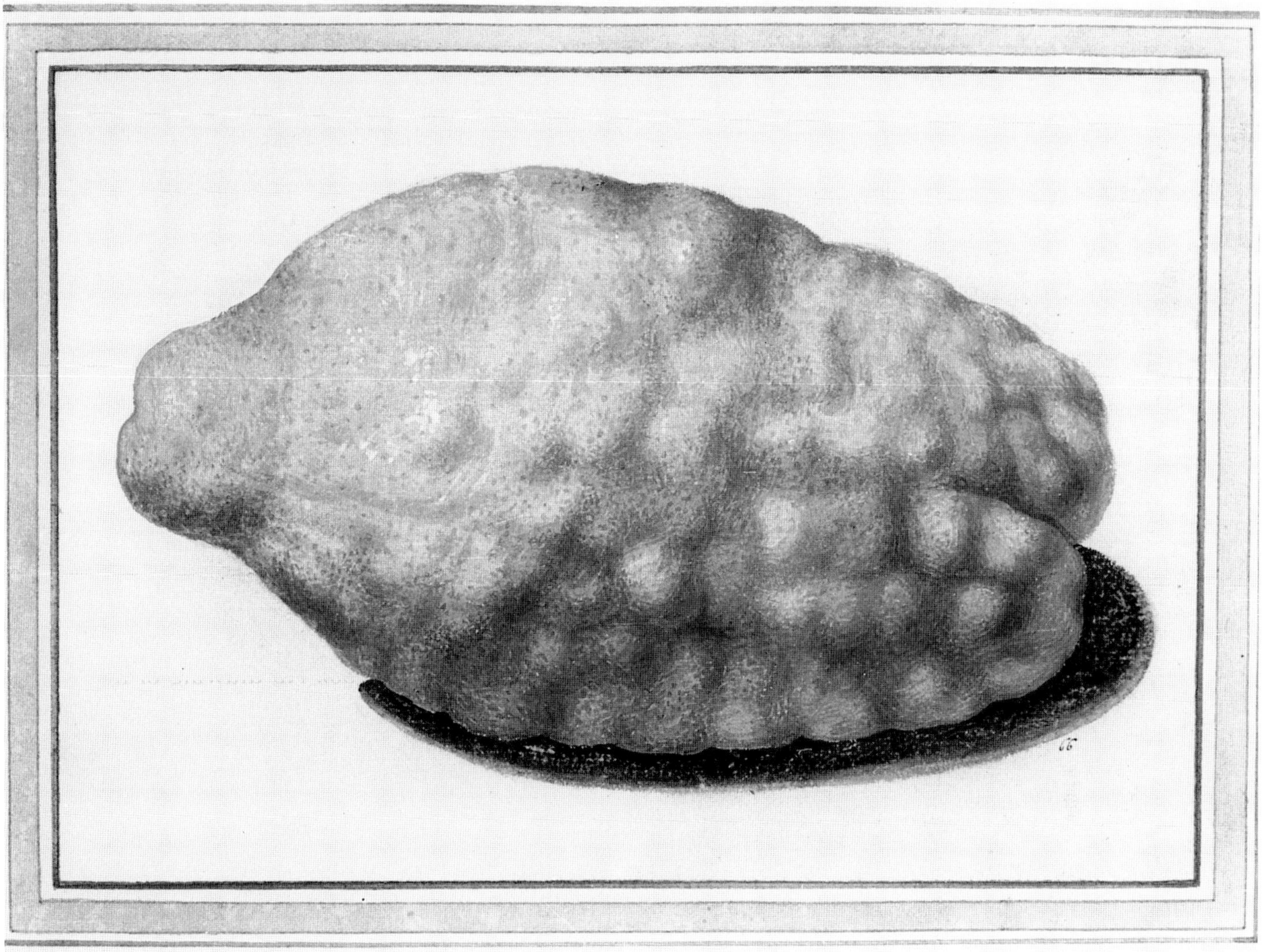

7

8. Citron, *Citrus medica* L.: whole fruit

VINCENZO LEONARDI (fl. 1621–46)

Location unknown

Watercolour and bodycolour, with gum heightening, over black chalk

245 × 128 mm (joint measurement with **9**. Sheet cut and rejoined). Trimmed through shadow at right

ANNOTATION: *E*

MOUNT SHEET: mounted with **9**, within the same wash border

Not reproduced in the *Hesperides*, this drawing shows a large, ellipsoid, pale yellow, smooth and finely pitted fruit. A thick stalk is inserted on its evenly rounded base. The drawing of the half-fruit (**9**) shows a thin peel and remarkably thick mesocarp. The segments are rudimentary and seedless. The thickness of the mesocarp, the very rudimentary segments and the general shape of the fruit bear certain similarities to the *Lumia expers medullae* illustrated on p. 329 of the *Hesperides*, although the present specimen seems to be a citron rather than any of the varieties Ferrari described as *lumie* (See **59–62**).

8 and **9** presumably once formed part of the same sheet, were separated, and then mounted together again at the time of the rearrangement of the drawings after entering the collection of George III. The two drawings are known only from a 35 mm colour slide, made by the Courtauld Institute in 1967. The only measurement taken was that of the two drawing sheets together.

PROVENANCE: dal Pozzo; Albani; George III, by descent to George V; London art market (Mendelson)

9. Citron, *Citrus medica* L.: half-fruit

VINCENZO LEONARDI (fl. 1621–46)

Location unknown

Watercolour and bodycolour, with gum heightening, over black chalk

245 × 128 mm (joint measurement with **8**. Sheet cut and rejoined)

MOUNT SHEET: mounted with **8**, within the same wash border

See **8**

PROVENANCE: dal Pozzo; Albani; George III, by descent to George V; London art market (Mendelson)

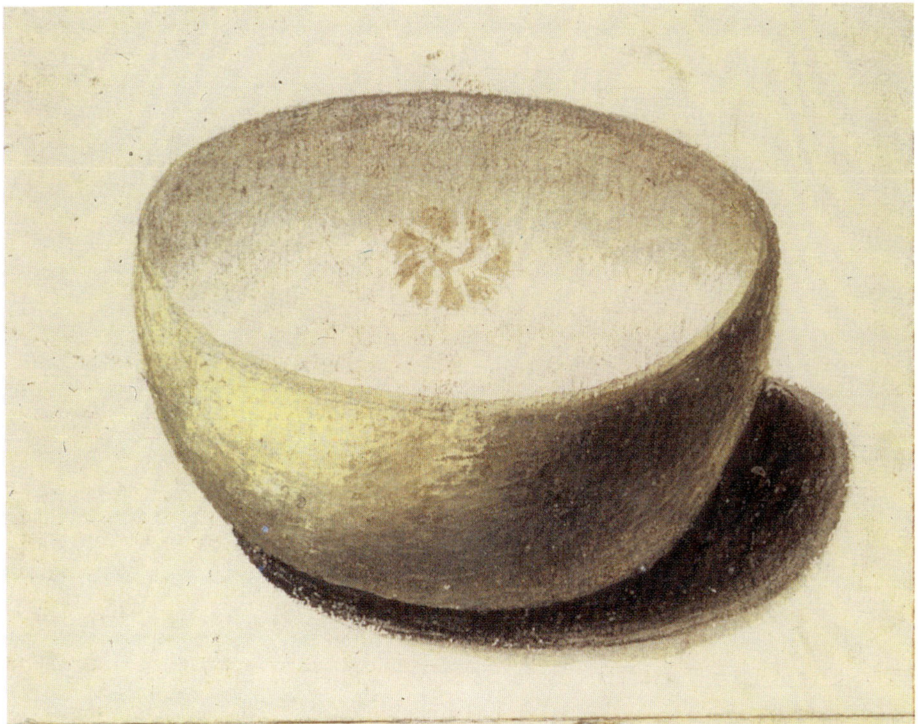

9

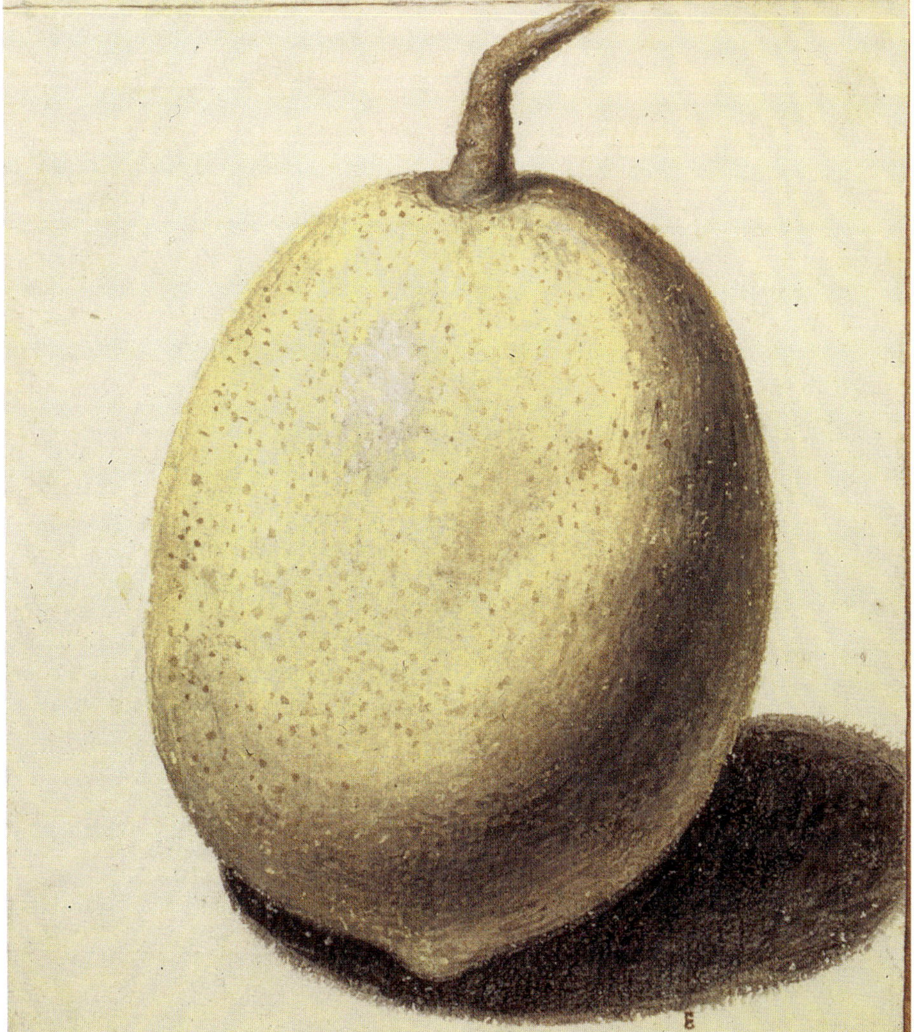

8

10. Citron, *Citrus medica* L.: leafed twig and fruit

?VINCENZO LEONARDI (fl. 1621–46)

Location unknown

Watercolour and bodycolour over black chalk

152 × 188 mm

ANNOTATION: [below stalk] *63*

MOUNT SHEET: 585 × 432 mm. Mounted on same sheet as **11**

Not represented in the *Hesperides*, this drawing shows a yellow, ellipsoid fruit with a smooth rind, a rounded base and a nippled apex. Two elliptical leaves and a leaflet are inserted on a thornless twig ending with the fruit. The drawing is mounted on the same sheet as **11**.

PROVENANCE: dal Pozzo; Albani; George III, by descent to George V; London art market (Mendelson); James R. Herbert Boone; Sotheby's (New York), 16 September 1988, lot 149

11. Citron, *Citrus medica* L.: half-fruit

?VINCENZO LEONARDI (fl. 1621–46)

Location unknown

Watercolour and bodycolour, with gum heightening, over black chalk

90 × 86 mm

ANNOTATION: *65*

MOUNT SHEET: mounted on same sheet as **10**

This drawing, not reproduced in the *Hesperides*, shows a half-citron with a thick rind surrounding small, pale yellow and apparently seedless segments. See **10** for the whole fruit.

PROVENANCE: dal Pozzo; Albani; George III, by descent to George V; London art market (Mendelson); James R. Herbert Boone; Sotheby's (New York), 16 September 1988, lot 149

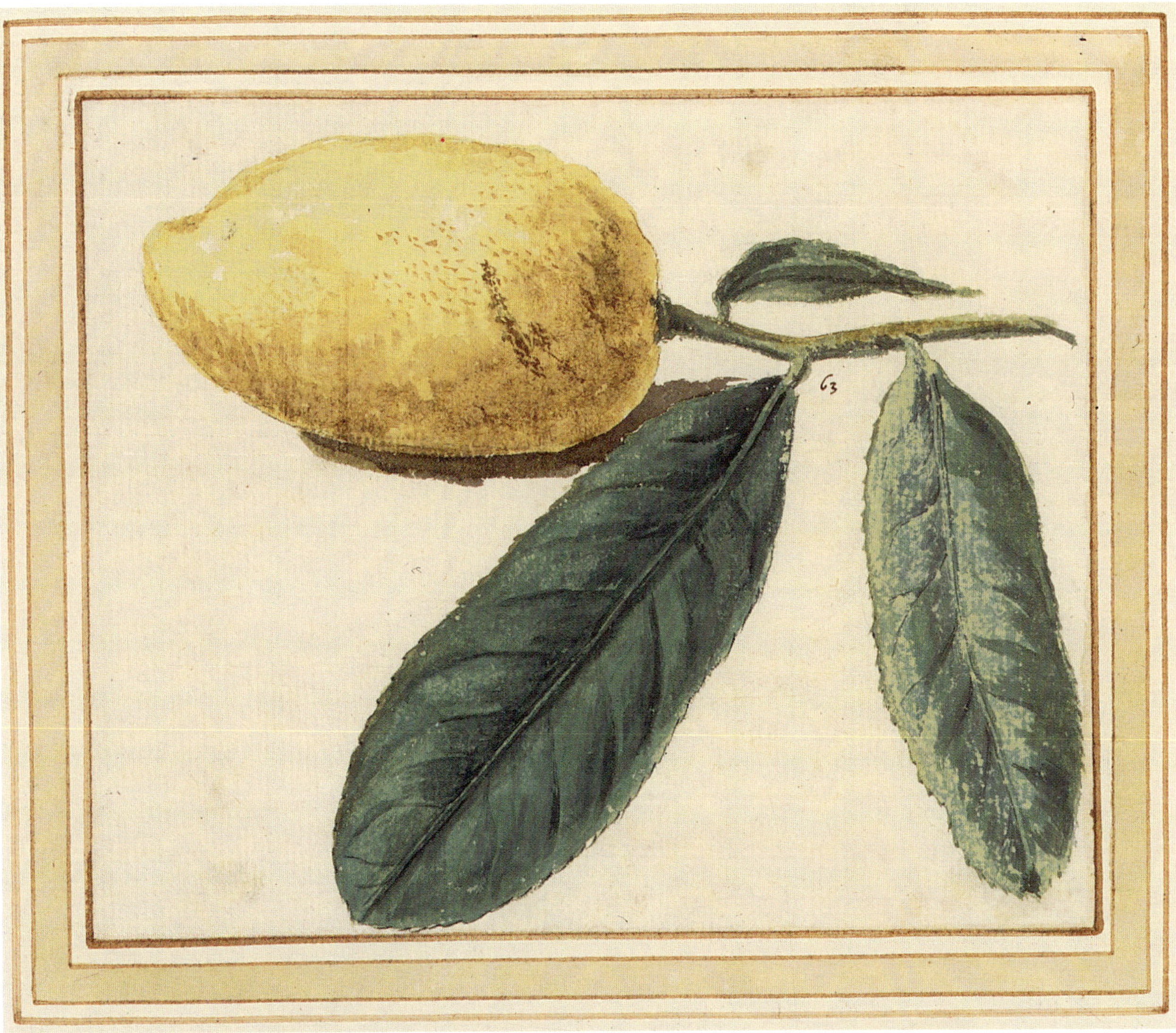

10

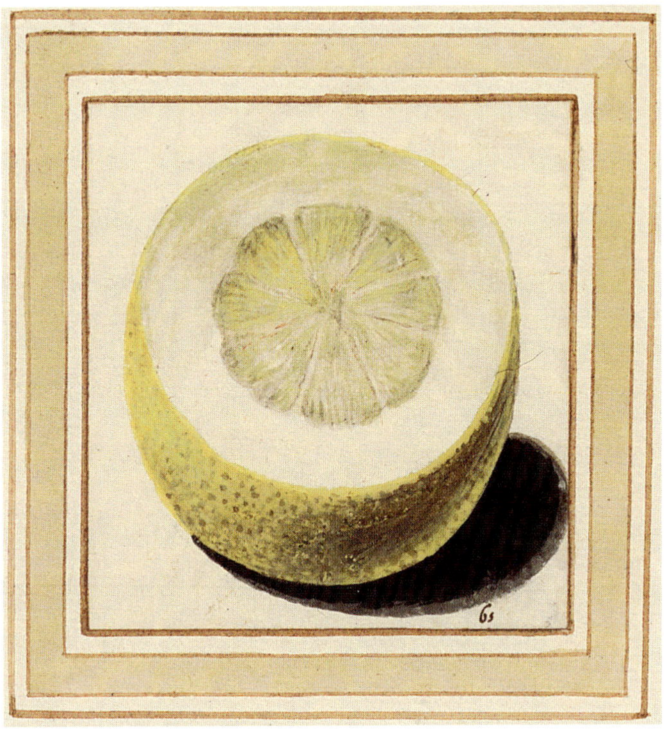

11

12. Citron, *Citrus medica* L.: leafed twig and fruit

?VINCENZO LEONARDI (fl. 1621–46)

New York art market

Watercolour and bodycolour over black chalk; *pentimenti* in black chalk and watercolour

190 × 174 mm. *Watermark*: none recorded

ANNOTATION: *31*

Not represented in the *Hesperides*, the drawing shows an obovate, apically rounded, bumpy citron, and two elliptical leaves with wingless petioles on a thorny twig. The drawing was mounted on the same sheet as **13** at the time of the 1988 sale, but the two have since been separated.

PROVENANCE: dal Pozzo; Albani; George III, by descent to George V; London art market (Mendelson); James R. Herbert Boone; Sotheby's (New York), 16 September 1988, lot 142; W. M. Brady and Co. Inc., New York

13. Citron, *Citrus medica* L.: half-fruit

?VINCENZO LEONARDI (fl. 1621–46)

Private collection, U.S.A.

Watercolour and bodycolour over black chalk

90 × 105 mm. Trimmed through shadow at right. *Watermark*: none recorded

MOUNT SHEET: see **12**

Not reproduced in the *Hesperides*, this drawing shows a half-fruit, with a thick white rind surrounding a firm, seeded, pale yellow flesh.

The drawing must originally have been on the same sheet as the whole fruit (see **12**), the two later separated and then mounted together in the 18th century when they entered the collection of George III. After the Sotheby's sale in 1988 the mount sheet was cut in two and the drawings entered different collections.

PROVENANCE: dal Pozzo; Albani; George III, by descent to George V; London art market (Mendelson); James R. Herbert Boone; Sotheby's (New York), 16 September 1988, lot 142; W. M. Brady and Co. Inc., New York

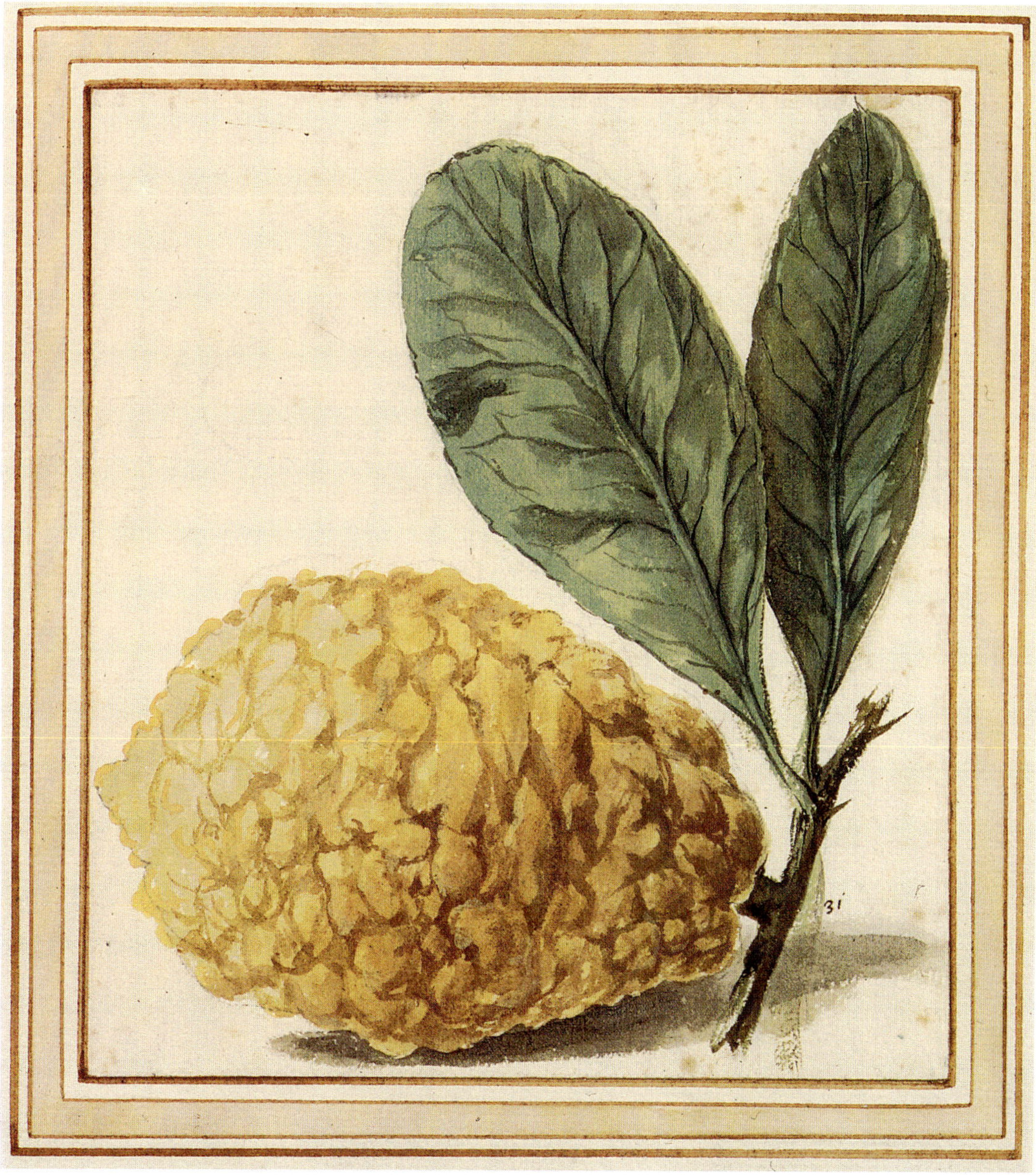

12

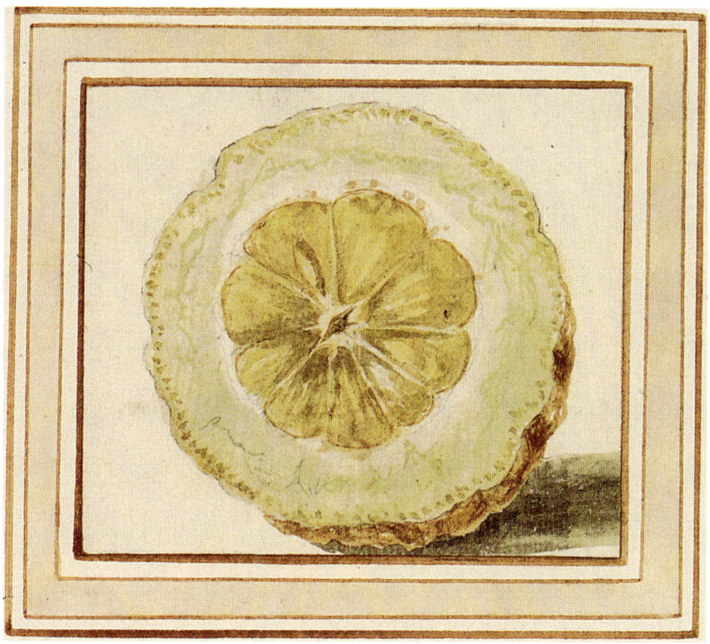

13

14. Citron, *Citrus medica* L.: whole fruit and thorny twig

?VINCENZO LEONARDI (fl. 1621–46)

Location unknown

Watercolour and bodycolour and pen and ink

217 × 276 mm. Badly foxed. Sheet trimmed around leaf, upper right. *Watermark*: none recorded

Not reproduced in the *Hesperides*. A large yellow citron with irregular bumps, and three elliptical leaflets not typical of this species. No drawing of the half-fruit survives among the known drawings in the Paper Museum.

PROVENANCE: dal Pozzo; Albani; George III, by descent to George V; London art market (Mendelson); James R. Herbert Boone; Sotheby's (New York), 16 September 1988, lot 141

15. Citron, *Citrus medica* L.: whole fruit and thorny twig

?VINCENZO LEONARDI (fl. 1621–46)

Location unknown

Watercolour and bodycolour over black chalk; *pentimenti* in black chalk

285 × 215 mm. Foxed and stained. Trimmed at right. *Watermark*: none recorded

ANNOTATION: [lower right] *83*

Not reproduced in the *Hesperides*, this drawing shows a large, dark yellow citron hanging from a very thorny limb, and a narrow, elliptical leaf with a wingless, reduced petiole. The fruit looks very much like the *Citrium pyriforme … pondere libras quinque* ('Pear-shaped citron, weighing five pounds') in Aldrovandi I.P. III, fol. 166 and in Montalbani 1668, p. 528. No drawing of the half-fruit survives among the known drawings in the Paper Museum.

PROVENANCE: dal Pozzo; Albani; George III, by descent to George V; London art market (Mendelson); James R. Herbert Boone; Sotheby's (New York), 16 September 1988, lot 141

REFERENCES: Aldrovandi I.P. III, fol. 166; Montalbani 1668, p. 528; Baldini 1989, p. 153f.; Baldini 1990, p. 66

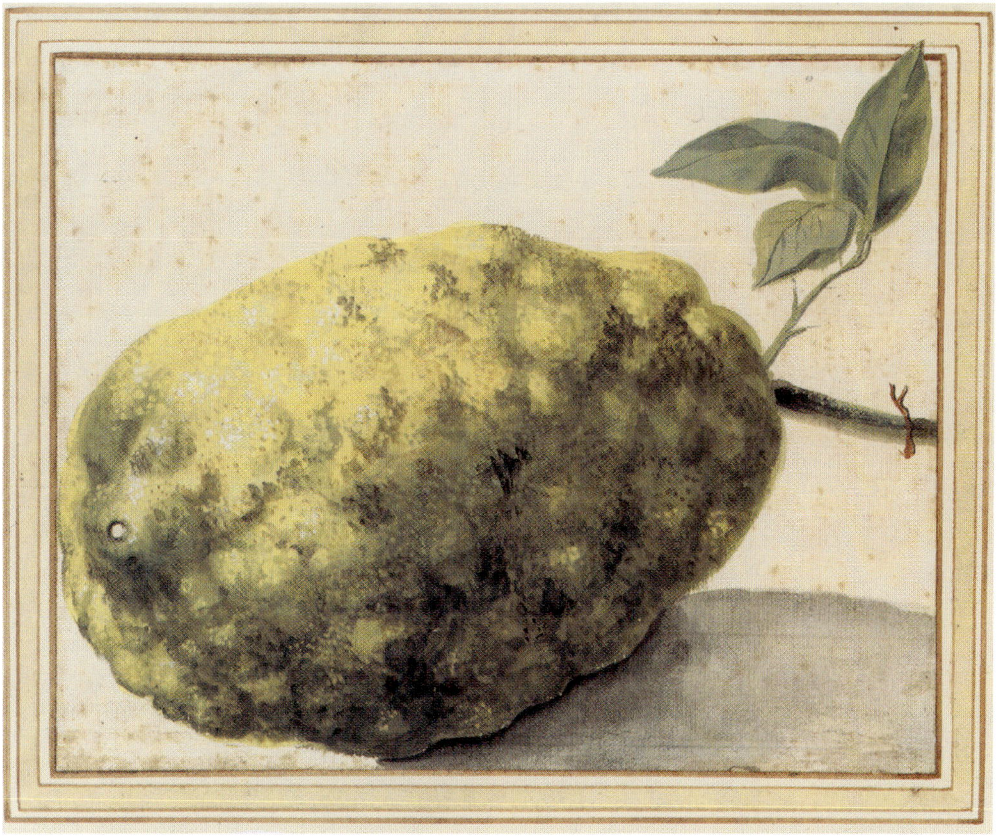

14

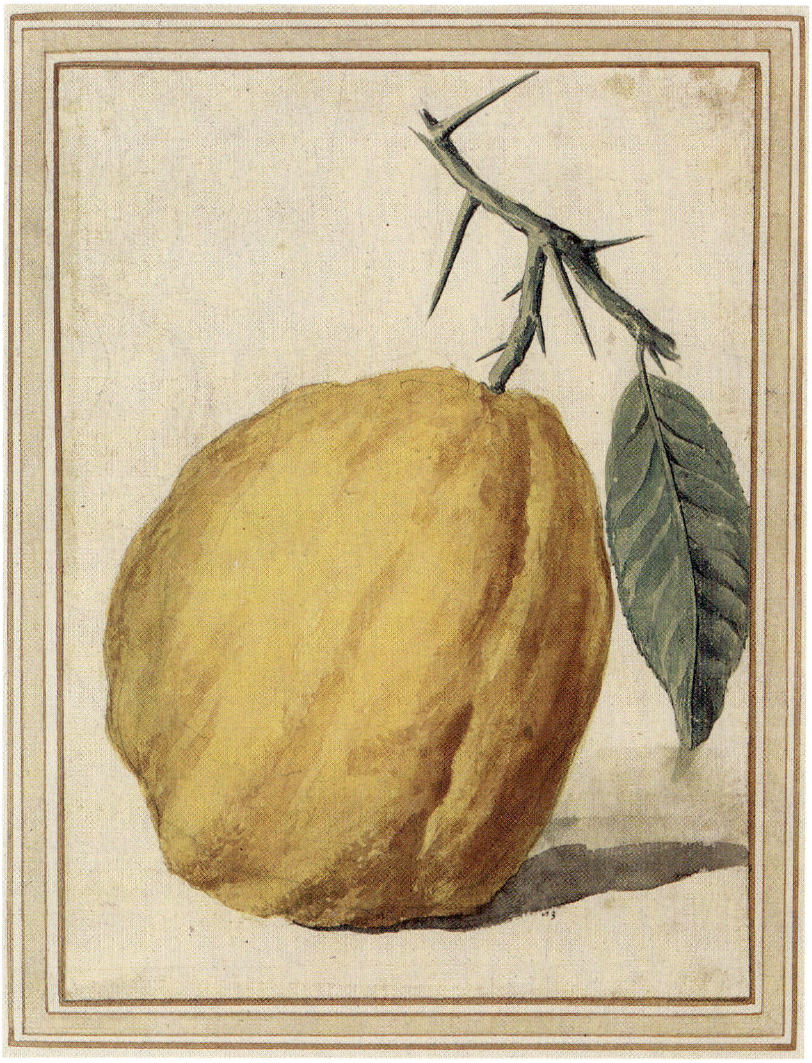

15

16. Citron, *Citrus medica* L.: rough whole fruit

?VINCENZO LEONARDI (fl. 1621–46)

Windsor, RL 19323

Watercolour and bodycolour with gum heightening

197 × 198 mm

ANNOTATION: [lower left] *B*

MOUNT SHEET: 586 x 434 mm. Badly water-stained along right edge, elsewhere marked and stained. *Watermark*: D & C Blauw

The specimen is an oblate, whole fruit with a bumpy and ribbed rind. The apex of the fruit has clearly been deeply depressed and furrowed. It looks like the Sicilian cultivar named *Cedro Vozza Vozza,* whose ancestors Deng *et al.* 1995 recognized as being the lemon and the pummelo.

PROVENANCE: dal Pozzo; Albani; George III, by descent to George V; London art market (Mendelson); Rex Nan Kivell; Her Majesty The Queen

Neither this nor **17** were reproduced in Ferrari's Hesperides. They must once have formed part of the same sheet before being separated and then finally reunited when they entered the collection of George III (by being mounted together on the same mount sheet).

17. Citron, *Citrus medica* L.: rough half-fruit

?VINCENZO LEONARDI (fl. 1621–46)

Windsor, RL 19323A

Watercolour and bodycolour with gum heightening

181 × 198 mm

ANNOTATION: [lower left] *C*

MOUNT SHEET: 586 × 434 mm. Badly water-stained along right edge, elsewhere marked and stained. *Watermark*: D & C Blauw

Longitudinal half of the fruit shown in **16**, showing three loose segments surrounded by a rather thick rind. See **16**.

PROVENANCE: dal Pozzo; Albani; George III, by descent to George V; London art market (Mendelson); Rex Nan Kivell; Her Majesty The Queen

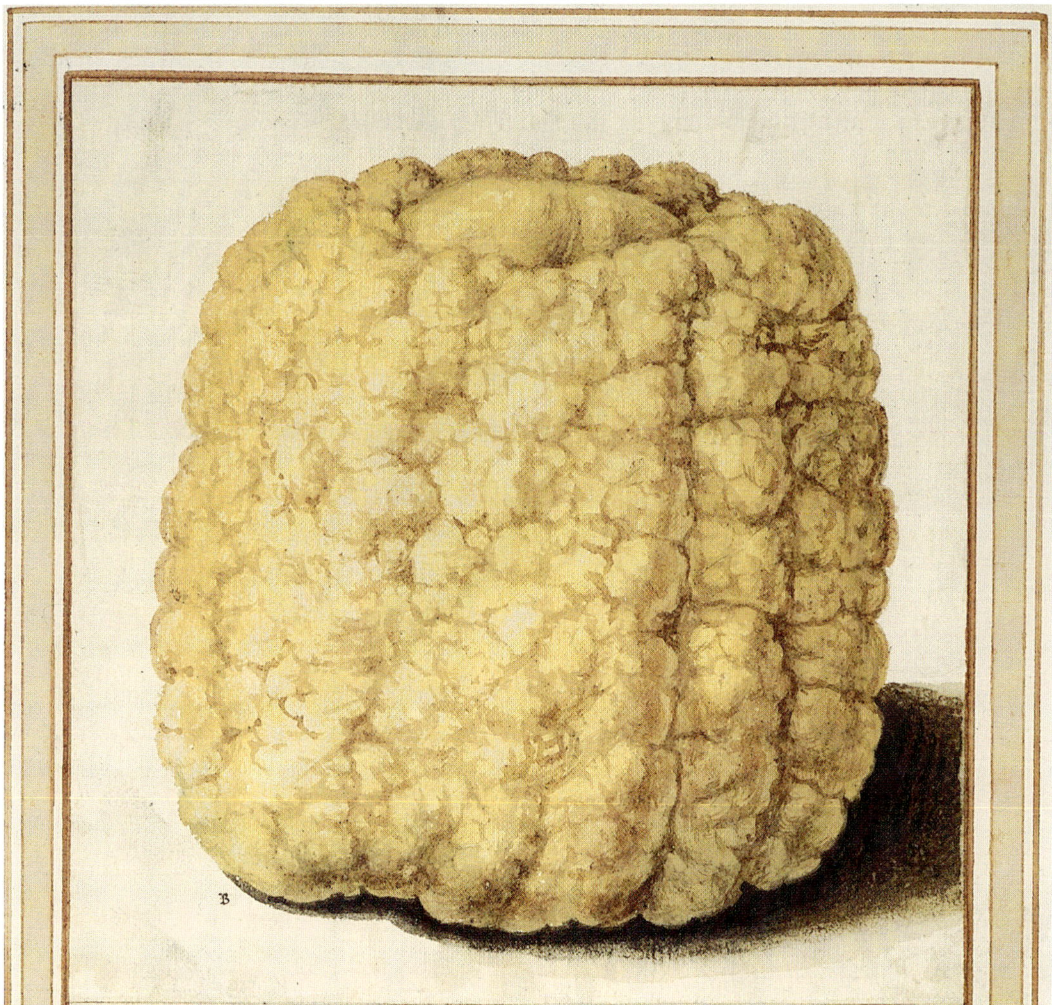

16

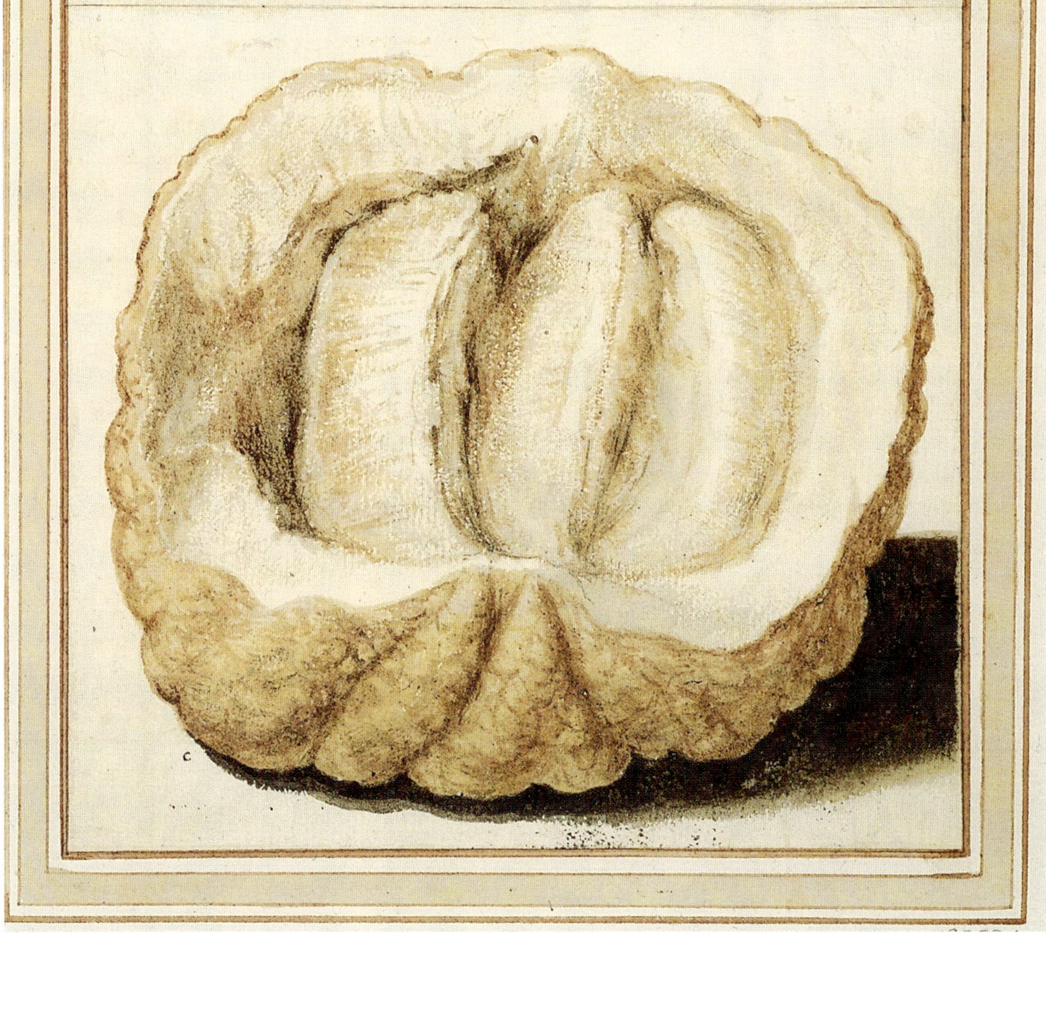

17

18. Lemon, *Citrus limon* (L.) Burm. f.: flowering twig

VINCENZO LEONARDI (fl. 1621–46)
Private collection, Italy
Watercolour and bodycolour over black chalk
276 × 189 mm. *Watermark*: none recorded
ANNOTATION: [below stem] *ddd*

This is the preparatory drawing for the etching of the branch — with flower-buds, flowers and leaves — of a lemon tree on p. 189 of Ferrari's *Hesperides*. The etched illustration is entitled *Limoniae flores*, inscribed on the characteristic scroll that adorns each of the botanical illustrations in the *Hesperides*. It carries the monogram of Dominique Barrière on the base of the stalk itself.

Just as in the case of Books II and IV on citrons and oranges respectively, the engraving of the flowers of the tree precedes all the engravings of the fruits. In chapter I of the third book (entitled *Arethusa sive Malum Limonium*), Ferrari gives a brief and rather generic description of the lemon tree. He then goes on to comment that while writers such as Pontanus divided the lemons into three kinds (*genera*), Clusius into four, and Giovanni Battista della Porta only into two, he had found very many varieties, which he planned to enumerate chapter by chapter (*capitulatim*).

The specimen illustrated here is a thornless branch bearing clusters of flowers, nearly all still in bud, with reddish tinted petals. The leaves are elliptical and serrate, with wingless petioles.

Some everbearing lemons, such as the so-called *Femminello* (the most important cultivar in Sicily), blossom and crop more than once a year, giving rise to differently shaped fruits depending on flowering season (see p. 282).

PROVENANCE: dal Pozzo; Albani; George III, by descent to George V; London art market (Mendelson); James R. Herbert Boone; Sotheby's (New York), 16 September 1988, lot 134

REFERENCES: Ferrari 1646, p. 187

ENGRAVINGS: Ferrari 1646, p. 189

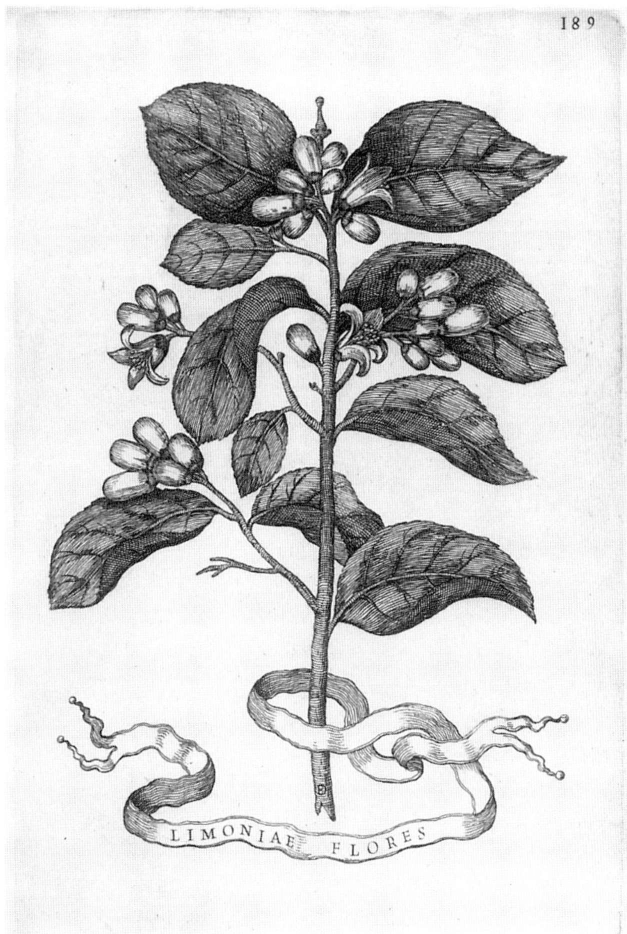

Comp. fig. 18: Ferrari 1646 p. 189

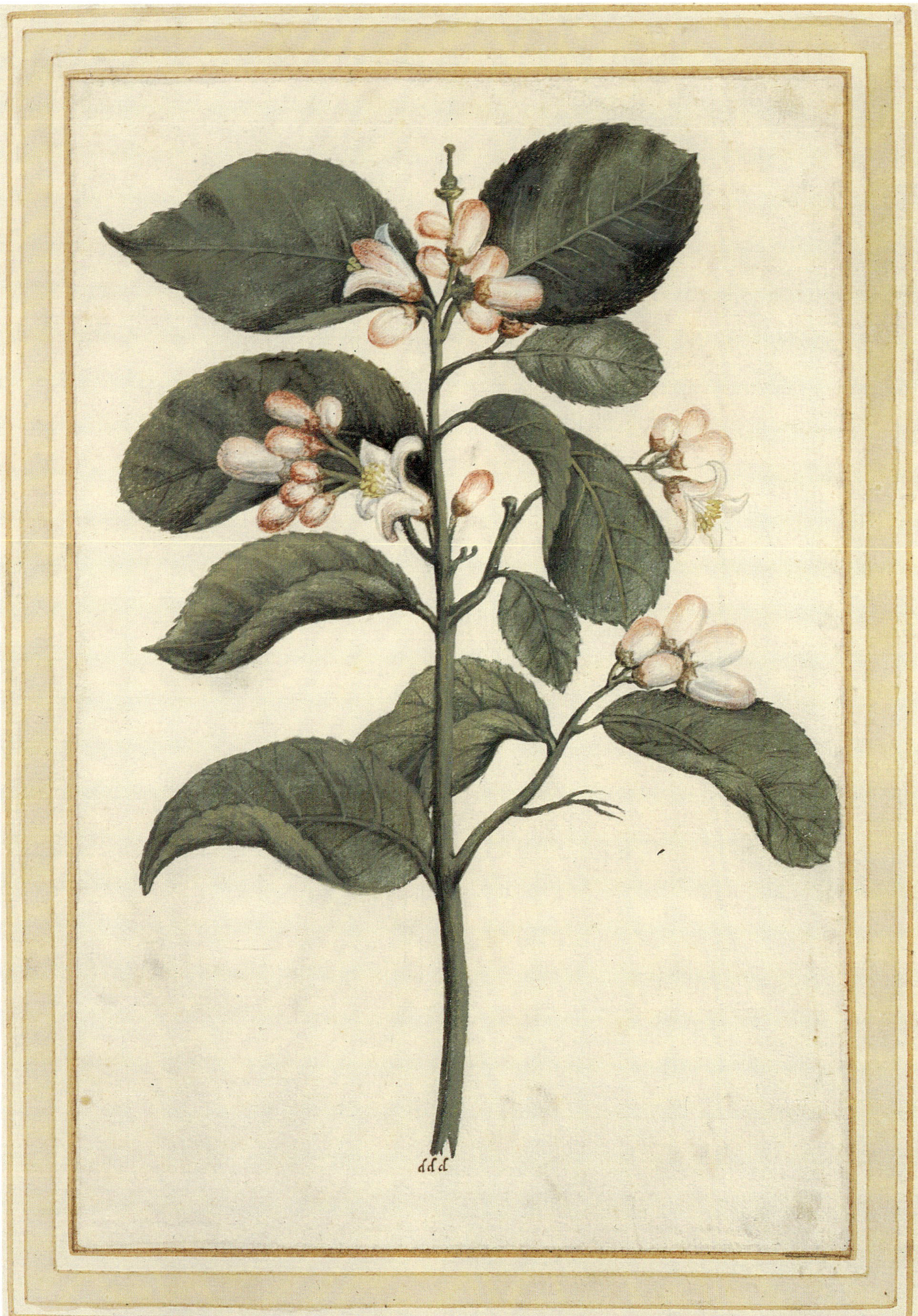
ddd

19. Lemon, *Citrus limon* (L.) Burm. f.: whole and half-fruit

VINCENZO LEONARDI (fl. 1621–46)

Windsor, RL 19335

Watercolour and bodycolour over black chalk, with incised outlines

173 × 239 mm. Cream prepared paper. Survival of 5 mm overlay of original 17th-century mount at right edge

ANNOTATION: [below stalk] *VV*

MOUNT SHEET: 592 × 435 mm. Stained top right corner and marked elsewhere. *Watermark*: D & C Blauw

This is the preparatory drawing for the engraving of the common lemon (*Limon vulgaris*) on p. 193 of Ferrari's *Hesperides*. In his discussion on p. 191, Ferrari commented on the fact that this lemon was used for seasoning food, and that its acidic juice made it rather less satisfactory for eating. Then he proceeded to note some of the varied features within this particular group: some were oval, others more spherical; some were paler, others of a deeper yellow; some had a firmer flesh than others; and the juice was not always as sour in some examples as in others. After thus listing the 'differences', (but not suggesting that such differences had anything to do with speciation), Ferrari observed that there were many similarities too. None are any less than obvious (thin, toughish, tasteless skin, sour juice, a few longish seeds, etc.), although it is typical of him to include among the similarities a few details related to matters of cultivation, such as the fact that this kind of lemon was tolerant of the cold, and did not require special cultural practices.

The specimen represented here shows a long, ellipsoid, apiculate fruit with a long neck and pale yellowish, smooth skin. The cross-section reveals a rather thin rind and a juicy flesh with some seeds. The leaves are serrate and narrowly elliptical, the petioles wingless.

PROVENANCE: dal Pozzo; Albani; George III, by descent to George V; London art market (Mendelson); Rex Nan Kivell; Her Majesty The Queen

REFERENCES: Ferrari 1646, p. 191

ENGRAVINGS: Ferrari 1646, p. 193

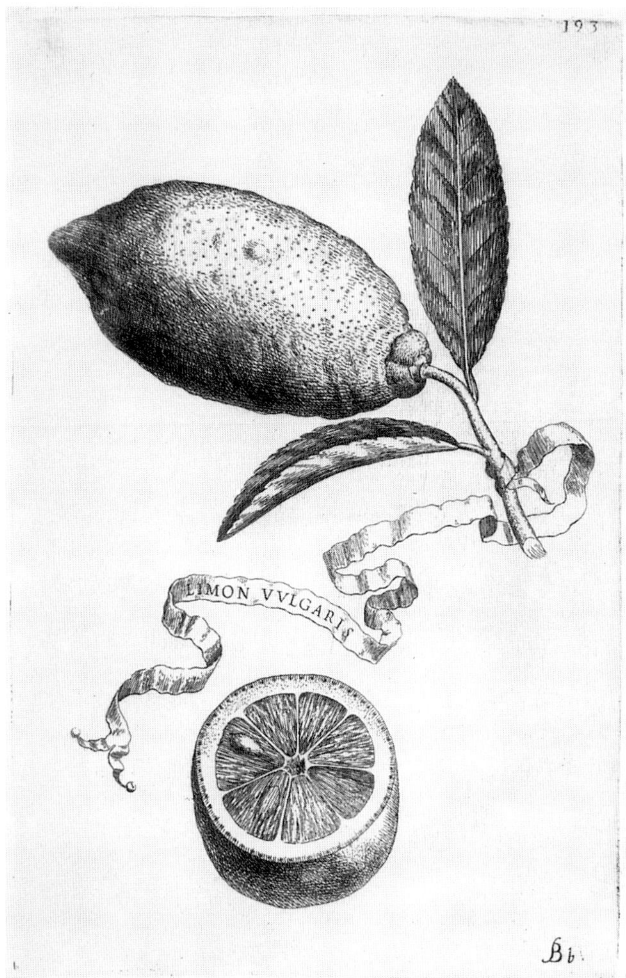

Comp. fig. 19: Ferrari 1646 p. 193

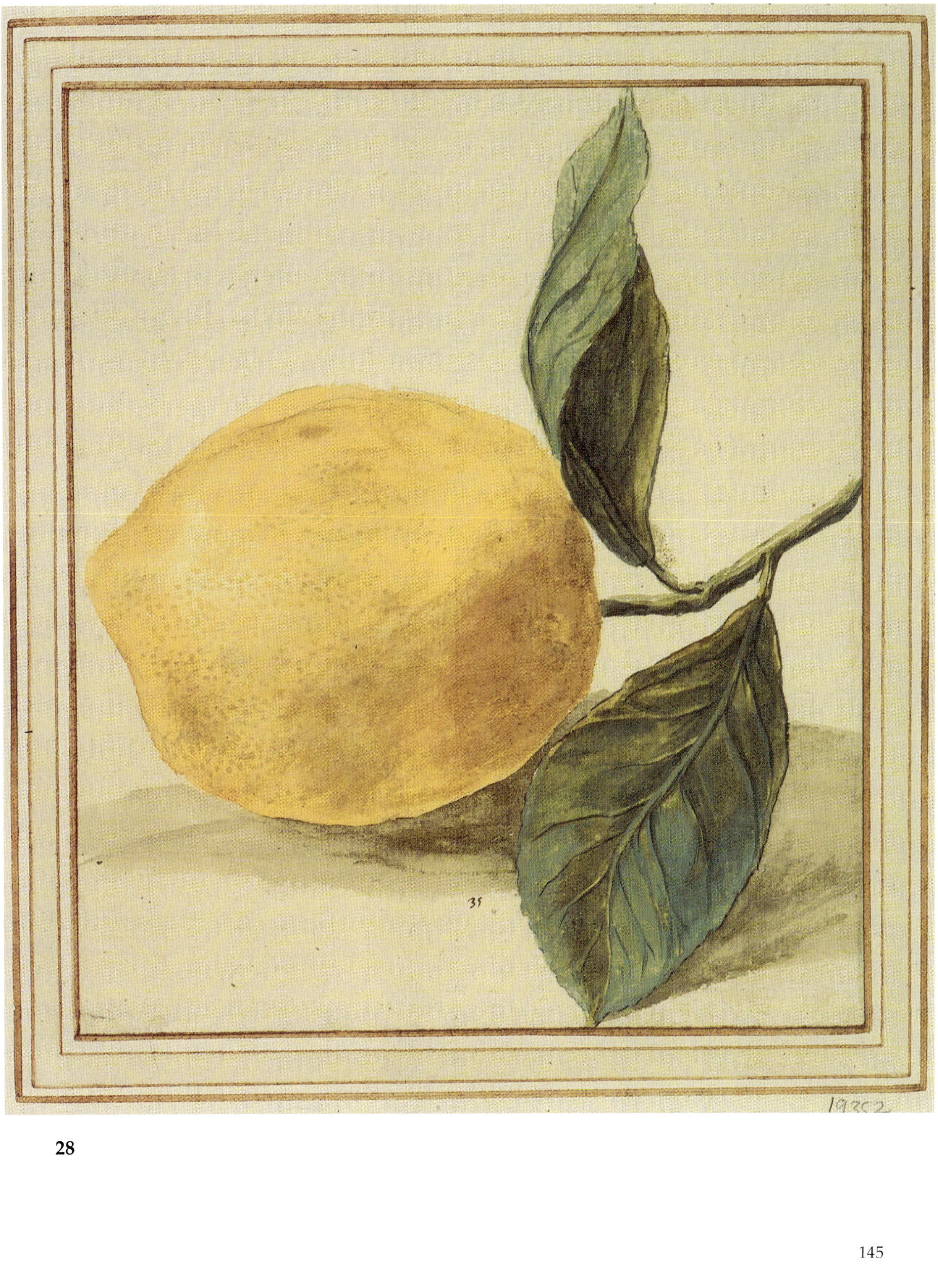

28

29. Lemon, *Citrus limon* (L.) Burm. f.: whole fruit

?VINCENZO LEONARDI (fl. 1621–46)

Location unknown

Watercolour and bodycolour over black chalk; *pentimenti* in black chalk

131 × 213 mm. Trimmed at right ?and left. *Watermark*: none recorded

ANNOTATION: [centre] *50*

MOUNT SHEET: 585 × 432 mm. Mounted on same sheet as **30**

Not engraved in Ferrari's *Hesperides*. The fruit is obovate, obtuse at the apex and covered by a wrinkled, bumpy, yellow rind. Two wide, elliptical leaves with wingless petioles are also shown. *Pentimenti* show different contours for fruit and a possible outline for the scroll.

PROVENANCE: dal Pozzo; Albani; George III, by descent to George V; London art market (Mendelson); James R. Herbert Boone; Sotheby's (New York), 16 September 1988, lot 150

30. Lemon, *Citrus limon* (L.) Burm. f.: half-fruit

?VINCENZO LEONARDI (fl. 1621–46)

Location unknown

Watercolour over black chalk

83 × 101 mm. Trimmed through shadow at right *Watermark*: none recorded

ANNOTATION: [below fruit] *51*

MOUNT SHEET: mounted on same sheet as **29**

This drawing must originally have been on the same drawing sheet as the whole fruit (see **29**), the two later separated, and then in the 18th century mounted together when they entered the collection of George III. Like the drawing of the whole fruit, this one was not reproduced in Ferrari's *Hesperides* either. The fruit has a moderately thick rind and nine yellow segments with many seeds.

PROVENANCE: dal Pozzo; Albani; George III, by descent to George V; London art market (Mendelson); James R. Herbert Boone; Sotheby's (New York), 16 September 1988, lot 150

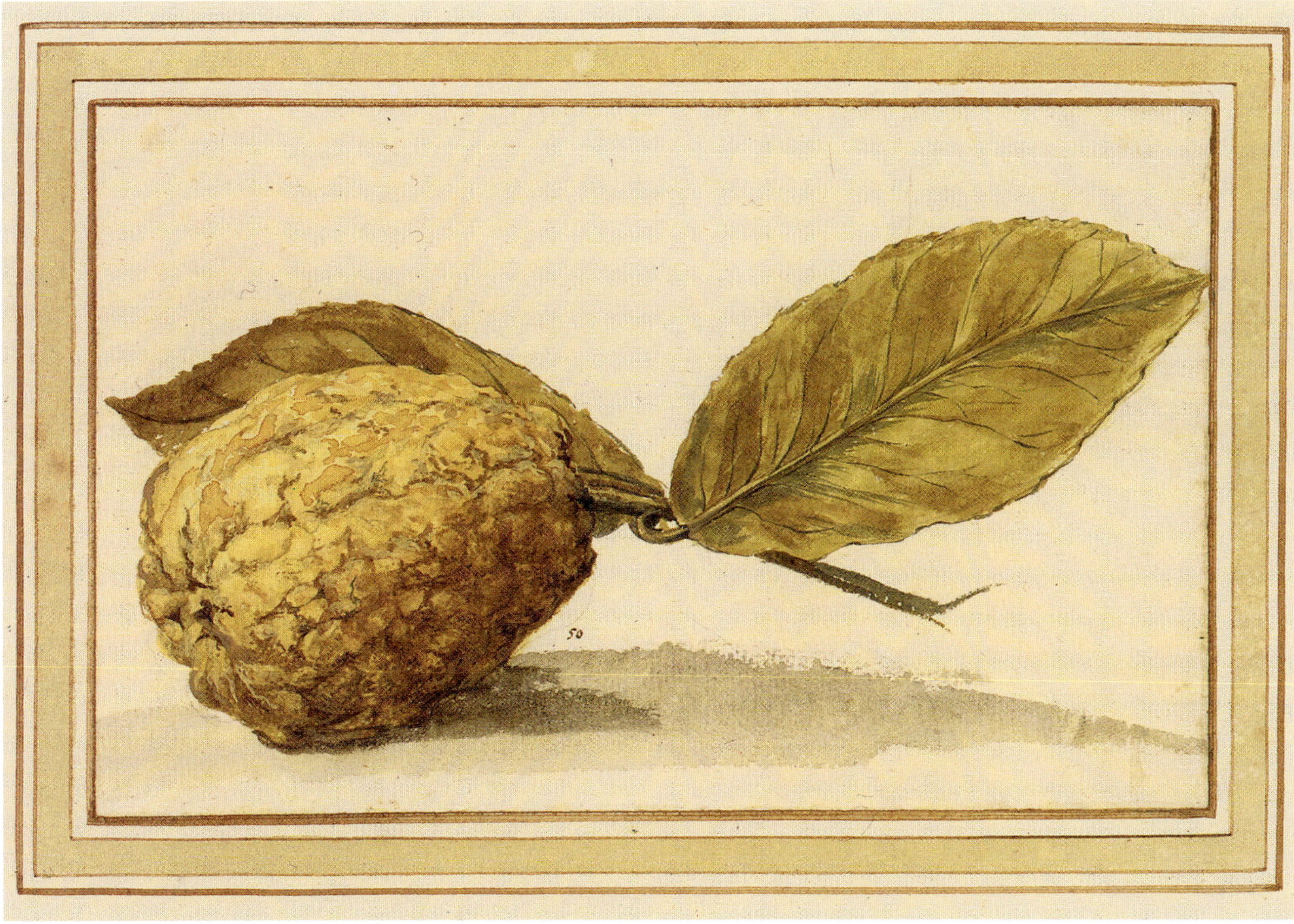

29

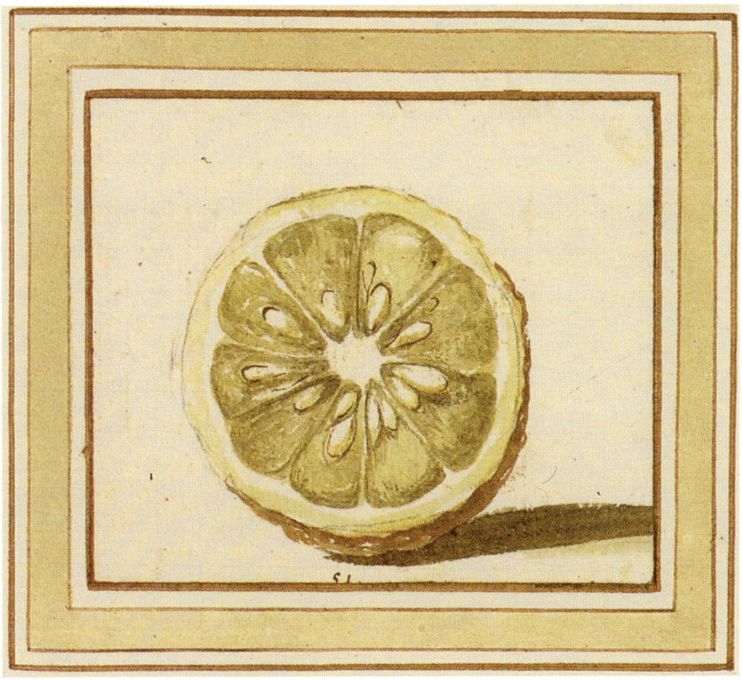

30

31. Sour orange *Citrus aurantium* L.: flowering twig and fruits

VINCENZO LEONARDI (fl. 1621–46)

Private collection, England

Watercolour and bodycolour, with gum heightening, over black chalk; unfinished scroll in black chalk

290 × 188 mm. Foxed

ANNOTATION: *aaa*

As in the case of the books devoted to citrons and lemons in the *Hesperides* (see **18**), that on oranges opens with an illustration of a flowering branch. But the book on oranges also contains an illustration of a double-flowered specimen (*Aurantium flore duplici*) on p. 391, engraved after the present drawing. Unlike most of the other surviving drawings for the *Hesperides*, it shows the scroll on which the name of the specimen would eventually be engraved on the final plate (but see also **20**, **29** and **49**). The fruit itself is attached to a small branch with a flower cluster still intact; the leaves are elliptical with broadly winged petioles, while, at the side, a cross-section of the fruit is also shown. Ferrari's definition of the branch as double-flowered (*flore duplici*) depends on the supernumerary petals which can easily be observed on the fully open flowers. In oranges, as in most citrus fruit, blossoms can occur when fruits are still on the tree. Since the Renaissance, orange flowers, called *zagara* in Italian, were picked and distilled to extract essential oils for perfumes.

Ferrari discusses this type of orange on p. 387, along with the *Aurantium crispo folio*, so called for the dense cluster of small curly leaves around each shoot and fruit. He was so taken with the latter ('no orange', he observed, 'pleases the eyes and the palate more fully') that he barely commented on the double-flowered type at all, noting only its shape, the colour and quantity of its flesh (*medulla croceum, peramplum & seminosum*), and its 'medium' taste, halfway between sharp and sweet.

Ordinary double-flowered oranges of this kind are also represented in the paintings of Bartolomeo Bimbi, such as Villa Med. C. 594.

PROVENANCE: dal Pozzo; Albani; George III, by descent to George V; London art market (Mendelson)

LITERATURE: Freedberg 1993, pp. 181, 183, pl. 110

REFERENCES: Ferrari 1646, p. 387; Bimbi, Villa Med. C. 594; Volkamer *Nürnberg. Hesp.* I, p. 201f., pl. 202b; Micheli *Enumeratio*, p. 30f., pls. 16–7; Gallesio 1811, p. 129f.; unpublished drawing on loose, unnumbered folio, GAG; Baldini 1996b, p. 20, pls. 16–17

ENGRAVINGS: Ferrari 1646, p. 391

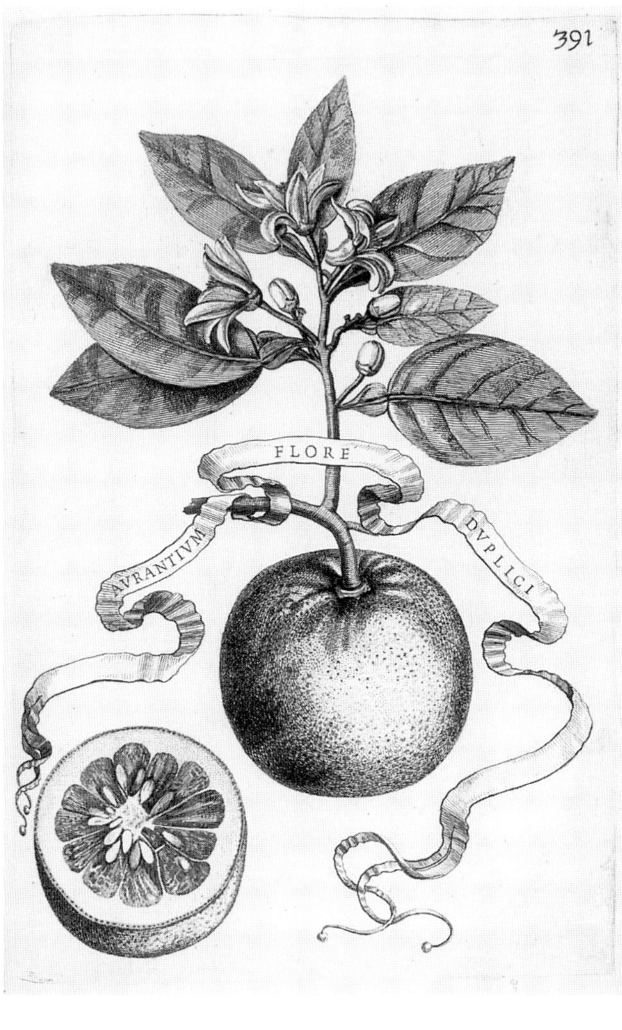

Comp. fig. 31: Ferrari 1646 p. 391

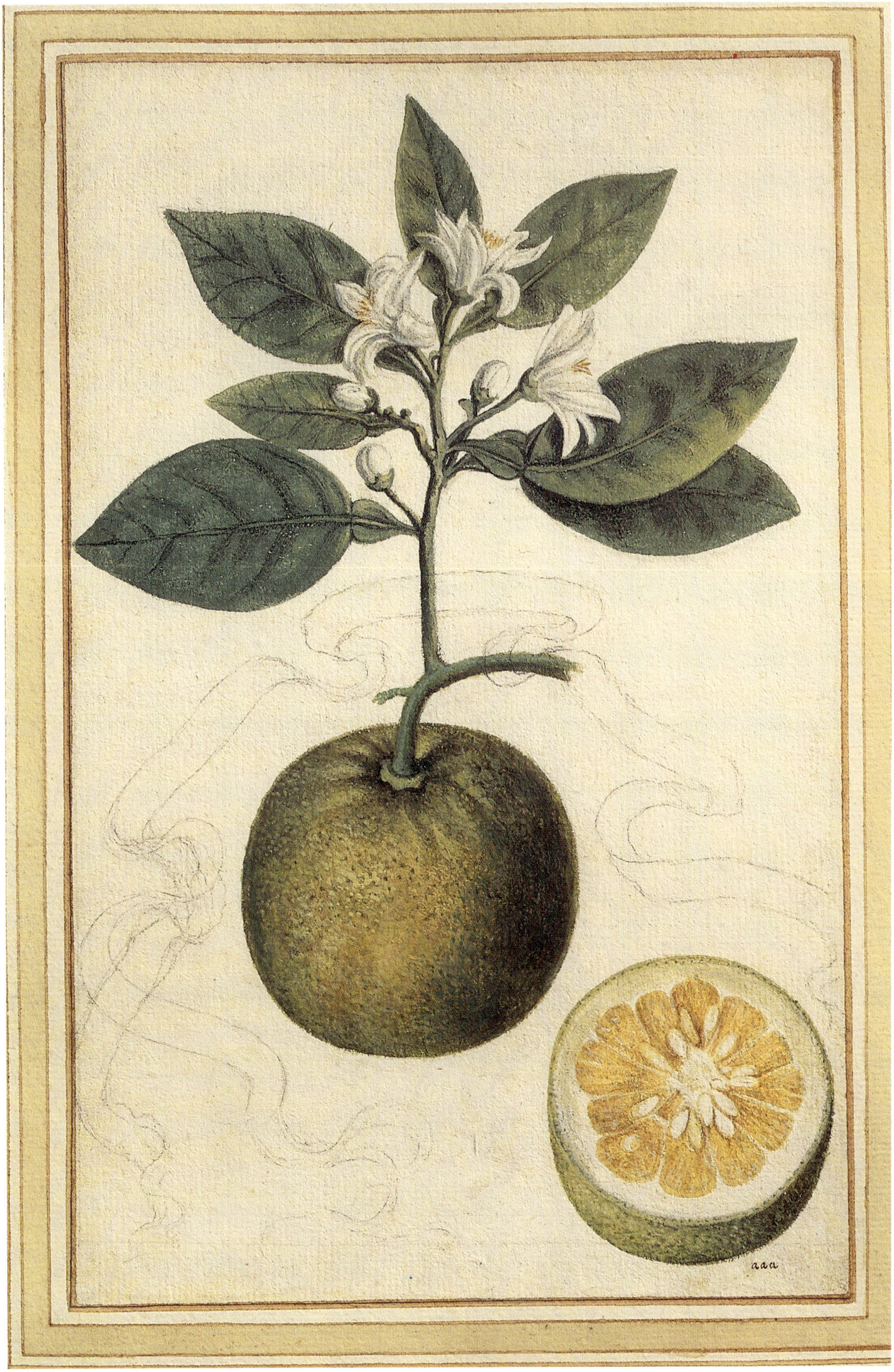
aaa

32. Sweet orange, *Citrus sinensis* (L.) Osb.: whole fruit

VINCENZO LEONARDI (fl. 1621–46)

Windsor, RL 19331

Watercolour and bodycolour, with gum heightening, over black chalk

93 × 127 mm. Cream prepared paper

MOUNT SHEET: 587 × 434 mm. Mounted on same sheet as **33**. Tears at left edge, repaired on verso. *Watermark*: D & C Blauw

The pretty Portuguese orange, Ferrari recalls, was sent from Lisbon to be cultivated in the gardens of the Pii and the Barberini. This is corroborated by the wonderfully detailed account in Archivio Dal Pozzo MS 39, fol. 51, of the arrival of this type in Rome. Ferrari illustrates it on p. 427, and just as in the present drawing, it shows a flattened sweet orange with a sunken apex and a smooth, finely pitted peel. But in the *Hesperides* three elliptical, pointed leaves with winged petioles were added to this example of what Ferrari entitled the *Aurantium Olysiponense*. In fact, the sweet oranges of Italy were generally called *portogalli*, recalling their Iberian origins. But as Ferrari himself noted in his chapter on them (p. 425), many authorities asserted that they had originally come from China, and thus were often called Chinese oranges as well. He also observed that their leaves were strongly perfumed, and their flesh so gleamingly yellow that it seemed to produce an almost golden juice. This had a pleasantly sharp sweetness which was often good for sweetening the breath. For Peiresc's comments on the Chinese orange, see p. 51.

PROVENANCE: dal Pozzo; Albani; George III, by descent to George V; London art market (Mendelson); Rex Nan Kivell; Her Majesty The Queen

REFERENCES: Archivio Dal Pozzo MS 39, fols. 51–52v; Ferrari 1646, p. 425; Cupani 1696, p. 180; Bimbi, Villa Med. C. 594; Micheli *Enumeratio*, p. 22, pl. 7; Gallesio 1811, p. 149; Risso/Poiteau 1818, p. 36, pl. 5; unpublished drawing on loose, unnumbered folio, GAG; Baldini 1996b, p. 27, pl. 28

ENGRAVINGS: Ferrari 1646, p. 427

33. Sweet orange, *Citrus sinensis* (L.) Osb.: half-fruit

VINCENZO LEONARDI (fl. 1621–46)

Windsor, RL 19332

Watercolour and bodycolour over black chalk

94 × 127. Cream prepared paper

ANNOTATION: *DD*

MOUNT SHEET: mounted on same sheet as **32**

This is the preparatory drawing for the half-fruit of the *Aurantium Olysiponense* illustrated on p. 427 of the *Hesperides* and discussed in the previous entry. It has a thin rind and an orange seeded flesh divided into twelve segments. In Ferrari's plate two elliptical leaves with winged petioles have been added.

PROVENANCE: dal Pozzo; Albani; George III, by descent to George V; London art market (Mendelson); Rex Nan Kivell; Her Majesty The Queen

REFERENCES: Ferrari 1646, p. 425; Cupani 1696, p. 180; Micheli *Enumeratio*, p. 22, pl. 7; Gallesio 1811, p. 149; Risso/Poiteau 1818, p. 36, pl. 5

ENGRAVINGS: Ferrari 1646, p. 427

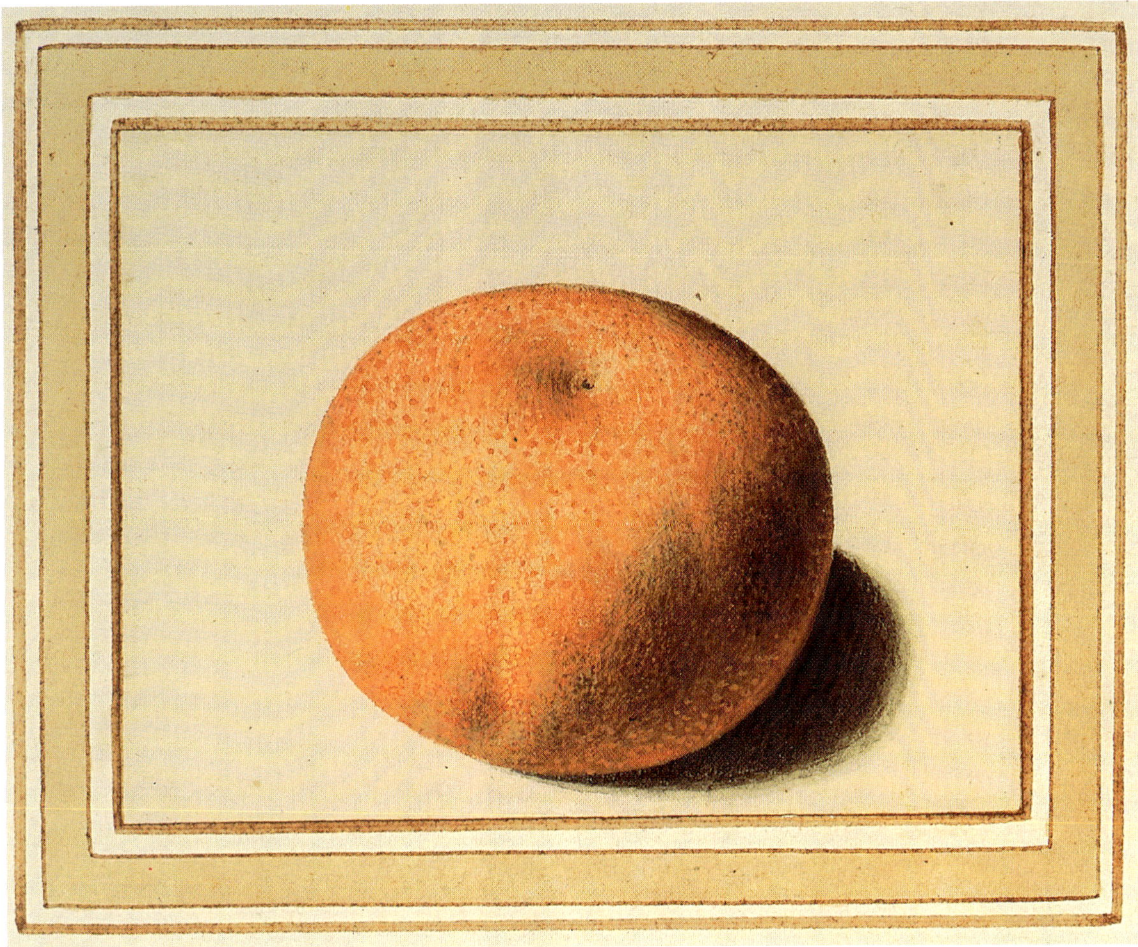

32

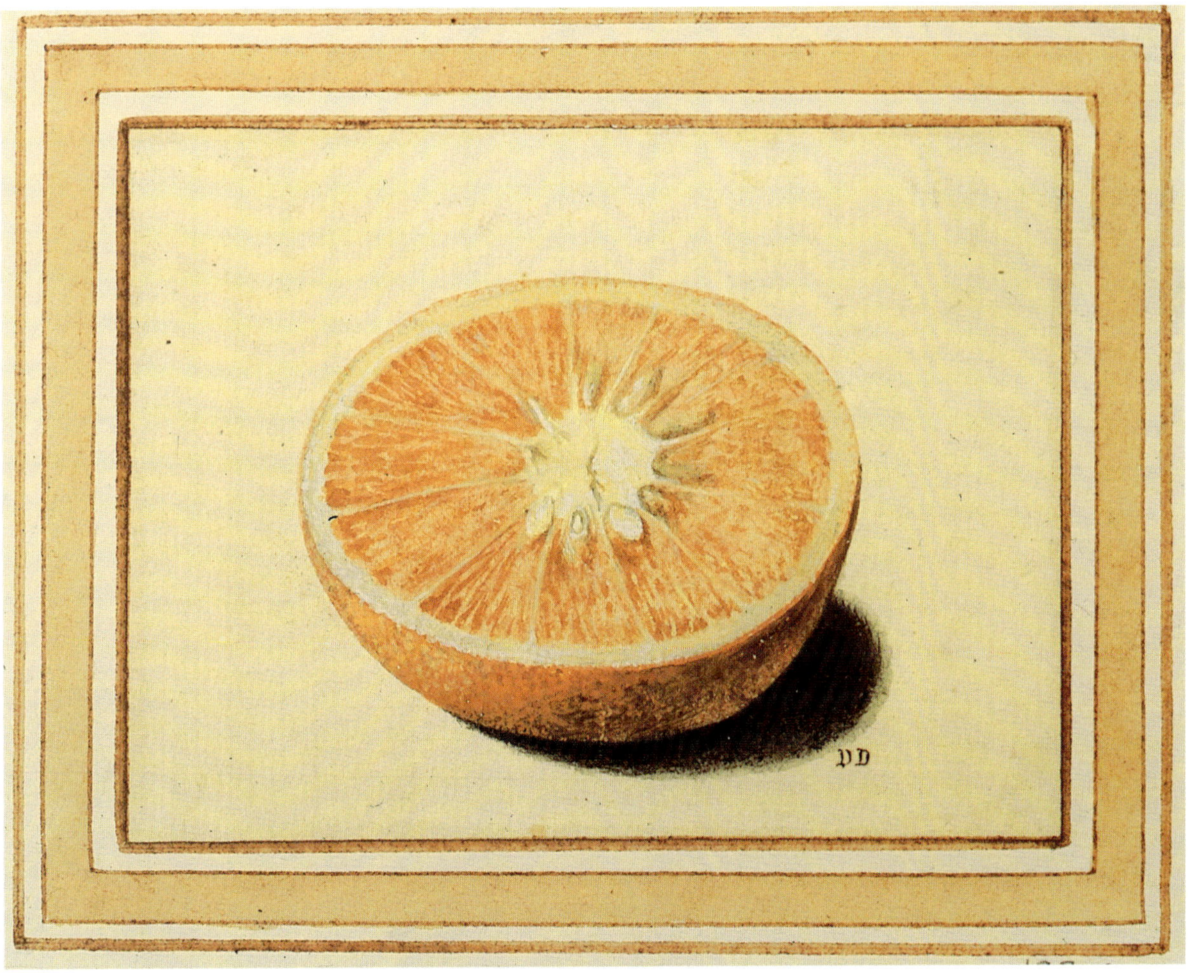

33

34. Myrtle-leaf orange, *Citrus myrtifolia* Raf.: two whole fruits and leaves

VINCENZO LEONARDI (fl. 1621–46)

London, Victoria and Albert Museum, E. 734-1949

Watercolour and bodycolour over black chalk, with gum heightening and incised outlines

64 × 90 mm

ANNOTATION: *19*

MOUNT SHEET: 585 × 435 mm.
Annotation: [verso: in pencil, in a 20th-century hand] *Mrs de Beer £8.8.0*

Mounted together with **35** and **36**, on the same sheet. This drawing served as the preparatory design for the orange branch shown at the bottom of p. 433 of Ferrari's *Hesperides*. As the inscription on this page makes clear, the fruit shown here is an *Aurantium sinense*, as opposed to the *Aurantium dulci cortice* shown in much larger scale at the top of the sheet. This last, in fact, is a pummelo hybrid, for which the preparatory drawings also survive (see **63** and **64**). Just as in the case of the present drawing, it too is mounted together with the drawing of its section on the same sheet. Since the section of the present *Aurantium sinense* (see the following entry) was etched together with the section of the *Aurantium dulci cortice* on the next page in the *Hesperides* (p. 435), and since it too was shown on a much smaller scale than the section of the pummelo hybrid, one must assume that these representations of the *Aurantium sinense* were added as an afterthought, for reasons not known to us now. The fact that the drawings of the whole fruit and sections of both the *Aurantium sinense* and the *Aurantium dulci cortice* are separate but mounted together on the same sheets suggests some manipulation on the part of the etcher of the plates to accommodate what must have been a late decision to fill the space at the bottom of the pages intended at one stage or another to show only the *Aurantium dulci cortice*. Perhaps it was felt that these drawings of the *Aurantium sinense* were too pleasant or showed too important a species or variety not to be illustrated.

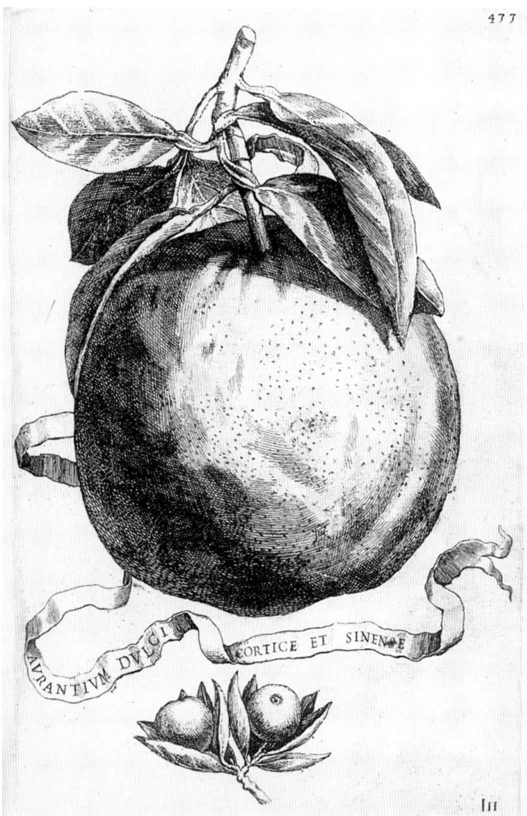

Comp. fig. 34: Ferrari 1646 p. 433

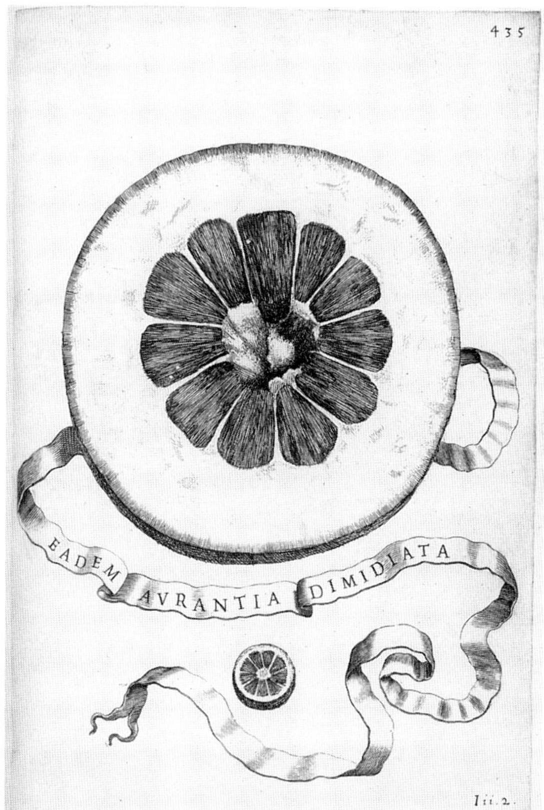

Comp. fig. 35: Ferrari 1646 p. 435

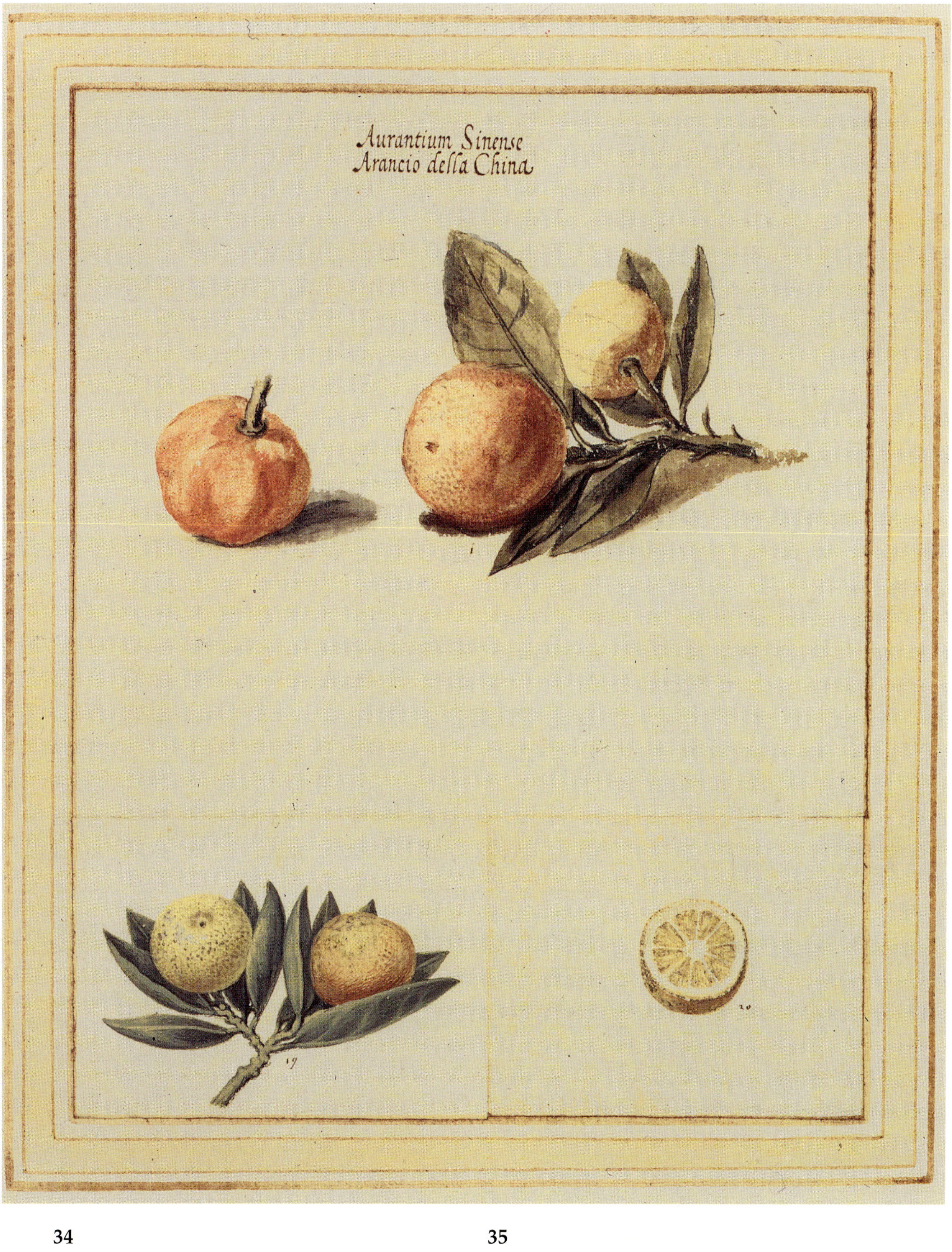

36

34 35

In any event, Ferrari describes the *Aurantium sinense* in typically close detail. It generally has nine or ten segments, has no seeds, is nicely golden, and has a pleasantly acidic flavour. Ferrari notes that the Jesuit Alvaro Semedo had made it clear that this type of fruit did not actually originate in China, but rather in India, and Goa more specifically. In prescient anticipation of its modern taxonomy, Ferrari notes that the density of its small myrtle-like leaves caused it sometimes to be called the myrtle-leafed orange. He concludes with a short note to the effect that the same Semedo had recorded a true Chinese orange, which was a somewhat smaller tree, about the size of olive trees from Valencia.

Myrtle-leaf oranges or *chinotti* have very short internodes and small leaves and fruits. This ornamental species was probably brought to Mediterranean countries from China in the 16th century. Pictures and descriptions of it appear in Aldrovandi I.P. (*Mala aurantia ex China*), in Bimbi (*Arancio della China*), in Montalbani (*Mala Aurantia Sinensis*), while significant references are also to be found — needless to say — in the *Notizie diverse* gathered by Cassiano (e.g. the *Arancetti della China* to be found both in the garden of the Archduchess in Florence and in the Aldobrandini gardens in Rome (Archivio Dal Pozzo MS 39, fol. 48).

PROVENANCE: dal Pozzo; Albani; George III, by descent to George V; London art market (Mendelson); Mrs Helen de Beer

REFERENCES: Aldrovandi I.P.VII, fol. 68; Ferrari 1646, p. 430; Montalbani 1668, p. 489; Bimbi, Villa Med. C. 594; Volkamer *Nürnberg. Hesp.* I, p. 207f., pl. 207; Micheli *Enumeratio*, p. 43, pl. 29; Gallesio 1811, pp. 132, 134; Risso/Poiteau 1818, p. 104, pl. 50; unpublished drawing on loose, unnumbered folio, GAG; Gallesio 1839, p. 11; De Noter 1926, p. 173; Baldini 1996b, p. 21, pl. 19

ENGRAVINGS: Ferrari 1646, p. 433

35. Myrtle-leaf orange, *Citrus myrtifolia* Raf. : half-fruit

VINCENZO LEONARDI (fl. 1621–46)

London, Victoria and Albert Museum, E. 734-1949

Watercolour and bodycolour over black chalk, with gum heightening and incised outlines

64 × 83 mm

ANNOTATION: *20*

MOUNT SHEET: 585 × 435 mm.
Annotation: [verso: in pencil, in a 20th-century hand] *Mrs de Beer £8.8.0*

This is Vincenzo Leonardi's small but meticulous preparatory drawing for the section of the *Aurantium sinense* shown at the bottom of p. 435 in Ferrari's *Hesperides* and tucked within the lower loop of the scroll (the much larger and self-evidently more telling illustration of a section of the *Aurantium dulci cortice* is shown above). It seems even more of an afterthought — or a late addition — than the whole fruit of the *Aurantium sinense* on the previous plate. For Ferrari's comments on this fruit, and for speculation about the possible late addition of the *Aurantium sinense* to the *Hesperides* illustrations, see **34**.

PROVENANCE: dal Pozzo; Albani; George III, by descent to George V; London art market (Mendelson); Mrs Helen de Beer

REFERENCES: Aldrovandi I.P. VII, fol. 68; Ferrari 1646, p. 430; Montalbani 1668, p. 489; Bimbi, Villa Med. C. 594; Volkamer *Nürnberg. Hesp.* I, p. 207f., pl. 207; Micheli *Enumeratio*, p. 43, pl. 29; Gallesio 1811, pp. 132, 134; Risso/Poiteau 1818, p. 104, pl. 50; unpublished drawing on loose, unnumbered folio, GAG; Gallesio 1839, p. 11; De Noter 1926, p. 173; Baldini 1996b, p. 21, pl. 19

ENGRAVINGS: Ferrari 1646, p. 435

36. Myrtle-leaf orange, *Citrus myrtifolia* Raf.: three whole fruits and leaves

SEVENTEENTH-CENTURY ITALIAN

London, Victoria and Albert Museum, E. 734-1949

Watercolour and bodycolour over black chalk

154 × 174 mm

ANNOTATION: [top, in a 17th-century italic hand] *Aurantium Sinense / Arancio della China*; [in a different ink, below central fruit] *i*

MOUNT SHEET: 585 × 435 mm.
Annotation: [verso, mount: in pencil, in a 20th-century hand] *Mrs de Beer £8.8.0*

This is the topmost of the three drawings mounted on the same 18th-century sheet (for the others see the two preceding entries). It is clearly by a different, inferior hand to the others, and is certainly not by Vincenzo Leonardi (as they presumably are). Its rather more cursory technique and its lack of gum arabic aligns it with the group of anonymous drawings in the corpus of citrus drawings collected by Cassiano (see p. 59).

Despite the fact that it is also a drawing of an *Aurantium sinense*, as the inscription makes clear, it was not chosen for reproduction in Ferrari's *Hesperides*. It is, however, unique amongst the surviving citrus drawings in that it carries an inscription written in the same neat scribal hand that occurs on many of the drawings and in the indices of the bound volumes from the collection of Cassiano dal Pozzo. These inscriptions may already have been added in the lifetime of Federico Cesi (*d.* 1630), from whose estate Cassiano acquired so many of his drawings (see p. 34); but they are more likely attributable to a scribe of Cassiano's.

It is very possible that this drawing, like the drawings of fossils still preserved at Windsor (see B. III), came from Cesi's collection, and that it therefore antedates the two drawings by Leonardi on the same sheet.

PROVENANCE: dal Pozzo; Albani; George III, by descent to George V; London art market (Mendelson); Mrs Helen de Beer

REFERENCES: Aldrovandi I.P. VII, fol. 68; Ferrari 1646, p. 430; Montalbani 1668, p. 489; Bimbi, Villa Med. C. 594; Volkamer *Nürnberg. Hesp.* I, p. 207f., pl. 207; Micheli *Enumeratio*, p. 43, pl. 29; Gallesio 1811, pp. 132, 134; Risso/Poiteau 1818, p. 104, pl. 50; unpublished drawing on loose, unnumbered folio, GAG; Gallesio 1839, p. 11; De Noter 1926, p. 173; Baldini 1996b, p. 21, pl. 19

37. Sweet orange, *Citrus sinensis* (L.) Osb.: whole and half-fruits, leaf and segment

SEVENTEENTH-CENTURY ITALIAN

Private collection

Watercolour and bodycolour over graphite

284 × 235 mm

ANNOTATION: [bottom centre] *560*; [in a different ?17th-century hand, graphite, numbering each fruit] *1 - 5*

The five drawings on this single sheet — none of them reproduced in the *Hesperides* — show (1) a whole fruit, (2) a cross-section, (3) a longitudinal section, (4) a segment, and (5) an elliptical leaf with a moderately winged petiole. Except in the case of the leaf, the cast shadows are unusually pronounced. No other surviving drawing shows such an arrangement of whole fruit, sections, segment and leaf. The finely pitted orange peel, the thin rind, the colour of the segment and the oval shape of the leaf are all consistent with the taxonomy of the sweet orange or *Citrus sinensis* (L.) Osb.

PROVENANCE: dal Pozzo; Albani; George III, by descent to George V; London art market (Mendelson)

REFERENCES: Archivio Dal Pozzo MS 39, fols. 51–52v; Ferrari 1646, p. 425; Cupani 1696, p. 180; Micheli *Enumeratio*, p. 22, pl. 7; Gallesio 1811, p. 149; Risso/Poiteau 1818, p. 36, pl. 5; unpublished drawing on loose, unnumbered folio, GAG; Baldini 1996b, p. 27, pl. 28

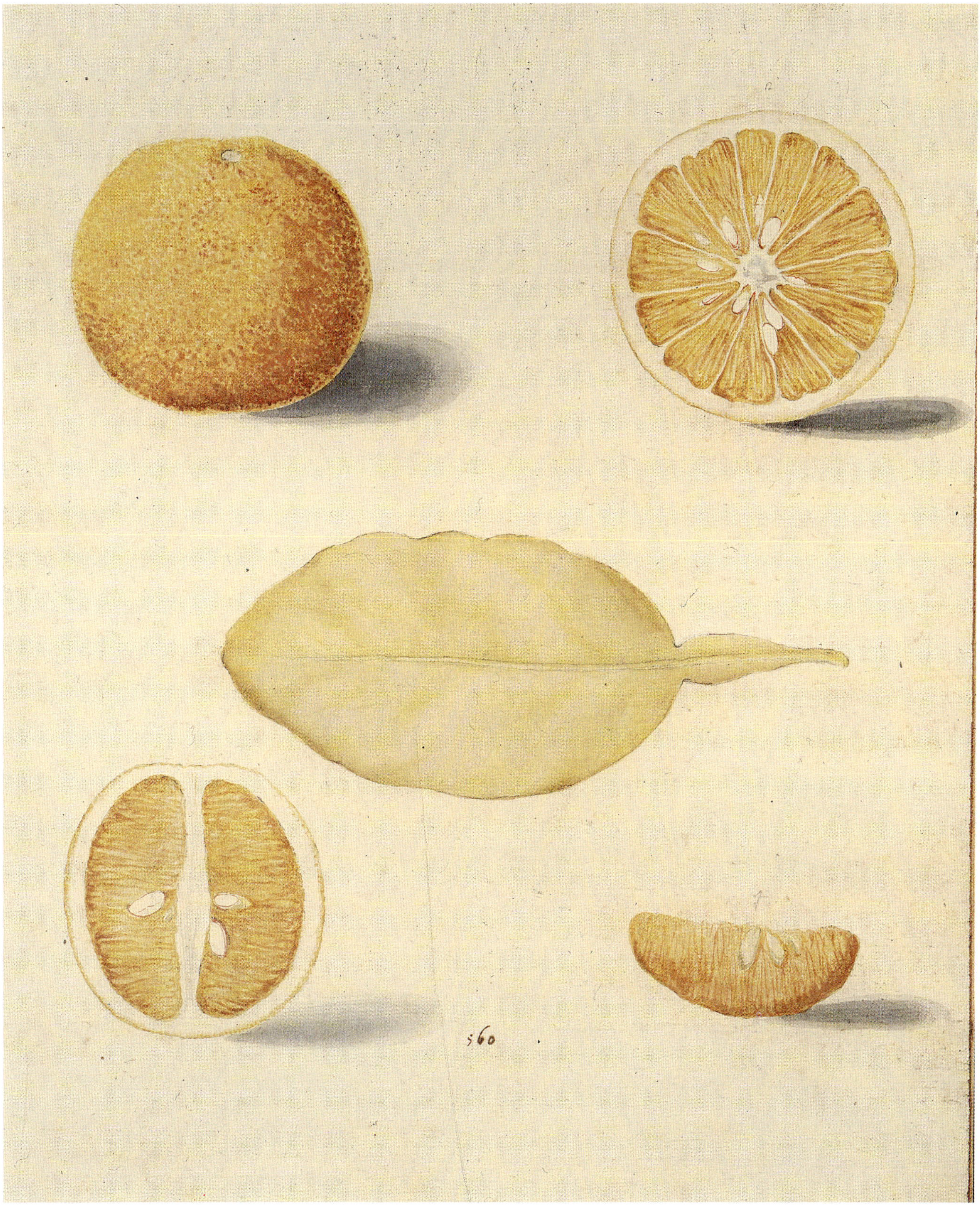

37

38. Pummelo, *Citrus grandis* (L.) Osb.: whole fruit and leaf

VINCENZO LEONARDI (fl. 1621–46)

Location unknown

?Watercolour and bodycolour, with gum heightening

Measurements unknown

ANNOTATION: [bottom centre] *hhh*

This is the magnificent preparatory drawing for the engraving which Ferrari calls the *Aurantium maximum* on p. 439 of the *Hesperides*. The fruit is in fact a pummelo, which Ferrari mistakenly groups with oranges. In his long chapter on this species (p. 437f.) Ferrari enthuses at great length about its size, and offers many comments on its taste. The chapter begins with the observation that the orange species is remarkable for the variety of sizes of fruit it can produce, from dwarf specimens to those which are the size not just of large gourds, but even of playing-balls. The fruits of the *Aurantium maximum* can be gigantic, even elephantine, sometimes swelling up to the size of five palms. Nothing about the original tree would lead one to suspect the production of such huge specimens, he says; but give them two or three years and they reach their full size. By this time, however, only their skin is sweet; the flesh itself becomes increasingly acid and turns blackish yellow. But the skin has not a trace of bitterness to it. Flower and leaf can be very large too: indeed, one such leaf, sent to Ferrari from Naples, was thirty inches long and eight inches wide! Typically, such plants had ten or more thorns per branch, with no seeds in the fruit, and were of a deep yellow colour. As usual, Ferrari goes on to speak of gardens and people he knew: a particularly magnificent specimen, for example, towered over the gardens of Carlos de Cardeñas, Marquis of Laynez, in Naples, whither it had been imported from Genoa. Ferrari's close friend, the excellent doctor and plant pharmacologist Mario Schipani had applied the name of *Hipponeranzion* to this species, and divided it into two kinds — the one a sessile plant with a sweet flesh and spotted yellow leaves, the other with a rather unpleasant-tasting flesh but with shorter, thicker and uniformly coloured leaves.

The drawing shows a very large, globose, light yellow fruit with a large, broad, orbicular leaf and a widely winged petiole behind it. Rumpf (1750, p. 34) called this fruit the *Pompelmoes-boom*, *Limo decumanus* or *Lomen cassomba*; Gallesio (1811, p. 139) the *Citrus aurantium indicum fructu maximo citrato* or *Pomo d'Adamo*; Volkamer (*Nürnberg. Hesp.* I, p. 189) the *Pompelmoes*, and De Noter (1926, pp. 100-03) the *Citrus aurantium decumana* with different varieties: *chadeck*, *vulgaris*, *crispata*, and *racemosa*.

This drawing is known only from an image on 35 mm miniature film, made by the Courtauld Institute in the late 1950s.

PROVENANCE: dal Pozzo; Albani; George III, by descent to George V; London art market (Mendelson)

REFERENCES: Ferrari 1646, p. 437f.; Volkamer *Nürnberg. Hesp.* I, p. 181f., pls. 182a-b; Micheli *Enumeratio*, p. 13f., pl. 1; Rumpf 1750, p. 34; Gallesio 1811, pp. 161–4; Risso/Poiteau 1818, p. 130, pls. 61–3; unpublished drawing on loose, unnumbered folio, GAG; Gallesio 1839, p. 12; De Noter 1926, pp. 101–03; Baldini 1996b, p. 22, pl. 20

ENGRAVINGS: Ferrari 1646, p. 439

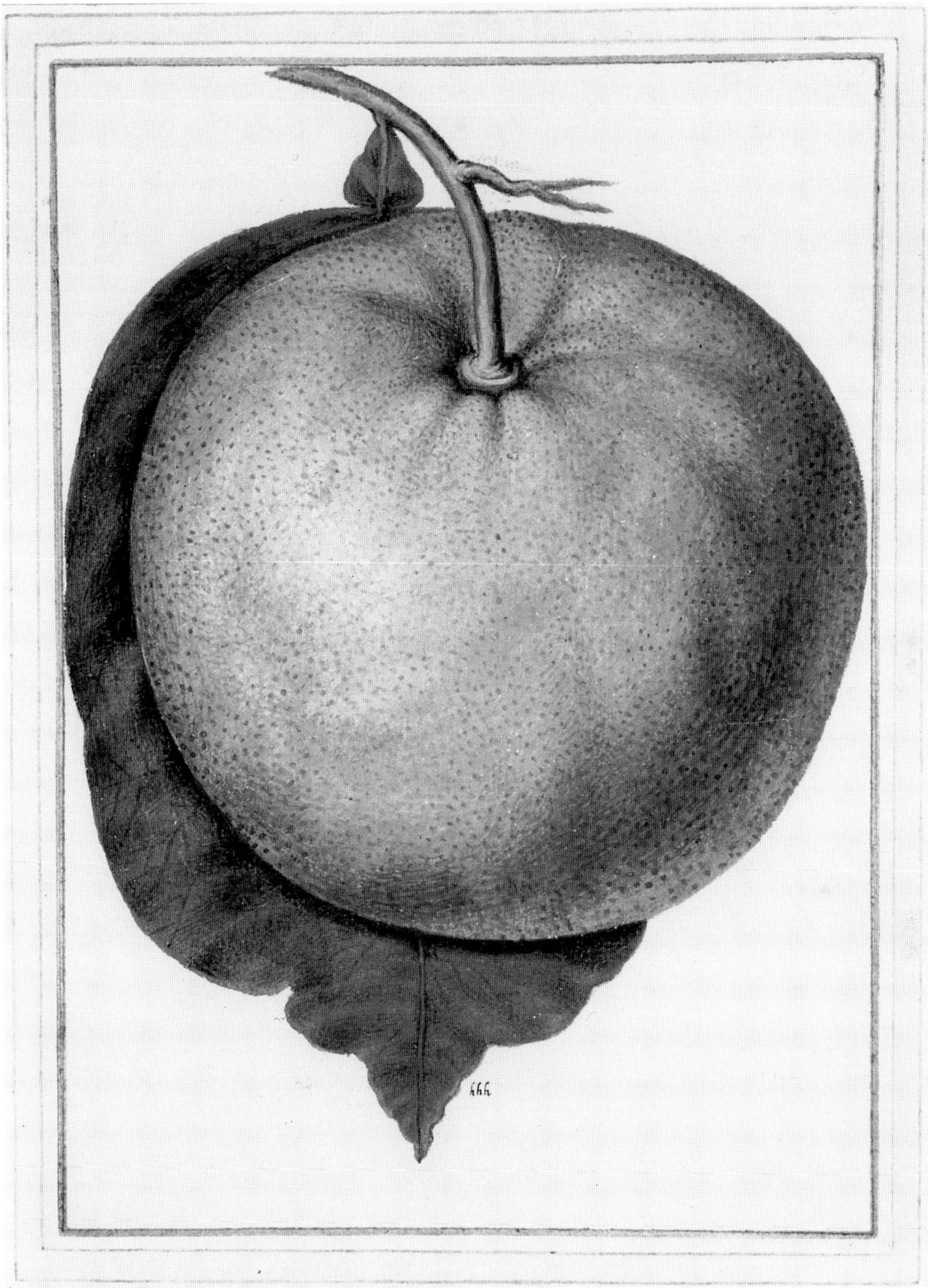

38

39. Pummelo, *Citrus grandis* (L.) Osb.: half-fruit

VINCENZO LEONARDI (fl. 1621–46)
Location unknown
Watercolour and bodycolour over black chalk
181 × 178 mm. White ?prepared paper. *Watermark*: unknown
MOUNT SHEET: mounted with **40**, within the same wash border

Engraved on p. 441 of the *Hesperides*, this drawing shows the half-fruit (180 mm in diameter) of the *Aurantium maximum*, with a very thick mesocarp and a yellowish-orange flesh, divided (as in Ferrari's engraving) into fourteen small seedless segments.

This drawing and **36**, although drawn on separate sheets, were mounted together (the leaf upside-down) in the 18th century when they entered the collection of George III.

PROVENANCE: dal Pozzo; Albani; George III, by descent to George V; London art market (Mendelson); James R. Herbert Boone; Sotheby's (New York), 16 September 1988, lot 155

REFERENCES: Ferrari 1646, p. 437f.; Volkamer *Nürnberg. Hesp.* I, p. 181f., pls. 182a-b; Micheli *Enumeratio*, p. 13f., pl. 1; Rumpf 1750, p. 34; Gallesio 1811, pp. 161–4; Risso/Poiteau 1818, p. 130, pls. 61–3; unpublished drawing on loose, unnumbered folio, GAG; Gallesio 1839, p. 12; De Noter 1926, pp. 101–03; Baldini 1996b, p. 22, pl. 20

ENGRAVINGS: Ferrari 1646, p. 441

40. Pummelo, *Citrus grandis* (L.) Osb.: leaf

VINCENZO LEONARDI (fl. 1621–46)
Location unknown
Watercolour and bodycolour over pen and ink
274 × 179 mm. *Watermark*: unknown
ANNOTATION: *FF*
MOUNT SHEET: mounted with **39**, within the same wash border

This leaf is virtually the same one which is seen behind the whole fruit (**38**) in both the drawing and the engraving (p. 439 of the *Hesperides*). Featuring a remarkably winged petiole (discussed by Ferrari on p. 438) and a broad, orbicular blade, the leaf serves to emphasize the dimensions of the fruit (**38** and **39**).

PROVENANCE: dal Pozzo; Albani; George III, by descent to George V; London art market (Mendelson); James R. Herbert Boone; Sotheby's (New York), 16 September 1988, lot 155

REFERENCES: Ferrari 1646, p. 437f.; Volkamer *Nürnberg. Hesp.* I, p. 181f., pls. 182a-b; Micheli *Enumeratio*, p. 13f., pl. 1; Rumpf 1750, p. 34; Gallesio 1811, pp. 161–4; Risso/Poiteau 1818, p. 130, pls. 61–3; unpublished drawing on loose, unnumbered folio, GAG; Gallesio 1839, p. 12; De Noter 1926, pp. 101–03; Baldini 1996b, p. 22, pl. 20

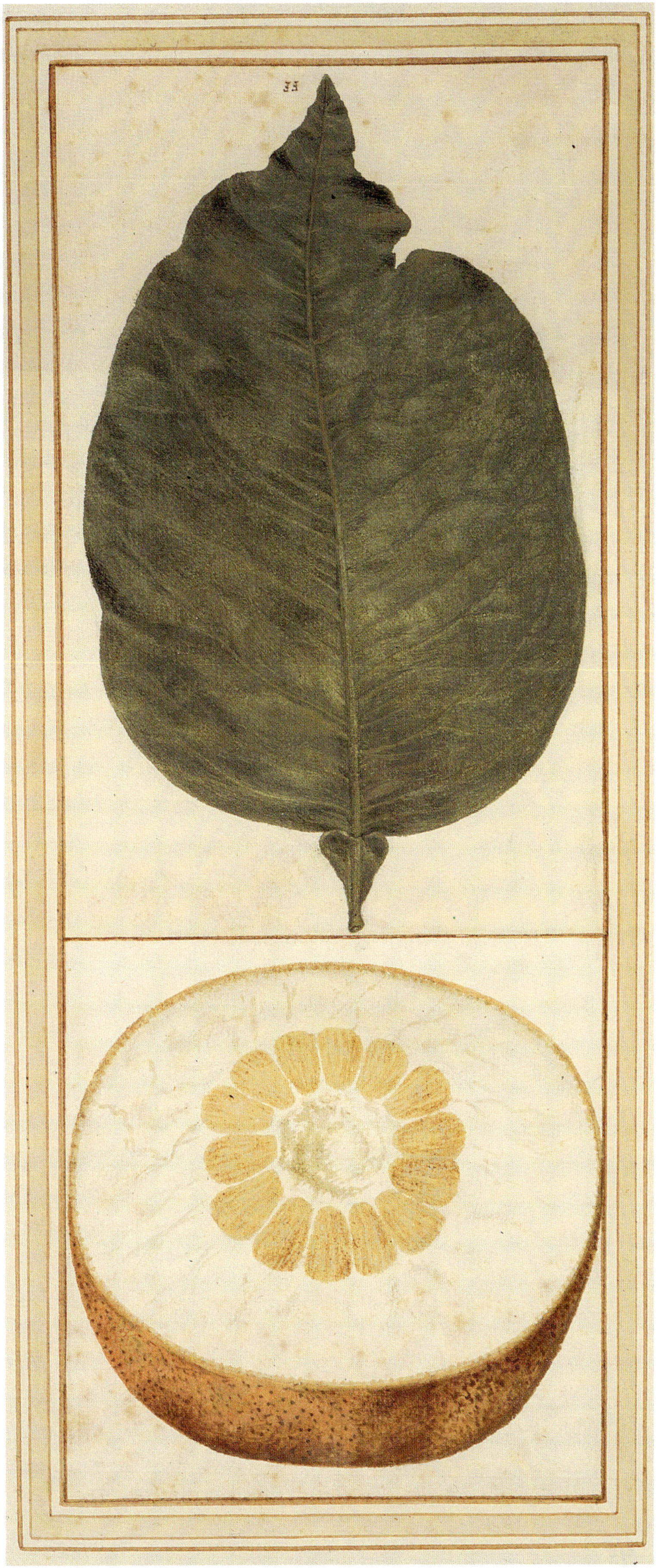

40

39

41. Pummelo, *Citrus grandis* (L.) Osb.: whole and half-fruit

VINCENZO LEONARDI (fl. 1621–46)

Windsor, RL 19333

Watercolour and bodycolour, with gum heightening, over black chalk

344 × 205 mm. Cream prepared paper. Survival of 5 mm overlay of original 17th-century mount at right-hand side

ANNOTATION: [lower right] *EE*

MOUNT SHEET: 596 × 435 mm

This drawing was not reproduced in Ferrari's *Hesperides*. At the top a large orange-coloured pummelo hangs from the end of a twig with a small green fruitlet to one side, and a large orbicular leaf, with a broadly winged petiole, on the other. The section has a thick rind with a partly brownish mesocarp and many small segments. No seeds are in evidence.

PROVENANCE: dal Pozzo; Albani; George III, by descent to George V; London art market (Mendelson); Rex Nan Kivell; Her Majesty The Queen

REFERENCES: Ferrari 1646, p. 437f.; Volkamer *Nürnberg. Hesp.* I, p. 181f., pls. 182a-b; Micheli *Enumeratio*, p. 13f., pl. 1; Rumpf 1750, p. 34; Gallesio 1811, pp. 161–4; Risso/Poiteau 1818, p. 130, pls. 61–3; unpublished drawing on loose, unnumbered folio, GAG; Gallesio 1839, p. 12; De Noter 1926, pp. 101–03; Baldini 1996b, p. 22, pl. 20

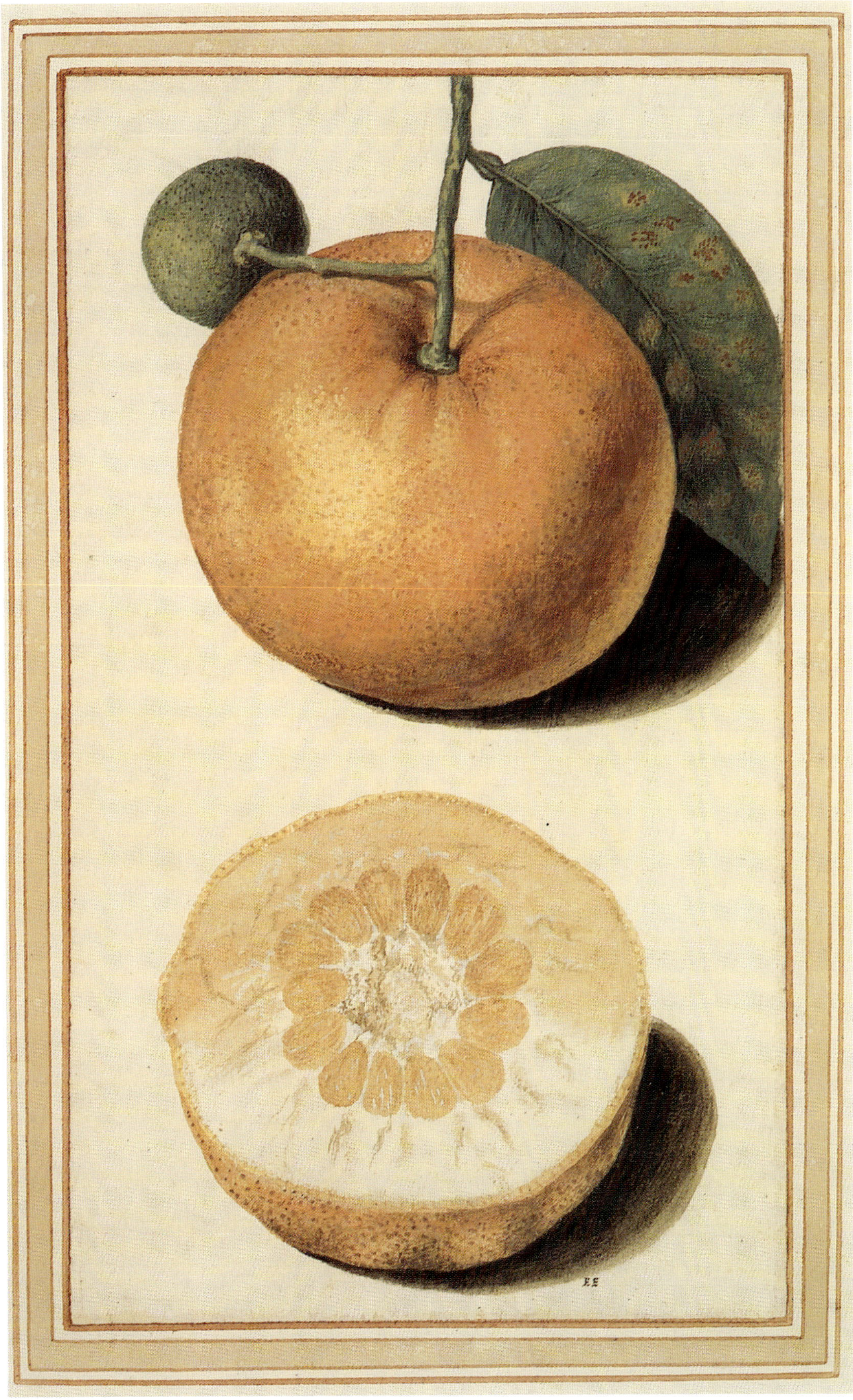

41

42. Pummelo, *Citrus grandis* (L.) Osb.: whole fruit

VINCENZO LEONARDI (fl. 1621–46)

Location unknown

Watercolour and bodycolour, with gum heightening, over black chalk

149 × 165 mm. Trimmed at right and lower edges. Badly stained and foxed

Not reproduced in the *Hesperides*, this drawing shows a large, globose, yellow pummelo with a smooth, roughly pitted peel and a depressed, radially furrowed base; an occasional pimple and some shallow longitudinal ribs protrude from the rind. The specimen is similar to the other pummelos in the present group of drawings from Cassiano's Paper Museum (see **38** and **41**). No drawing of the half-fruit survives among the known drawings in the Paper Museum.

This drawing is known only from a 35 mm colour slide, made by the Courtauld Institute in 1967.

PROVENANCE: dal Pozzo; Albani; George III, by descent to George V; London art market (Mendelson)

REFERENCES: Ferrari 1646, p. 437f.; Volkamer *Nürnberg. Hesp.* I, p. 181f., pls. 182a-b; Micheli *Enumeratio*, pp. 13f., pl. 1; Rumpf 1750, p. 34; Gallesio 1811, pp. 161–4; Risso/Poiteau 1818, p. 130, pls. 61–3; unpublished drawing on loose, unnumbered folio, GAG; Gallesio 1839, p. 12; De Noter 1926, pp. 101–03; Baldini 1996b, p. 22, pl. 20

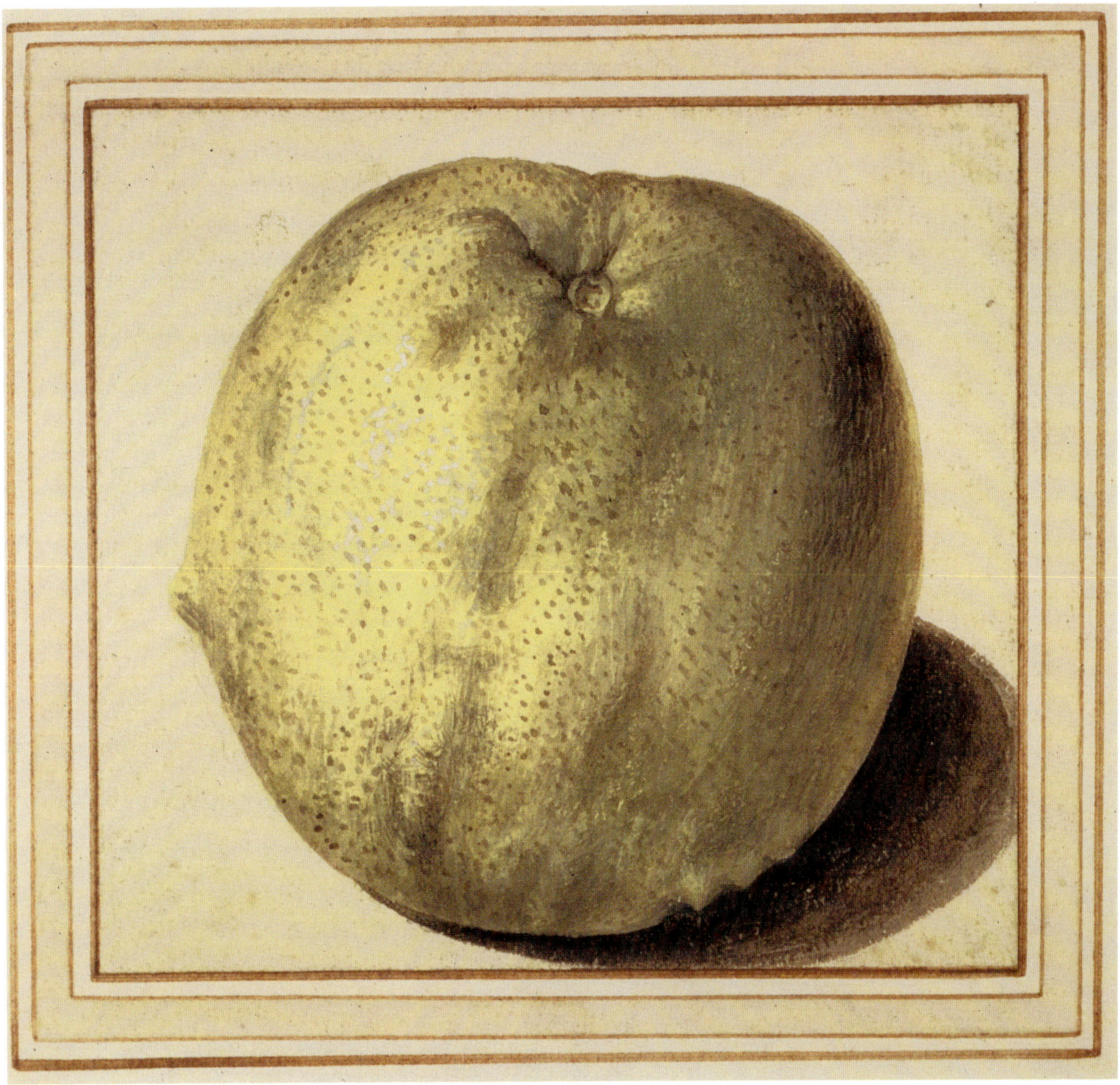

42

43. Citron-lemon, *Citrus limonimedica* Lush.: whole and half-fruit

VINCENZO LEONARDI (fl. 1621–46)

Windsor, RL 19336

Watercolour and bodycolour, with gum heightening, over black chalk; leaf rubbed at top

315 × 206 mm. Badly stained. Trimmed through stalk at right. *Watermark:* Figure (cut and unidentifiable)

ANNOTATION: [lower right] *gg*

MOUNT SHEET: c. 585 × 434 mm. Badly water-stained. *Watermark*: D & C Blauw

This is the preparatory drawing for the *Limon Liguriae ceriescus* illustrated on p. 199 of Ferrari's *Hesperides*. Ferrari discusses it in the third chapter of his book on lemons (p. 195f.), along with other lemons from the Ligurian coast, viz., the *Limon Sancti Remi* and the ordinary Genoese lemon, *Limon ligusticus*, popularly known as *pusilla pila*. This particular citrus also comes from San Remo, but its fruit is much smaller than that of the *Limon Sancti Remi* discussed both by Ferrari and Cassiano (Archivio Dal Pozzo MS 39, fol. 80). On p. 195 of the *Hesperides* Ferrari noted that the popular name of the fruit called *ceriescus* (presumably the Latin equivalent of a now lost Italian form describing a citron-like lemon) arose from its close similarity, both internally and externally, to a citron. This *semperferens* cultivar flowers and fruits at least four times a year (in May, July, August and September). It matures early, because of the benign winters on the Ligurian coast. Often it looks somewhat like a *lumia* (see **58–62**) and can weigh up to 7 *librae* (more than 3 kg). The *Limon ceriescus* was a popular fruit and was exported to Germany and to Southern France, where the whole fruit was eaten, just as in the case of pears.

The drawing shows a citron-shaped, elliptical leaf with a wingless petiole and a roundish, nippled, seedless fruit with a rather thick albedo.

PROVENANCE: dal Pozzo; Albani; George III, by descent to George V; London art market (Mendelson); Rex Nan Kivell; Her Majesty The Queen

REFERENCES: Ferrari 1646, p. 195; Volkamer *Nürnberg. Hesp.* I, p. 153f.; Gallesio 1811, p. 110; Risso/Poiteau 1818, p. 177, pl. 85; De Noter 1926, p. 133f.

ENGRAVINGS: Ferrari 1646, p. 199

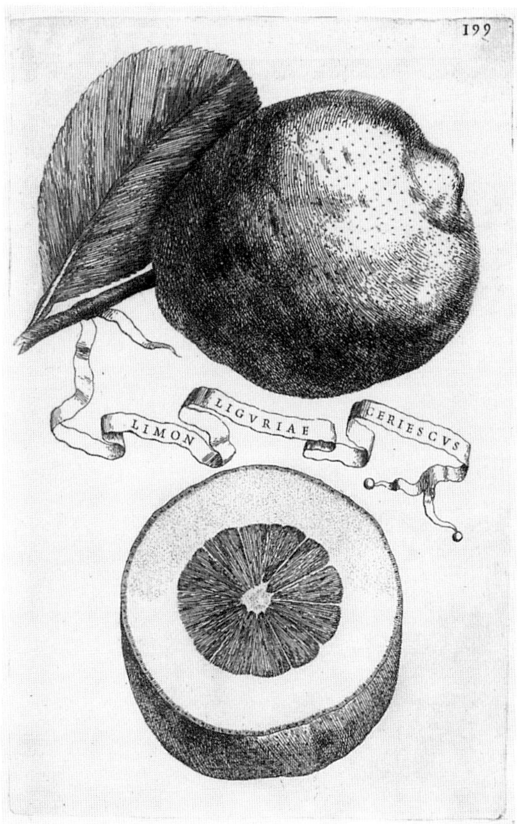

Comp. fig. 43: Ferrari 1646 p. 199

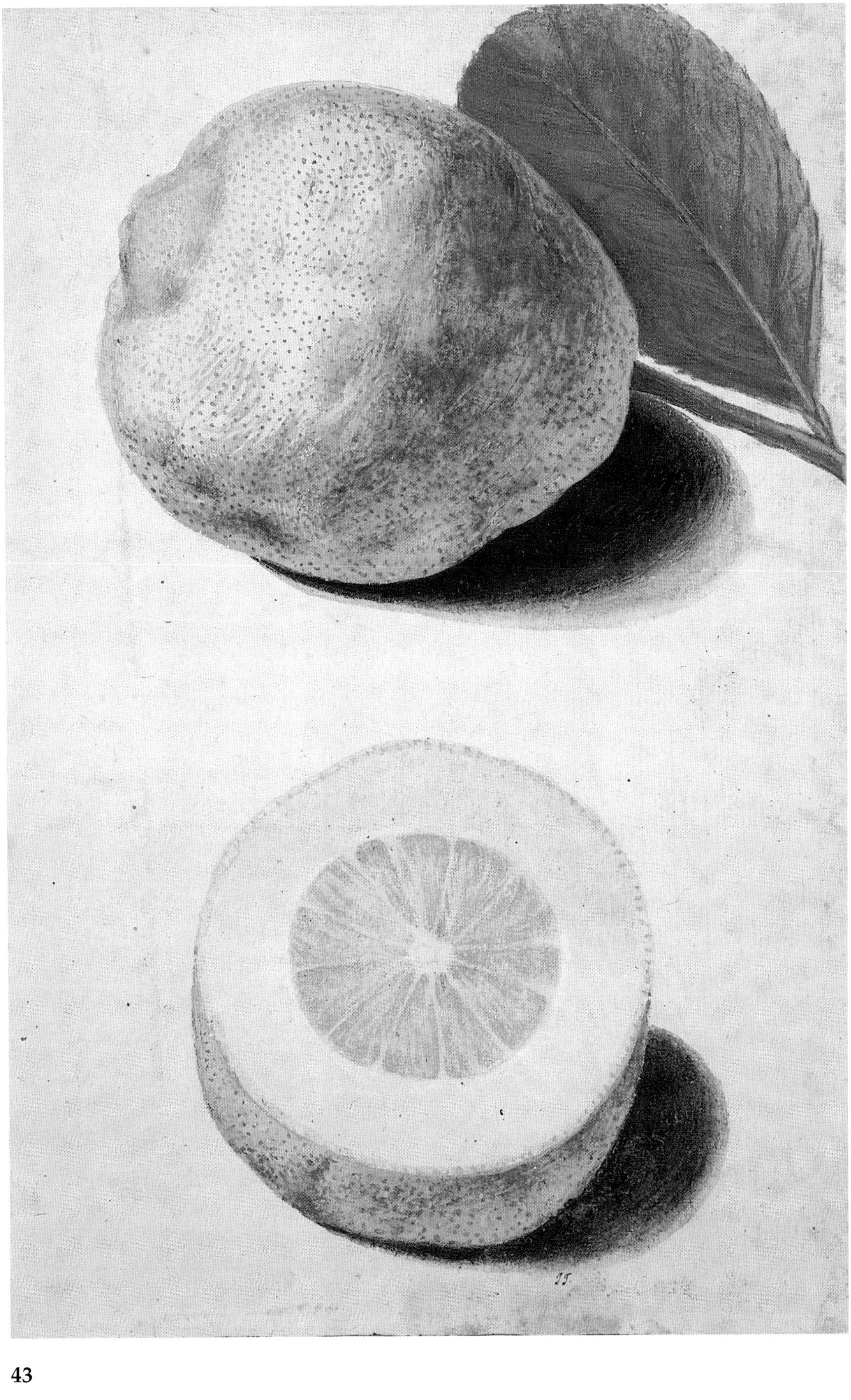

43

44. Citron-lemon, *Citrus limonimedica* Lush.: whole and half-fruit

VINCENZO LEONARDI (fl. 1621–46)

Private collection, England

Watercolour and bodycolour, with gum heightening, over black chalk

239 × 202 mm. Badly foxed

ANNOTATION: *37*

MOUNT SHEET: 404 × 310 mm. *Watermark*: D & C Blauw

This is the preparatory drawing for the *Limon incomparabilis* illustrated on p. 223 of Ferrari's *Hesperides*. As Ferrari notes on p. 221, the name for this fruit refers to its incomparable taste, unequalled by any other lemon. For the rest, Ferrari's description is typically careful, noting size, shape and texture; when it comes to his comment on the thickness of the leaves, however, he testifies to his botanical knowledge by comparing them to those of the 'American Laurel' (?*Laurocerasus caroliniana* Roem, or wild orange). He also notes that a tree was sent, along with other splendid plants, from Gaeta to the gardens of the Villa Mattei in Rome, where it acquired its name. It was then propagated by grafting in other *pomaria* in Rome.

In the present drawing a large, yellow, ellipsoid citron-lemon hangs from a leafy twig; the half-fruit shows a thin, yellow and moderately bumpy peel, a rather thick mesocarp, and a few seeds, just as in Ferrari's description. The leaves are elliptical and serrate with wingless petioles.

PROVENANCE: dal Pozzo; Albani; George III, by descent to George V; London art market (Mendelson); Hazlitt, Gooden and Fox, London; Kyburg Gallery, London

REFERENCES: Ferrari 1646, p. 221; Volkamer *Nürnberg. Hesp.* II, p. 98, pl. 99

ENGRAVINGS: Ferrari 1646, p. 223

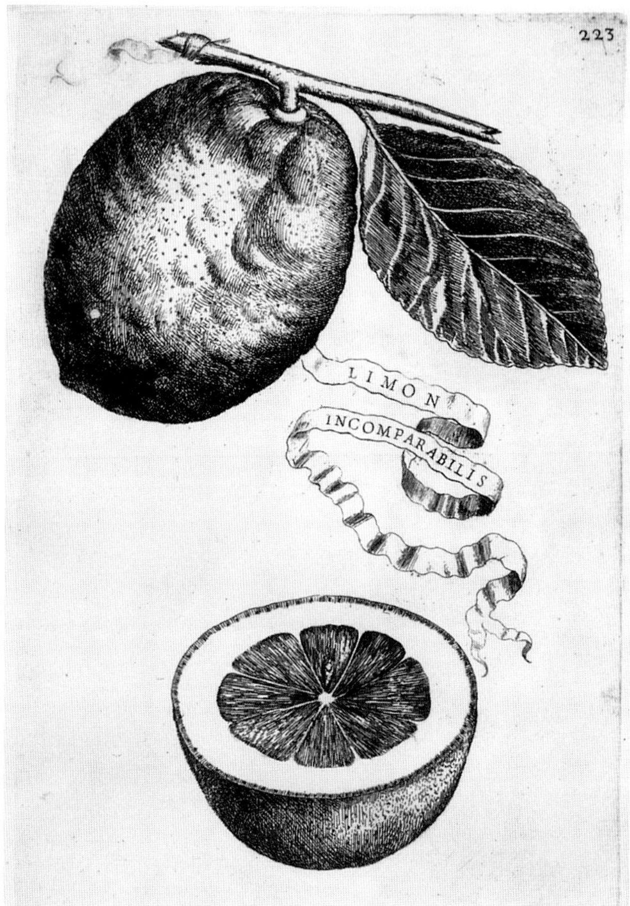

Comp. fig. 44: Ferrari 1646 p. 223

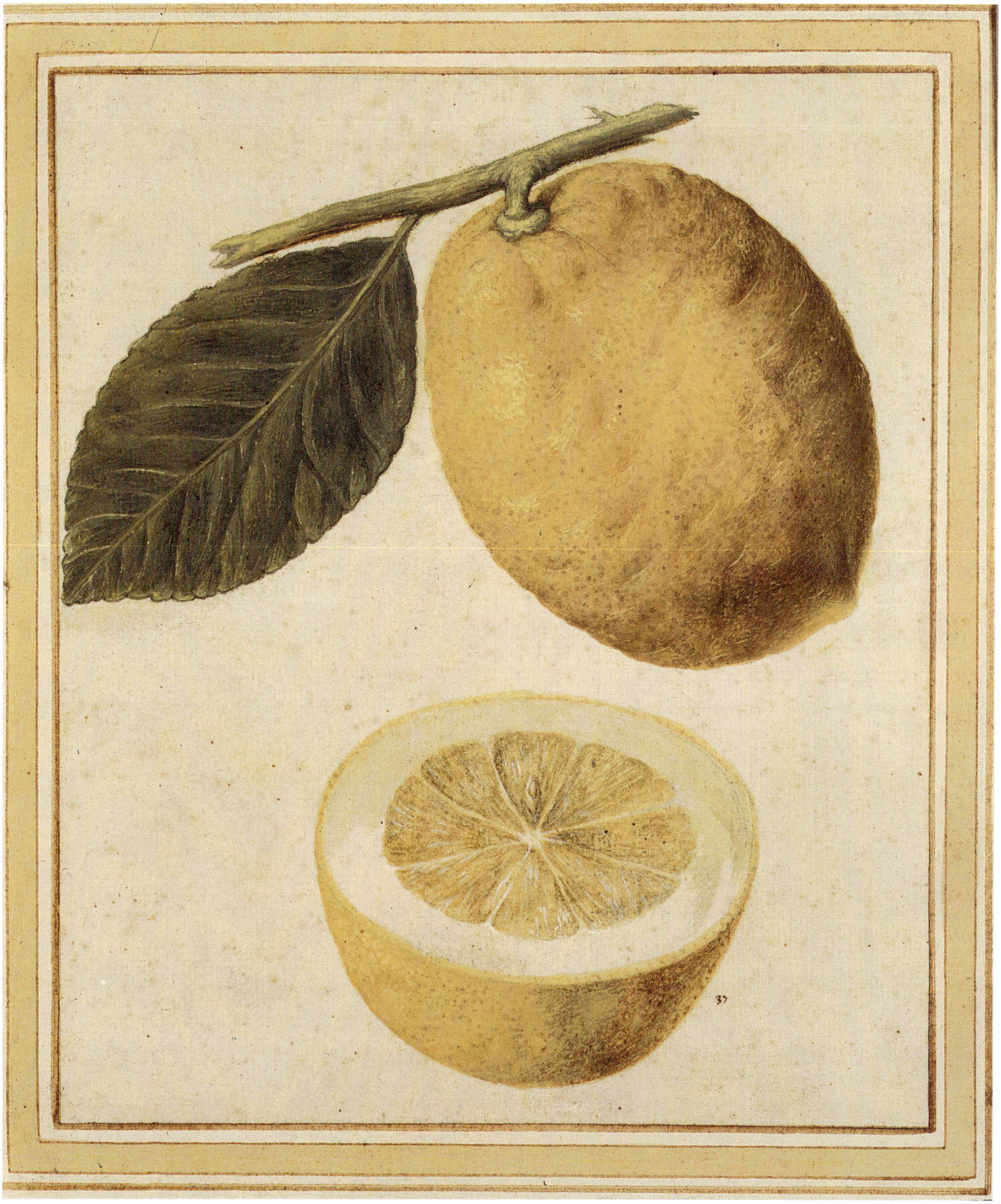

44

45. Citron-lemon, *Citrus limonimedica* Lush.: leafy twig and fruits

?VINCENZO LEONARDI (fl. 1621–46)

Location unknown

Watercolour and bodycolour over black chalk

187 × 193 mm. Trimmed through shadow at right edge

ANNOTATION: *56*

Unlike almost all the other drawings related to pages in the *Hesperides*, this one, which shows the fruit of the *Peretta sive Spatafora duplex* on the lower part of p. 233 of the *Hesperides*, is not reversed (but see also **56**). Aside from the three leaves on either side of the central fruit, and the one just above it and to the left, the drawing also shows a number of other leaves and a flower bud not represented in the etching. The drawing does not show the two whole fruit and the section on the top half of the plate, above the scroll. Nor does it show the half-fruit curiously placed at the bottom of the plate. What exactly all this signifies for the status of the present drawing and its relationship to the final etching is not clear.

The illustration in the *Hesperides* shows two fruiting clusters and two seedless half-fruits of the pear-shaped lemon known as the *Peretta sive Spatafora duplex*. Ferrari explains that the name of this particular cultivar came from the old Sicilian Spatafora family, whose grove housed it. There were in fact three types of *Peretta*, according to Ferrari: the *minor*, the *maior*, and the *striatus*. The smaller type was called the Florentine; the whole fruit could be eaten, tasting rather like a pear, but it did not do the stomach any good. The larger type could in fact grow to a great size; it came from Reggio Calabria, looked beautiful, and was abundant the whole year round. The striated type was much rarer: it came from Naples, but had recently been sent to the Martelletti gardens in Rome, and fruit of this kind were thus sometimes called Martelletti. But it was so pear-like in appearance and taste, he said, that it hardly qualified as a lemon. A fourth type was even further removed from the normal appearance of a lemon. It had a very thin skin indeed, no seeds, a pale colour, and a rather acid taste — and so on and so forth, in the endlessly descriptive manner that is so typical of Ferrari.

The fruiting cluster of Cassiano's drawing shows elliptical, narrow leaves with wingless petioles. The fruits are small, coupled, obovate to pyriform, yellow and hollowed at their apex where a rudimentary style persists.

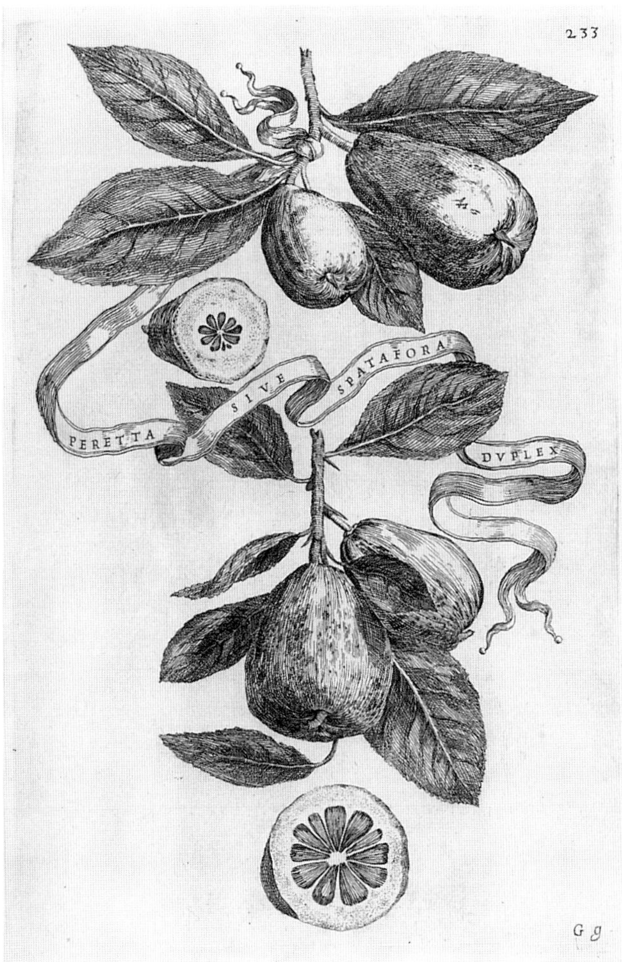

Comp. fig. 45: Ferrari 1646 p. 233

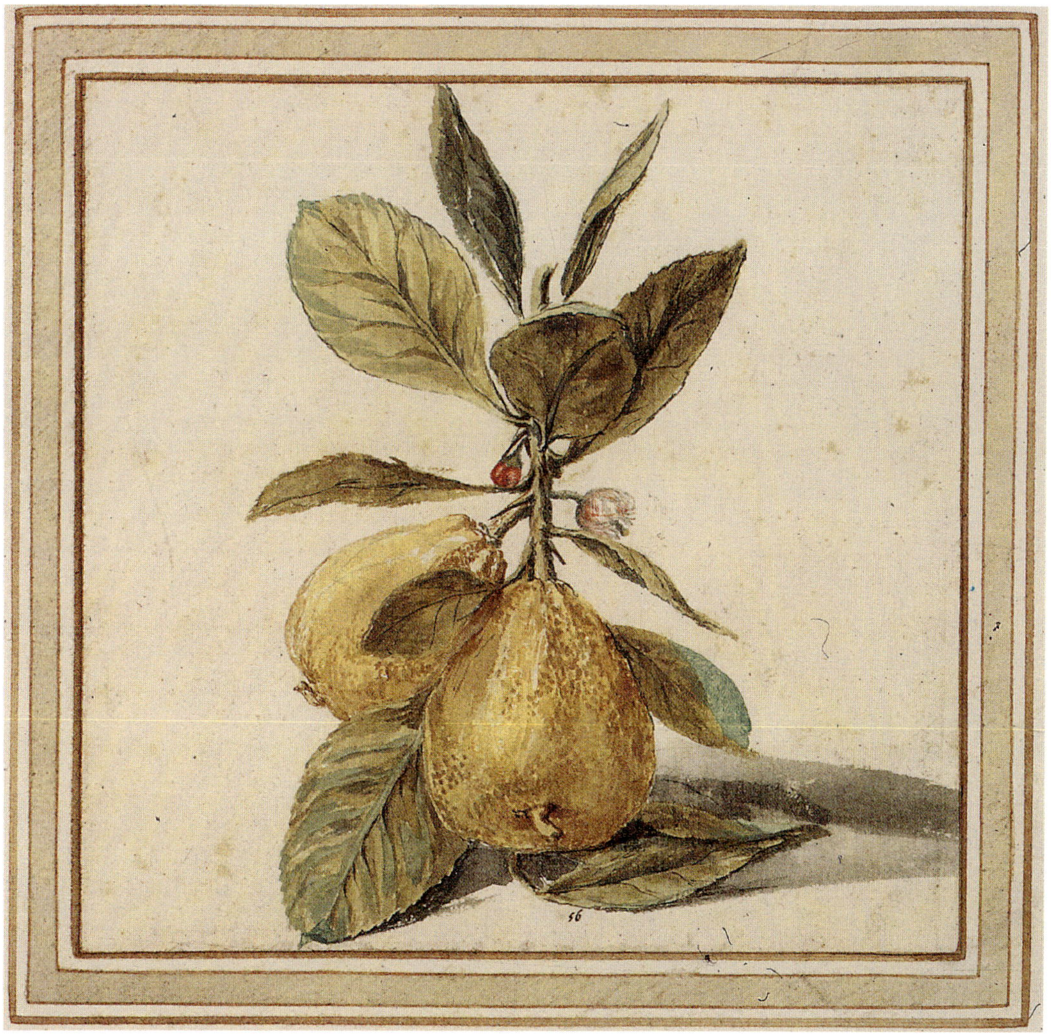

45

In Volkamer's plate 140a a *Limon peretto Spada fora incanellato* (i.e. furrowed) is shown together with a smooth *Perettin Personzin*. Among Micheli's manuscripts a drawing of *Peretta di San Domenico del giardino degli Strozzi*) can be found, apparently the same as the *Limonier Peretta Spatafore* described by Risso and Poiteau and the *Citrus Medica Peretta minor*, *Limone Peretta di San Domenico* by Targioni Tozzetti. Continella and Tribulato (1992, p. 444) described a citron-lemon type named *Piretto* quite similar to the specimen of Cassiano's drawing: DNA analysis (RAPD) has shown that the citron and lemon are the ancestors of this old variety (Deng *et al.* 1995, p. 78).

This drawing is known only from a 35 mm colour slide made by the Courtauld Institute in 1967.

PROVENANCE: dal Pozzo; Albani; George III, by descent to George V; London art market (Mendelson)

REFERENCES: Archivio Dal Pozzo MS 39, fols. 56–56v; Ferrari 1646, p. 231; Volkamer *Nürnberg. Hesp.* I, p. 139, pls. 140a, 140b; Micheli *Enumeratio*, pl. 187; Risso/Poiteau 1818, p. 172, pl. 82; Targioni Tozzetti 1825, p. 50, pl. 42; Continella/Tribulato 1992, p. 444; Deng *et al.* 1995, p. 78

ENGRAVINGS: Ferrari 1646, p. 233

46. Citron-lemon, *Citrus limonimedica* Lush.: whole and half-fruit

VINCENZO LEONARDI (fl. 1621–46)

Private collection, England

Watercolour and bodycolour, with grey wash, over black chalk

250 × 214 mm. Trimmed through shadow at right, where sheet protrudes onto mount sheet. Stained and foxed.
Watermark: none recorded

ANNOTATION: [below whole fruit] *1*

This is the preparatory drawing for the *Limon Spatafora Rheginus* illustrated on p. 241 of Ferrari's *Hesperides.* The name derives from the old Sicilian family of Spatafora, and from the town of Reggio Calabria. Ferrari notes that this 'lemon' is different from the varieties of pear-shaped lemons — popularly known as the *Peretta* or the *Spatafora* — which he discusses in his previous chapter (11) and illustrates on pp. 233, 235 and 237 (see **45**). His description, as always, is concise and to-the-point, but it does not add significantly to what can be judged from the illustration. The skin is pitted and roughish, lightly corrugated, and with ridges and raised tubercles. There are generally twelve segments, somewhat dry and sharp. What is perhaps most exceptional about this cultivar, it seems, is the huge size of the tree, with its deep green leaves and very white flowers.

This drawing shows an ellipsoidal, yellow whole fruit, its half-fruit, and an orbicular broad leaf with wingless petioles. The whole fruit has a rough skin; the half shows a rather thick mesocarp and eleven juicy segments with many seeds. According to Continella/Tribulato 1992 (p. 443), a cultivar named *Limone cedrato Spadafora* is still present sporadically in north-eastern Sicily.

PROVENANCE: dal Pozzo; Albani; George III, by descent to George V; London art market (Mendelson); James R. Herbert Boone; Sotheby's (New York), 16 September 1988, lot 148; Jonathan Harris, London

REFERENCES: Ferrari 1646, p. 239; Bimbi, Villa Med. C. 597; Volkamer *Nürnberg. Hesp.* II, p. 128f., pl. 130; Risso/Poiteau 1818, p. 172; Continella/Tribulato 1992, p. 443

ENGRAVINGS: Ferrari 1646, p. 241

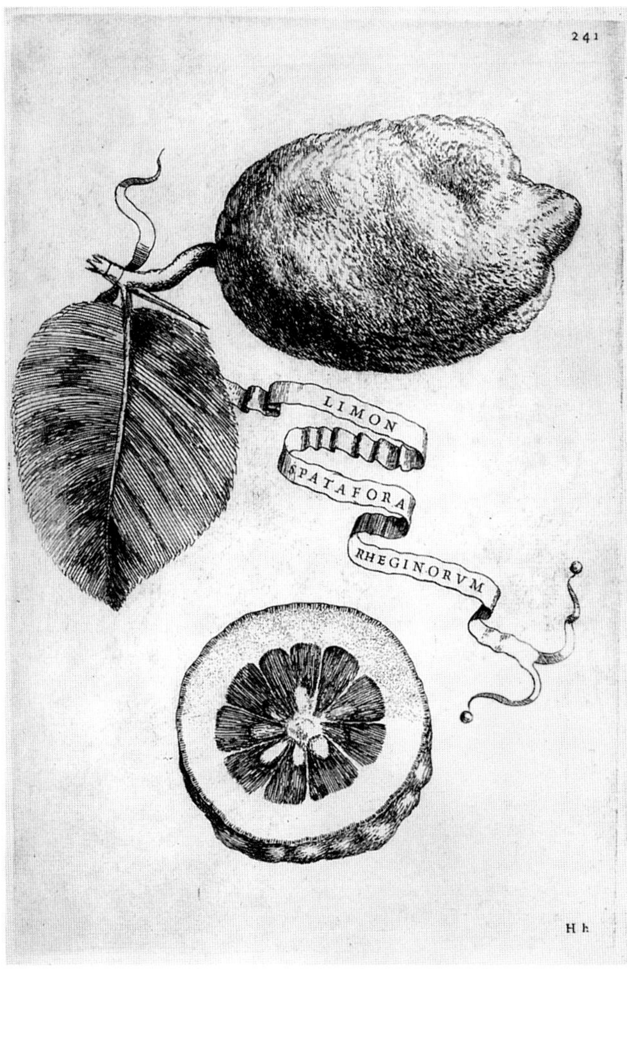

Comp. fig. 46: Ferrari 1646 p. 241

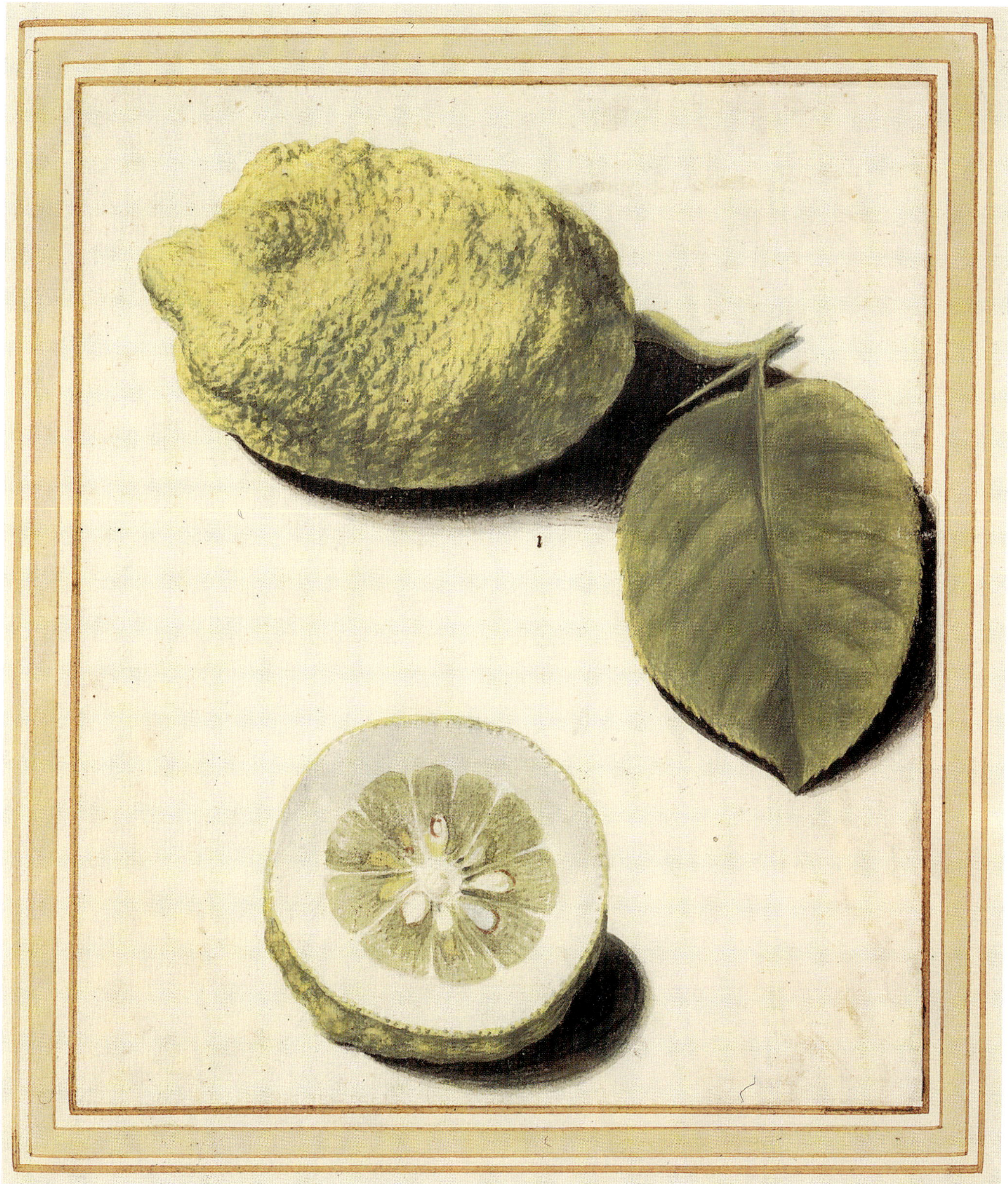

46

47. Citron-lemon, *Citrus limonimedica* Lush.: fruit cluster and half-fruit

VINCENZO LEONARDI (fl. 1621–46)

Private collection, Italy

Watercolour and bodycolour over black chalk; rubbed at top

364 × 270 mm. Badly stained. *Watermark*: none recorded

ANNOTATION: [below half-fruit] *HH*

This is the preparatory drawing for the *Limon racemosus* on p. 243 of Ferrari's *Hesperides*. 'With constant playfulness, the lemon sometimes simulates the appearance of the vine', wrote Ferrari at the head of his small chapter on this cultivar. The branches produce clusters of fruit — hence the comparison. Sometimes there are even five fruit on each cluster. The fruits, he says, incline to orange, although it does not appear so from the present rather fine drawing. The tree is very similar to that of the *lumia* from Valencia. Ferrari noted that he himself had gathered this kind of fruit, with the result that his description was unusually accurate. As in the case of the *Pomum Paradisi* discussed below in **56**, the cross-section of this fruit shows a very thick mesocarp (nearly 100% of the fruit), while segments and seeds are lacking. The *Limon racemosus* thus looks more like a citron-lemon rather than a lemon, as its name would suggest. Ferrari judged it 'moderately sweet'.

PROVENANCE: dal Pozzo; Albani; George III, by descent to George V; London art market (Mendelson); James R. Herbert Boone; Sotheby's (New York), 16 September 1988, lot 152

REFERENCES: Ferrari 1646, p. 239; Volkamer *Nürnberg. Hesp.* II, p. 75f., pls. 76–7; Risso/Poiteau 1818, p. 183, pl. 90

ENGRAVINGS: Ferrari 1646, p. 243

Comp. fig. 47: Ferrari 1646 p. 243

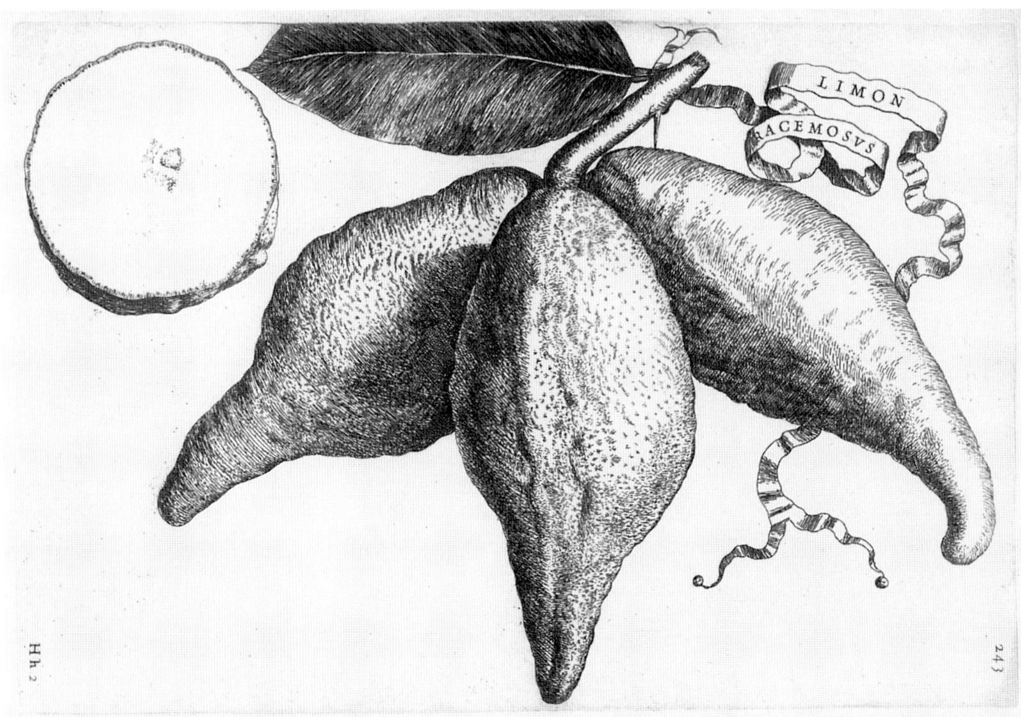

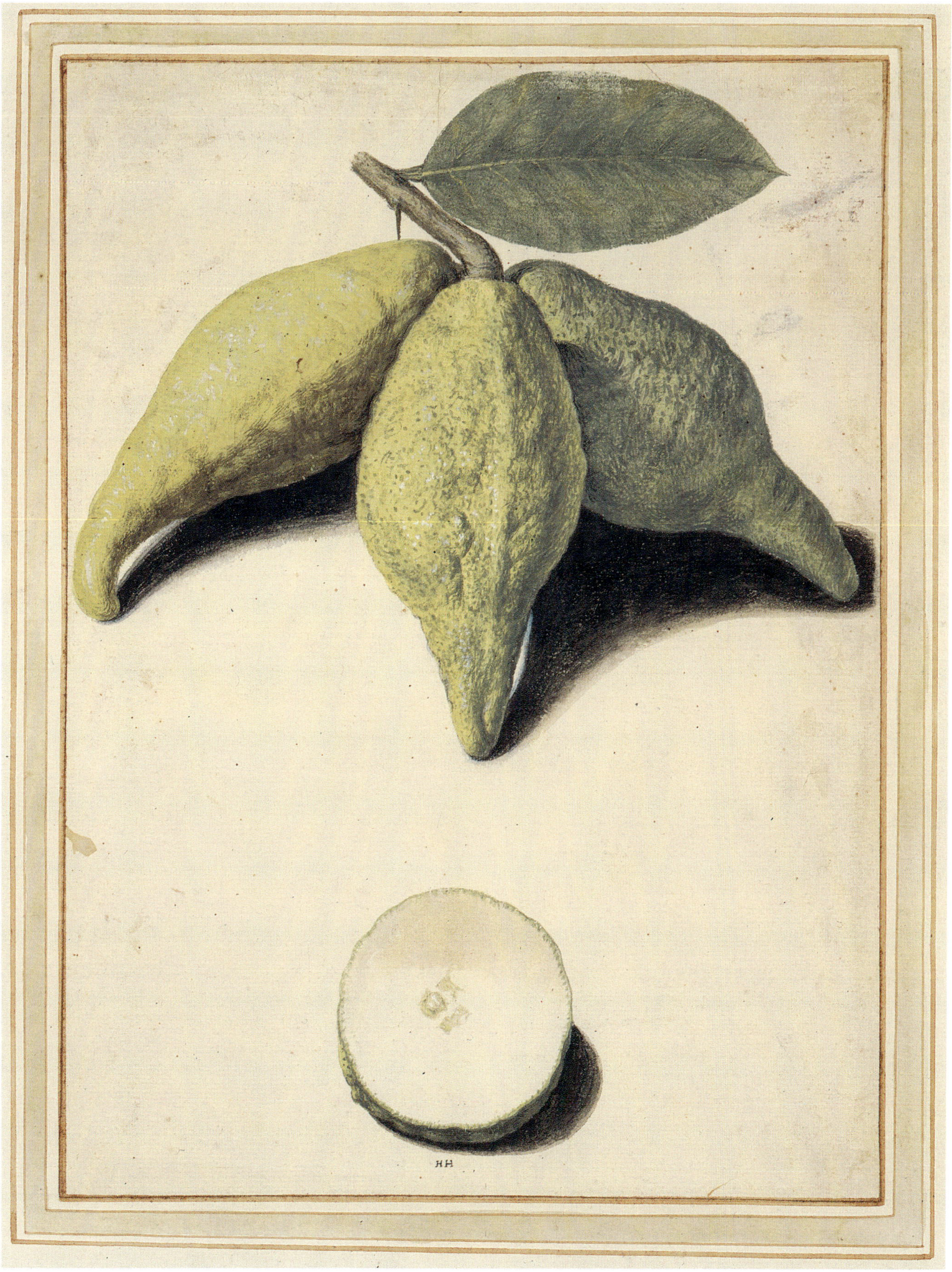

47

48. Citron-lemon, *Citrus limonimedica* Lush.: whole and half-fruit

?VINCENZO LEONARDI (fl. 1621–46)

Windsor, RL 19364

Watercolour and bodycolour over black chalk

262 × 195 mm. Cream prepared paper. Survival of 5 mm overlay of original 17th-century mount at left-hand side. Lower left corner streaked after mounting. *Watermark*: Figure, same group as Figure 60 (cut)

ANNOTATION: *80*

MOUNT SHEET: 587 × 434 mm. Stained and marked, with streaks across lower left. *Watermark*: D & C Blauw

This is the preliminary drawing for the first of Ferrari's citrated lemons, of the group which he calls the smoother kind. These were known as *cedrini* in Pietrasanta; others grew in the region of Florence. According to Ferrari, the citrated lemons derived their name from the fact that their aroma and other characteristics showed similarities to the citron group. Generally such fruits were oblong, nippled, and often pregnant with other fruits. This phenomenon was one which greatly interested Ferrari, and which inspired the central 'fables' of the *Hesperides* (see **72**, **81** and **90**, and pp. 62, 74).

The present drawing, used on p. 265 of Ferrari's *Hesperides*, showing the *Limon citratus primae notae laevior*, is of a large, ellipsoidal, long-nippled fruit with a smooth yellow peel. The fruit hangs from a thorny twig featuring two wide, narrow and serrate leaves with wingless petioles. The half-fruit has a rather thick mesocarp and yellow seedless flesh. This citron-lemon could conceivably be the *Cédratier de Rome* or the *Cédratier de Florence* described by Risso and Poiteau (1818, pp. 202–4), or the *Cedratello di Firenze* described by Targioni Tozzetti (1825, p. 26) and Gallesio (1811, p. 103).

PROVENANCE: dal Pozzo; Albani; George III, by descent to George V; London art market (Mendelson); Rex Nan Kivell; Her Majesty The Queen

REFERENCES: Ferrari 1646, p. 263f.; Volkamer *Nürnberg. Hesp.* I, p. 123, pl. 124; Gallesio 1811, p. 103; Risso/Poiteau 1818, pp. 202–4, pls. 102 and 104; Targioni Tozzetti 1825, p. 26, pl. 25; unpublished drawing on loose, unnumbered folio, GAG; Baldini 1996b, p. 12, pl. 2; Freedberg 1996a

ENGRAVINGS: Ferrari 1646, p. 265

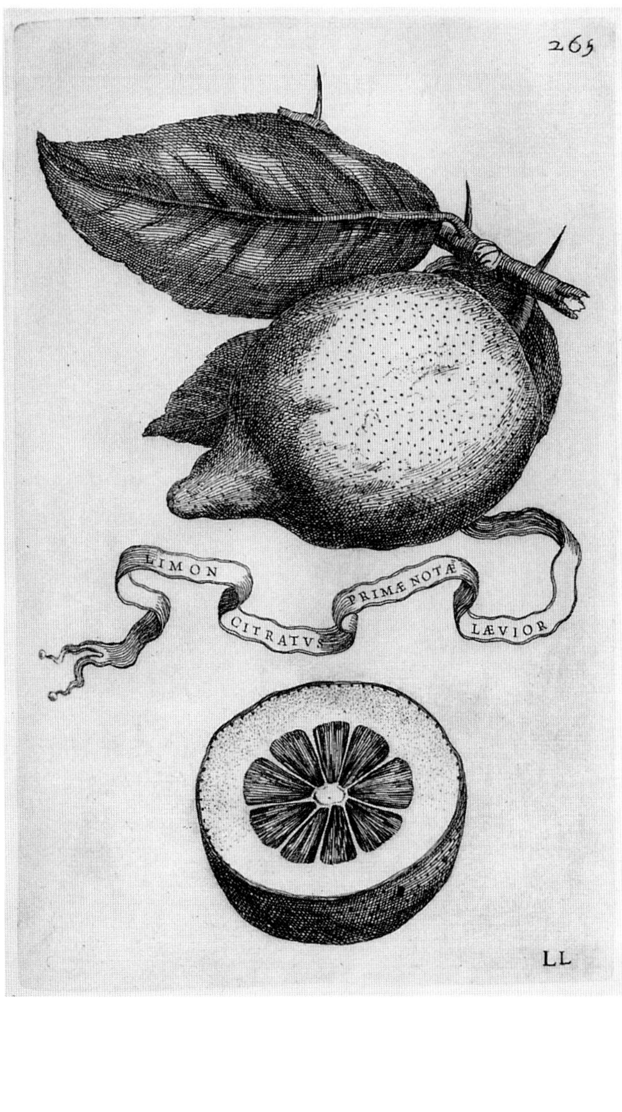

Comp. fig. 48: Ferrari 1646 p. 265

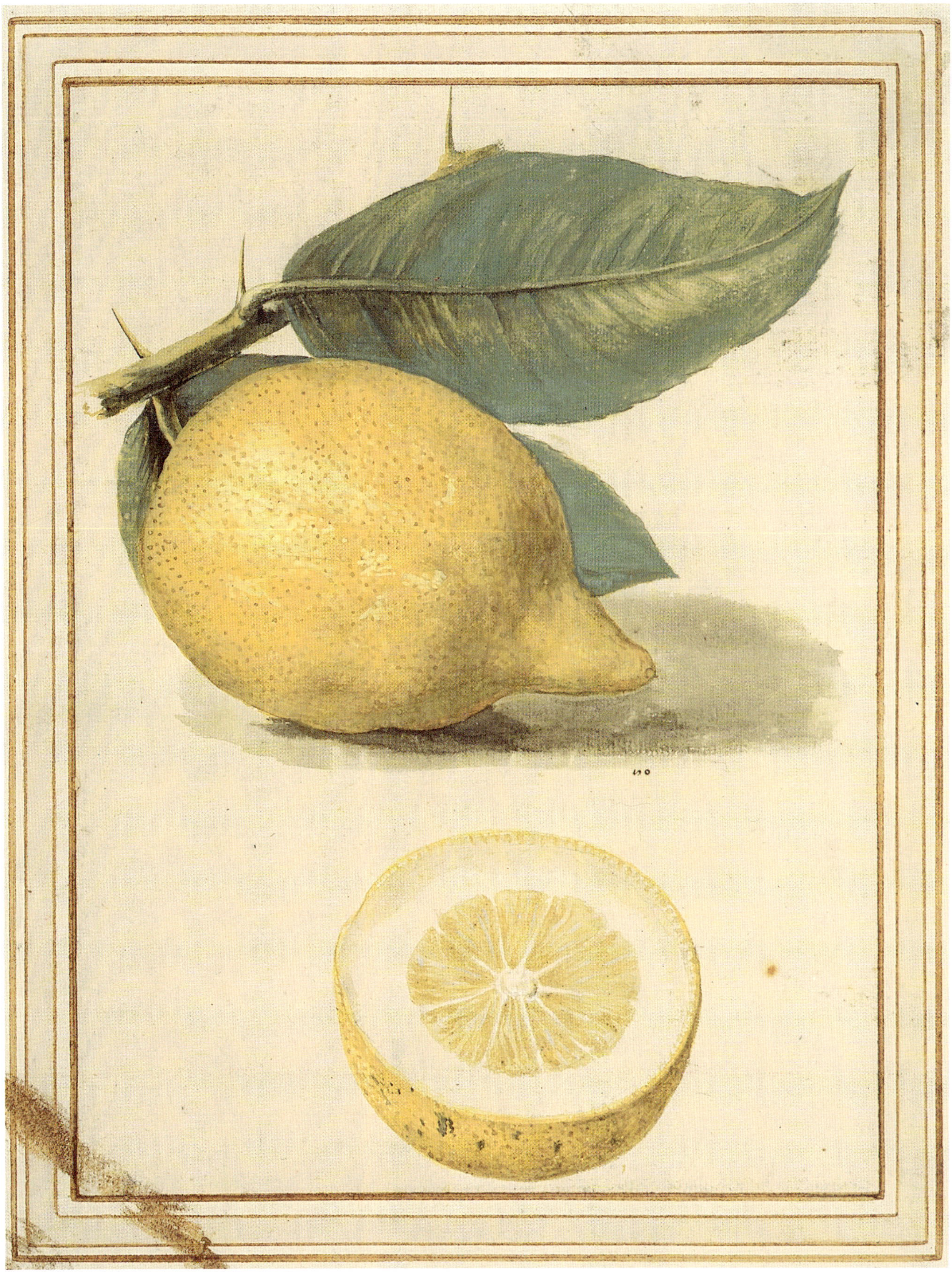

49. Citron-lemon, *Citrus limonimedica* Lush.: whole fruit and thorny, leafy twig

?VINCENZO LEONARDI (fl. 1621–46)

Private collection, U.S.A.

Watercolour and bodycolour over black chalk; scroll in black chalk

173 × 186 mm. Trimmed at lower edge, right edge ?and top.
Watermark: Figure ?60 (cut)

ANNOTATION: *81*

MOUNT SHEET: 585 × 432 mm

Unlike the group of citrated lemons discussed in the previous entry, this drawing, for the *Limon citratus primae notae scaber* illustrated in Ferrari's *Hesperides* p. 267, shows the rough kind of citrated lemon. It shows an irregularly shaped, furrowed, nipple-less yellow fruit and three elliptic-lanceolate leaves with wingless petioles. Ferrari wrote that its fruit was generally wrinkled (*verrucosus*) and often occurred in pregnant form. Its tree was able to flower all year long (*semperflorens*).

This is one of the few drawings in the surviving corpus of drawings made for the *Hesperides* to show the elegant scroll that accompanied each of the drawings in the book itself (for other examples, see **20**, **29** and **31**), providing a decorative space for the name given to each fruit by Ferrari.

PROVENANCE: dal Pozzo; Albani; George III, by descent to George V; London art market (Mendelson); James R. Herbert Boone; Sotheby's (New York), 16 September 1988, lot 142; W.M. Brady and Co., Inc.

REFERENCES: Ferrari 1646, pp. 263–4

ENGRAVINGS: Ferrari 1646, p. 267

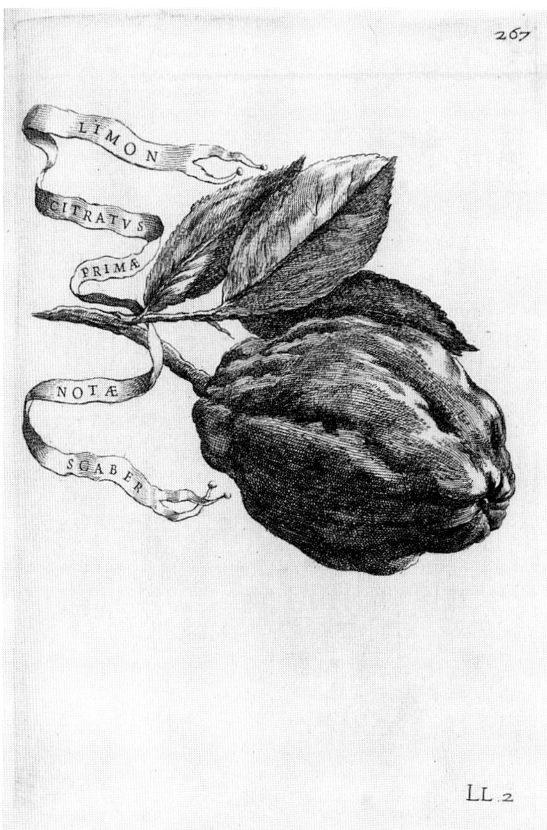

Comp. fig. 49: Ferrari 1646 p. 267

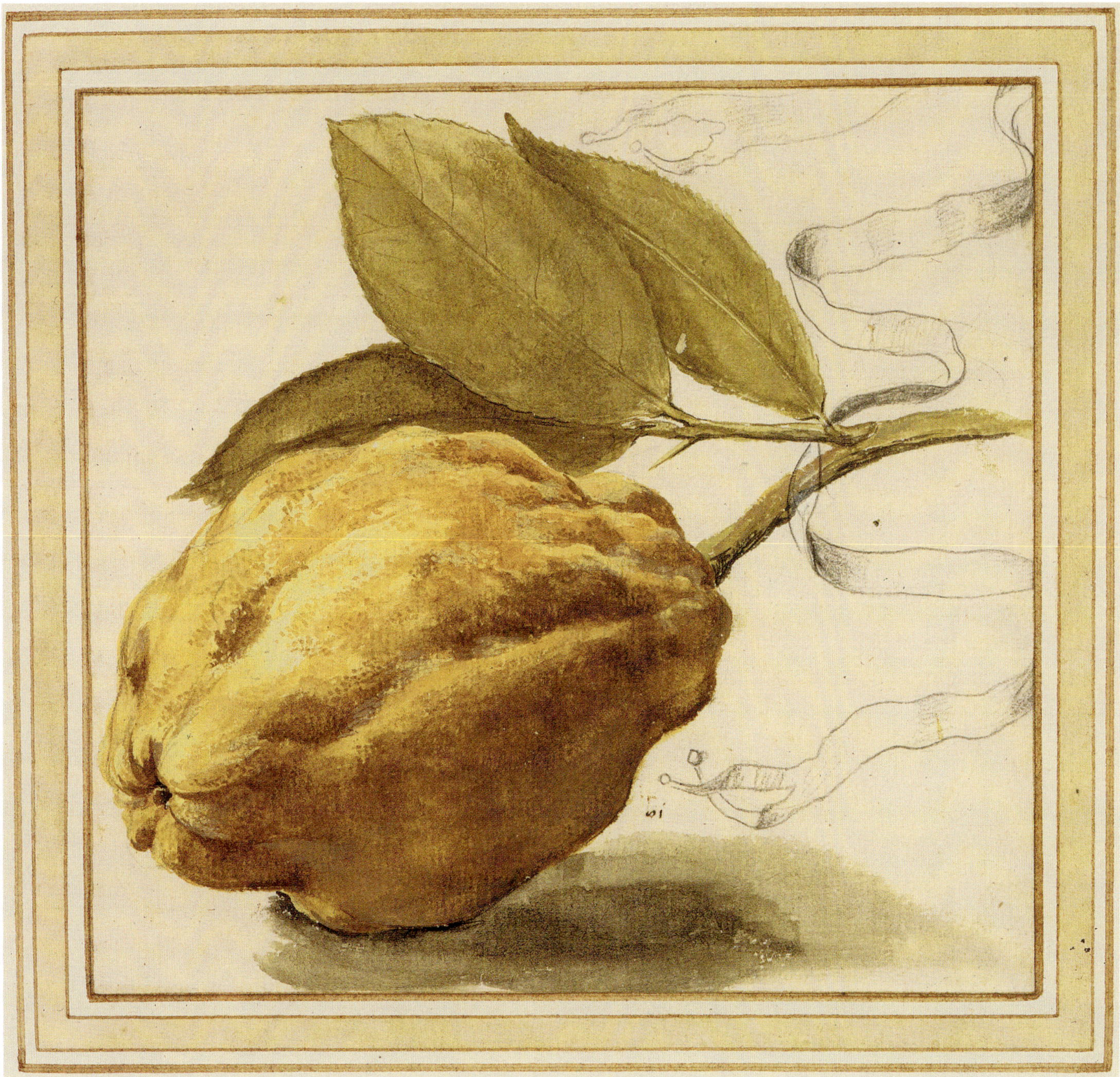

49

50. Citron-lemon, *Citrus limonimedica* Lush.: whole fruit

VINCENZO LEONARDI (fl. 1621–46)

Windsor, RL 19342

Watercolour and bodycolour, with gum heightening, over black chalk

105 x 139 mm. Cream prepared paper

MOUNT SHEET: 586 × 435 mm. Mounted on same sheet as **51**. Stained and marked, red ?ink streak across lower right. *Watermark*: D & C Blauw

This is the preparatory drawing for the whole ellipsoid fruit of the *Limon citratus Amalphitanus* illustrated on p. 281 of Ferrari's *Hesperides*. Ferrari briefly describes it in his chapter on the second category of citrated lemons, the *Limon citratus secundae notae* (p. 279). It takes its name from the town of Amalfi, and differs only slightly in size from the first group of citron-lemons. It is sessile, and has a yellow and rather rough skin. Its bitter taste and smell are milder than the others; although it has very similar leaves and the same number of seeds, the tree itself is thornier. To modern botanists, in fact, the differences are very slight.

After his unusually short discussion of this type, Ferrari moves on to examples of citron-lemons (or what he thinks are citron-lemons) which he knew in Rome.

PROVENANCE: dal Pozzo; Albani; George III, by descent to George V; London art market (Mendelson); Rex Nan Kivell; Her Majesty The Queen

REFERENCES: Ferrari 1646, p. 279; Risso/Poiteau 1818, p. 189, pl. 93; De Noter 1926, p. 140f.

ENGRAVINGS: Ferrari 1646, p. 281

51. Citron-lemon, *Citrus limonimedica* Lush.: half-fruit

VINCENZO LEONARDI (fl. 1621–46)

Windsor, RL 19341

Watercolour and bodycolour, with gum heightening, over black chalk

103 × 121 mm. Cream prepared paper

ANNOTATION: [below fruit] *xxiii*

MOUNT SHEET: mounted on same sheet as **50**

This is the half-fruit of the fruit described in the preceding entry. It is thus the preparatory drawing for the lower part of the illustration of the *Limon citratus Amalphitanus* illustrated on p. 281 of the *Hesperides*. It shows a half-fruit with a relatively thick rind and ten seedless, yellow segments.

PROVENANCE: dal Pozzo; Albani; George III, by descent to George V; London art market (Mendelson); Rex Nan Kivell; Her Majesty The Queen

REFERENCES: Ferrari 1646, p. 279; Risso/Poiteau 1818, p. 189, pl. 93; De Noter 1926, p. 140f.

ENGRAVINGS: Ferrari 1646, p. 281

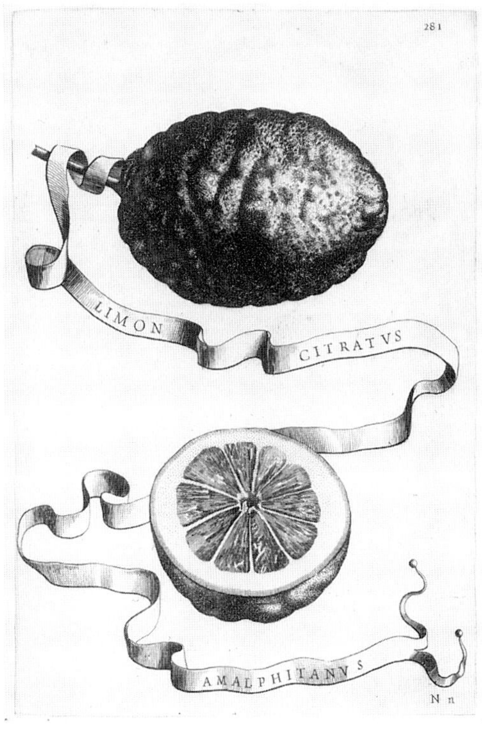

Comp. fig. 50, 51: Ferrari 1646 p. 281

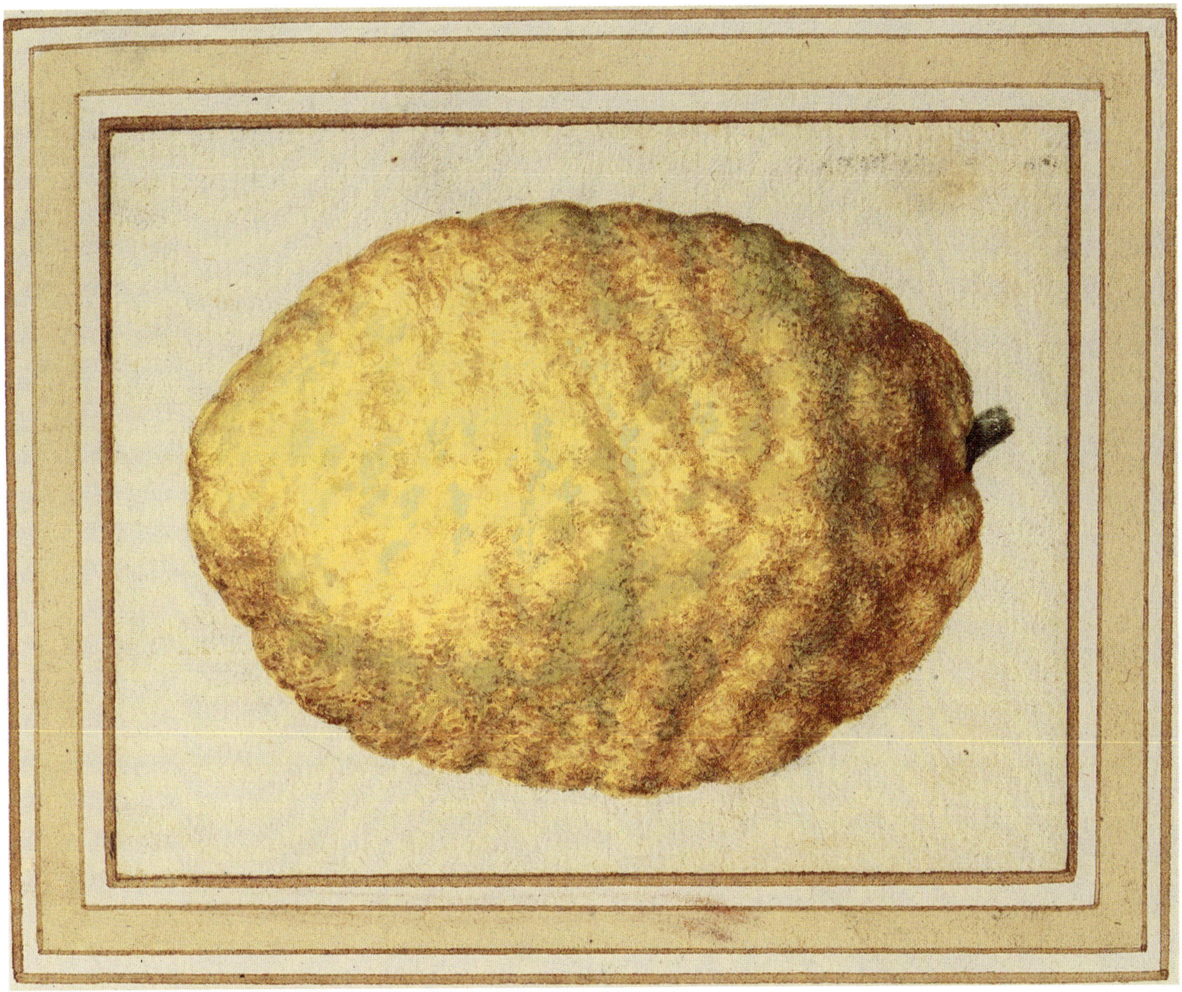

50

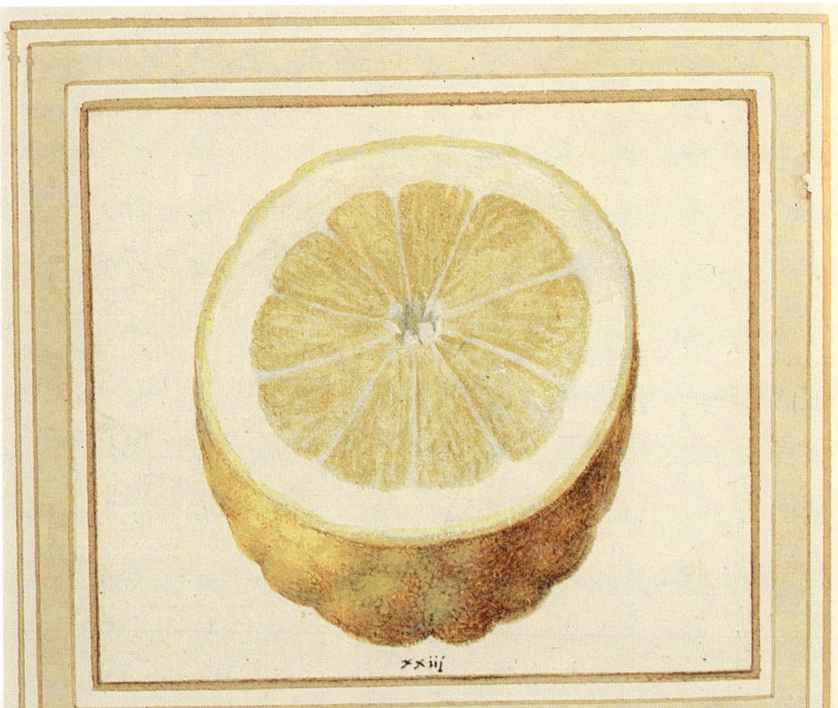

51

52. Citron-lemon, *Citrus limonimedica* Lush.: whole fruit and leaf

VINCENZO LEONARDI (fl. 1621–46)
Location unknown
Watercolour and bodycolour ?over black chalk
152 × 154 mm. Foxed and stained. *Watermark*: none recorded
MOUNT SHEET: 585 × 432 mm. Mounted on same sheet as **53**

This is the preparatory drawing for the *Limon citratus in coenobio Teresiano trans Tyberim* shown on p. 283 of Ferrari's *Hesperides*. It is a small, oblong, pointed citron-lemon *secundae notae*. Ferrari gives no particular reason for including this citron-lemon, grown by the Discalced Carmelite Fathers of Santa Teresa in Trastevere, although he describes it with his usual minute care — smooth skin, moderately golden, slender, a little sweet, the flesh somewhat hard, of diluted sulphurous colour, with about nine segments, very few and very small seeds. This particular specimen is a yellow, cylindrical-apiculate fruit with an elliptical and remarkably toothed leaf (a fact commented on by Ferrari as well), with a wingless petiole.

The specimen looks similar to the modern cultivar *Interdonato*, which currently attracts interest because of its tolerance to 'mal secco' disease (*Phoma tracheiphila*).

PROVENANCE: dal Pozzo; Albani; George III, by descent to George V; London art market (Mendelson); James R. Herbert Boone; Sotheby's (New York), 16 September 1988, lot 150

REFERENCES: Ferrari 1646, pp. 279–80; Bimbi, Villa Med. C. 597; Risso/Poiteau 1818, p. 189, pl. 93; De Noter 1926, p. 140f.

ENGRAVINGS: Ferrari 1646, p. 283

53. Citron-lemon, *Citrus limonimedica* Lush.: half-fruit

VINCENZO LEONARDI (fl. 1621–46)
Location unknown
Watercolour and bodycolour over black chalk
99 × 98 mm. *Watermark*: none recorded
ANNOTATION: *H*
MOUNT SHEET: mounted on same sheet as **52**

This is the drawing for the half-fruit of the citron-lemon grown in the monastery of Santa Teresa in Trastevere and illustrated on p. 283 of the *Hesperides*. It is now mounted on the same sheet as the drawing of the whole fruit (**52**); both probably once formed part of the same sheet, were separated, and then mounted together in the 18th-century rearrangement of the drawings.

The cross-sectioned fruit shows a moderately thick mesocarp and juicy yellow flesh. There are no seeds in its segments.

PROVENANCE: dal Pozzo; Albani; George III, by descent to George V; London art market (Mendelson); James R. Herbert Boone; Sotheby's (New York), 16 September 1988, lot 150

REFERENCES: Ferrari 1646, p. 279; Risso/Poiteau 1818, p. 189, pl. 93; De Noter 1926, p. 140f.

ENGRAVINGS: Ferrari 1646, p. 283

54. Citron-lemon, *Citrus limonimedica* Lush.: flowering twig and fruits

VINCENZO LEONARDI (fl. 1621–46)

Windsor, RL 19372

Watercolour and bodycolour, with gum heightening, over graphite

258 × 174 mm. Cream prepared paper

ANNOTATION: *ggg*

MOUNT SHEET: 587 × 434 mm. Small tears and stains.
Watermark: D & C Blauw.
Annotation: [verso: pencil, in a 20th-century hand] *7*

The last of the four fruits discussed by Ferrari in his chapter on citrated lemons *secundae notae* is the small type that could once be seen in the garden of Ottavio Roberti, close to the Porta Flaminia in Rome. It usually weighed about nine ounces, was lightly striped and oblong. At first its cortex was reddish, then green, then a palish yellow. It had a strong aroma and was often wrinkly. Its flesh was very acid. Its leaf was that of a citron, but narrower; its flower was more like that of a *citrinum*. The fruits themselves were wild (*silvestres*), in the sense that they were both very sour and very hard and small. These citron-lemons were usually propagated by seed and not by grafting.

The present preparatory drawing for the *Limon citratus silvestris*, illustrated on p. 287 of the *Hesperides*, is one of the most attractive in the surviving corpus of Cassiano's citrus drawings. Unusually, it also shows the flower along with two fruits, the branch and five leaves. Different stages of fruit development are also represented. The ripe whole fruit is cylindrical-apiculate, with an evenly rounded base and a smooth yellow peel. The half-fruit shows a rather thick mesocarp and some seeds. The leaves are elliptical; their petioles are wingless.

PROVENANCE: dal Pozzo; Albani; George III, by descent to George V; London art market (Mendelson); Rex Nan Kivell; Her Majesty The Queen

LITERATURE: Freedberg 1989b, pp. 54–60, pl. 22

REFERENCES: Ferrari 1646, p. 280

ENGRAVINGS: Ferrari 1646, p. 287

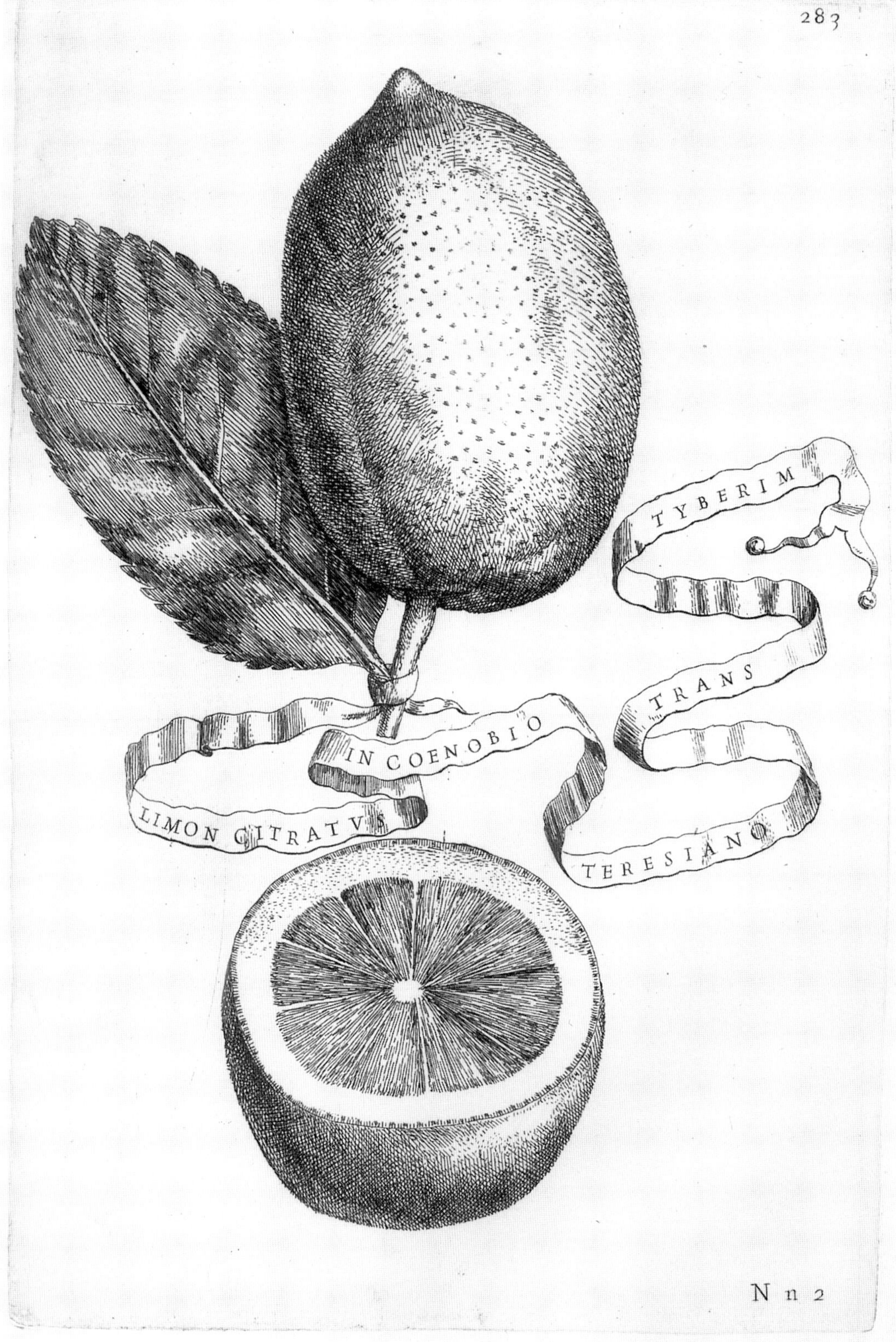

Comp. fig. 52, 53: Ferrari 1646 p. 283

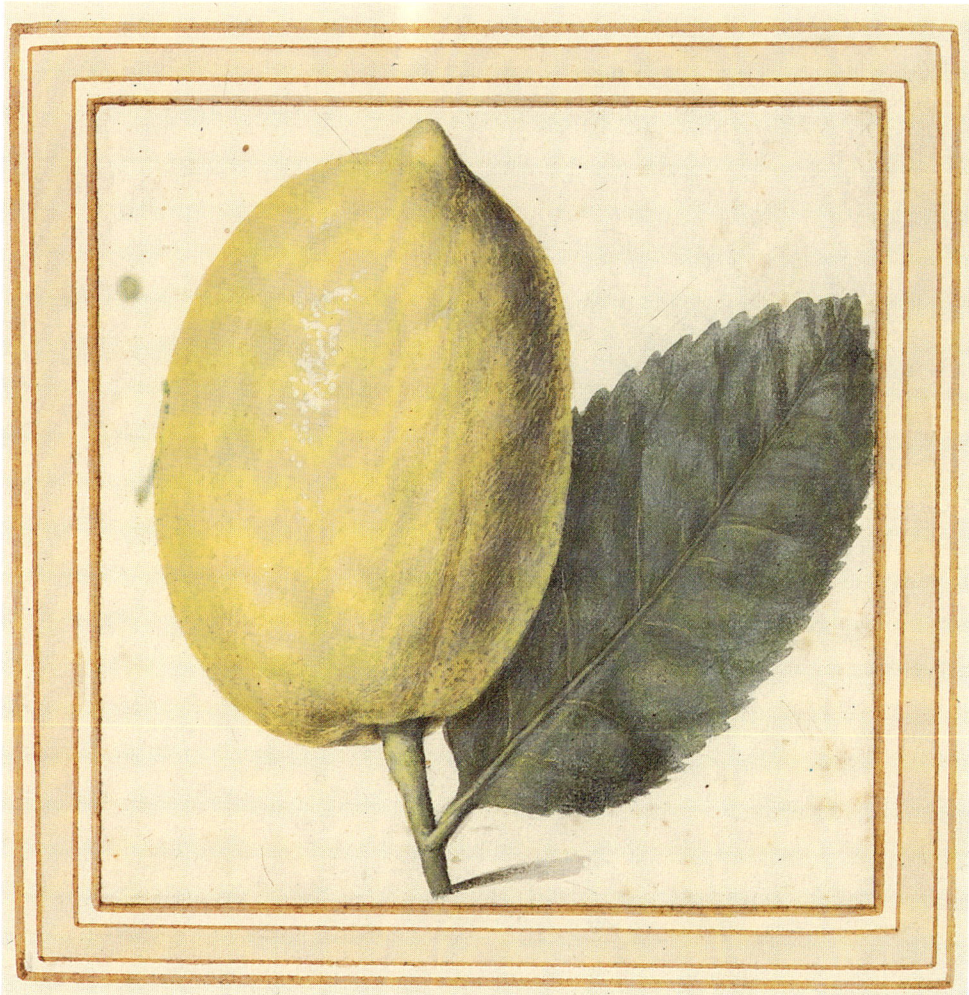

52

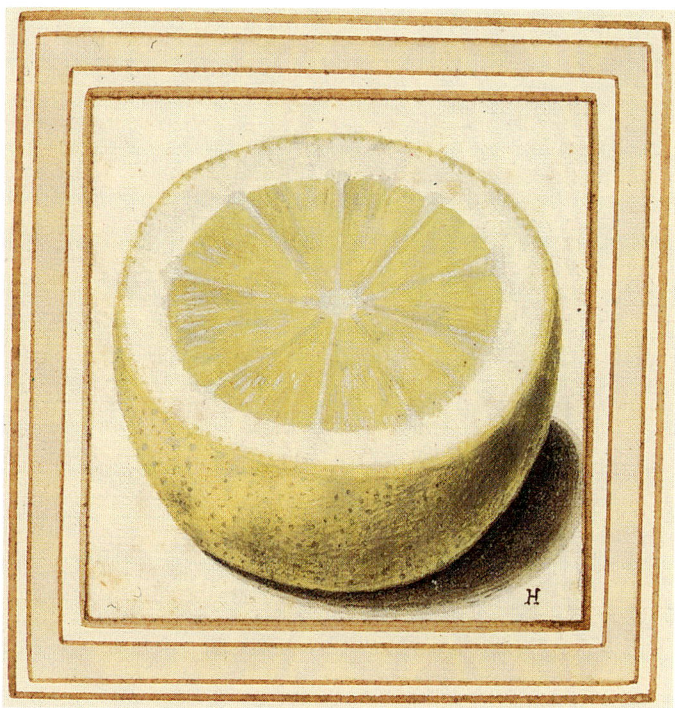

53

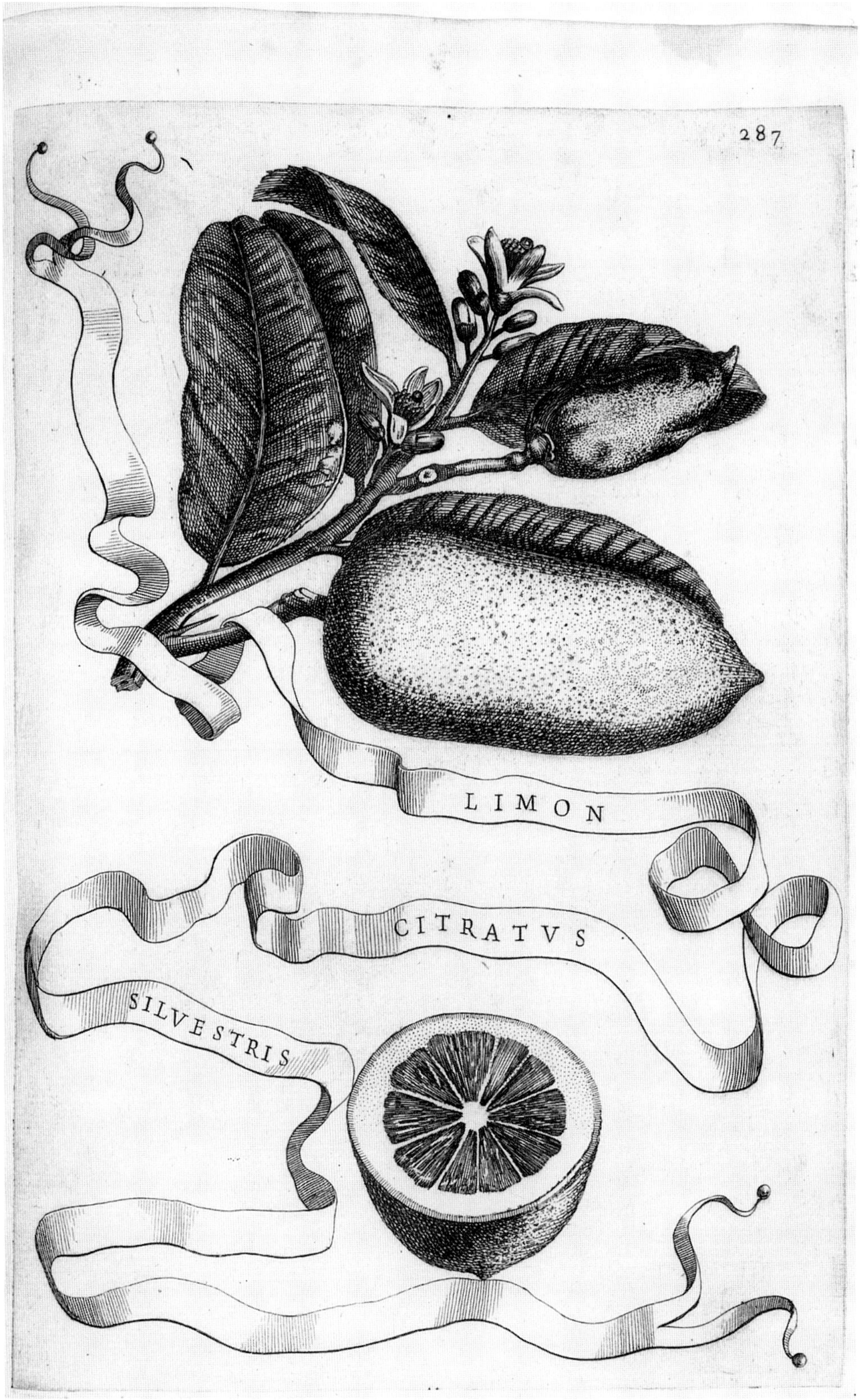

Comp. fig. 54: Ferrari 1646 p. 287

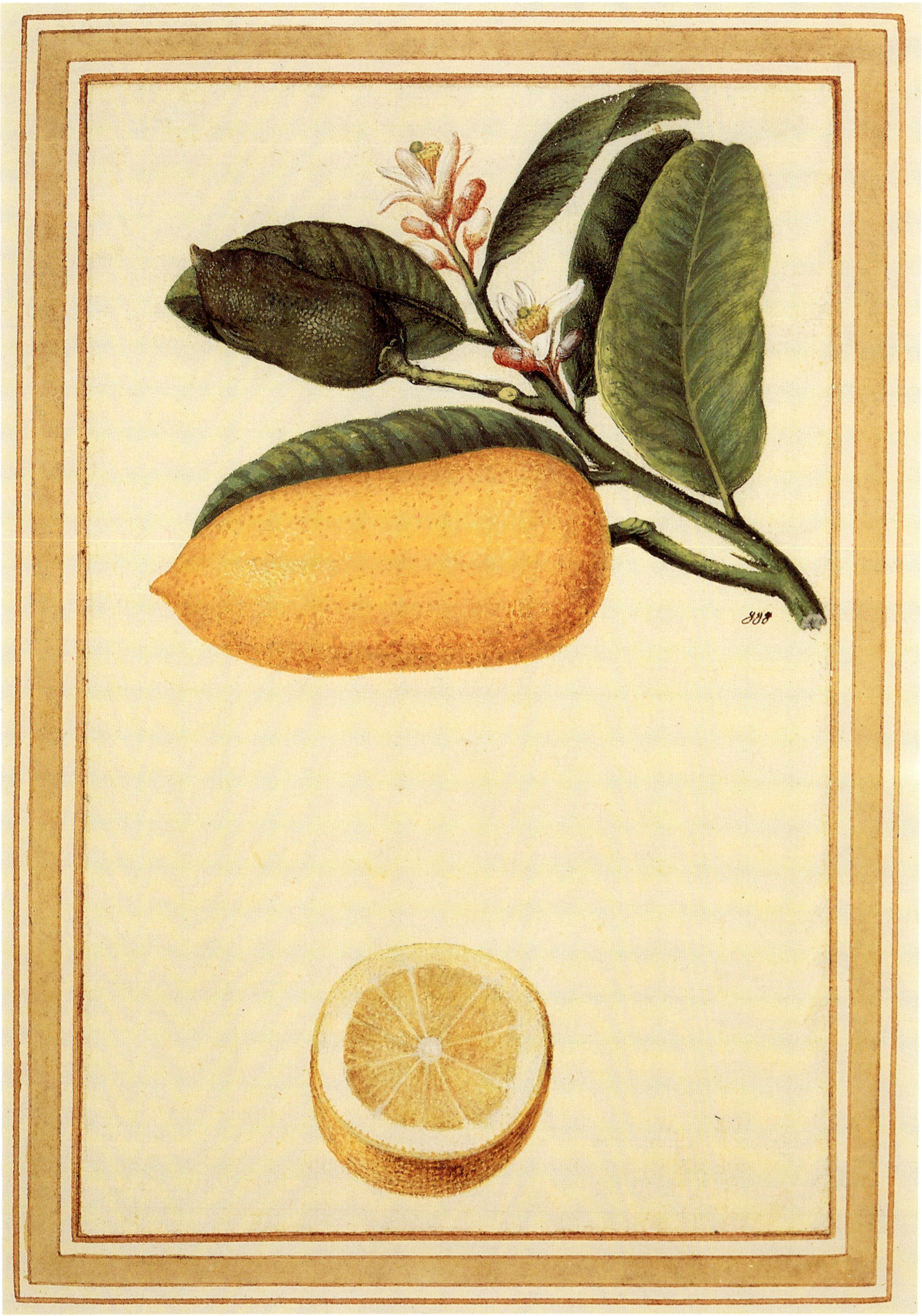

55. Citron-lemon, *Citrus limonimedica* Lush.: whole and half-fruit

?VINCENZO LEONARDI (fl. 1621–46)

Windsor, RL 19353

Watercolour and bodycolour, with gum heightening, over black chalk

302 × 195 mm. Cream prepared paper. Survival of 5 mm overlay of original 17th-century mount at top edge. *Watermark:* Figure, same group as Figure 60 (cut)

ANNOTATION: *39*

MOUNT SHEET: 585 × 435 mm. Stained and marked, with crease at top right corner and tear at right-hand side. *Watermark*: D & C Blauw

On pp. 289–90 of the *Hesperides* Ferrari writes about five or six different kinds of *ponzini*, the name which in Italy was given, as Ferrari puts it, to rather large round lemons with an extremely pleasing taste (see also pp. 52, 63). The *ponzinus* grown in the gardens of the Mattei in Rome was almost round, rather like a *lumia*, slightly papillate, with a most pleasant taste. It was especially sweet and tender in June. As usual, Ferrari makes detailed colouristic and textural observations — a smoothish skin, growing from pale yellow to deeper yellow. The flesh was divided into around nine segments, with few seeds. The tree was much more like that of a *lumia*, although it had fewer thorns. There were some who thought that this *ponzinus* was raised from a sweet lime grafted onto a lemon, thus suggesting a morphogenetic effect of the rootstock on the scion.

This preparatory drawing for the *Limon Ponzinus in hortis Matthaeiorum* was used for the illustration of the fruit on p. 295 of Ferrari's *Hesperides*. The whole fruit is represented as ellipsoid-globose and with a yellow, finely papillate skin; the half-fruit shows a thick mesocarp and a little flesh with seeds. The leaves are rather large and wide, broadly elliptical and serrate, with wingless petioles.

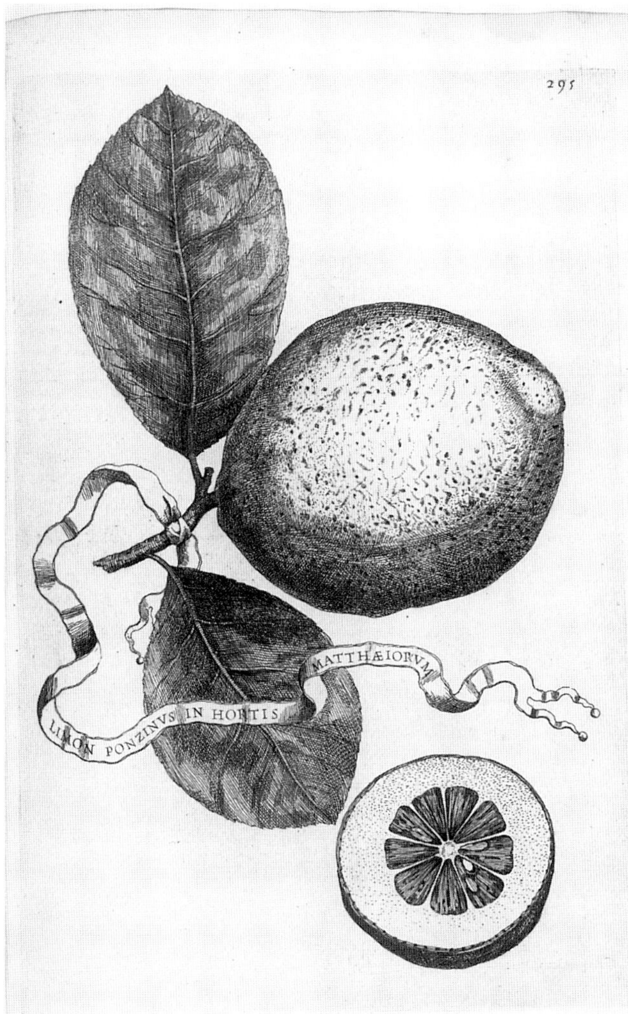

Comp. fig. 55: Ferrari 1646 p. 295

PROVENANCE: dal Pozzo; Albani; George III, by descent to George V; London art market (Mendelson); Rex Nan Kivell; Her Majesty The Queen

REFERENCES: Archivio Dal Pozzo MS 39, fols. 92–92v; Ferrari 1646, p. 289; De Noter 1926, p. 127

ENGRAVINGS: Ferrari 1646, p. 295

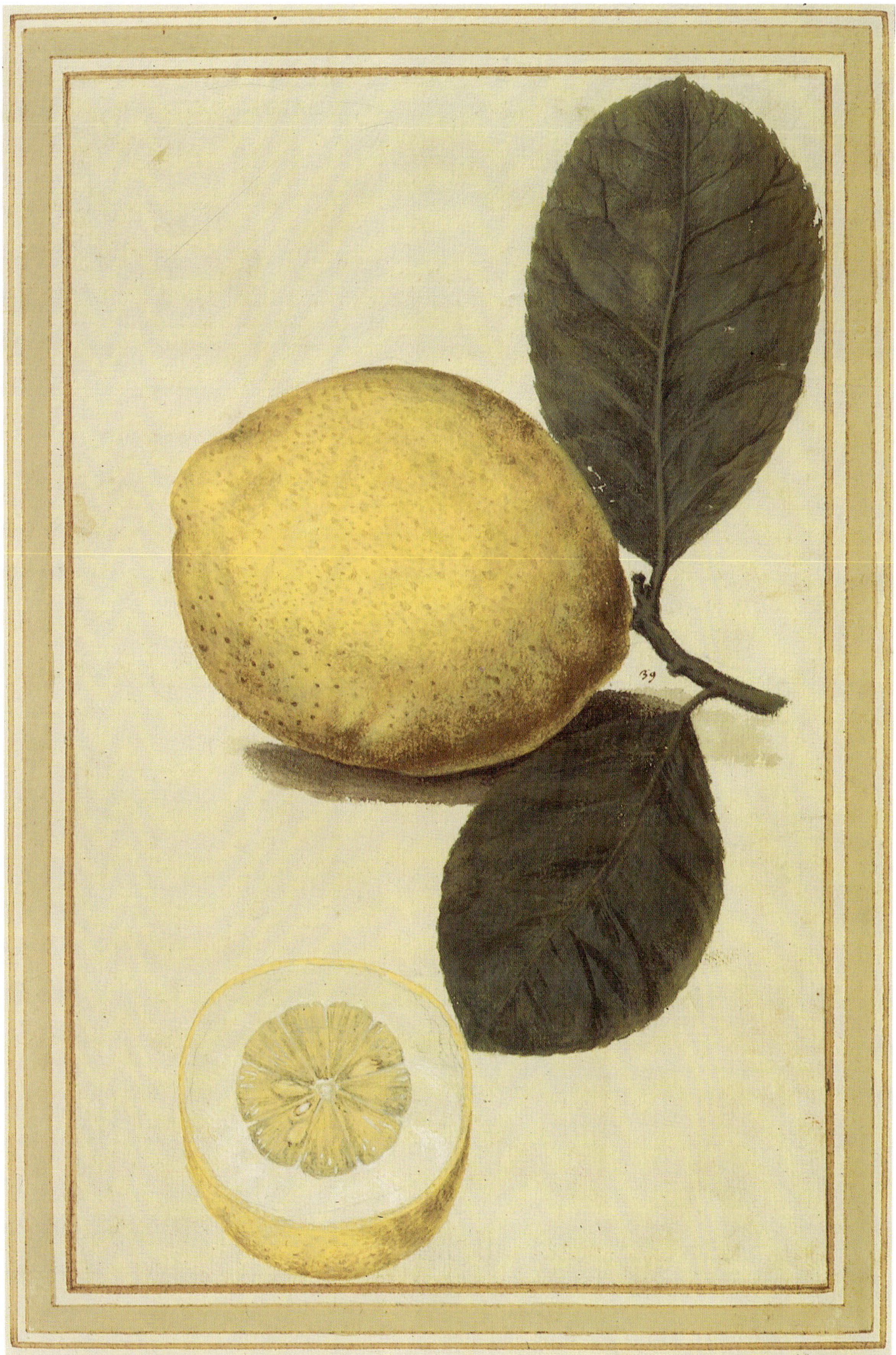

56. Citron-lemon, *Citrus limonimedica* (Lush.): whole fruit and leafy twig

?VINCENZO LEONARDI (fl. 1621–46)

Location unknown

Watercolour and bodycolour over black chalk

141 × 169 mm. Trimmed at right and through leaf at top. *Watermark*: none recorded

ANNOTATION: [below stalk] *26*

MOUNT SHEET: 585 × 432 mm. Mounted on same sheet as **57**

The *Pomum Paradisi* or 'Apple of Paradise' was thus called, according to Ferrari on p. 305 of the *Hesperides*, because of its marvellously sweet taste. But for some very small differences, the present drawing is almost exactly like the *Pomum Paradisi* reproduced on p. 307 of the *Hesperides*. It is, however, in the same sense as the engraving, which suggests that this drawing may be a copy.

This particular specimen is ellipsoidal and apiculate, with a smooth yellow peel and a stalk inserted on a twig supporting a narrow, lanceolate leaf with wingless petioles.

PROVENANCE: dal Pozzo; Albani; George III, by descent to George V; London art market (Mendelson); James R. Herbert Boone; Sotheby's (New York), 16 September 1988, lot 149

REFERENCES: Archivio Dal Pozzo MS 39, fol. 95; Ferrari 1646, p. 305; Bimbi, Villa Med. C. 594; Volkamer *Nürnberg. Hesp.* I, p. 162f; Gallesio 1811, p. 116f.; Risso/Poiteau 1818, p. 183, pl. 90; unpublished drawing on loose, unnumbered folio, GAG; De Noter 1926, p. 139f.; Baldini 1996b, p. 17, pl. 12

ENGRAVINGS: Ferrari 1646, p. 307

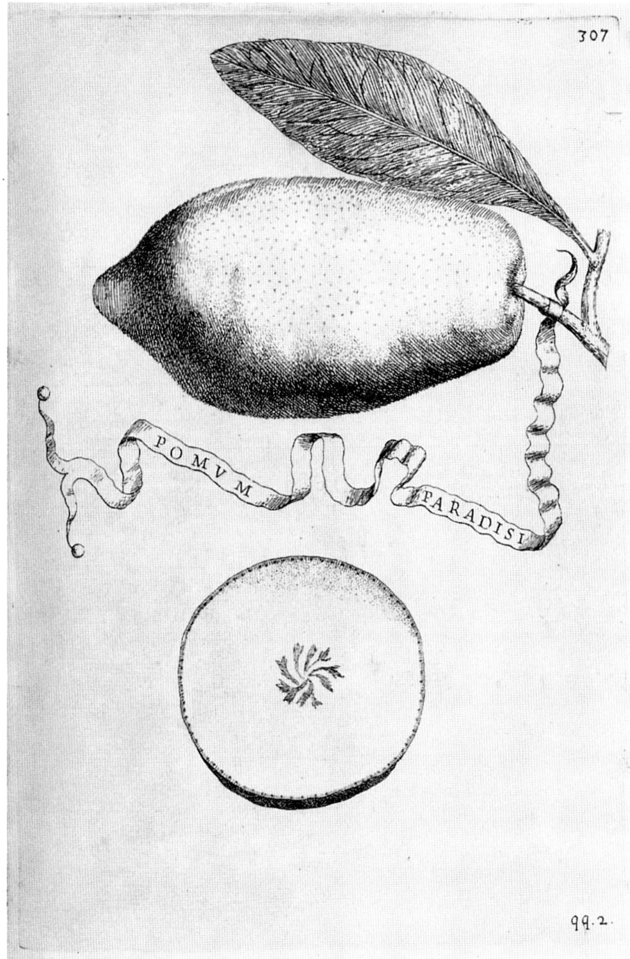

Comp. fig. 56, 57: Ferrari 1646 p. 307

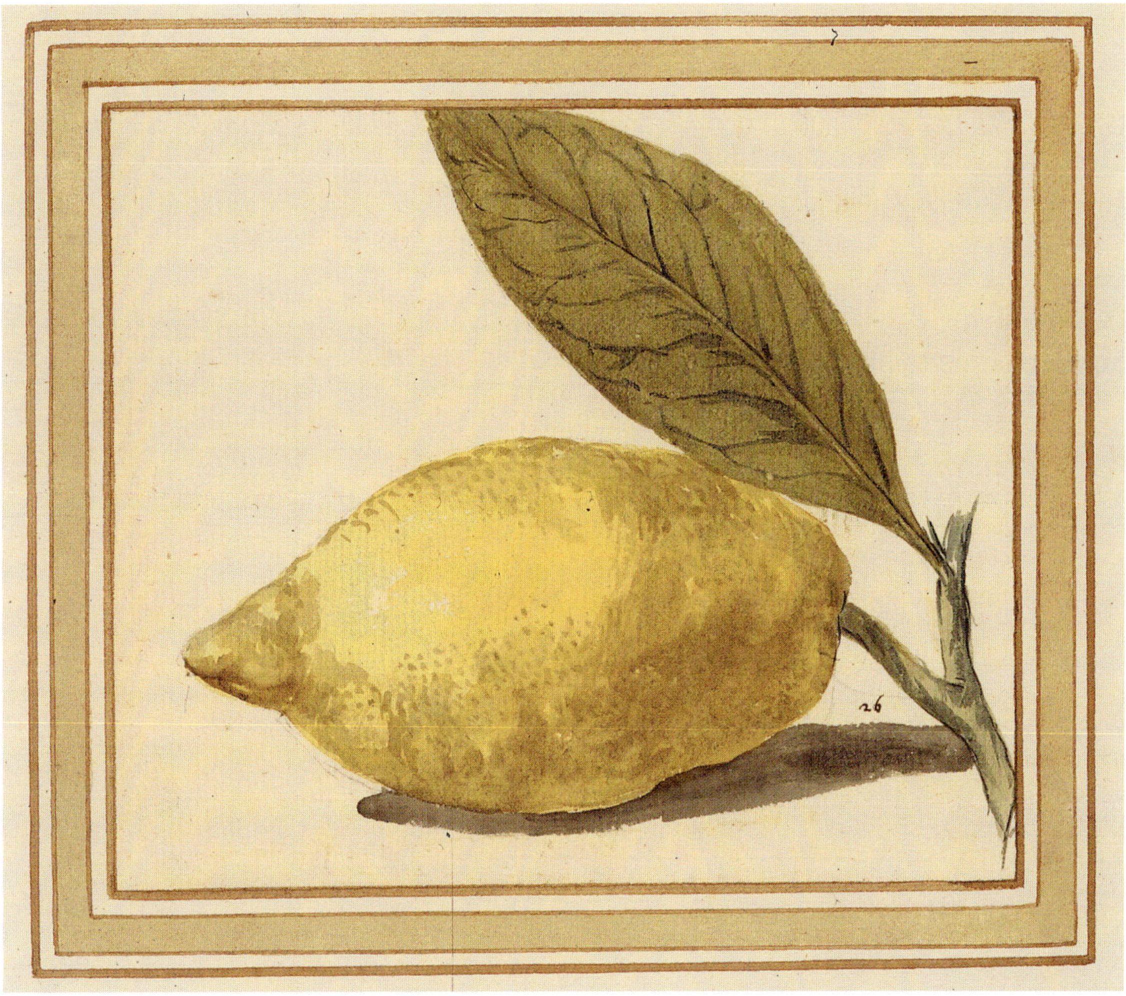

56

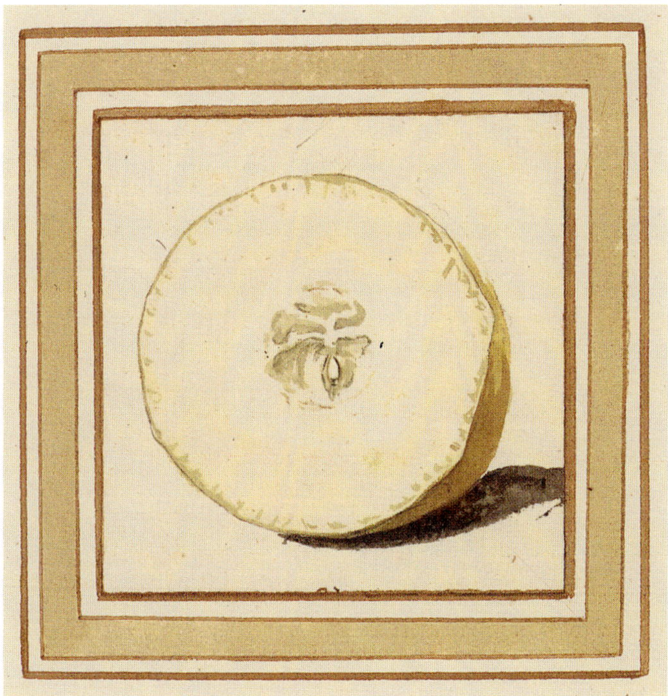

57

57. Citron-lemon, *Citrus limonimedica* (Lush.): half-fruit

?VINCENZO LEONARDI (fl. 1621–46)

Location unknown

Watercolour and bodycolour over black chalk

86 × 88 mm. Trimmed through shadow at right.
Watermark: none recorded

ANNOTATION: [bottom centre, cut] ?27

MOUNT SHEET: mounted on same sheet as **56**

This drawing, showing the half-fruit of the *Pomum Paradisi* on p. 307 of Ferrari's *Hesperides*, must originally have been on the same drawing sheet as the whole fruit (see **56**), the two later separated, and then in the 18th century mounted together when they entered the collection of George III. It represents a half-fruit completely lacking in flesh and seeds, and featuring a very thick mesocarp; therefore it looks like a citron-lemon rather than a lemon. As is clear from the drawing as well as the etching, the fruit seems to be arranged — as pointed out — in the shape of a wheel or rose. The little membranes dividing it up, Ferrari observed, were so pretty that they seemed to be made 'not by the artifice of nature, but by the art of the brush.' The insistence on the parallel with art — or on the essentially artistic quality of God's creations — is comparatively common in Ferrari.

In Archivio Dal Pozzo MS 39 (fol. 95), it is noted that the *Pomo Paradiso* can also be called a *limone tutta scorza*, just as indicated by the present drawing. Here it is noted that the example came from the well-known Roman garden of Tranquillo Romauli, with whom Ferrari seems to have been especially well acquainted, as he is referred to a number of times both in the *De Florum Cultura* and in the *Hesperides* (see also Freedberg 1989b, p. 44).

PROVENANCE: dal Pozzo; Albani; George III, by descent to George V; London art market (Mendelson); James R. Herbert Boone; Sotheby's (New York), 16 September 1988, lot 149

LITERATURE: Freedberg 1989b, p. 44, note 61; Freedberg 1996b

REFERENCES: Archivio Dal Pozzo MS 39, fol. 95; Ferrari 1646, pp. 305, 307; Bimbi, Villa Med. C. 594; Volkamer *Nürnberg. Hesp.* I, p. 162f.; De Noter 1926, pp. 139f.; Gallesio 1811, p. 116f.; Risso/Poiteau 1818, p. 183, pl. 90; unpublished drawing on loose, unnumbered folio, GAG; Baldini 1996b, p. 17, pl. 12

ENGRAVINGS: Ferrari 1646, p. 307

58. Citron-lemon, *Citrus limonimedica* (Lush.): whole and half-fruit

?VINCENZO LEONARDI (fl. 1621–46)

Private collection, U.S.A.

Watercolour and bodycolour, with gum heightening, over black chalk

277 × 200 mm. Badly stained. *Watermark*: none recorded

ANNOTATION: [below whole fruit] *L*

Many are the types of citrus fruit, observes Ferrari on p. 309, which have been called 'Adam's Apples'; but the one to which he wants to attach the name comes from Reggio Calabria (see Archivio Dal Pozzo MS 39, fols. 15–16v, for a long description of those from San Remo as well). It is bigger than the majority of lemons, with a very strong yellow colour. But it is not really to be counted amongst the prized lemons; indeed, claims Ferrari, it is rather a lowly food. The Neapolitans and some people from Reggio Calabria call it a Valencian *lumia*. Whether it derives its name from its sweet taste, or whether from what appear to be old bite marks (the bite of Adam in Paradise), Ferrari does not decide. Although he had some hesitation, Ferrari admits, about whether to include this fruit with the lemons (rather than with the oranges), he had decided to group it with the lemons, on the grounds that it shared slightly more characteristics with them than with oranges. This, in fact, was the most popular of the Neapolitan lemons, and was sometimes just called a lemon. Pontanus (Giovanni Pontano) had devoted several important lines to the *Pomum Adami* in his long poem called *De hortis Hesperidum*, where he divided it into three types. Pedro Pintor of Valencia, Alexander VI's doctor, had taught the Neapolitan apothecaries to make a drug out of hyacinths and the bitter juice of this 'lemon'. Mario Schipani, Cassiano's friend, had continued to make this mixture when he was Protomedicus of the city magistrates of Naples. He too had proven that the word 'lemon' was not specifically Neapolitan. In general, however, the *Pomum Adami* was better for looking at than for other purposes. Sometimes it produced numerous 'pregnant' fruits. You would think, concluded Ferrari, that the fruit got its name not just from Adam, but from this very superabundance (but see also **90**, **91** and **92**).

The present drawing, then, is a preparatory study for the *Pomum Adami Rheginum* (whole and half fruit) illustrated on p. 311 of the *Hesperides*, as always following the long morphological, agricultural, and anatomical description provided by Ferrari. The drawing shows an ellipsoid fruit with a flattened, apical nipple and a coarsely papillose, yellow skin. The half-fruit has a rather thick rind and pale yellow flesh with some seeds. The leaf is elliptical and serrate, the petiole slightly winged.

PROVENANCE: dal Pozzo; Albani; George III, by descent to George V; London art market (Mendelson); private collection, U.S.A.; Sotheby's (New York), 10 January 1995, lot 165

REFERENCES: Ferrari 1646, p. 309

ENGRAVINGS: Ferrari 1646, p. 311

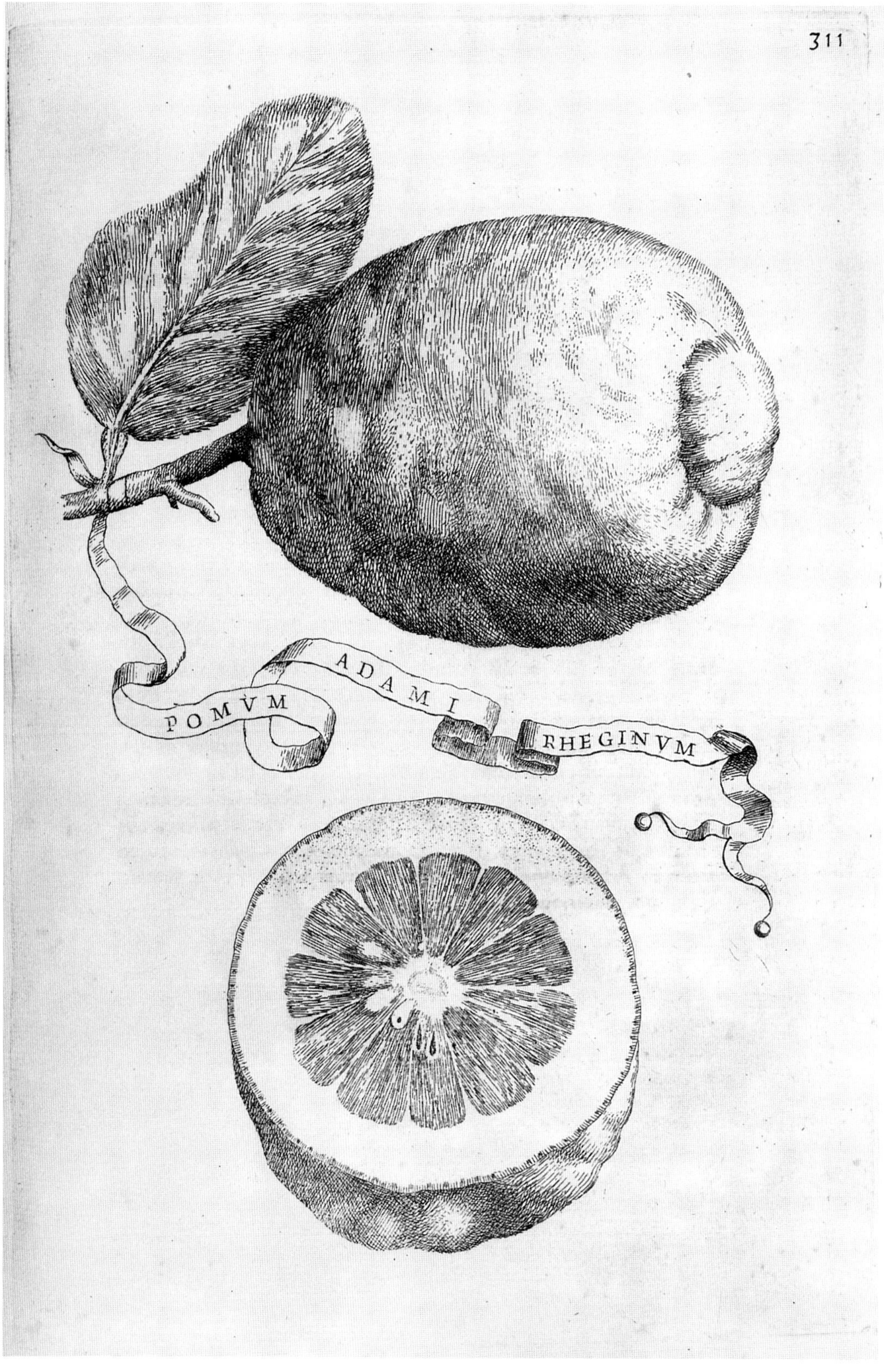

Comp. fig. 58: Ferrari 1646 p. 311

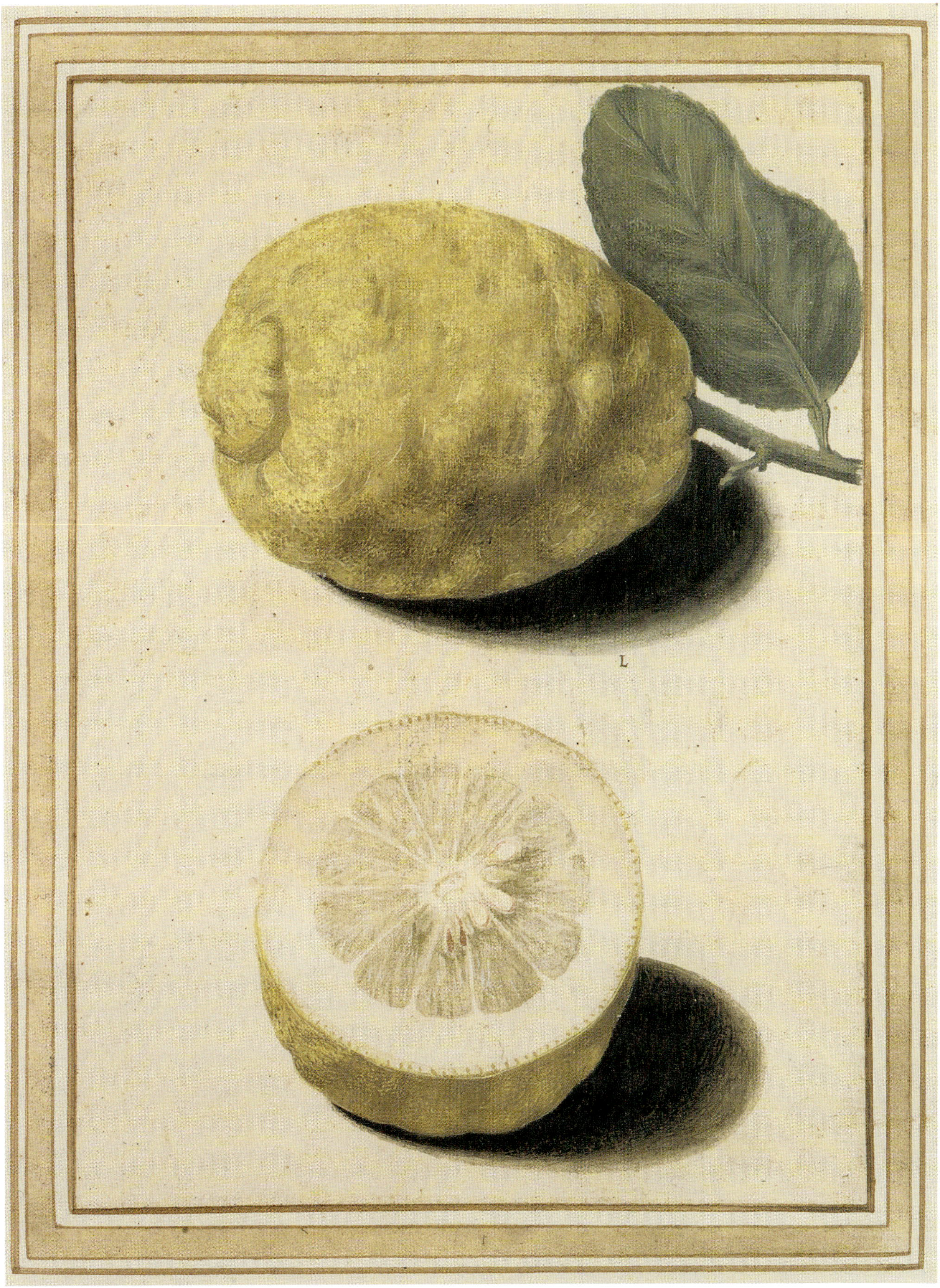
L

59. Lumia, *Citrus limon* (L.) Burm., f. × *Citrus sinensis* (L.) Osb.: whole fruit

VINCENZO LEONARDI (fl. 1621–46)

Location unknown

Watercolour and bodycolour, with gum heightening, over black chalk

141 × 147 mm. *Watermark*: none recorded

ANNOTATION: *N*

MOUNT SHEET: 585 × 432 mm. Mounted on same sheet as **60**

Ferrari, like Volkamer after him, used the term *lumia* to indicate certain hybrids of lemons and sweet oranges: *Media inter aurantiorum et limonum poma incedit lumia* (Gallesio 1811, p. 84); Gallesio, on the other hand, regarded the *lumia* as a hybrid of the citron and the sweet orange. In Sicily, however, the word *lumia* is currently used to indicate lemons in general. Ferrari describes three kinds of fruit which he regards as clearly falling into this group: from Valencia, Genoa, and finally those which are called the 'lumia of St. Dominic' or the 'salt-lumia'. He also discusses a group of four or five different citrus types from Reggio Calabria which are also called *lumia* by the inhabitants of that region. Characteristically Ferrari goes to great lengths not just to explain why they might thus be named, but also to place them into their correct groups. Thus there are those which are really 'apples of paradise' (the *Pomum Paradisi*) and one which is more likely a sweet lime. Of the first group, there was the fruit which the people of Reggio called the Jerusalem *lumia*; of the second were all those fruits which the Spanish called *lumie* but which the Italians took to be limes. Writers like Matthioli, Dodonaeus, and even Clusius sometimes called the *Pomum Paradisi* a kind of *lumia*. At the end of his rather pedantic discussion of all such varieties from Reggio Calabria, Ferrari wryly notes that there are almost as many names for citrus fruit as there are people there (p. 319).

The *Lumia Divi Dominici sive salis* represented here and used as the basis for the illustration on p. 323 got its name *sive salis* because it was usually eaten with some salt. There is also a related type that is a spurious *Spatafora* in the shape of a pear (see also **61**), illustrated on p. 325. These *lumiae* were often used to dispel nauseous tastes, especially those arising from food cooked with oil. Thus it was a good idea to season cabbages and other vegetables cooked in oil with a little lumia of this kind; it was also a useful accompaniment for roast meat in general.

The particular specimen shown here is a large, sub-globose, nippled fruit, with an elliptical, serrate leaf with an inconspicuously winged petiole. It looks a little like the *Lumia cedrata di quarta classe* represented by Bimbi in his painting of citrus fruit in the Villa Medicea (C. 594).

In 1631 Peiresc wrote rather anxiously to Cassiano that 'as for the *Lumia* of San Domenico, we certainly have never seen it; but I would like to see the drawing of flower and fruit, and even better the plant itself, if it could be found for a reasonable price' (see p. 46). Alas, Peiresc (who died in 1637) would not live to see the publication of the *Hesperides*, but perhaps Cassiano had in the meantime been able to send him either a drawing or a specimen itself, just as Peiresc had so insistently suggested.

PROVENANCE: dal Pozzo; Albani; George III, by descent to George V; London art market (Mendelson); James R. Herbert Boone; Sotheby's (New York), 16 September 1988, lot 151

REFERENCES: Archivio Dal Pozzo MS 39, fols. 88 and 89v; Ferrari 1646, pp. 317–19; Bimbi, Villa Med. C. 594; Volkamer *Nürnberg. Hesp.* II, p. 149, pl. 150; Gallesio 1811, p. 84

ENGRAVINGS: Ferrari 1646, p. 323

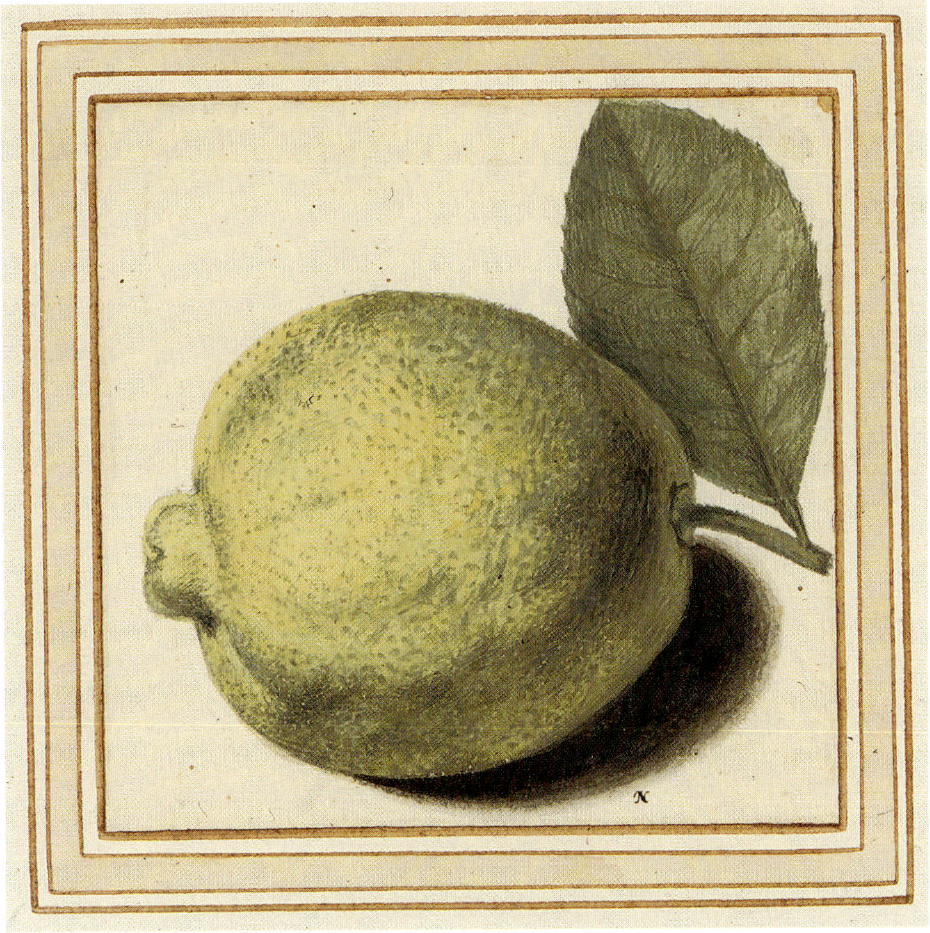

59

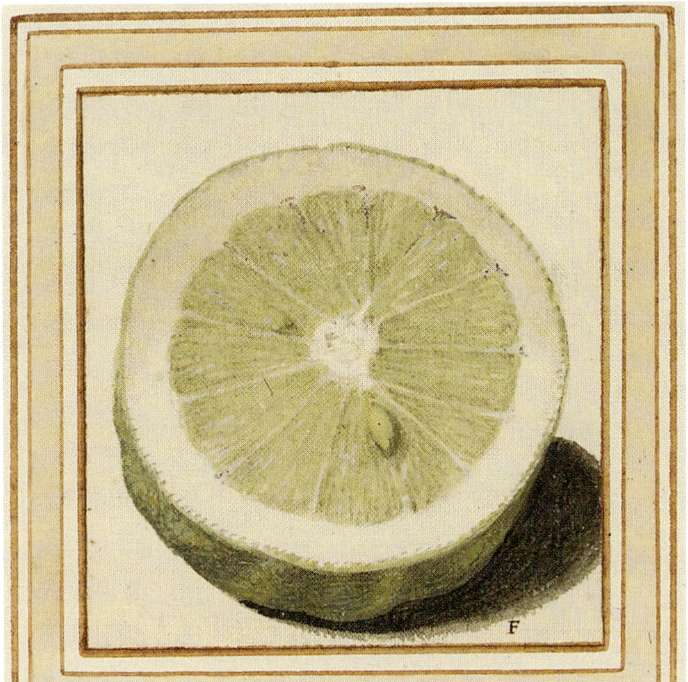

60

60. Lumia, *Citrus limon* (L.) Burm. f. × *Citrus sinensis* (L.) Osb.: half-fruit

VINCENZO LEONARDI (fl. 1621–46)

Location unknown

Watercolour and bodycolour, with gum heightening, over black chalk

115 × 107 mm. *Watermark*: none recorded

ANNOTATION: *F*

MOUNT SHEET: mounted on same sheet as **59**

This drawing, showing the half-fruit of the *Lumia Divi Dominici sive salis*, must originally have been on the same drawing sheet as the whole fruit (see **59**), the two later separated, and then in the 18th century mounted together when they entered the collection of George III. However, it does not correspond very closely to the relevant half-fruit as illustrated by Ferrari on the bottom half of p. 323. It has a thinner rind and more segments, with at least one seed.

PROVENANCE: dal Pozzo; Albani; George III, by descent to George V; London art market (Mendelson); James R. Herbert Boone; Sotheby's (New York), 16 September 1988, lot 151

REFERENCES: Archivio Dal Pozzo MS 39, fols. 88 and 89v; Ferrari 1646, pp. 317–19; Volkamer *Nürnberg. Hesp.* II, p. 149, pl. 150; Gallesio 1811, p. 84

ENGRAVINGS: Ferrari 1646, p. 323

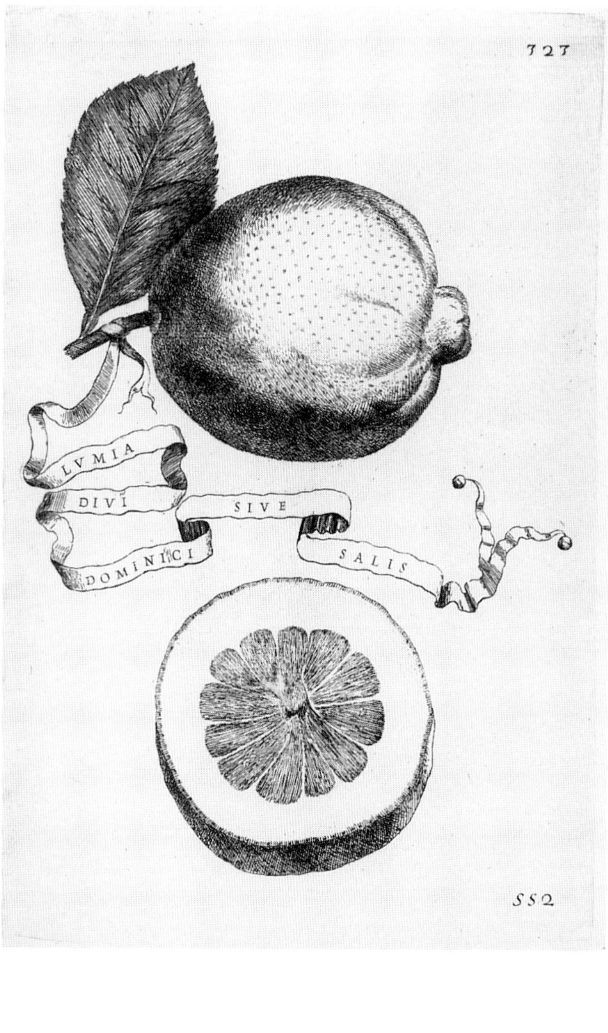

Comp. fig. 59, 60: Ferrari 1646 p. 323

61. Lumia, *Citrus limon* (L.) Burm. f. × *Citrus sinensis* (L.) Osb.: whole fruit

?VINCENZO LEONARDI (fl. 1621–46)
Location unknown
Watercolour and bodycolour over [?] black chalk
181 × 193 mm. *Watermark*: Figure ?60 (cut)
ANNOTATION: [below fruit] *54*
MOUNT SHEET: 585 × 432 mm. Mounted on same sheet as **62**

The second of the two *lumiae* generally eaten with salt was the variant which looked like a pear and derived its name from the Sicilian Spatafora family, hence its name *Lumia salis sive Spatafora notha*. In Cassiano's compilation on citrus fruit (Archivio Dal Pozzo MS 39, fol. 88v) the similarity with the appearance of a pear is discussed at some length. The present drawing served as the basis for the illustration of this variant in Ferrari's *Hesperides*, p. 325. It shows a globose, nippled fruit and two lanceolate leaves with wingless petioles on a thornless twig. The flowers shown in Ferrari's illustration are not present here. Amongst other details about this variant, Ferrari also notes the extreme fragility of the trees, which are easily damaged.

PROVENANCE: dal Pozzo; Albani; George III, by descent to George V; London art market (Mendelson); James R. Herbert Boone; Sotheby's (New York), 16 September 1988, lot 149

REFERENCES: Archivio Dal Pozzo MS 39, fols. 87–8; Ferrari 1646, pp. 317–19

ENGRAVINGS: Ferrari 1646, p. 325

62. Lumia, *Citrus limon* (L.) Burm. f. × *Citrus sinensis* (L.) Osb.: half-fruit

?VINCENZO LEONARDI (fl. 1621–46)
Location unknown
Watercolour and bodycolour, over [?] black chalk
75 × 85 mm. Foxed and stained. *Watermark*: none recorded
ANNOTATION: *57*; [bottom centre] *1*
MOUNT SHEET: mounted on same sheet as **61**

This drawing must originally have been on the same drawing sheet as the whole fruit (see **61**), the two later separated, and then in the 18th century mounted together when they entered the collection of George III. This is the preparatory drawing for the half-fruit of the *Lumia salis sive Spatafora notha* illustrated on p. 325 of Ferrari's *Hesperides*. It shows a cross-section with a relatively thick rind and ten apparently seedless segments.

PROVENANCE: dal Pozzo; Albani; George III, by descent to George V; London art market (Mendelson); James R. Herbert Boone; Sotheby's (New York), 16 September 1988, lot 149

REFERENCES: Archivio Dal Pozzo MS 39, fols. 87–8; Ferrari 1646, pp. 317–19

ENGRAVINGS: Ferrari 1646, p. 325

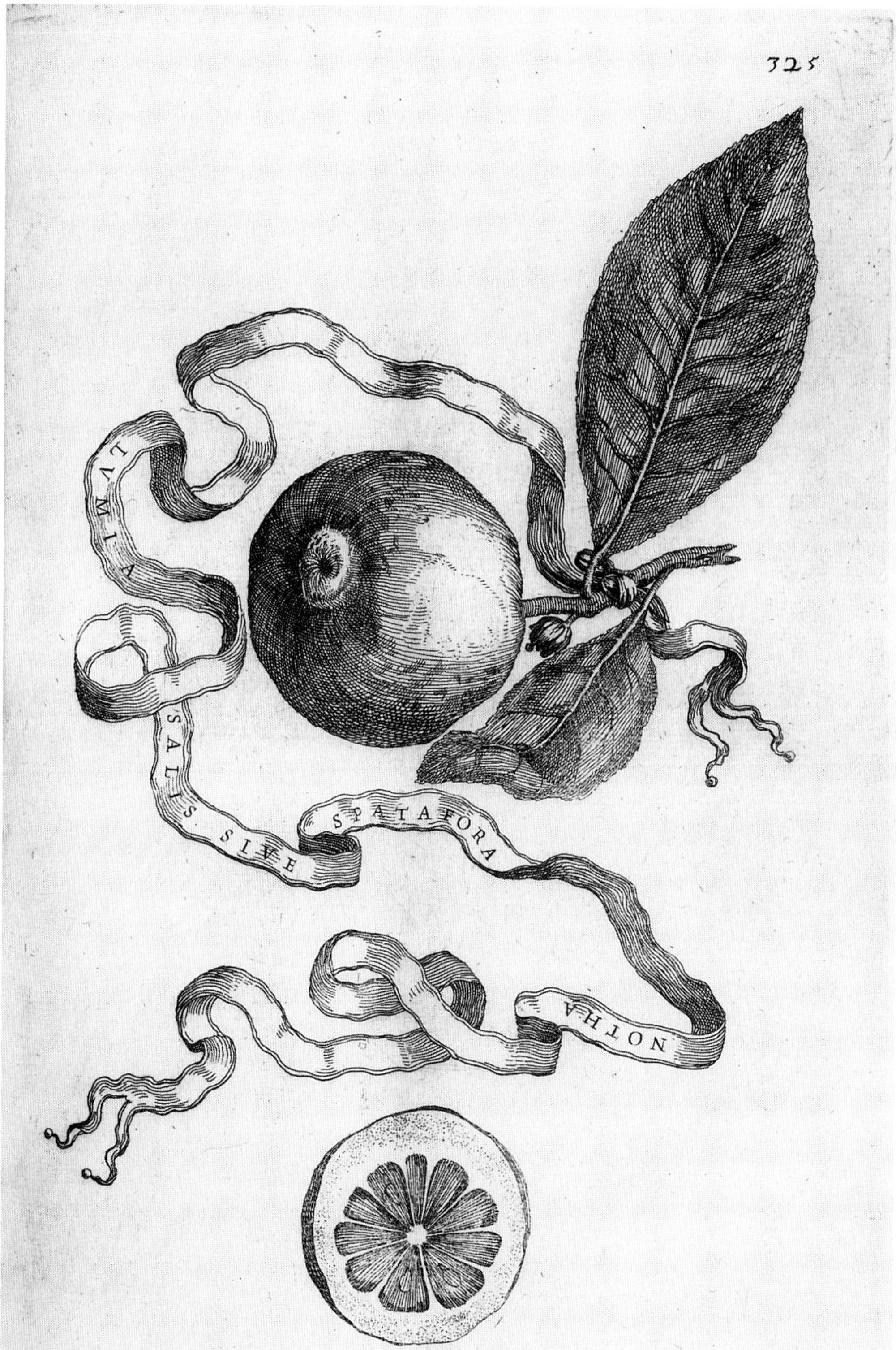

Comp. fig. 61, 62: Ferrari 1646 p. 325

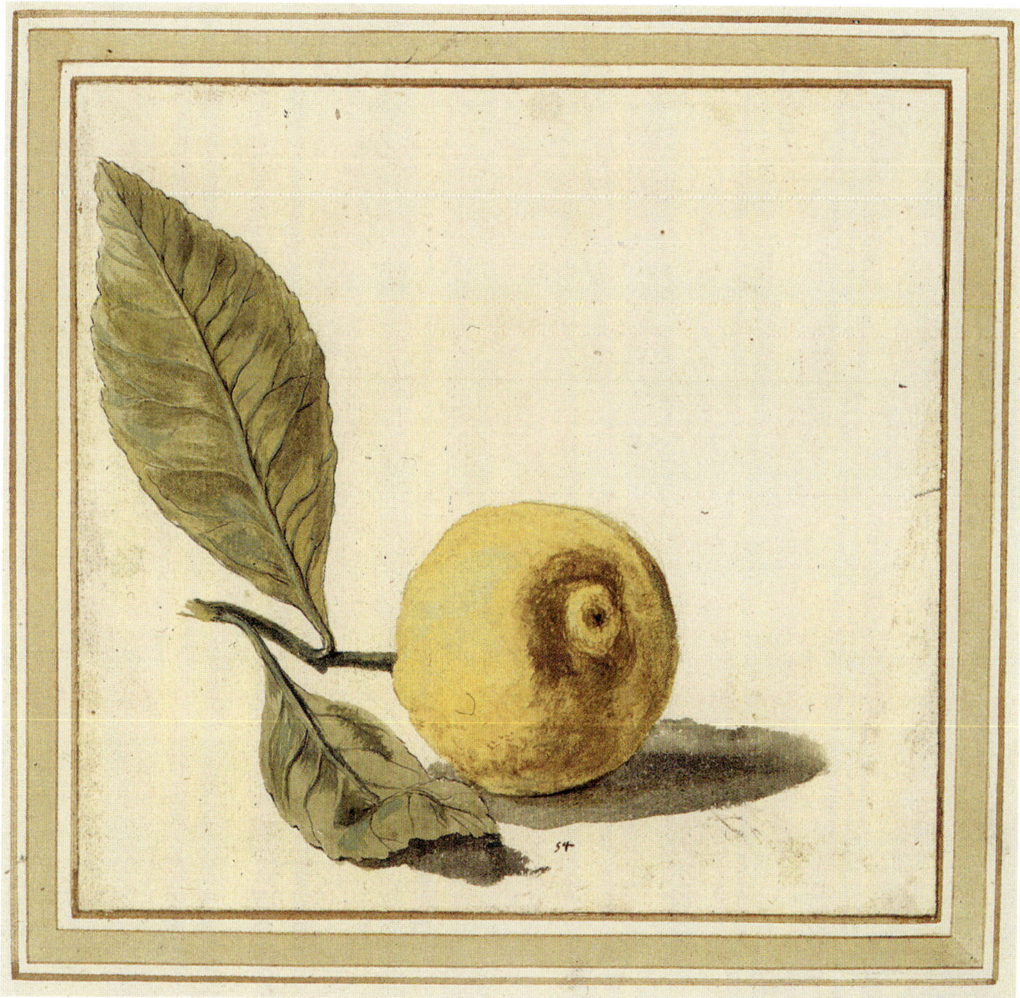

61

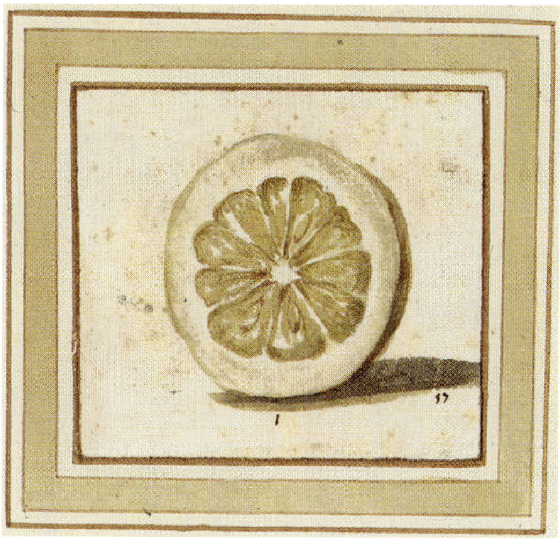

62

63. Pummelo, *Citrus grandis* (L.) Osb. hybrid: whole fruit and leafy twig

VINCENZO LEONARDI (fl. 1621–46)

Private collection, Scotland

Watercolour and bodycolour, with gum heightening, over black chalk

230 × 205 mm. Foxed and stained. Trimmed through shadow at right

ANNOTATION: [below] *10*

MOUNT SHEET: mounted with **64**, within the same wash border

This drawing served as the basis for the whole fruit of the *Aurantium dulci cortice et sinense* shown on p. 433 of Ferrari's *Hesperides* (for the myrtle-leaf orange shown at the bottom of p. 433, see **34–36**). The point about this 'orange', as Ferrari's taxonomy makes very clear, is that it differs from all other oranges in having a sweet rind. But it is not actually an orange at all. The thickness of its rind, its colour and its size make it much more likely to be a pummelo hybrid. Ferrari comments on the lasting dark-green colour of the leaves of this type, of their toughness and relative smallness. He notes the abundance and 'laxity' of its flowers. He comments, as always, on the colours of its interior and its slightly bitter taste. He points out the unusually large number of seeds. In Sicily and in Calabria it is called an *Aurantium pomum*, in Genoa an *Adami pomum*, or an Adam's apple (cf. also Archivio Dal Pozzo MS 39, fol. 15). In Reggio Calabria it has no status at all, but is used as a kind of ball in popular games, sometimes stuffed with grass or tow. While Hippomenes defeated Atalanta with the third throw of the ball, the people of Reggio used a whole variety of citrus projectiles in their games. But these are cheap ones, unlike the golden one picked up by Atalanta. The mythological/ethnographic parallel is one that appealed to Ferrari and is a common feature of his writing.

The drawing shows a large globose deep-yellow fruit suspended from a twig with seven small narrow, lanceolate leaves with slightly winged petioles.

In Archivio Dal Pozzo MS 39 there are several pages devoted to the *Cedro dell'Agro dolce (e colorito nella scorza di ranciato come gl'Aranci) di Reggio di Calavria*, which in many respects conforms to Ferrari's description of the present fruit. In this passage specific mention is made not just of the fact that it had been sent to Father Ferrari along with other specimens of fruit from Reggio, but that it had also been drawn or painted for him (Archivio Dal Pozzo MS 39, fols. 67–67v; see also fols. 56–56v for a long discussion of the *Arancio della scorza dolce*).

This drawing and **64** must originally have been part of the same sheet, then cut into two, and finally mounted together on the same sheet when they entered the collection of George III.

PROVENANCE: dal Pozzo; Albani; George III, by descent to George V; London art market (Mendelson)

REFERENCES: Ferrari 1646, p. 430f.; Gallesio 1811, p. 136

ENGRAVINGS: Ferrari 1646, p. 433

63

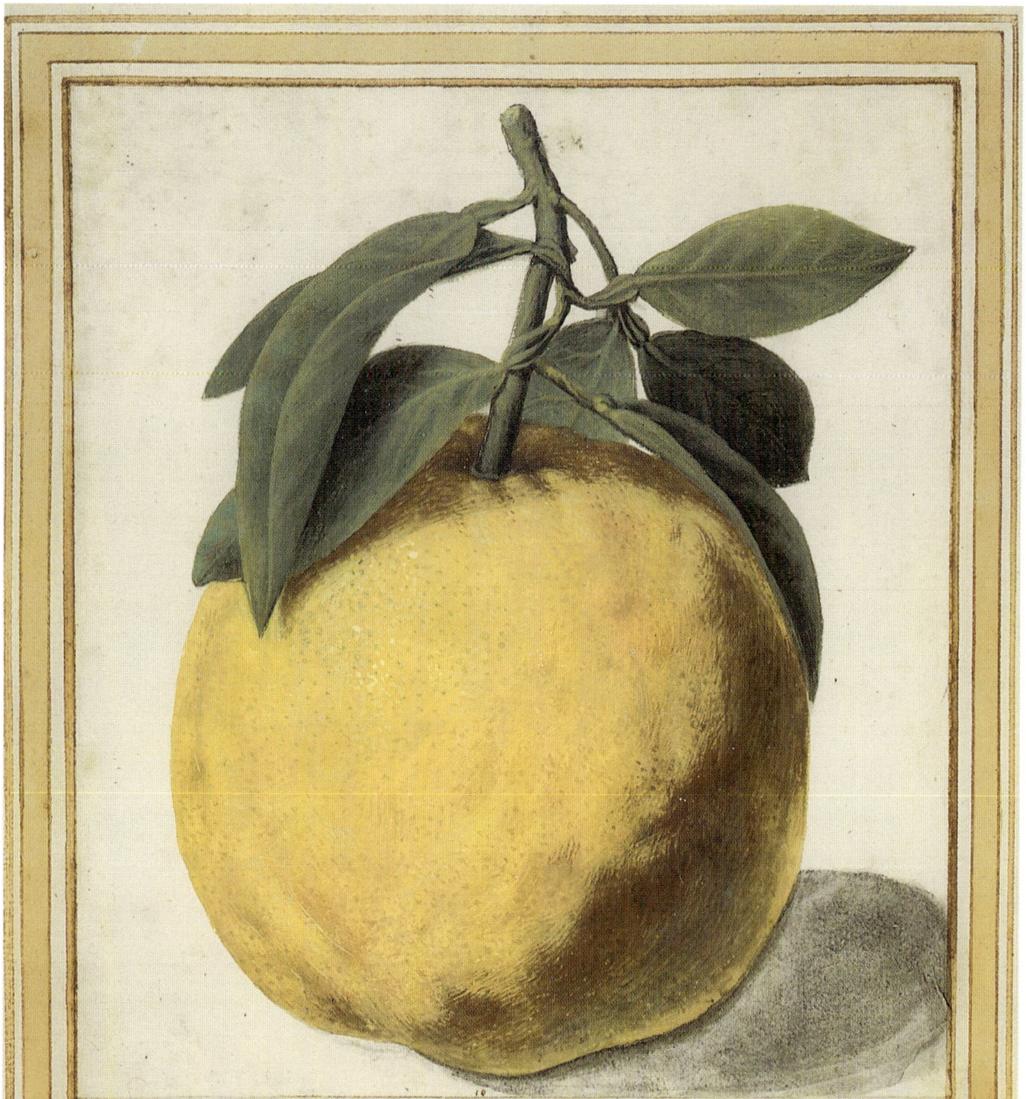

64

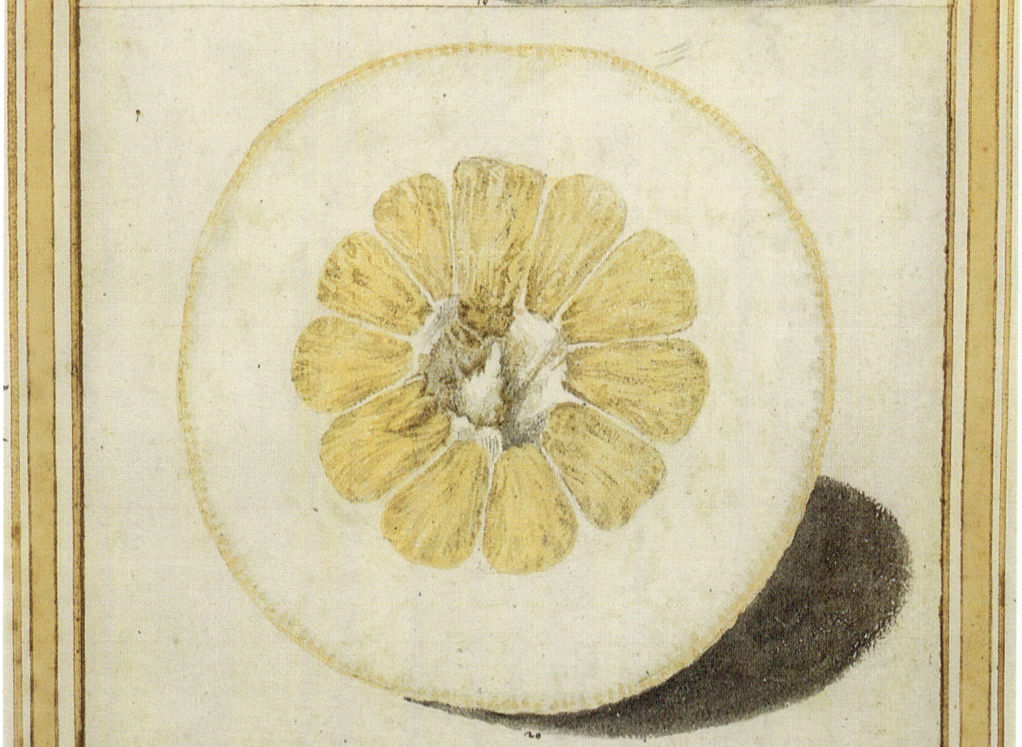

64. Pummelo, *Citrus grandis* (L.) Osb. hybrid: half-fruit

VINCENZO LEONARDI (fl. 1621–46)

Private collection, Scotland

Watercolour and bodycolour, over black chalk

175 × 205 mm. Foxed and stained

ANNOTATION: [below] *20*

MOUNT SHEET: mounted with **63**, within the same wash border

This drawing of the cross-section of the *Aurantium dulci cortice et sinense* was reproduced on p. 435 of Ferrari's *Hesperides*. The half-fruit has a thick mesocarp and small yellow segments.

See **63**.

PROVENANCE: dal Pozzo; Albani; George III, by descent to George V; London art market (Mendelson)

REFERENCES: Ferrari 1646, p. 430f.; Gallesio 1811, p. 136

ENGRAVINGS: Ferrari 1646, p. 435

65. Citron-lemon, *Citrus limonimedica* Lush.: whole fruit and leafy twig

?VINCENZO LEONARDI (fl. 1621–46)

Windsor, RL 19348

Watercolour over black chalk; *pentimenti* in black chalk and watercolour

187 × 209 mm. Foxed. Survival of 5 mm overlay of original 17th-century mount at lower edge and right-hand side.
Watermark: Crown 38 (cut)

ANNOTATION: 22

MOUNT SHEET: 434 × 586 mm. Very badly stained and torn at right-hand side, elsewhere stained and marked. *Watermark*: D & C Blauw

Not reproduced in Ferrari's *Hesperides*, the drawing is partially unfinished, and a rough outline of two further leaves may be discerned on the right of the sheet. One of these leaves in black chalk shows traces of watercolour as well. The drawing shows a relatively large, ellipsoid, smooth, yellow citron-lemon, depressed at the base and somewhat obtuse at its apex. Two large elliptical leaves with wingless petioles are inserted on the twig, which is continuous with the fruit stalk.

PROVENANCE: dal Pozzo; Albani; George III, by descent to George V; London art market (Mendelson); Rex Nan Kivell; Her Majesty The Queen

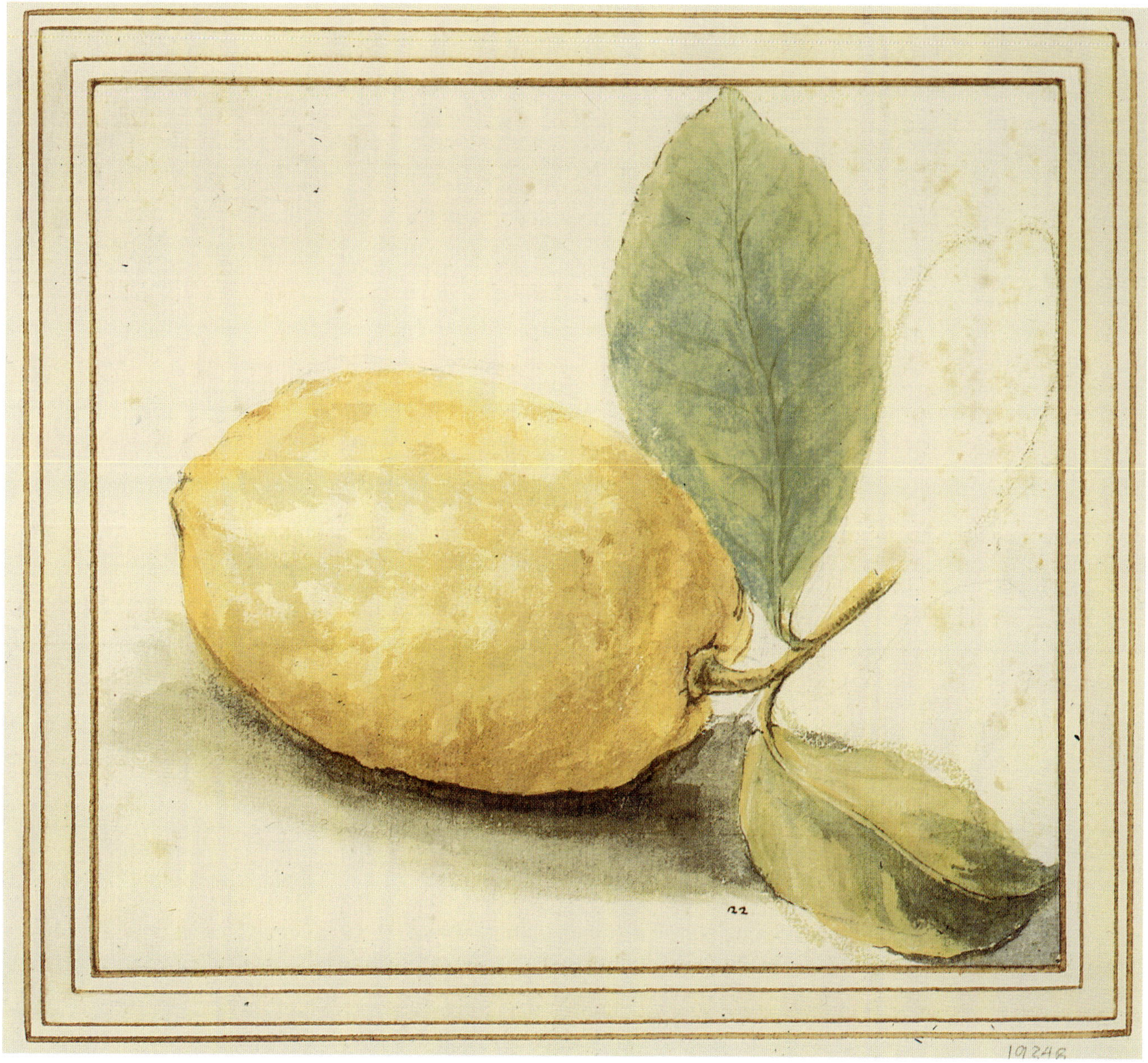

65

66. Citron-lemon, *Citrus limonimedica* Lush.: half-fruit

?VINCENZO LEONARDI (fl. 1621–46)

Windsor, RL 19339

Watercolour with gum heightening, over black chalk

149 × 161 mm. Slight foxing

ANNOTATION: *000*

MOUNT SHEET: 231 × 274 mm. Some staining at upper edge. *Watermark*: D & C Blauw

Not reproduced in Ferrari's *Hesperides*, this drawing is evidently by the same hand as the drawing in the previous entry and presumably the half-fruit of the specimen shown there. Its diameter and colour are the same. This irregular cross-section has a rather thick rind and small yellow seedless segments.

PROVENANCE: dal Pozzo; Albani; George III, by descent to George V; London art market (Mendelson); Rex Nan Kivell; Her Majesty The Queen

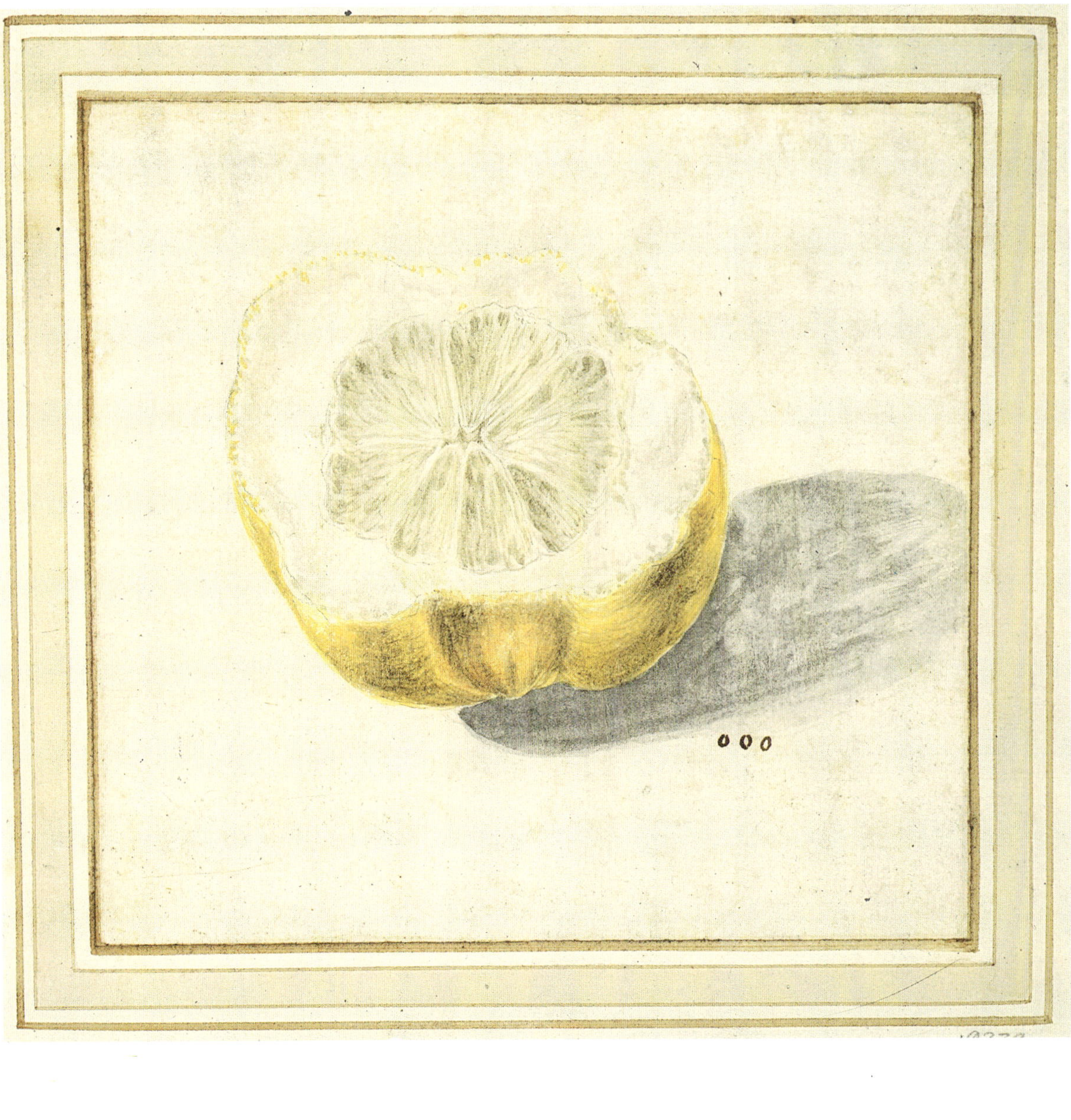

66

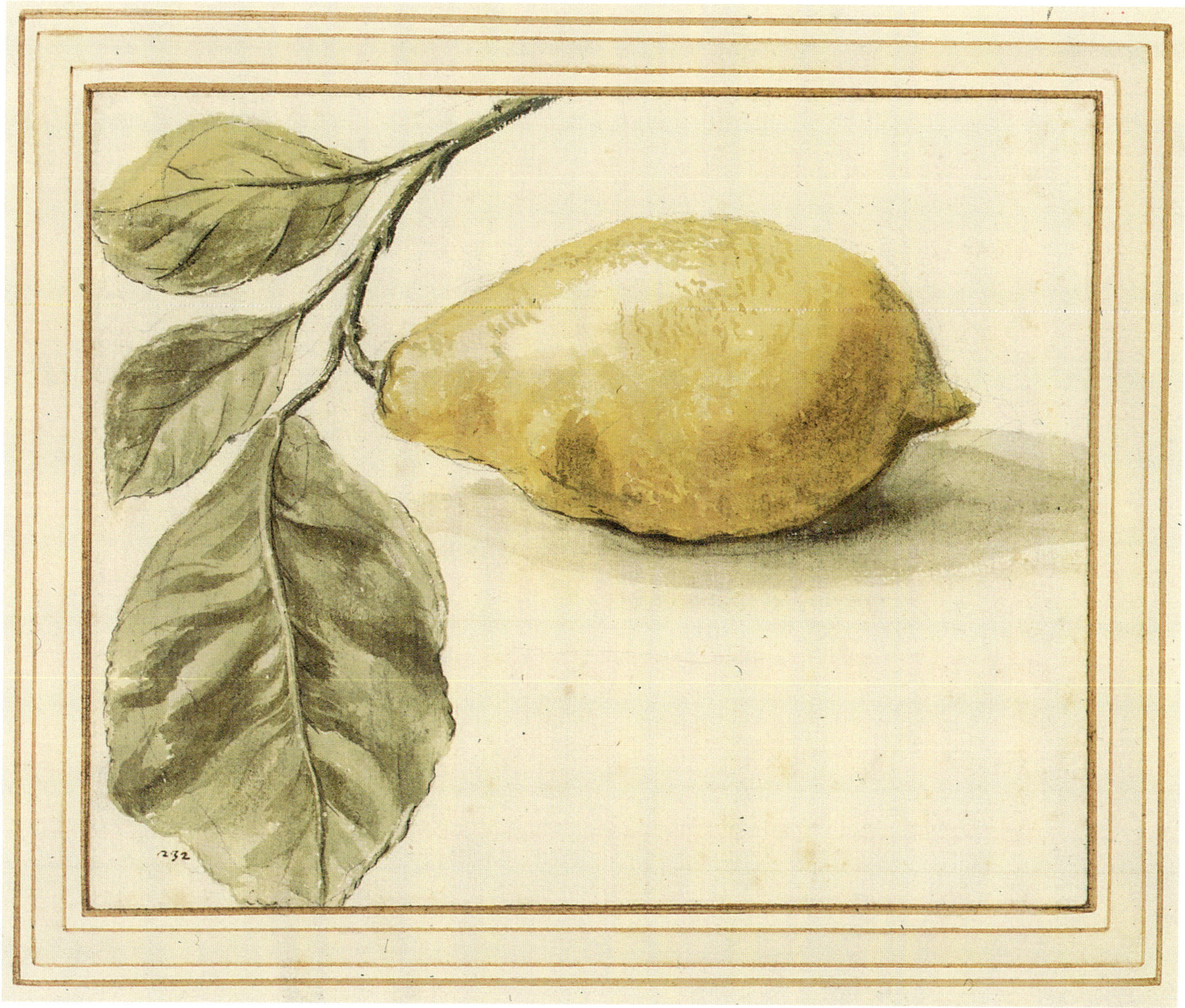

67

67. Citron-lemon, *Citrus limonimedica* (Lush.): whole fruit and leaves

?VINCENZO LEONARDI (fl. 1621–46)

Location unknown

Watercolour over [?] black chalk; *pentimenti* in black chalk

163 × 207 mm. Trimmed at bottom edge. *Watermark*: none recorded

ANNOTATION: [lower left] *232*

MOUNT SHEET: 585 × 432 mm. Mounted on same sheet as **68**

This drawing is not reproduced in Ferrari's *Hesperides*. A whole, ellipsoid citron-lemon with a prominent, pointed apex and a long-necked base. The peel is yellow and rather smooth. Three orbicular or elliptical leaves with wingless petioles are inserted on the twig.

PROVENANCE: dal Pozzo; Albani; George III, by descent to George V; London art market (Mendelson); James R. Herbert Boone; Sotheby's (New York), 16 September 1988, lot 151

68. Citron-lemon, *Citrus limonimedica* Lush.: half-fruit

?VINCENZO LEONARDI (fl. 1621–46)

Location unknown

Watercolour over [?] black chalk

92 × 102 mm. Trimmed through shadow at right.
Watermark: none recorded

ANNOTATION: [bottom centre] *24.i*

MOUNT SHEET: mounted on same sheet as **67**

Not reproduced in Ferrari's *Hesperides*, this drawing must originally have been on the same drawing sheet as the whole citron-lemon (see **67**). The two must have been later separated, and then in the 18th century mounted together when they entered the collection of George III. It shows a round half-fruit with a thick rind and a few small yellow segments with large seeds.

PROVENANCE: dal Pozzo; Albani; George III, by descent to George V; London art market (Mendelson); James R. Herbert Boone; Sotheby's (New York), 16 September 1988, lot 151

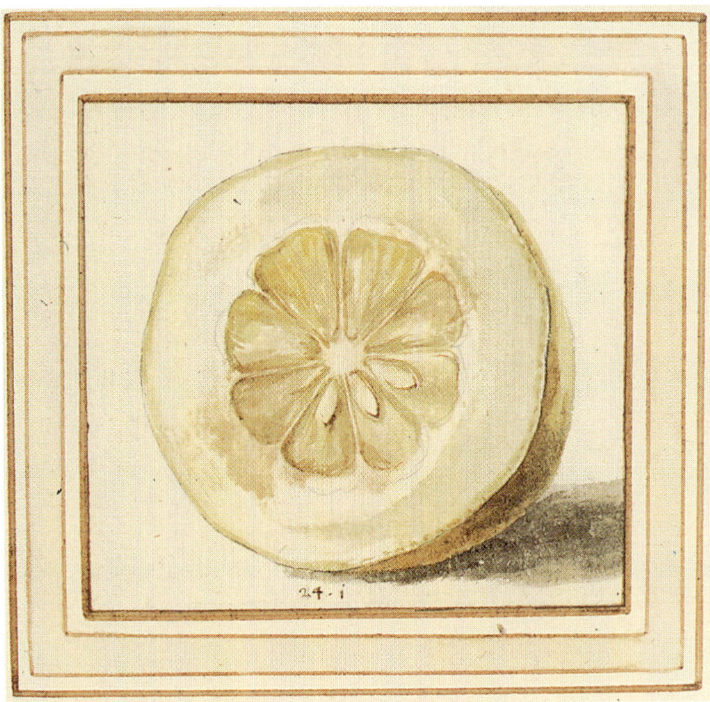

68

69. Citron-lemon, *Citrus limonimedica* Lush.: whole fruit and leafy twig

?VINCENZO LEONARDI (fl. 1621–46)

Windsor, RL 19351

Watercolour and bodycolour over black chalk

149 × 188 mm. Survival of 5 mm overlay of original 17th-century mount at edges

ANNOTATION: [below stem] *30*

MOUNT SHEET: 434 × 584 mm. *Watermark*: D & C Blauw

Not reproduced in Ferrari's *Hesperides*, this drawing shows a long, ellipsoid citron-lemon with a papillose yellow rind, a small protruding apical nipple and an evenly rounded base. The twig has two orbicular leaves restricted at the base with long inconspicuously winged petioles. There is no representation of the half-fruit amongst the surviving drawings.

PROVENANCE: dal Pozzo; Albani; George III, by descent to George V; London art market (Mendelson); Rex Nan Kivell; Her Majesty The Queen

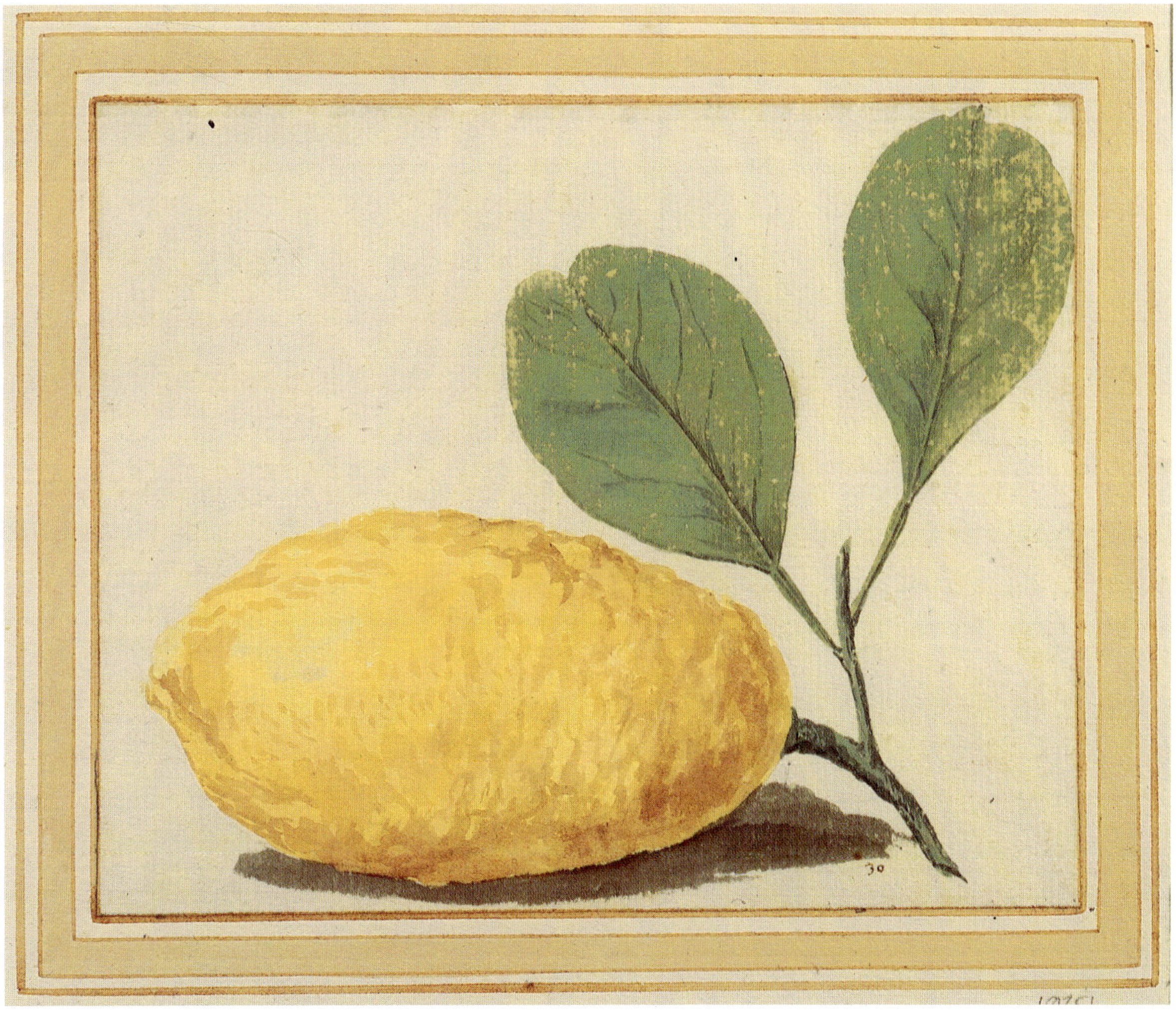

69

70. Citron-lemon, *Citrus limonimedica* Lush.: leafy twig, whole and half-fruit

VINCENZO LEONARDI (fl. 1621–46)

London, Victoria and Albert Museum, E. 733-1949

Watercolour and bodycolour (parts oxidised on cross-section), with gum heightening, over black chalk

274 × 208 mm

ANNOTATION: [below half-fruit] *nnn*

MOUNT SHEET: 585 × 435 mm.
Annotation: [verso: pencil, in a 20th-century hand] *Mrs de Beer*

This drawing is not reproduced in Ferrari's *Hesperides*. The leaves of this specimen are elliptical, with wingless petioles, but are manifestly of different widths; they look like those of citrons. The whole fruit is globose, lightly nippled and laterally furrowed with a yellow, wrinkled, and papillose peel. The half-fruit has a thick rind; segments are few, small and seeded. Since the taxonomy of this specimen remains unclear it has been ascribed tentatively to the heterogeneous group of citron-lemons.

PROVENANCE: dal Pozzo; Albani; George III, by descent to George V; London art market (Mendelson); Mrs Helen de Beer

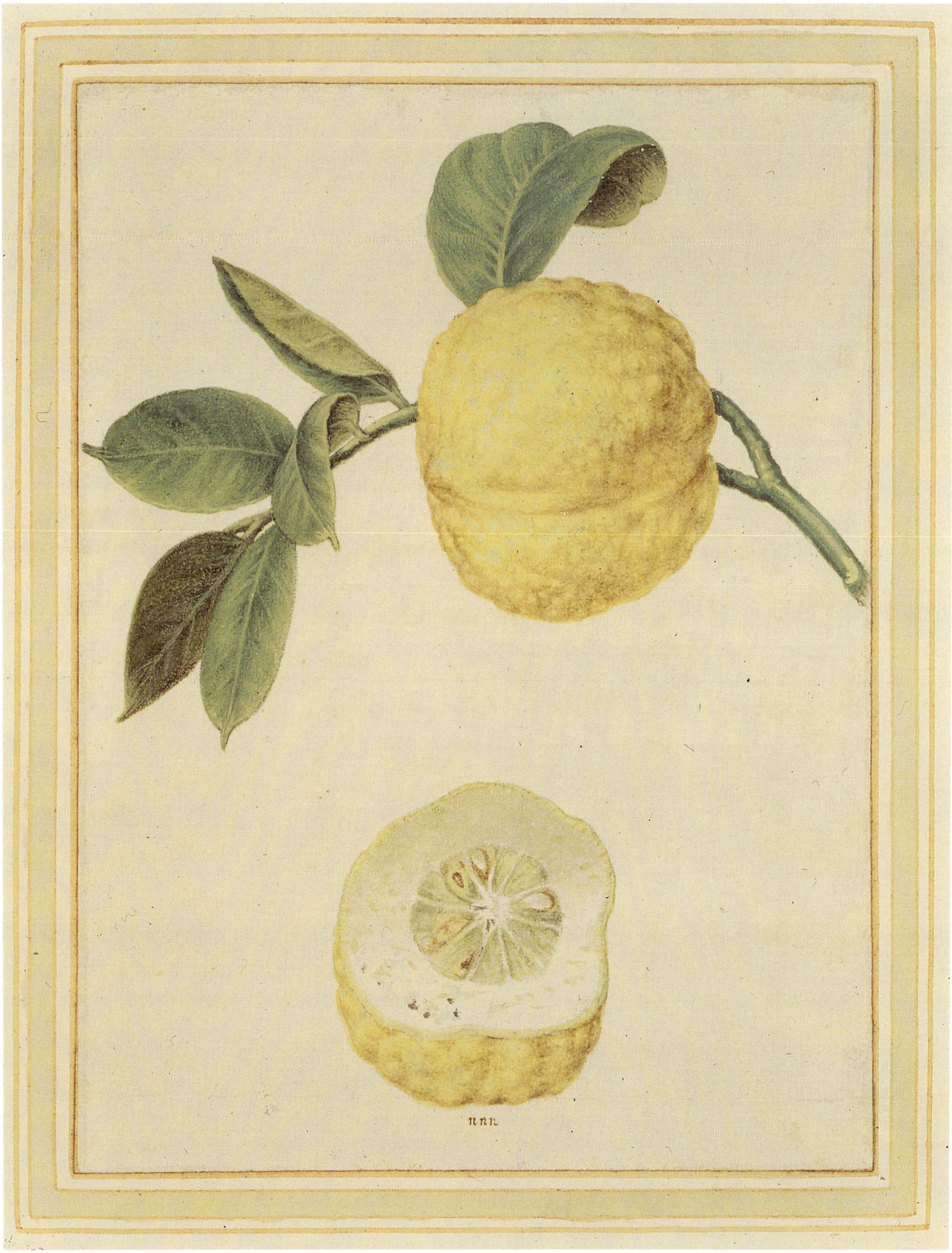
nnn

71. Citron × sour orange, *Citrus medica* L. × *Citrus* aurantium L.: whole and half-fruits

SEVENTEENTH-CENTURY ITALIAN

Location unknown

Watercolour and bodycolour over black chalk

321 × 424 mm. Trimmed at right. Foxed and stained. *Watermark*: Crown 35 (cut)

ANNOTATION: [bottom centre] *752*

This is an usually large drawing, not reproduced in Ferrari's *Hesperides*. It shows four specimens of the same fruit, represented on what seem to be stone steps, again an unusual element in the drawings of citrus fruit collected by Cassiano (see also **98** and **107**). Below, the fruit is seen whole from two sides, while crosswise and longitudinal sections are depicted above. This arrangement is unique in the surviving corpus of citrus drawings. The whole fruits look similar to Gallesio's *Cedro della China* and to Cassiano's *Citrus medica Pomum Adami citratum* in Targioni Tozzetti, who classified it amongst the citron hybrids. A *Pomo d'Adamo cedrato* is still sporadically present in Sicily (Continella/Tribulato 1992, p. 446). Since a secondary fruitlet is included at the top of the axis of the fruit on the upper right, it is clear that this is a 'navel' specimen. The rind is fairly thick, deep yellow, papillose and remarkably wrinkled. Gallesio (1811, pp. 100–02) suspected that the ancestors of this hybrid could be the citron and sour orange; Deng *et al.* (1995, p. 78) have confirmed this phylogenetic hypothesis by RAPD testing.

PROVENANCE: dal Pozzo; Albani; George III, by descent to George V; London art market (Mendelson); James R. Herbert Boone; Sotheby's (New York), 16 September 1988, lot 156

REFERENCES: Micheli *Enumeratio*, pl. 55; Gallesio 1811, pp. 100–2; Targioni Tozzetti 1825, p. 16, pl. 15; unpublished drawing on loose, unnumbered folio, GAG; Gallesio 1839, p. 10; Continella/Tribulato 1992, p. 446; Deng *et al.* 1995, p. 78; Baldini 1996b, p. 17, pl. 12

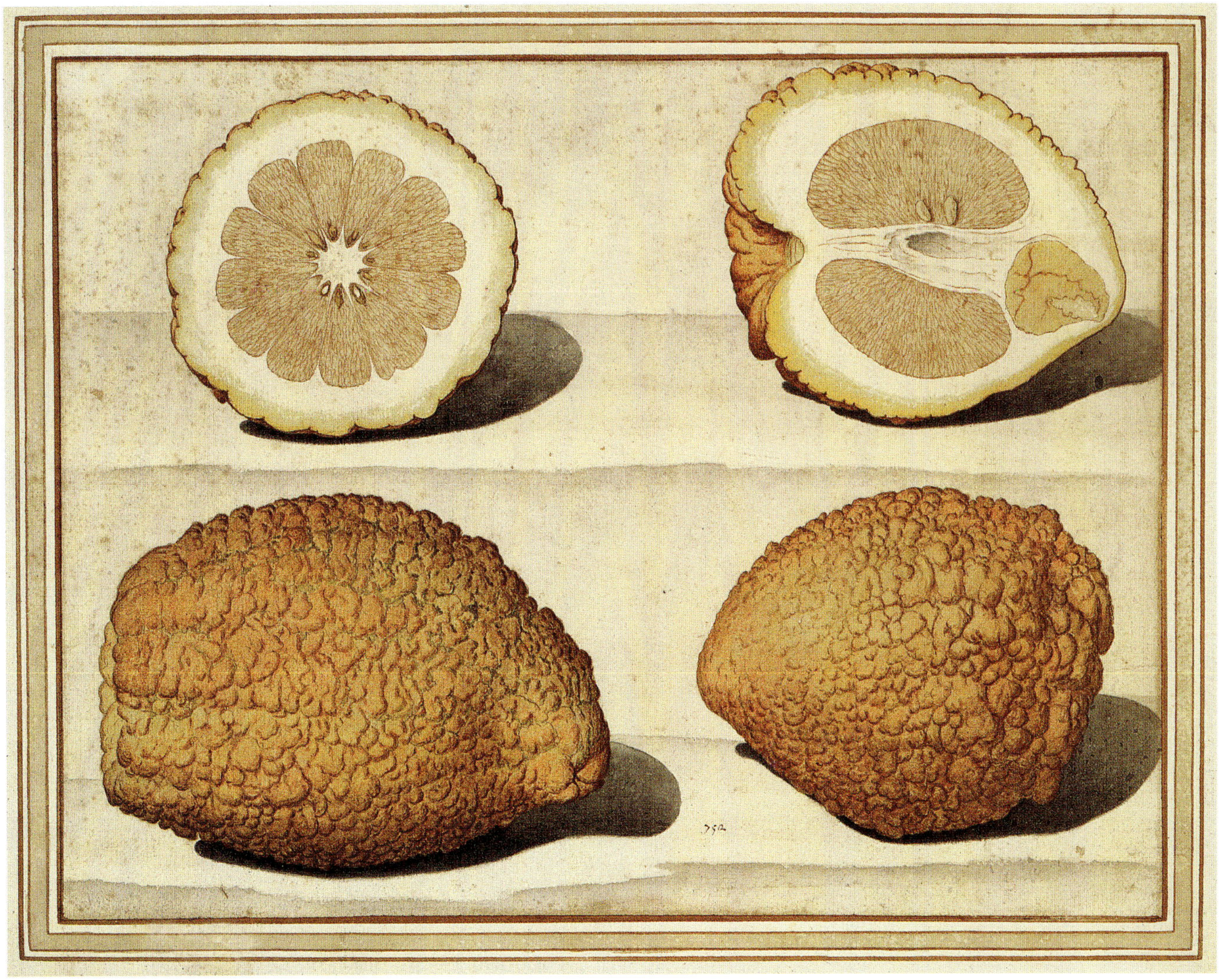

71

72. Citron, *Citrus medica* L.: misshapen whole fruit

VINCENZO LEONARDI (fl. 1621–46)

Private collection, England

Watercolour and bodycolour, with gum heightening, over black chalk

310 × 217 mm

ANNOTATION: *60*

This is the preparatory drawing for the *Malum citreum digitatum seu multiforme* illustrated on p. 77 of Ferrari's *Hesperides*. It is thus the first of the 'malformed' fruits which so fascinated Ferrari to be illustrated in the book. 'Nature plays nowhere more freely than in citrus fruit', he commented on p. 75. 'It so wonderfully and so sweetly makes various forms that it seems less to exercise the art of fruit than the art of abbreviation' [*siglinam*]. There are those, he observed, which expand into a digitated hand, others which swell into two breasts. Some seem to take on the form of the beak of a bird (here he may well have been thinking of plates such as the *Limon prosopaetos* on p. 515 of Montalbani's 1668 edition of Aldrovandi's *Dendrologia*, (see Fig. 26), others the crest of a rooster. Still others are horned like a deer or an ox. But why, asks Ferrari, should he enumerate the inexhaustible number of such forms? And he proceeds to describe this particular example in characteristic detail.

The multiform digitated citron can weigh up to 12 libbre. It is better to look at than to taste. The tree can live about forty years at most. Its forms are more extravagant as it grows older. It is regarded especially highly in Reggio Calabria; but it is not woven into hedges there, as in the case of the ordinary citron. It is prized simply because of the way it looks. In this manner Ferrari continues his eulogistic description of this type of citron. Its tree is taller than that of the ordinary citrus, its trunk thicker, its branches denser, its foliage thicker, longer, and greener. Its cortex is paler. It has sharp thorns, and frequently pricks.

Following his compact description of this citron, Ferrari speculates about a variety of possible reasons for its curious form. Could it be, he asks, because of the weakness of the seed that the fruit aborts? Or because a particular kind of fertilization results in several fruits (he refers to them as foetuses), some of which remain unformed, while others turn into graceful shapes, always, of course, slightly aborted? But then he stops himself, and says that the time has come to stop groping for philosophical explanations (*Et hactenus quidem philosophicas inter coniecturas caecutisse sit satis*). And he embarks on the first of his long poetic narratives, all of his own invention, in which he offers a wholly different form of explanation for the appearance of digitated citrus fruit. On page 89, in his illustration of the story of the unhappy fate of the youthful singing prodigy Harmonillus (see p. 74, Figs. 15, 28), Andrea Sacchi has shown the young man in the process of being transformed into a citron tree. His thumbs begin to sprout — one on each side — specimens of what are clearly the present *Malum citreum digitatum multiforme*. The little finger of his left hand, now clearly transformed into the branches of a citrus tree, seems to sprout another digitated fruit, here closest of all to the digitated variety of the *limon a rivo seu rio* (see *Hesperides* p. 215, and **81–88**).

De Noter called this citron *Cédratier-poncire digité*. It is a large bumpy fruit, apically split into three protruding appendages caused by the *Aceria sheldoni* mite (see p. 281), or even by endogenous floral disorders like ovary extrusions or staminal concretion.

PROVENANCE: dal Pozzo; Albani; George III, by descent to George V; London art market (Mendelson)

LITERATURE: Freedberg 1989a, p. 34; Freedberg 1989b, pp. 55–7, pls. 20–1

REFERENCES: Ferrari 1646, pp. 75–6; Montalbani 1668, p. 515 (see Fig. 26, p. 70); Micheli *Enumeratio*, p. 98, pl. 98; De Noter 1926, pp. 149–50; Baldini 1991, pp. 152–8

ENGRAVINGS: Ferrari 1646, p. 77

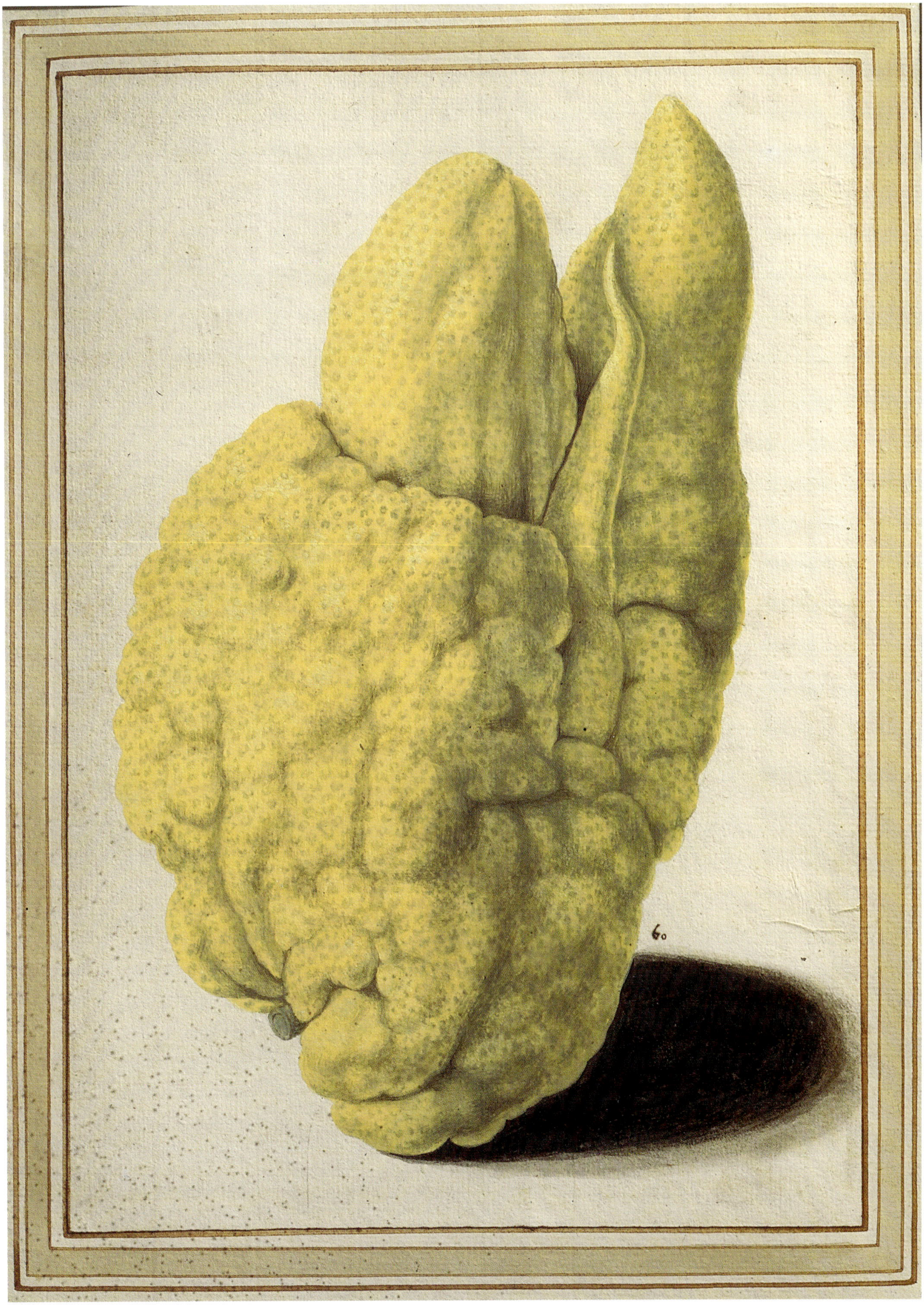
60

73. Citron, *Citrus medica* L.: misshapen half-fruit and seeds

VINCENZO LEONARDI (fl. 1621–46)

Windsor, RL 19347

Watercolour and bodycolour
over black chalk, with incised outlines

199 × 231 mm. Cream prepared paper

ANNOTATION:[below fruit] *61*

MOUNT SHEET: 435 × 586 mm. Some staining at edges and small tear at right-hand side. *Watermark*: D & C Blauw

This drawing shows the irregular cross-section and seeds (*Idem citreum dissectum eiusque semina*) of the digitated citron (*Malum citreum digitatum multiforme*) in Ferrari's *Hesperides*, p. 79, and discussed in the preceding entry. The rind is formed by a very thin flavedo and by a very thick white albedo. Segments are lacking. In spite of this, three small comma-shaped seeds are also shown, one of which is still inside the fruit.

PROVENANCE: dal Pozzo; Albani; George III, by descent to George V; London art market (Mendelson); Rex Nan Kivell; Her Majesty The Queen

LITERATURE: Freedberg 1989a, p. 34; Freedberg 1989b, pp. 55–7, pls. 20–1

REFERENCES: Ferrari 1646, pp. 75–6; Montalbani 1668, p. 515; Micheli *Enumeratio*, p. 98, pl. 98; De Noter 1926, pp. 149–50; Baldini 1991, pp. 152–8

ENGRAVINGS: Ferrari 1646, p. 79

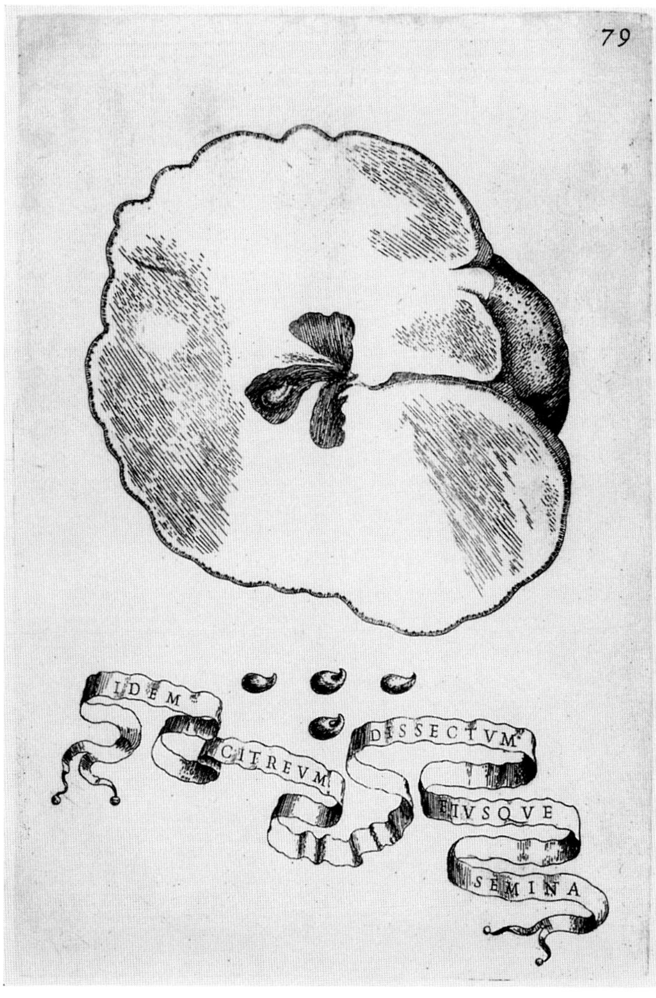

Comp. fig. 73: Ferrari 1646 p. 79

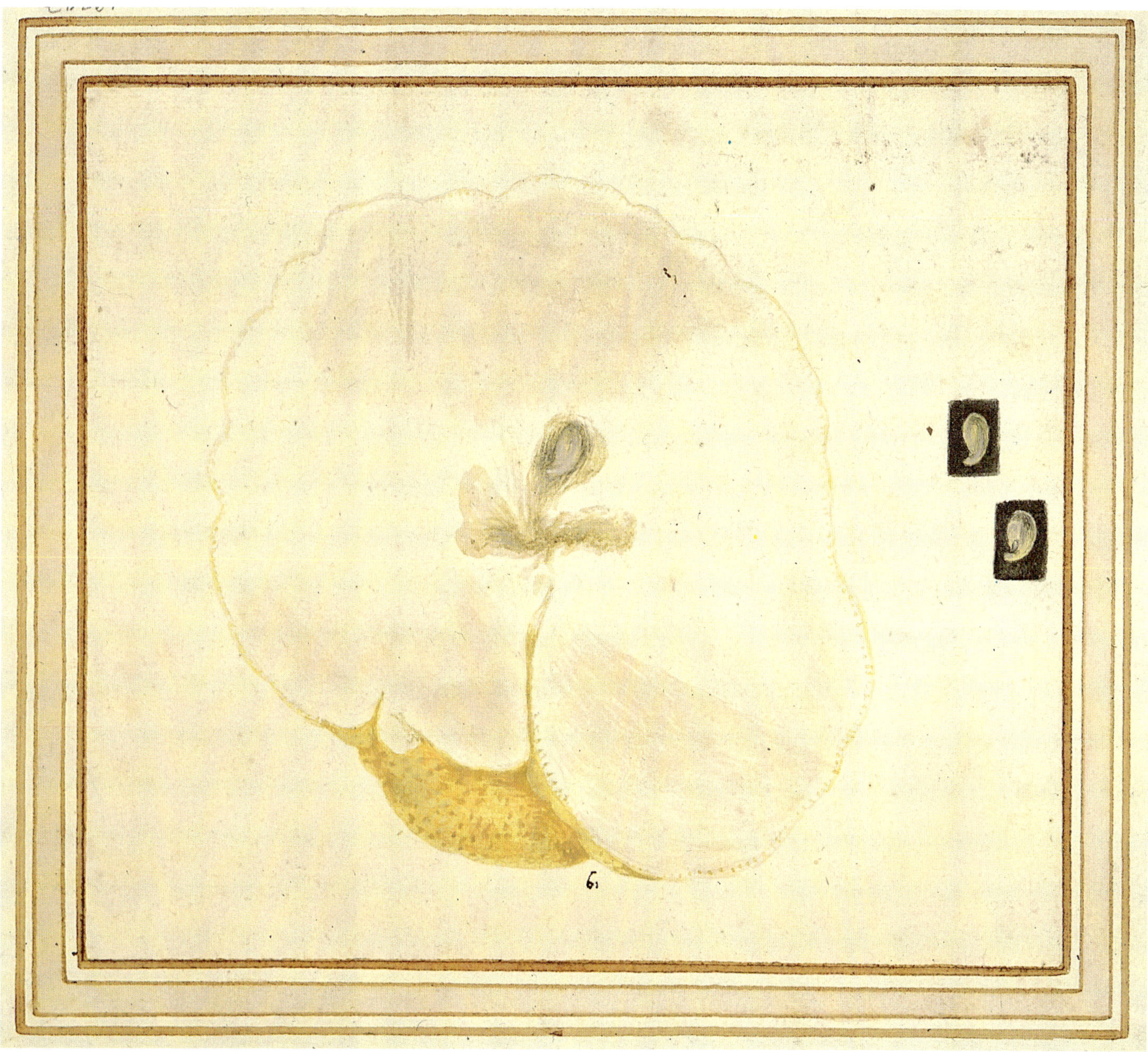

73

74 . Citron, *Citrus medica* L.: belted whole fruit

VINCENZO LEONARDI (fl. 1621–46)

Windsor, RL 19359

Watercolour and bodycolour, with gum heightening, over black chalk

175 × 238 mm. Cream prepared paper. Stained top right after mounting

ANNOTATION: [lower left] *73*

MOUNT SHEET: 435 × 586 mm. Water-damaged and stained. Large crease running top left to centre right. *Watermark*: D & C Blauw.
Annotation: [verso: pencil, in a 20th-century hand] *33*

This drawing does not correspond to any specific illustration in Ferrari's *Hesperides*, although its longitudinal ridge and the bumpiness of its skin does call to mind the similar characteristics of the *Malum citreum vulgare belluatum* illustrated on p. 65 of the *Hesperides*. It also resembles a striped citron picked in 1723 in the Grand Duke's Vagaloggia gardens near Florence for Bartolomeo Bimbi, who depicted it in a small painting which is now in the Museo Botanico, University of Florence (Bimbi, DBF 17). In all these citrons the irregularly bumpy yellow belt is marked by a green longitudinal rib attributable to a mericlinal autogenous cytochimera.

For other ribbed fruits, see not only the following entry, but also **96**, **100**, **103**, **104** and **107**.

PROVENANCE: dal Pozzo; Albani; George III, by descent to George V; London art market (Mendelson); Rex Nan Kivell; Her Majesty The Queen

REFERENCES: Ferrari 1646, p. 70f., pl. 65; Jonston 1662, pl. VI; Bimbi, DBF 17; Baldini 1982, p. 31; Tilney-Bassett 1986, p. 42; Baldini 1991, pp. 131–45

75. Citron *Citrus medica* L.: ribbed whole fruit

VINCENZO LEONARDI (fl. 1621–46)

Windsor, RL 19322

Watercolour and bodycolour, with gum heightening

117 × 155 mm. Cream prepared paper

ANNOTATION: *A*

MOUNT SHEET: 586 × 435 mm. Mounted on same sheet as **76**. Stained. Crease bottom right corner. *Watermark*: D & C Blauw.
Annotation: [verso: pencil, in a 20th-century hand] *£4.4.0.*

Not reproduced in Ferrari's *Hesperides*, the drawing shows an ellipsoid, apiculate yellow-greenish citron with a pointed apex and a narrow green rib running between the poles.

For other ribbed fruits, see also **74, 96, 100, 103, 104** and **107**.

PROVENANCE: dal Pozzo; Albani; George III, by descent to George V; London art market (Mendelson); Rex Nan Kivell; Her Majesty The Queen

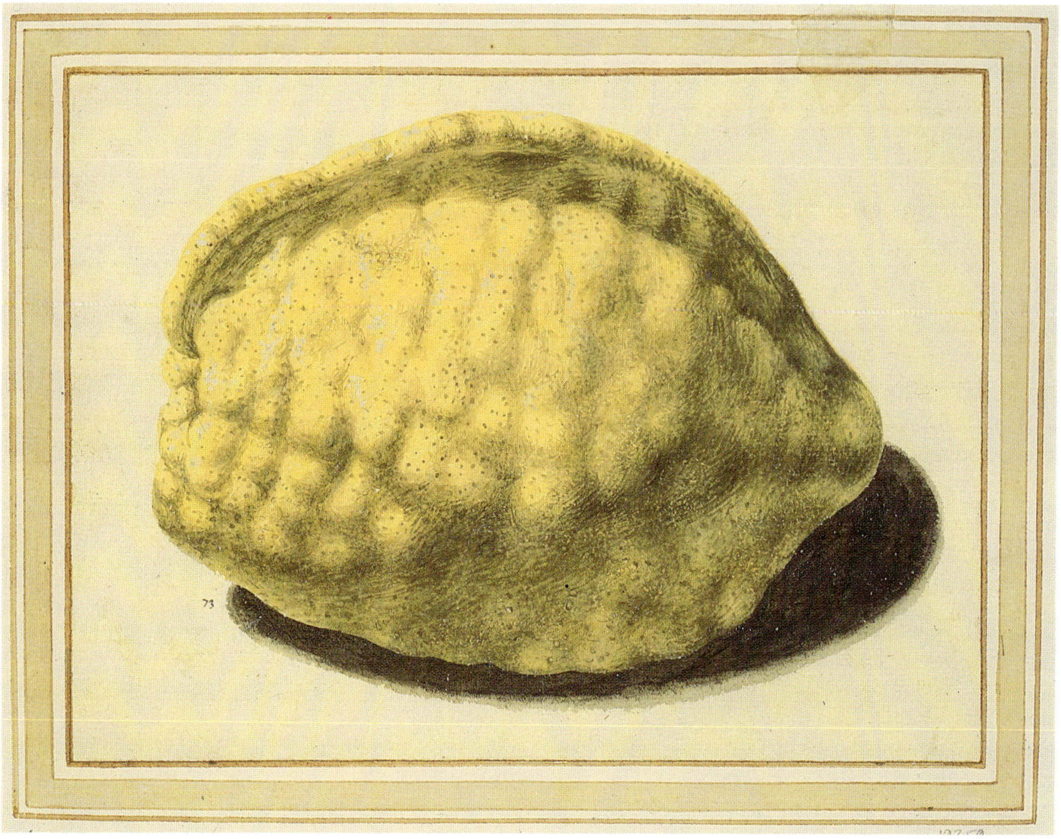

74

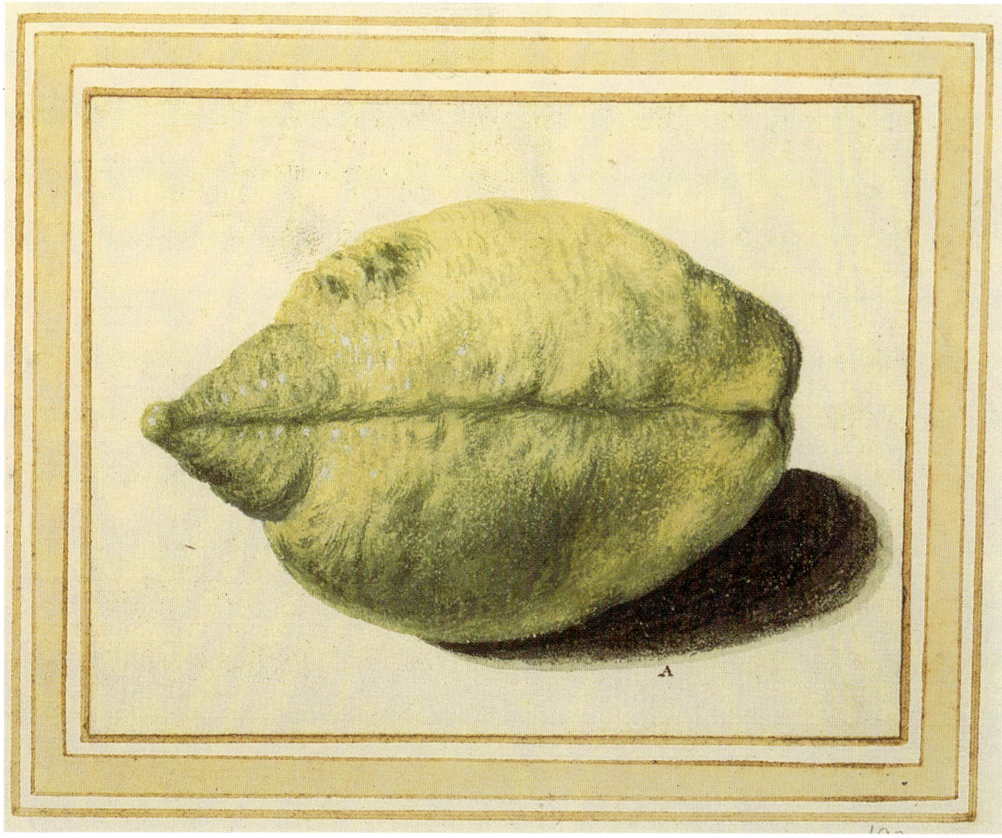

75

76. Citron, *Citrus medica* L.: ribbed half-fruit

VINCENZO LEONARDI (fl. 1621–46)
Windsor, RL 19321
Watercolour and bodycolour, with gum heightening
117 × 112 mm. Cream prepared paper
MOUNT SHEET: mounted on same sheet as **75**

Not reproduced in the *Hesperides*, this drawing must originally have been on the same drawing sheet as the whole fruit (see **69**), the two later separated, and then in the 18th century mounted together when they entered the collection of George III.

PROVENANCE: dal Pozzo; Albani; George III, by descent to George V; London art market (Mendelson); Rex Nan Kivell; Her Majesty The Queen

Although the rib trace does not appear on its cross-section, it could well be a half of the citron which is shown in **75**. It has a very thick mesocarp enclosing an inconspicuous, juiceless endocarp. Only a single segment contains a seed.

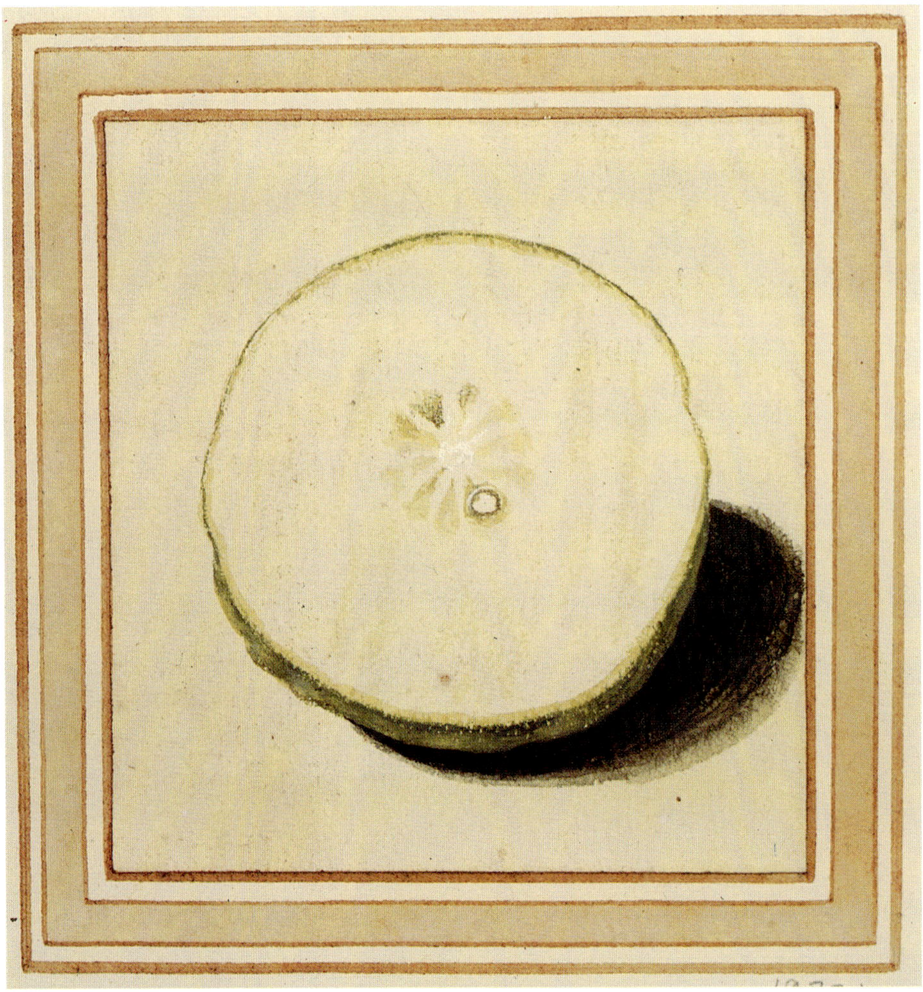

76

77. Citron, *Citrus medica* L.: whole and half-fruit with an apical pellet

VINCENZO LEONARDI (fl. 1621–46)

London art market

Watercolour and bodycolour, with gum heightening, over black chalk

388 × 279 mm. Foxed. *Watermark*: none recorded

ANNOTATION: *00*

Not reproduced in Ferrari's *Hesperides*, this drawing shows a large, yellow citron with a ball-shaped apical appendage, perhaps an hypertrophic nipple or a protruded secondary fruitlet. A lanceolate, serrate leaf, with a wingless petiole, is also represented. The half-fruit shows a very thick mesocarp surrounding small, yellow, seeded segments around a large axis. The whole fruit is somewhat similar to the apically belted citron in Risso/Poiteau, pl. 101 (*Cédratier de Salò*).

PROVENANCE: dal Pozzo; Albani; George III, by descent to George V; London art market (Mendelson); James R. Herbert Boone; Sotheby's (New York), 16 September 1988, lot 148

78. Citron, *Citrus medica* L.: crowned whole and half-fruit

VINCENZO LEONARDI (fl. 1621–46)

Location unknown

Watercolour and bodycolour, with gum heightening, over black chalk

357 × 204 mm. Foxed

ANNOTATION: [right of whole fruit] *T*

Not reproduced in Ferrari's *Hesperides*, this yellow citron has a 'crown' of irregular bumps around its apex ending in a very prominent nipple. The half-fruit shows a very thick mesocarp surrounding small segments with a few small seeds. This drawing is known only from a 35 mm colour slide made by the Courtauld Institute in 1967.

PROVENANCE: dal Pozzo; Albani; George III, by descent to George V; London art market (Mendelson)

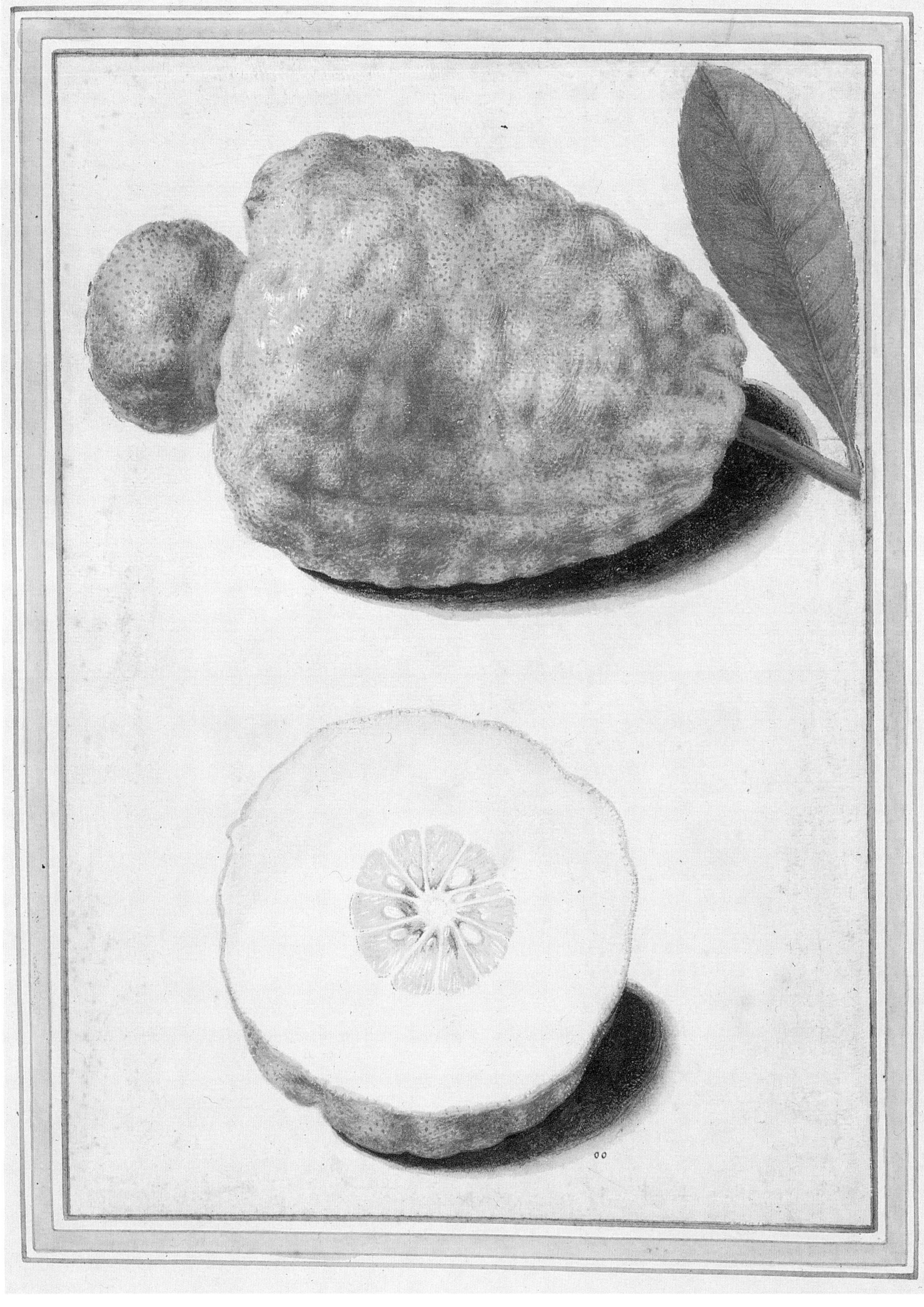

77

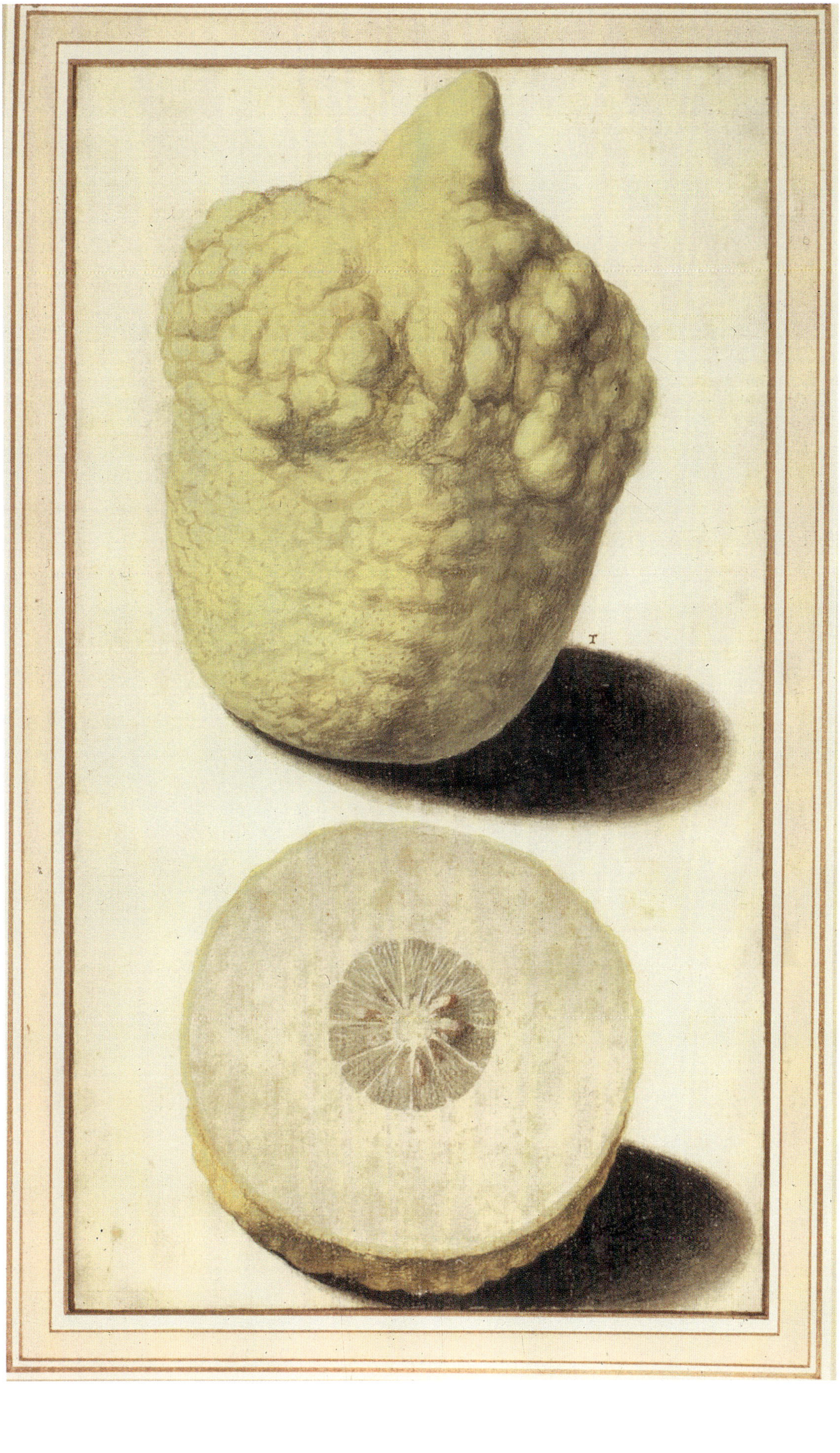

79. Citron, *Citrus medica* L.: misshapen whole fruit

VINCENZO LEONARDI (fl. 1621–46)

Windsor, RL 19324A

Watercolour and bodycolour, with gum heightening, over black chalk

114 × 139 mm. Cream prepared paper

ANNOTATION: *D*

MOUNT SHEET: 585 × 434 mm. Mounted with **80**, within the same wash border. Stained. Tear at top, repaired on verso, larger tear bottom right.
Watermark: D & C Blauw.
Annotation: [verso: pencil, in a 20th-century hand] *25*

This drawing and **80** once formed part of the same sheet; they were then separated, and were finally mounted together again on the same sheet in the 18th-century rearrangement of the drawings. They were not reproduced in Ferrari's *Hesperides*. The irregular bumpy apex of this fruit is presumably the result of a bud-mite injury. The rind is still green, but thick and roughly papillose. The seeds in the light yellow segments are large and numerous.

PROVENANCE: dal Pozzo; Albani; George III, by descent to George V; London art market (Mendelson); Rex Nan Kivell; Her Majesty The Queen

80. Citron, *Citrus medica* L.: misshapen half-fruit

VINCENZO LEONARDI (fl. 1621–46)

Windsor, RL 19324

Watercolour and bodycolour, over black chalk; blackened lead white highlights

112 × 139 mm. Cream prepared paper

MOUNT SHEET: mounted with **79**, within same wash border

See **79**

PROVENANCE: dal Pozzo; Albani; George III, by descent to George V; London art market (Mendelson); Rex Nan Kivell; Her Majesty The Queen

79

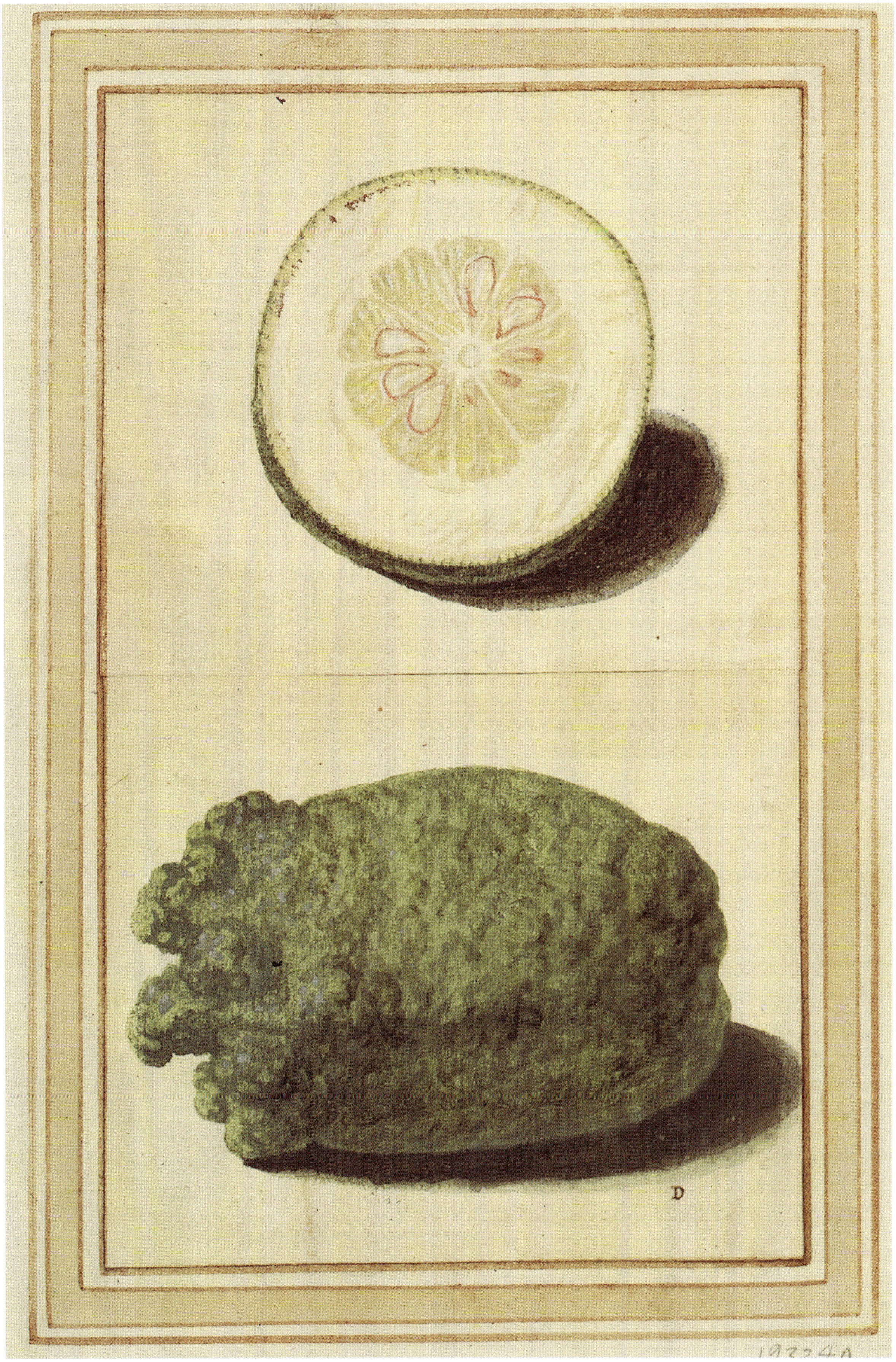

80

81. Lemon, *Citrus limon* (L.) Burm. f.: fingered whole fruit

VINCENZO LEONARDI (fl. 1621–46)

Windsor, RL 19379

Watercolour and bodycolour over black chalk

117 × 139 mm

MOUNT SHEET: 117 × 139 mm (trimmed to size of drawing sheet)
18th-century mount paper. Survival of 5 mm overlay of original
17th-century mount between drawing sheet and 18th-century mount

On the same page as the illustration of the ordinary *Limon a rivo seu rio* on p. 215 of the *Hesperides* (**20**), it was decided to include also, on the lower left, one teratological digitated specimen. The present drawing clearly served as the preparatory design for it. This particular kind of digitation is of the same order as four other drawings of digitated specimens also commissioned by Cassiano but it is of much higher quality than those now rather worn sheets (**82–85**). Ferrari was fascinated by the phenomenon of digitation, as he was by other teratological forms (see **72**). Thus, at the end of his chapter on the *Limon a rivo*, he notes that this particular type of lemon often produces digitated forms, and for all its arboreal origins, dares to imitate human forms. Several examples of such births — or rather misbirths — could be found: *Quin etiam nonnullae aliae limonum stirpes ludum hunc nonnunquam exhibent, abortu verius, quam partu, mirabiliores*. In this case the fruit seems to have been brought to Rome from Genoa, perhaps from a tree of *ponzini* (see **20**). The idea of a nature that played, even as it produced aborted fruits, was one that fascinated Ferrari and his contemporaries. He did not know that the digitation arose from chance infestation by the *Aceria sheldoni* mite affecting flower buds, or from genetic factors. In this chapter he simply concludes with a kind of apostrophe to a tree which so profusely — and so generously — produced specimens that seemed to be nothing less than human hands reaching out to one. No wonder then, that in his illustration of the story of the transformation of the young Harmonillus into a citron tree, with his own hands turning into citrons and digitated citrons (see also **72**), Andrea Sacchi also managed to include one fruit that looked almost exactly like the digitated *Limon a rivo seu rio*. No drawing of the half-fruit survives among the known drawings in the Paper Museum. Digitated fruits, popularly called *maravigghi* in Sicily, occasionally occur on lemon trees, but seldom on other citrus species.

It would appear that the five drawings commissioned by Cassiano to illustrate digitated fruits were originally inlaid into one 17th-century mount sheet, then removed from it after their arrival in the British Royal Collection and laid down onto 18th-century mount paper. After the 20th-century dispersal of this part of the Paper Museum (see p. 22) it would seem that this folio sheet suffered particular damage and the five drawings were therefore removed from it at some stage prior to their return to the Royal Library. Of the five drawings, only this one retains any part of the 18th-century mount.

Three other drawings of digitated lemons (**86**, **87** and **88**) also survive, in much better condition, and were presumably commissioned on a different occasion and perhaps for a different purpose. On the contrary, the corresponding drawings of *Aurantium distortum*, etched on p. 415 of the *Hesperides*, are lacking in Cassiano's Paper Museum.

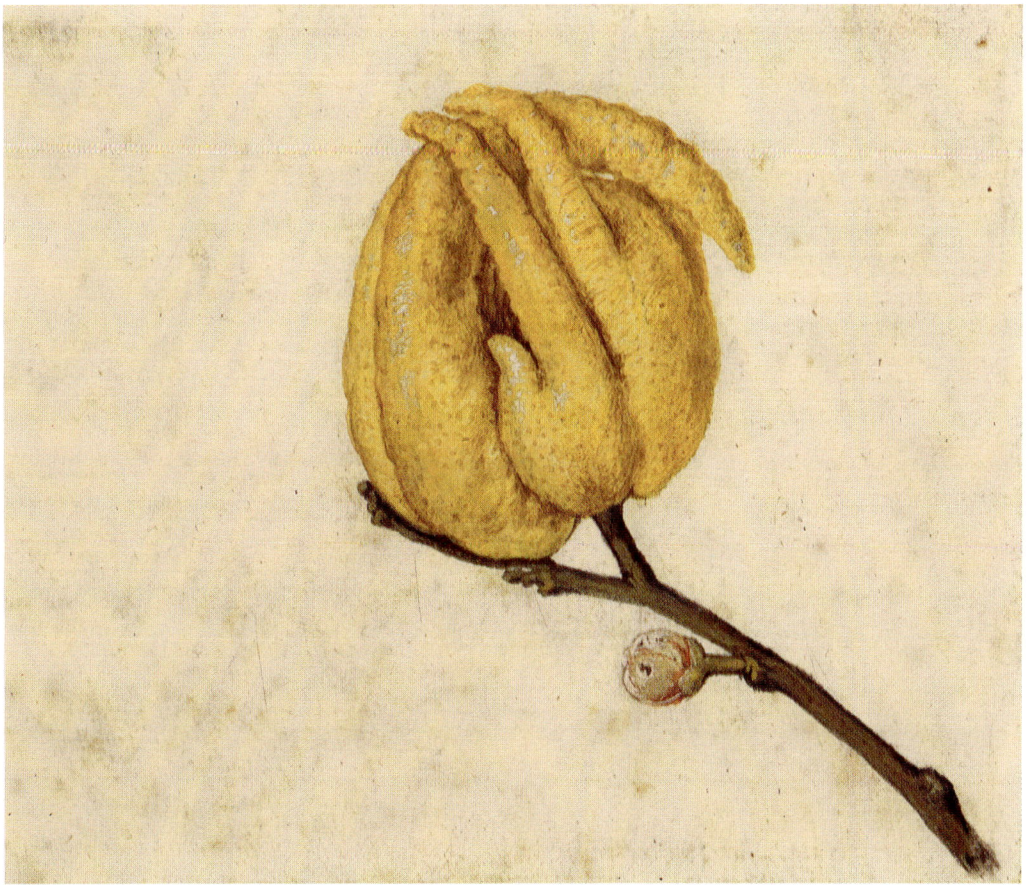

81

PROVENANCE: dal Pozzo; Albani; George III, by descent to George V; London art market (Mendelson); Rex Nan Kivell; Her Majesty The Queen

LITERATURE: Freedberg 1989a, p. 34; Freedberg 1989b, pp. 52–4; Freedberg 1993, p.180f.

REFERENCES: Aldrovandi MS 88, c.10; Ferrari 1646, p. 213; Jonston 1662, pl. VI; Montalbani 1668, p. 513f.; Cupani 1696, p. 130; Volkamer *Nürnberg. Hesp.* I, pl. 146b, II, p. 94f., pl. 96; Risso/Poiteau 1818, p. 192, pl. 96; unpublished drawing on unnumbered folio, GAG; De Noter 1926, p. 142; Ewing 1937, pp. 7–13; Boyce/Korsmeyer, 1941, pp. 745–56; Baldini 1991, pp. 152–8; 192; Baldini 1996b, p. 18, pl. 14

ENGRAVINGS: Ferrari 1646, p. 215 (See Comp. figs. 20, 21, p. 130)

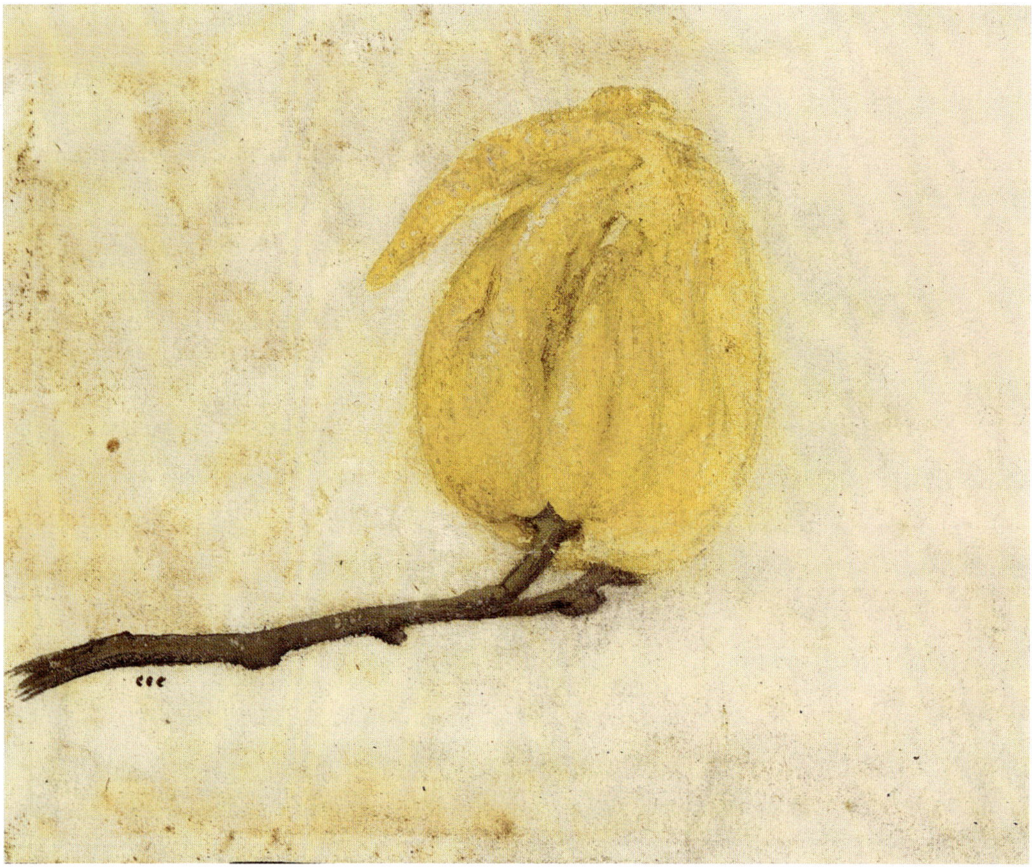

82

83

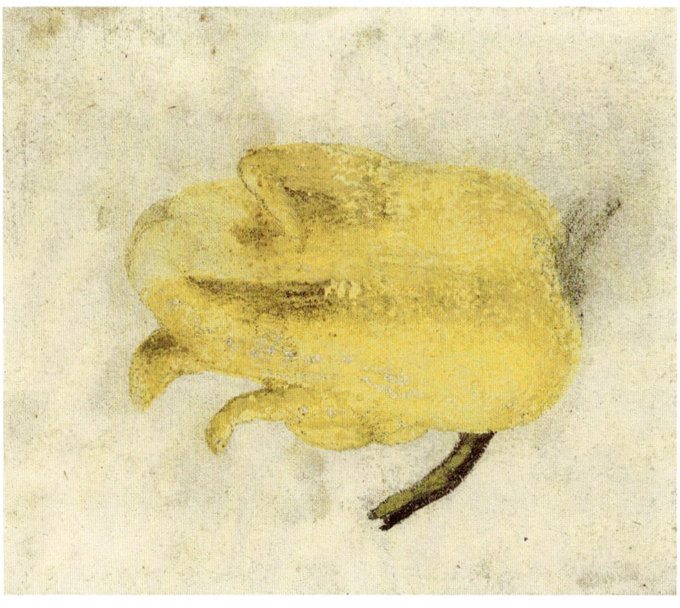

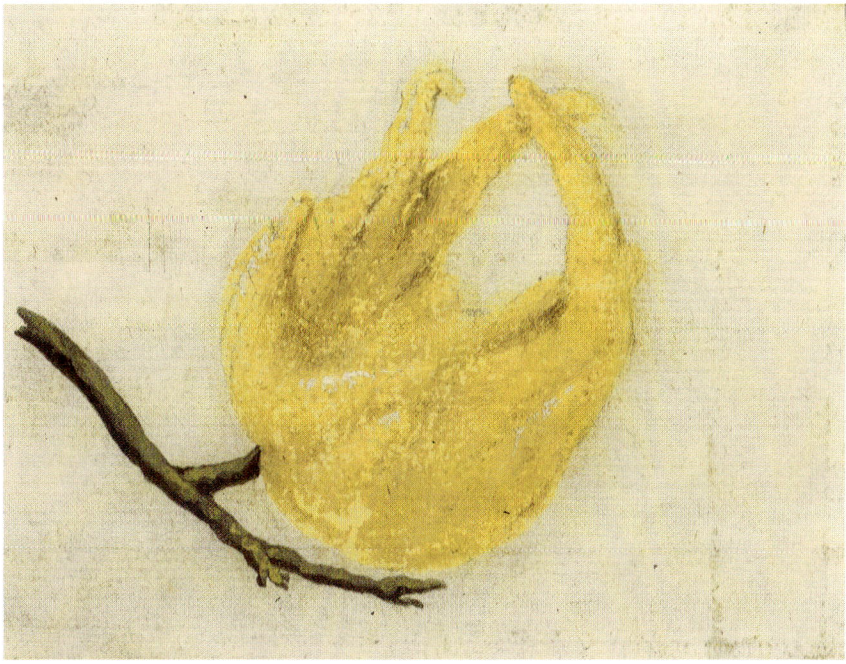

84

85

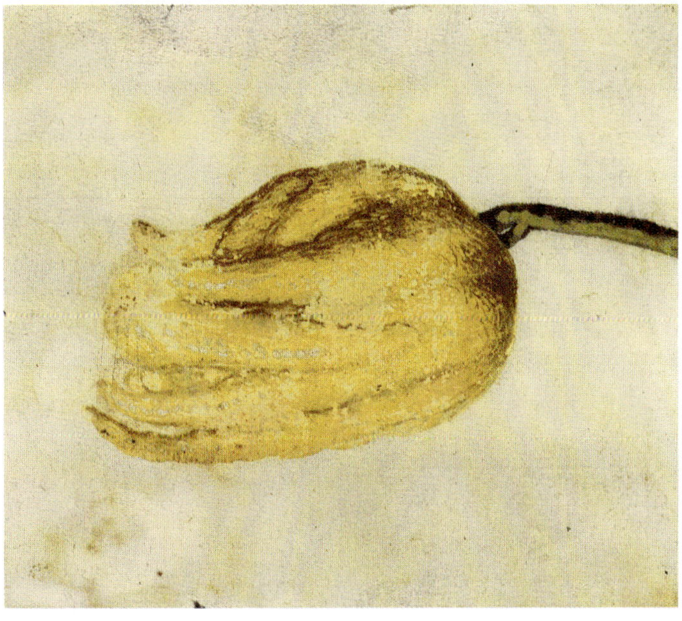

82. Lemon, *Citrus limon* (L.) Burm. f.: fingered whole fruits

VINCENZO LEONARDI (fl. 1621–46)

Windsor, RL 19380

Watercolour and bodycolour over black chalk, badly rubbed

115 × 140 mm. Marks of original 17th-century inlay (5 mm overlap) visible on verso. Edges torn and damaged

ANNOTATION: [below stalk] *ccc*

MOUNT SHEET: removed from mount; see **81**

For comments on this fruit, see **81**, **86**, and **88**. See **83–8** for the companion studies to this drawing.

PROVENANCE: dal Pozzo; Albani; George III, by descent to George V; London art market (Mendelson); Rex Nan Kivell; Her Majesty The Queen

LITERATURE: Freedberg 1989a, p. 34; Freedberg 1989b, pp. 52–4; Freedberg 1993, p.180f.

REFERENCES: Aldrovandi MS 88, c.10; Ferrari 1646, p. 213; Jonston 1662, pl. VI; Montalbani 1668, p. 513f.; Cupani 1696, p. 130; Volkamer *Nürnberg. Hesp.* I, pl. 146b, II, p. 94f., pl. 96; Risso/Poiteau 1818, p. 192, pl. 96; unpublished drawing on unnumbered folio, GAG; De Noter 1926, p. 142; Ewing 1937, pp. 7–13; Boyce/Korsmeyer, 1941, pp. 745–56; Baldini 1991, pp. 152–8; 192; Baldini 1996b, p. 18, pl. 14

83. Lemon, *Citrus limon* (L.) Burm. f.: fingered whole fruit

VINCENZO LEONARDI (fl. 1621–46)

Windsor, RL 19381

Watercolour and bodycolour over black chalk, badly rubbed

78 × 93 mm. Marks of original 17th-century inlay (5 mm overlap) visible on verso

MOUNT SHEET: removed from mount; see **81**

This drawing was not reproduced in Ferrari's *Hesperides*. Like the other drawings in this group (**81–5**), it is rubbed and in poorer condition than the other citrus drawings which came from Cassiano's collection. No drawing of the half-fruit survives among the known drawings in the Paper Museum.

PROVENANCE: dal Pozzo; Albani; George III, by descent to George V; London art market (Mendelson); Rex Nan Kivell; Her Majesty The Queen

LITERATURE: Freedberg 1989a, p. 34; Freedberg 1989b, pp. 52–4; Freedberg 1993, p.180f.

REFERENCES: Aldrovandi MS 88, c.10; Ferrari 1646, p. 213; Jonston 1662, pl. VI; Montalbani 1668, p. 513f.; Cupani 1696, p. 130; Volkamer *Nürnberg. Hesp.* I, pl. 146b, II, p. 94f., pl. 96; Risso/Poiteau 1818, p. 192, pl. 96; unpublished drawing on unnumbered folio, GAG; De Noter 1926, p. 142; Ewing 1937, pp. 7–13; Boyce/Korsmeyer, 1941, pp. 745–56; Baldini 1991, pp. 152–8; 192; Baldini 1996b, p. 18, pl. 14

84. Lemon, *Citrus limon* (L.) Burm., f.: fingered whole fruit

VINCENZO LEONARDI (fl. 1621–46)

Windsor, RL 19382

Watercolour and bodycolour over black chalk, badly rubbed

88 × 115 mm. Marks of original 17th-century inlay (5 mm overlap) visible on verso

MOUNT SHEET: removed from mount; see **81**

For comments on this drawing see **81**, **86**, and **88**.

PROVENANCE: dal Pozzo; Albani; George III, by descent to George V; London art market (Mendelson); Rex Nan Kivell; Her Majesty The Queen

LITERATURE: Freedberg 1989a, p. 34; Freedberg 1989b, pp. 52–4; Freedberg 1993, p.180f.

REFERENCES: Aldrovandi MS 88, c.10; Ferrari 1646, p. 213; Jonston 1662, pl. VI; Montalbani 1668, p. 513f.; Cupani 1696, p. 130; Volkamer *Nürnberg. Hesp.* I, pl. 146.b, II, p. 94f., pl. 96; Risso/Poiteau 1818, p. 192, pl. 96; unpublished drawing on unnumbered folio, GAG; De Noter 1926, p. 142; Ewing 1937, pp. 7–13; Boyce/Korsmeyer, 1941, pp. 745–56; Baldini 1991, pp. 152–8; 192; Baldini 1996b, p. 18, pl. 14

85. Lemon, *Citrus limon* (L.) Burm. f.: fingered whole fruit

VINCENZO LEONARDI (fl. 1621–46)

Windsor, RL 19383

Watercolour and bodycolour over black chalk, badly rubbed

80 × 93 mm. Marks of original 17th-century inlay (5 mm overlap) visible on verso. Tears and staining top and bottom

MOUNT SHEET: removed from mount; see **81**

For comments on this drawing see **81**, **86**, and **88**.

PROVENANCE: dal Pozzo; Albani; George III, by descent to George V; London art market (Mendelson); Rex Nan Kivell; Her Majesty The Queen

LITERATURE: Freedberg 1989a, p. 34; Freedberg 1989b, pp. 52–4; Freedberg 1993, p. 180f.

REFERENCES: Aldrovandi MS 88, c.10; Ferrari 1646, p. 213; Jonston 1662, pl. VI; Montalbani 1668, p. 513f.; Cupani 1696, p. 130; Lanzoni 1690; Volkamer *Nürnberg. Hesp.* I, pl. 146b, II, p. 94f., pl. 96; Risso/Poiteau 1818, p. 192, pl. 96; unpublished drawing on unnumbered folio, GAG; De Noter 1926, p. 142; Ewing 1937, pp. 7–13; Boyce/Korsmeyer, 1941, pp. 745–56; Baldini 1991, pp. 152–8, 192; Baldini 1996b, p. 18, pl. 14

86. Lemon, *Citrus limon* (L.) Burm. f.: fingered whole fruit

VINCENZO LEONARDI (fl. 1621–46)

Windsor, RL 19356

Watercolour and bodycolour, with gum heightening, over black chalk

149 × 139 mm

ANNOTATION: [below] *69*

MOUNT SHEET: 254 × 434 mm. Mounted with **87**, within the same wash border. Tears, one repaired on verso. Creased bottom left.
Watermark: D & C Blauw.
Annotation: [verso: pencil, in a 20th-century hand] *£8.8.0 78*

Neither this drawing nor **87** was chosen for reproduction in Ferrari's *Hesperides*. They represent the opposite lateral sides of a fingered, greenish, horn-shaped lemon presumably arising from flower buds injured by *Aceria sheldoni*. Abnormal fruits like this one were also collected and illustrated by Aldrovandi, and they appear among zoomorphic citrus such as the *Limon chirodes* and several other digitated fruits in Montalbani's edition of the *Dendrologia* and the *Curae Analiticae*. Once again, no drawing of the half-fruit survives among the known drawings in the Paper Museum.

These two drawings presumably once formed part of the same sheet, were then cut in two, and were reunited at the time of their rearrangement in the collections of George III.

PROVENANCE: dal Pozzo; Albani; George III, by descent to George V; London art market (Mendelson); Rex Nan Kivell; Her Majesty The Queen

LITERATURE: Freedberg 1989a, p. 35, pl. 24; Freedberg 1992b, p. 297

REFERENCES: Montalbani 1668, pp. 83, 513, 515 and 534; Montalbani 1671, p. 6; Baldini 1991, pp. 152–8; Baldini 1996b, p. 18, pl. 14

87. Lemon, *Citrus limon* (L.) Burm. f.: fingered whole fruit

VINCENZO LEONARDI (fl. 1621–46)

Windsor, RL 19357

Watercolour and bodycolour, with gum heightening, over black chalk

149 × 140 mm

ANNOTATION: [below] *69*

MOUNT SHEET: mounted with **86**, within the same wash border

See **86**

PROVENANCE: dal Pozzo; Albani; George III, by descent to George V; London art market (Mendelson); Rex Nan Kivell; Her Majesty The Queen

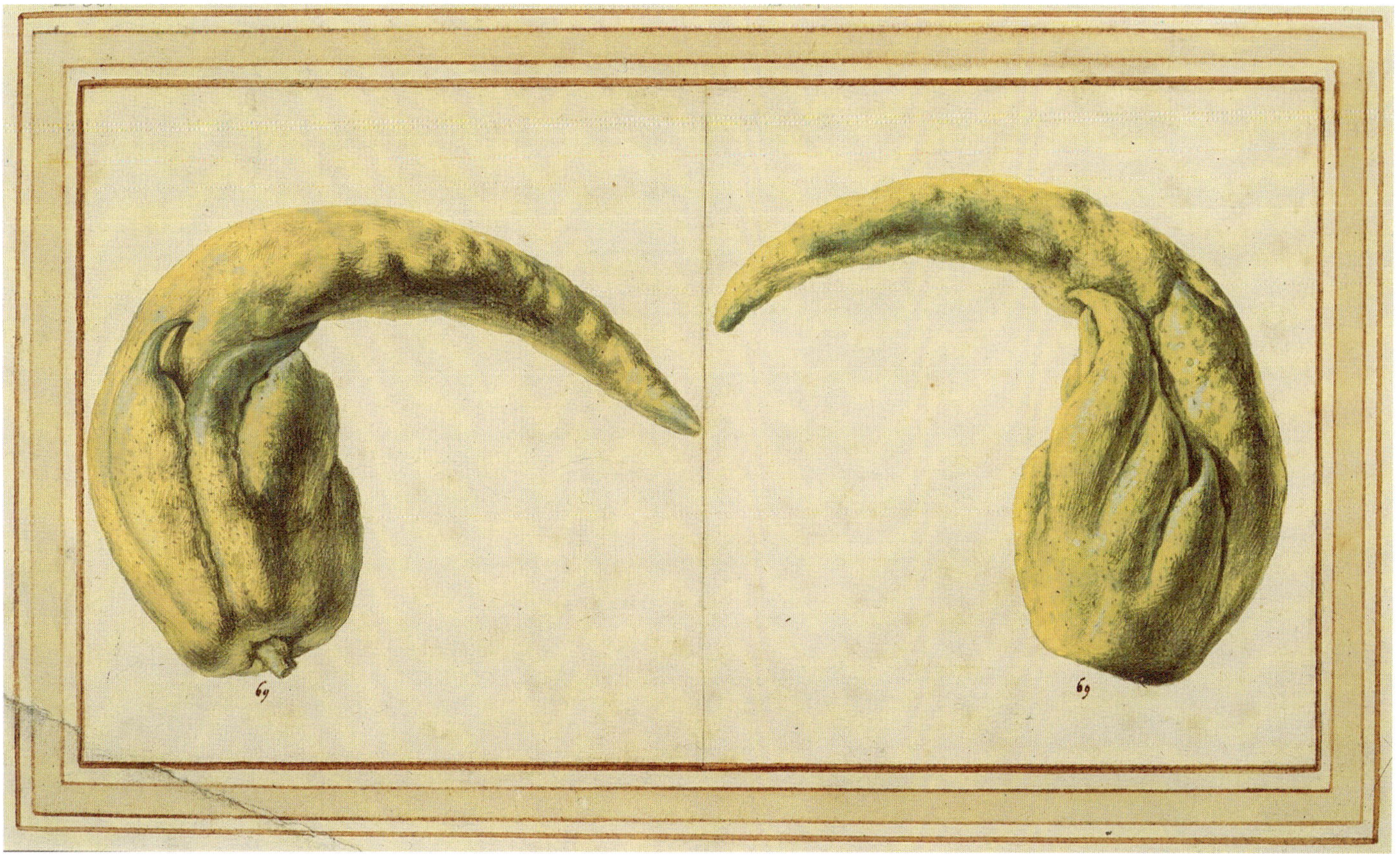

86 87

88. Lemon, *Citrus limon* (L.) Burm. f.: fingered whole fruit

VINCENZO LEONARDI (fl. 1621–46)

Windsor, RL 19358

Watercolour and bodycolour, with gum heightening, over black chalk

248 × 255 mm. Cream prepared paper. Drawing trimmed on left to protude slightly onto mount. *Watermark: Eagle ?15* (cut)

ANNOTATION: [below fruit] *70*

MOUNT SHEET: 326 × 345 mm. Stained. *Watermark*: D & C Blauw. *Annotation*: [verso: pencil, in a 20th-century hand] *86*

This spectacular drawing of a misshapen fruit was not reproduced in Ferrari's *Hesperides*. Nevertheless, its quality is so high and it is so carefully and meticulously drawn that one wonders — as in the case of several other drawings in the present corpus — whether it was not once intended for reproduction as well. Teratological specimens such as this, generally arising from flower buds damaged by *Aceria sheldoni*, differ from normal ones in that their apex protrudes irregularly as separate, large finger-like appendages. They usually have no pulp, and even if it is present it is generally very scanty. For Ferrari's 'explanation' of such fruits, see **72** and **81**, as well as p. 64. No drawing of the half-fruit of this particular specimen survives among the known drawings in the Paper Museum.

PROVENANCE: dal Pozzo; Albani; George III, by descent to George V; London art market (Mendelson); Rex Nan Kivell; Her Majesty The Queen

LITERATURE: Freedberg 1989a, p. 34; Freedberg 1989b, pp. 52–4; Freedberg 1993, p.180f.

REFERENCES: Aldrovandi MS 88, c.10; Ferrari 1646, p. 213; Jonston 1662, pl. VI; Montalbani 1668, p. 513f.; Cupani 1696, p. 130; Bimbi, Villa Med. C. 614; Volkamer *Nürnberg. Hesp.* I, pl. 146b, II, p. 94f.; pl. 96; Risso/Poiteau 1818, p. 192, pl. 96; De Noter 1926, p. 142; Baldini 1991, pp. 152–8 and 192; Baldini 1996b, p. 18, pl. 14

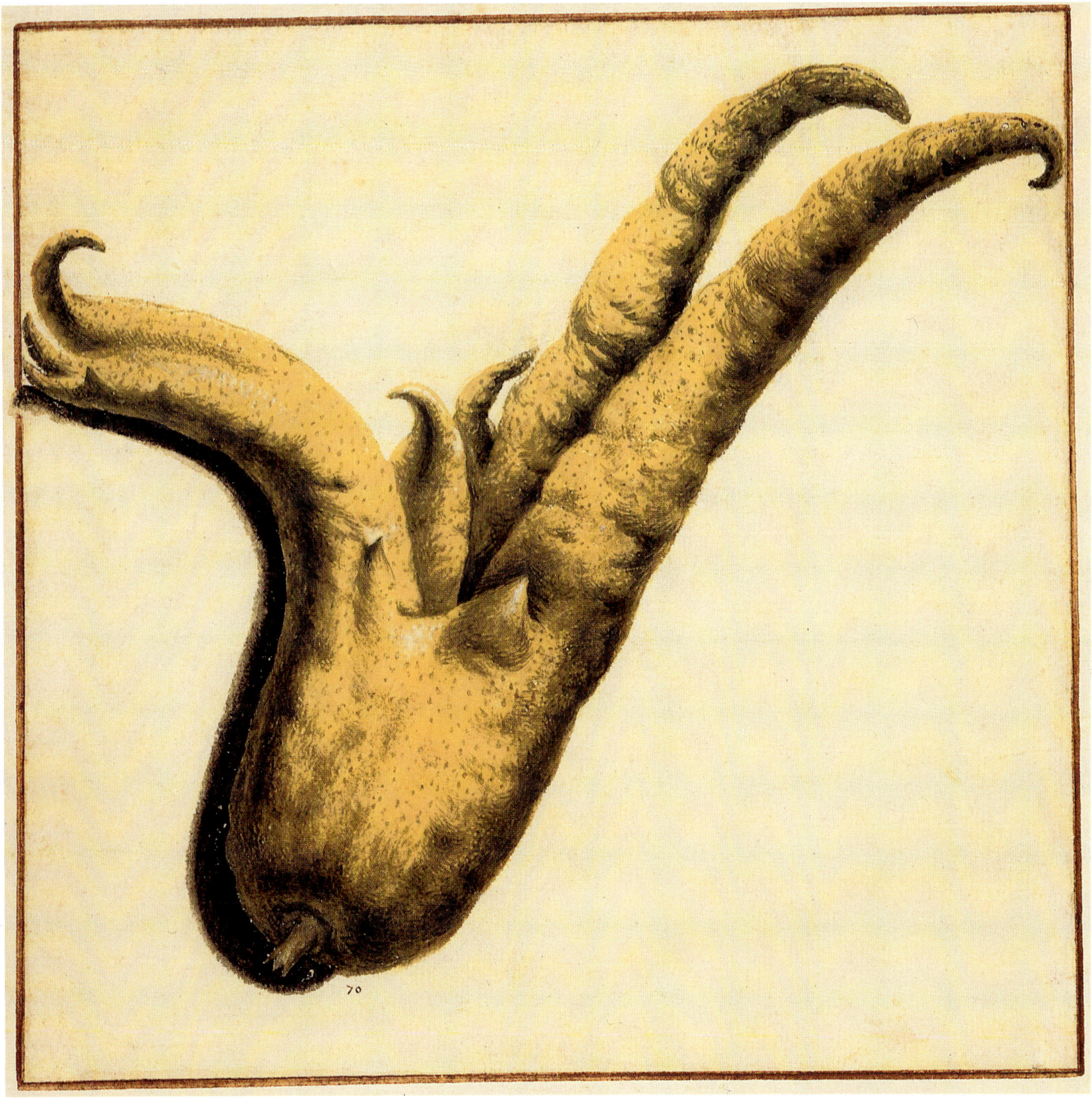

88

89. Citron-lemon, *Citrus limonimedica* Lush.: 'pregnant' whole fruit

VINCENZO LEONARDI (fl. 1621–46)

Windsor, RL 19363

Watercolour and bodycolour with gum heightening, over black chalk, with some incised lines

186 × 205 mm. Cream prepared paper. Damaged at right-hand edge

ANNOTATION: *77*

MOUNT SHEET: 587 × 435 mm. *Watermark*: D & C Blauw

In his very first chapter on citron-lemons, Ferrari observes that these fruits often produced pregnant specimens, that is, one fruit including another (see **48**, **71**, **90–94**). The phenomenon was one which interested Ferrari (and Cassiano) very deeply, as is also very clear from fol. 66 of the *Notizie diverse* prepared by Cassiano (see p. 51): *non si puo tra tutti gl'agrumi vedere frutto più degno d'ammirazione di questo del quale me vennero diverse volte inviate...* Ferrari gives instructions on how to cut such pregnant fruits, so that the inner fruits — much sweeter and tenderer — may be revealed, just as shown in the present drawing for p. 269 in Ferrari's *Hesperides*, showing a *Limon citratus alterum includens*. Here one can see, through a kind of artificial 'window' cut in the rind of this globose fruit, another fruit completely inside it. The leaf on the twig is elliptical and its petiole wingless. According to Ferrari this teratological specimen was the kind of fruit picked in a field outside Pietrasanta, and was a foetiferous, i.e. 'pregnant' kind of the *Limon citratus primae notae laevior* discussed in **48**.

PROVENANCE: dal Pozzo; Albani; George III, by descent to George V; London art market (Mendelson); Rex Nan Kivell; Her Majesty The Queen

LITERATURE: Freedberg 1989a, p. 24f., pl. 16; Freedberg 1992b, p. 300; Freedberg 1996a

REFERENCES: Aldrovandi MS 88, cc.4 and 8; Archivio Dal Pozzo MS 39, fol. 66; Ferrari 1646, p. 263f.; Jonston 1662, pl. XI; Volkamer *Nürnberg. Hesp.* II, p. 199f.; Baldini 1991, pp. 146–52; Baldini/Tagliaferri 1990, p. 25f.

ENGRAVINGS: Ferrari 1646, p. 269

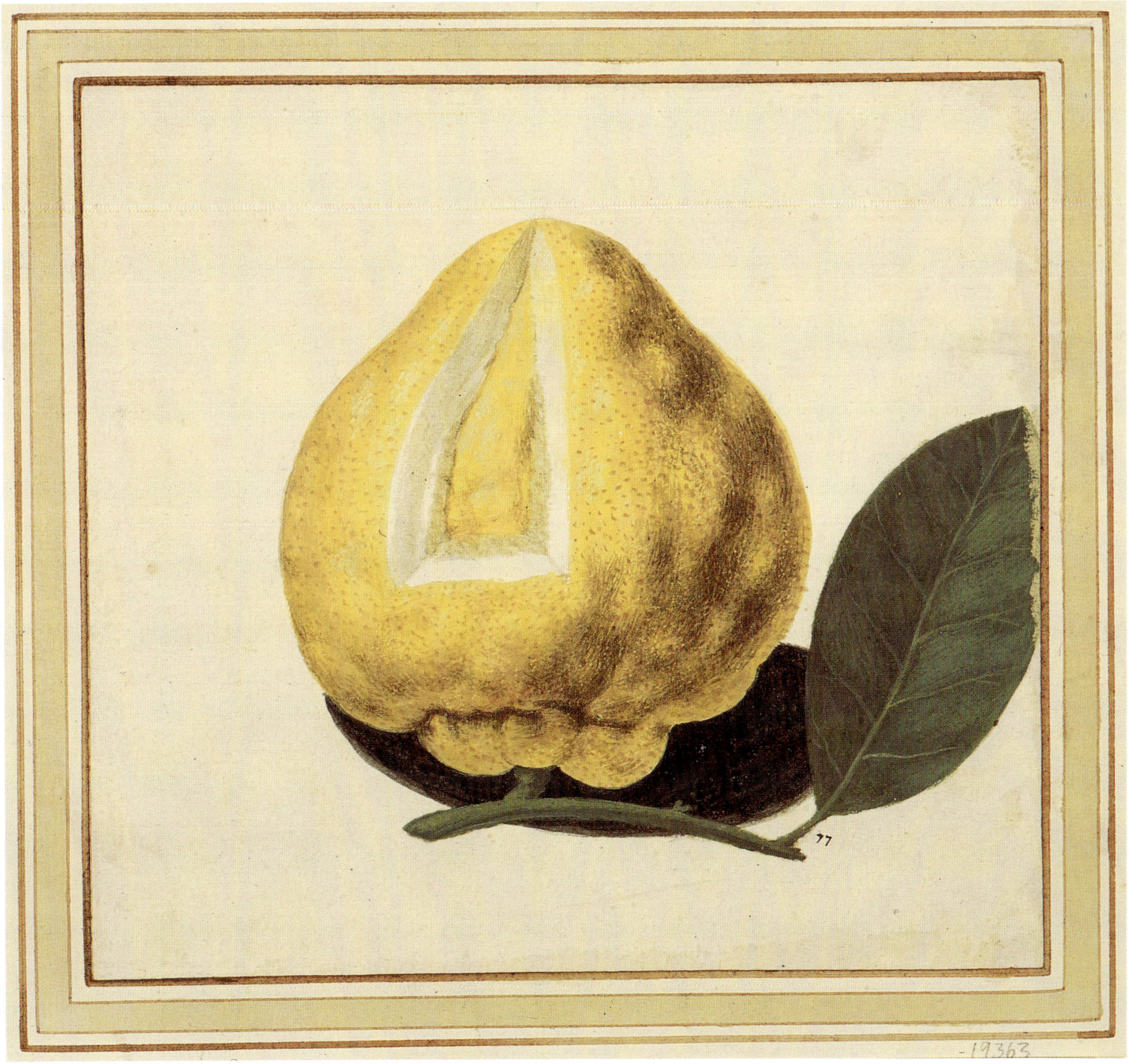

89

90. Citron-lemon, *Citrus limonimedica* Lush.: halves of a 'pregnant' fruit

VINCENZO LEONARDI (fl. 1621–46)

Windsor, RL 19362

Watercolour and bodycolour over black chalk, with incised outlines

185 × 288 mm. Cream prepared paper. *Watermark*: Crown 38 (cut)

ANNOTATION: [each view] *76*

MOUNT SHEET: 587 × 435 mm. Stained and marked. *Watermark*: D & C Blauw

This drawing shows two views of a section of the 'pregnant' citron-lemon discussed in the previous entry, but only the cross-section on the right was etched on p. 269 of Ferrari's *Hesperides*, below the etching of the whole fruit. Both sections here show the thick seedless mesocarp of the 'foetus', which fills nearly all the inside of the 'mother' fruit, surrounded by a thin layer of its residual segments. In some commercial orange cultivars, such as 'Washington navel', fruits regularly include a small, rudimentary fruitlet which can be seen through an apical small opening of the rind.

PROVENANCE: dal Pozzo; Albani; George III, by descent to George V; London art market (Mendelson); Rex Nan Kivell; Her Majesty The Queen

LITERATURE: Freedberg 1989a, p. 24f., pl. 17; Freedberg 1992b, p. 300

REFERENCES: Aldrovandi MS 88, cc.4 and 8; Archivio Dal Pozzo MS 39, fol. 66; Ferrari 1646, p. 263f.; Jonston 1662, pl. XI; Volkamer *Nürnberg. Hesp.* II, p. 199f.; Baldini 1991, pp. 146–52; Baldini/Tagliaferri 1990, p. 25f.

ENGRAVINGS: Ferrari 1646, p. 269

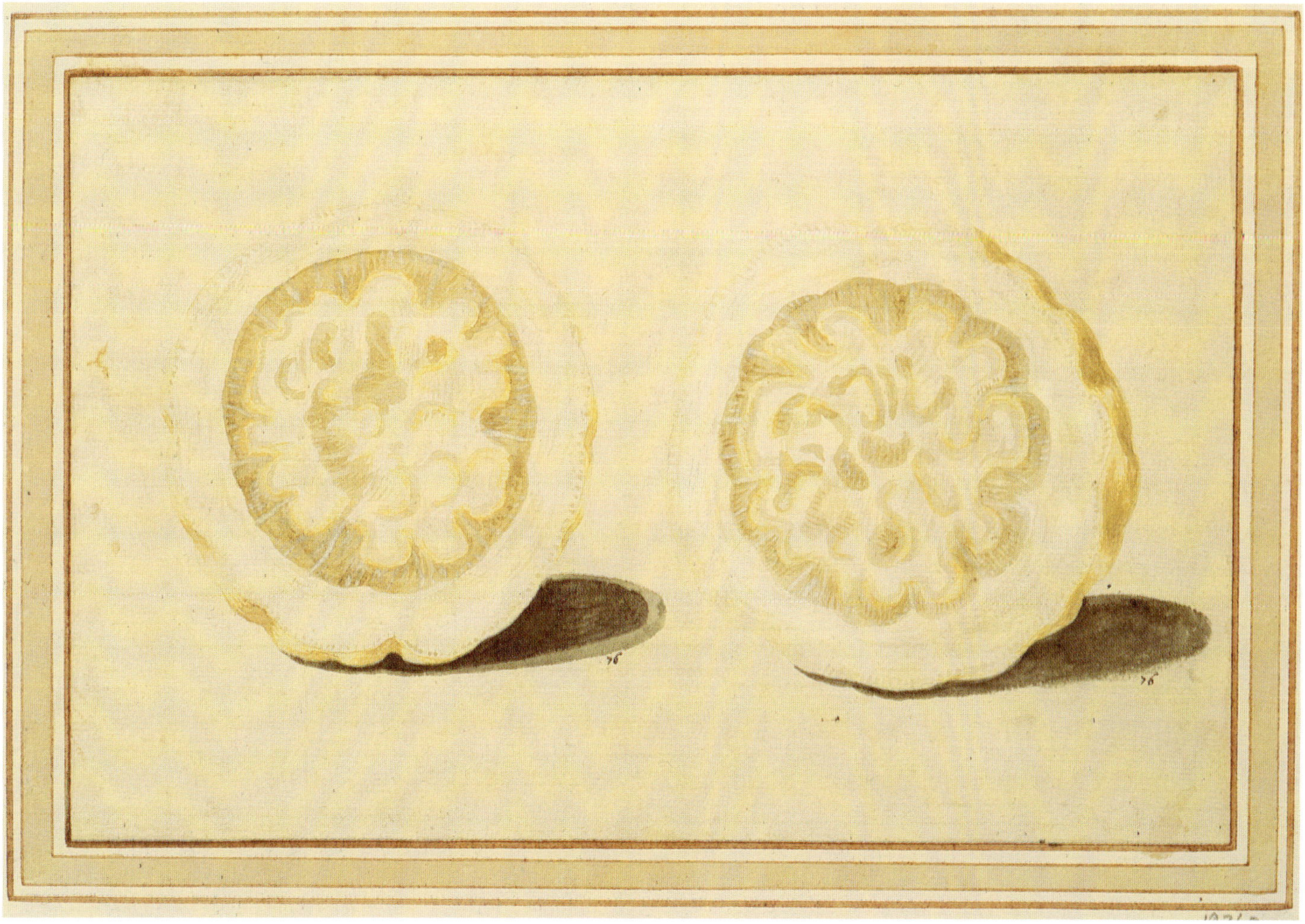

90

91. Citron-lemon, *Citrus limonimedica* Lush.: 'pregnant' whole fruit

VINCENZO LEONARDI (fl. 1621-46)

Private collection, Italy

Watercolour and bodycolour, with gum heightening and blackened lead white highlights, over black chalk

374 × 141 mm (joint measurement with **92**. Sheet cut and rejoined). Foxed and stained. *Watermark*: none recorded

ANNOTATION: *78*

MOUNT SHEET: mounted with **92**, within the same wash border

This drawing and **92** are Leonardi's models for the etchings entitled *Aliae formae citrati limonis alios includentis* on p. 271 of the *Hesperides*. They show a remarkable kind of 'pregnancy', as Ferrari himself noted. In his chapter devoted to this particular problem, he observed that although one could say that the seeds were fertilized by a living force of varying strength, nature, playful and playing, put together such instances of 'pregnancy'. He notes how a whole number of minute fruits may be seen through an opening in the top of the fruit, with the 'uterus', as it were, spontaneously opening to reveal the small fruits, just as in the case of modern navel oranges.

None of this offers much in the way of an explanation for the phenomena Ferrari describes so meticulously. But Ferrari, here as always, prefers to offer what he calls a poetical, rather than a philosophical (or what we might call 'scientific'), explanation for the causes of such phenomena. He proceeds to tell the story of how Tirsenia received the news of her son Harmonillus's death by transformation into a tree that carried digitated specimens. In the plate based on the story Ferrari made up, Romanelli shows Tirsenia receiving one of the fruits that had recently grown on the transformed Harmonillus and brought to her by the impossibly swift messenger, Ergastus (p. 277). And in the plate one does indeed see exactly the present fruit clasped by Tirsenia (herself now changing into a citron-lemon tree), in such a way that her fingers fit quite precisely into the first layer of opened segments, around the fruit included inside it. It is, in a way, a moving reminder of how Tirsenia herself once carried her son in her womb.

Although the *next* plate in the *Hesperides*, by Domenichino (Ferrari 1646, p. 421), was made to accompany a much later point in the extended poetic narrative that forms the poetic leitmotiv of the book as a whole, it too shows the topmost drawing of a pregnant citron-lemon, which is represented lying across the bow at the bottom of the illustration.

The specimens shown in this drawing and **92** are two split 'pregnant' citron-lemons with supplementary embryos inside, protruding from the split 'mother' fruit or able to be seen through a large opening (stylar hole) of the rind (**92**). No drawings of the half-fruits survive among the known drawings in the Paper Museum.

This and **92** once formed a single drawing which was cut in two, then both were mounted together in the 18th-century re-arrangement. They are perhaps two of the finest in the whole corpus of citrus drawings which once belonged to Cassiano.

PROVENANCE: dal Pozzo; Albani; George III, by descent to George V; London art market (Mendelson); James R. Herbert Boone; Sotheby's (New York), 16 September 1988, lot 147

LITERATURE: Freedberg 1992b, p. 300; Freedberg 1996a

REFERENCES: Aldrovandi MS 88; Ferrari 1646, p. 263f.; Jonston 1662, pl. X; Volkamer *Nürnberg. Hesp.* II, p. 199, pl. 200; Gallesio 1811, p. 169; unpublished drawing on loose, unnumbered folio, GAG; Baldini 1991, pp. 145–52; Baldini 1996b, p. 25, pl. 26

ENGRAVINGS: Ferrari 1646, p. 271

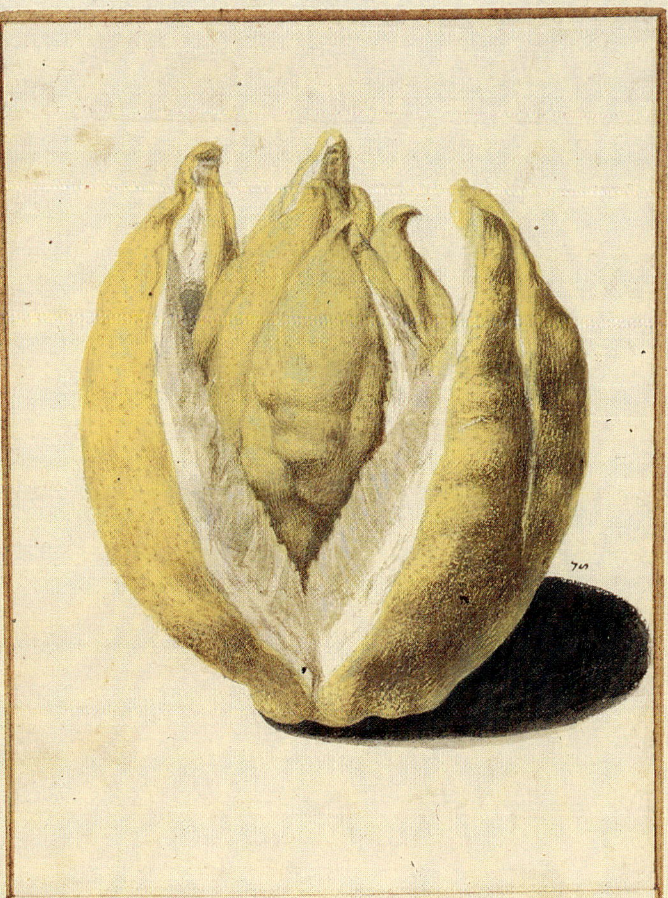

91

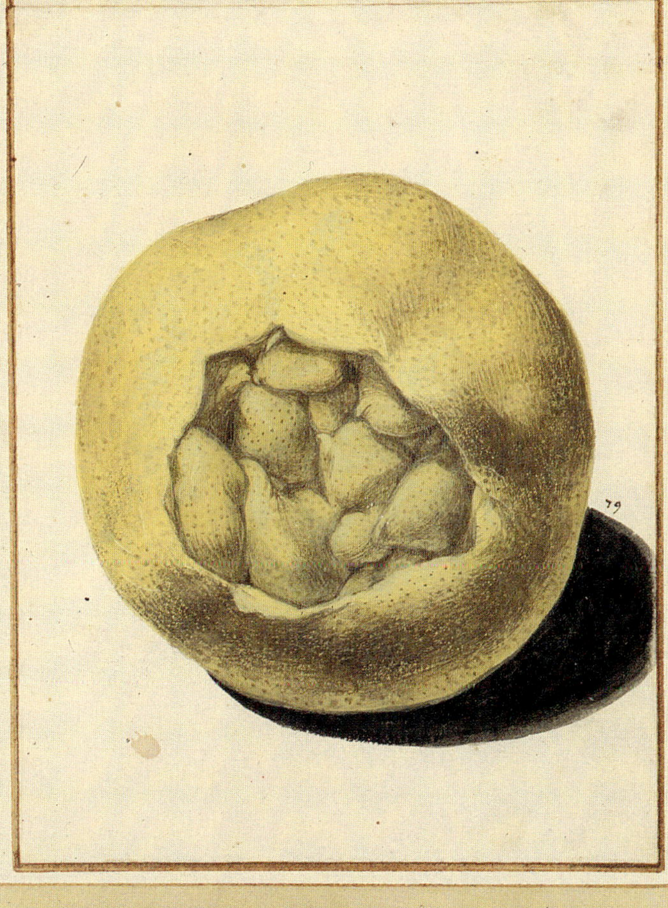

92

92. Citron-lemon, *Citrus limonimedica* Lush.: 'pregnant' whole fruit

VINCENZO LEONARDI (fl. 1621–46)

Private collection, Italy

Watercolour and bodycolour, with gum heightening and blackened lead white highlights, over black chalk

374 × 141 mm (joint measurement with **91**. Sheet cut and rejoined). Foxed and stained. *Watermark*: none recorded

ANNOTATION: *79*

MOUNT SHEET: mounted with **91**, within the same wash border

See **91**

PROVENANCE: dal Pozzo; Albani; George III, by descent to George V; London art market (Mendelson); James R. Herbert Boone; Sotheby's (New York), 16 September 1988, lot 147

ENGRAVINGS: Ferrari 1646, p. 271

93. Pummelo, *Citrus grandis* (L.) Osb., hybrid: 'pregnant' whole fruit

VINCENZO LEONARDI (fl. 1621–46)

Windsor, RL 19328

Watercolour and bodycolour, with gum heightening, over black chalk

109 × 122 mm. Cream prepared paper; survival of 5 mm overlay of original 17th-century mount at right-hand side

ANNOTATION: *M*

MOUNT SHEET: 585 × 435 mm. Mounted on same sheet as **94**. Tear at right-hand side. *Watermark*: D & C Blauw.
Annotation: [verso: pencil, in a 20th-century hand] *45, £7.7.0.*

At the end of his long chapter on the *Pomum Adami* (see **58**), Ferrari notes that the species often offers examples of what he here calls 'superfetation', just as in the case of the forms of citron-lemons which include one fruit within another. In all such cases it looks as if the embryos appear directly out of the open 'uterus'.

The present drawing served as the basis for the illustration of *Pomum Adami foetum* on p. 315 of the *Hesperides*. It shows a large cracked citrus fruit with an aborted fruit inside.

PROVENANCE: dal Pozzo; Albani; George III, by descent to George V; London art market (Mendelson); Rex Nan Kivell; Her Majesty The Queen

LITERATURE: Freedberg 1989a, p. 20f., pls. 13–14

REFERENCES: Ferrari 1646, p. 310, pl. 315; Jonston 1662, pl. XII; Volkamer *Nürnberg. Hesp.* II, p. 137f., pl. 141; Baldini 1991, pp. 145–52

ENGRAVINGS: Ferrari 1646, p. 315

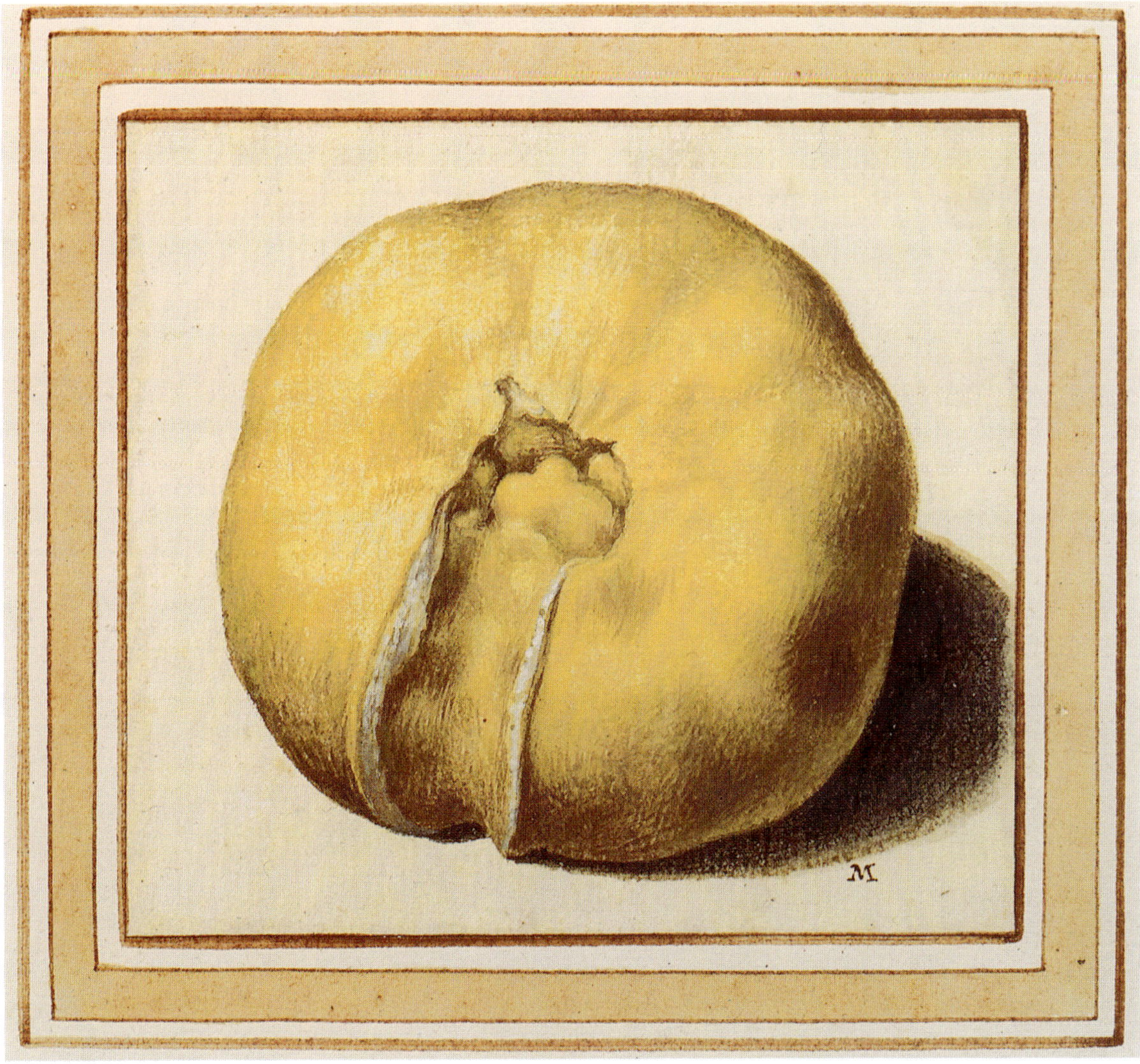

93

94. Pummelo, *Citrus grandis* (L.) Osb., hybrid: 'pregnant' half-fruit

VINCENZO LEONARDI (fl. 1621–46)
Windsor, RL 19327
Watercolour and bodycolour over black chalk
108 × 122 mm
MOUNT SHEET: mounted on same sheet as **93**

This is a fine drawing of the cross-section of the teratological Adam's Apple discussed in the previous entry and reproduced below it on p. 315 of Ferrari's *Hesperides*. One can clearly make out the small secondary fruit inside this cross-section.

PROVENANCE: dal Pozzo; Albani; George III, by descent to George V; London art market (Mendelson); Rex Nan Kivell; Her Majesty The Queen

LITERATURE: Freedberg 1989a, p. 20f., pls. 13–14

REFERENCES: Ferrari 1646, p. 310, pl. 315; Jonston 1662, pl. XII; Volkamer *Nürnberg. Hesp.*II, p. 137f., pl. 141; Baldini 1991, pp. 145–52

ENGRAVINGS: Ferrari 1646, p. 315

Comp. fig. 93, 94: Ferrari 1646 p. 315

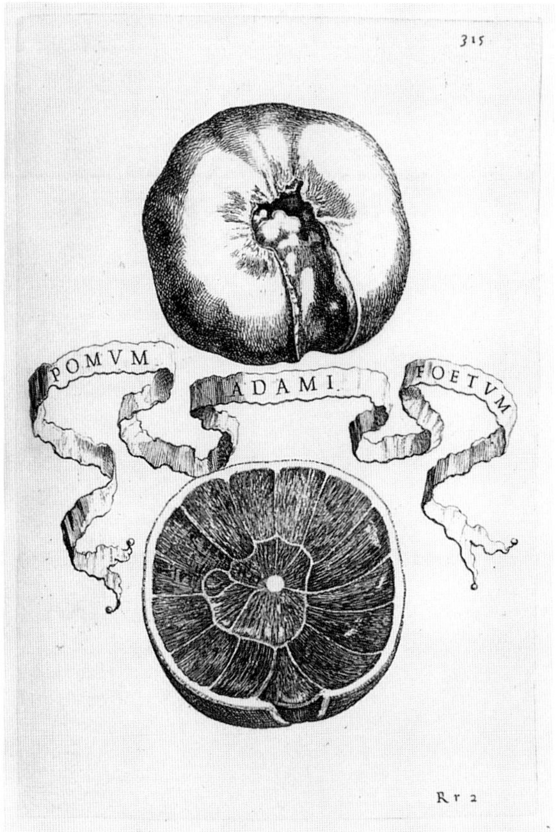

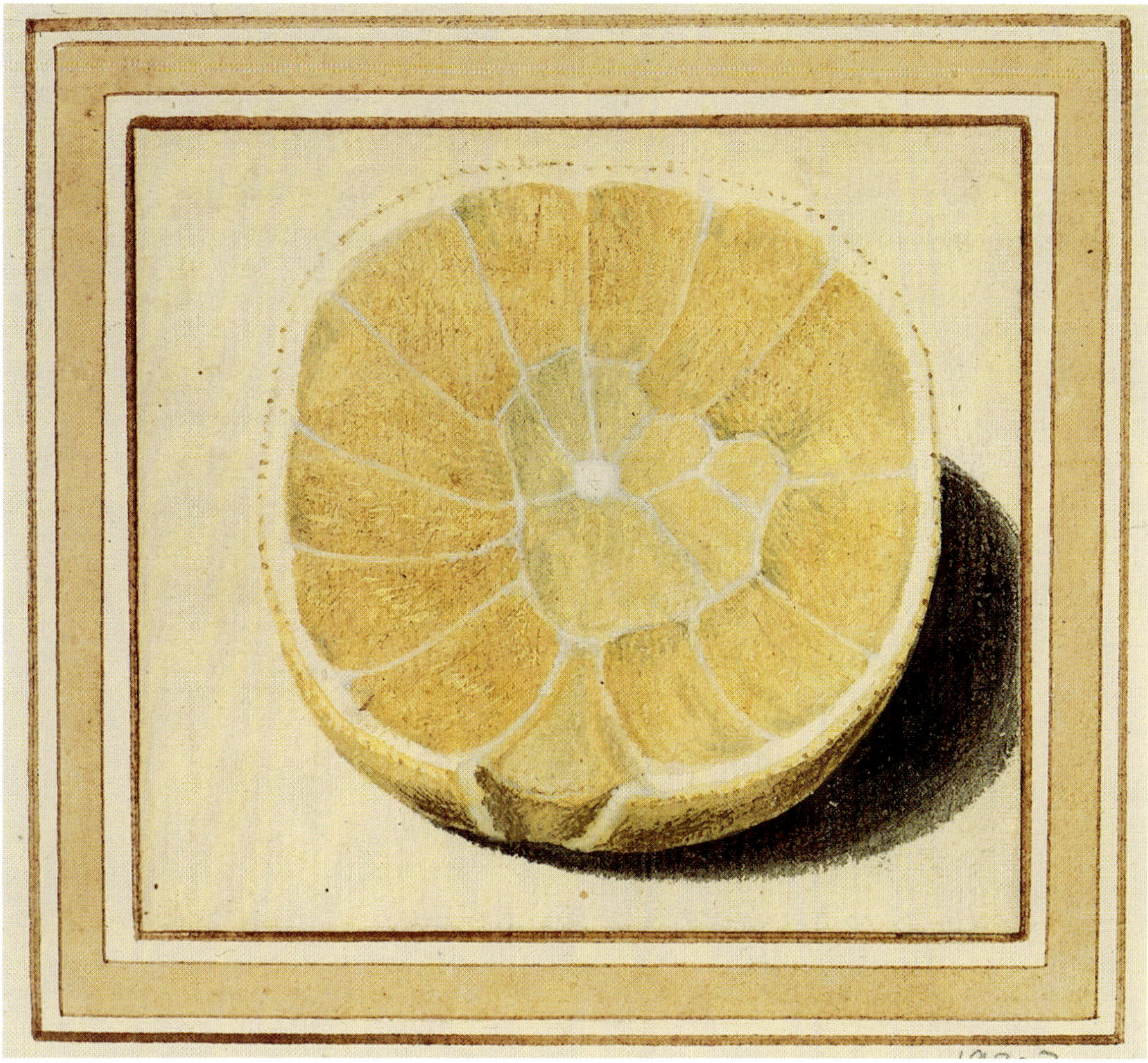

94

95. Sweet orange, *Citrus sinensis* (L.) Osb.: 'pregnant' whole and half-fruits

VINCENZO LEONARDI (fl. 1621–46)

Windsor, RL 21146

Watercolour and bodycolour with gum heightening, over black chalk; *pentimenti* in black chalk

320 × 202 mm

ANNOTATION: [bottom centre] *fff*

MOUNT SHEET: 555 × 400 mm. *Watermark*: D & C Blauw

In introducing his chapter on the *Aurantium foemina, sive foetiferum*, Ferrari notes that this fruit is similar to the type of lemon which includes others within it, such as the *Limon citratus alterum includens* (see **89–94** above), in that both give rise to aborted foetuses, *dum foetus quamvis abortivos invalido nisu parturit, non parit. Ita & hoc & proxime exhibendum, genere neutrale, a se ipso degenerat, vel in foeminam, vel in androgynum* (p. 403). The flavedo of this particular kind of orange has a strong aroma; its albedo is sweetish but its flesh pleasantly acid. Ferrari then goes on to describe in close detail the appearance of the 'navel' through the rind. Sometimes the navel remains green while the 'mother' grows yellow; but both parent and child age at the same pace. Ferrari attempts an 'explanation' of why the navel is so often divided, and why the mother produces a kind of double or twin offspring, suggesting that it has to do with the subtlety of the seminal substance, or some other congenital factor. His 'explanation', then, is typically vague and minimal — even more so than in the case of the pregnant citron-lemons or Adam's apple (**58**, **91–93**). It is also very unlike the long narratives which he invents to account for forms such as the pregnant citron-lemons (see **91** and **92**).

The present drawing is for the etching on p. 405 of the *Hesperides*. It shows four different views (from above, from below, a longitudinal and a transverse section) of a teratological navel orange with a fruitlet inside it. The disposition of the drawings on this sheet was slightly changed when it came to be reproduced in the *Hesperides*, since the whole fruits are there shown alongside one another, rather than one above the other, as here. This is, in fact, the only instance in the corpus where four drawings of the same fruit were shown — and, for that matter, reproduced — on one and the same sheet. Traces of black chalk underdrawing may still be seen for the section now on the lower left. This suggests that another arrangement was originally intended, perhaps with one drawing above the other, all aligned along a now still visible centre line, drawn in black chalk as well.

The leaves of this specimen are very small and lanceolate, with slightly winged petioles. As in the present example, navel fruits suffer rather frequently from a large open stylar hole and protruding aborted fruitlets, both of which phenomena Ferrari notes in his concise discussion of this fruit on p. 403. The resemblance between this fruit and **91** and **92** is striking.

PROVENANCE: dal Pozzo; Albani; George III, by descent to George V; London art market (Mendelson); Rex Nan Kivell; Her Majesty The Queen

LITERATURE: Freedberg 1989a, p. 32f., pl. 23; Freedberg 1989b, p. 63, pl. 23; Freedberg 1993, p. 184f., pl. 111

REFERENCES: Ferrari 1646, p. 403; Cupani 1696, p. 129; Volkamer *Nürnberg. Hesp.* II, p. 199f., pl. 200; Micheli *Enumeratio*, p. 55, pl. 42; Risso/Poiteau 1818, p. 78, pl. 33; Baldini 1991, pp. 145–52

ENGRAVINGS: Ferrari 1646, p. 405

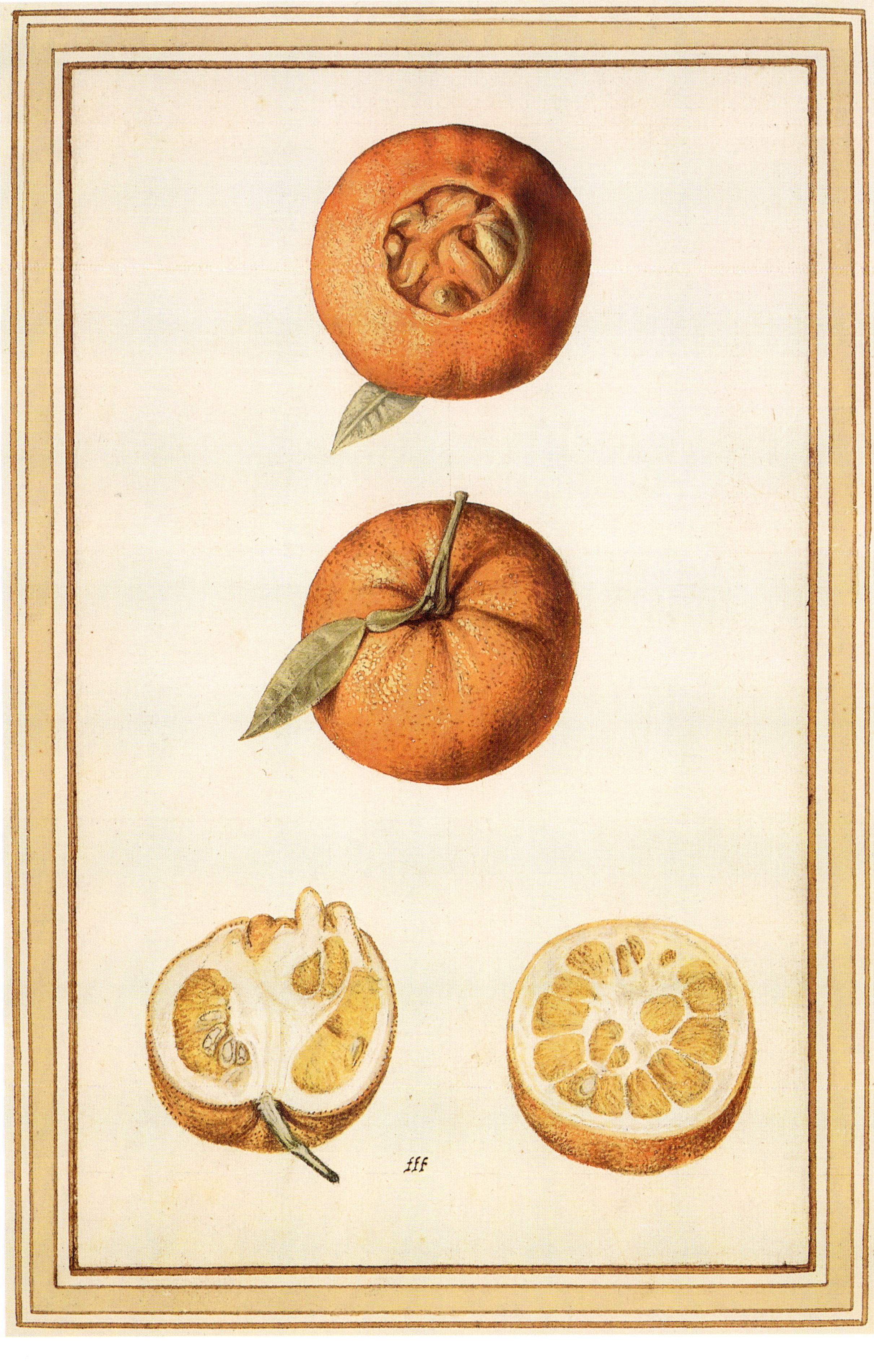

95

96. Lemon, *Citrus limon* (L.) Burm. f.: four misshapen fruits

?VINCENZO LEONARDI (fl. 1621–46)

Windsor, RL 19340

Watercolour and bodycolour over black chalk, with gum heightening on leaf; *pentimenti* in black chalk

194 × 192 mm. *Watermark:* Figure (cut and unidentifiable)

ANNOTATION: [bottom centre] *ii*

MOUNT SHEET: 585 × 434 mm. Stained. *Watermark*: D & C Blauw

The highly deformed shapes of these fingered, horned and ribbed yellow fruits are presumably the result of severe damage to the buds by the *Aceria sheldoni* mite. They seem to be lemons, but the elliptical and serrate leaf of the upper right specimen is clearly winged, a trait generally found in oranges and not in this species. No view of the half-fruits is shown. An unusual amount of preparatory drawing in black chalk is still visible in this drawing. The drawing was not reproduced in Ferrari's *Hesperides*.

PROVENANCE: dal Pozzo; Albani; George III, by descent to George V; London art market (Mendelson); Rex Nan Kivell; Her Majesty The Queen

REFERENCES: Micheli *Enumeratio*, p. 53, pl. 40

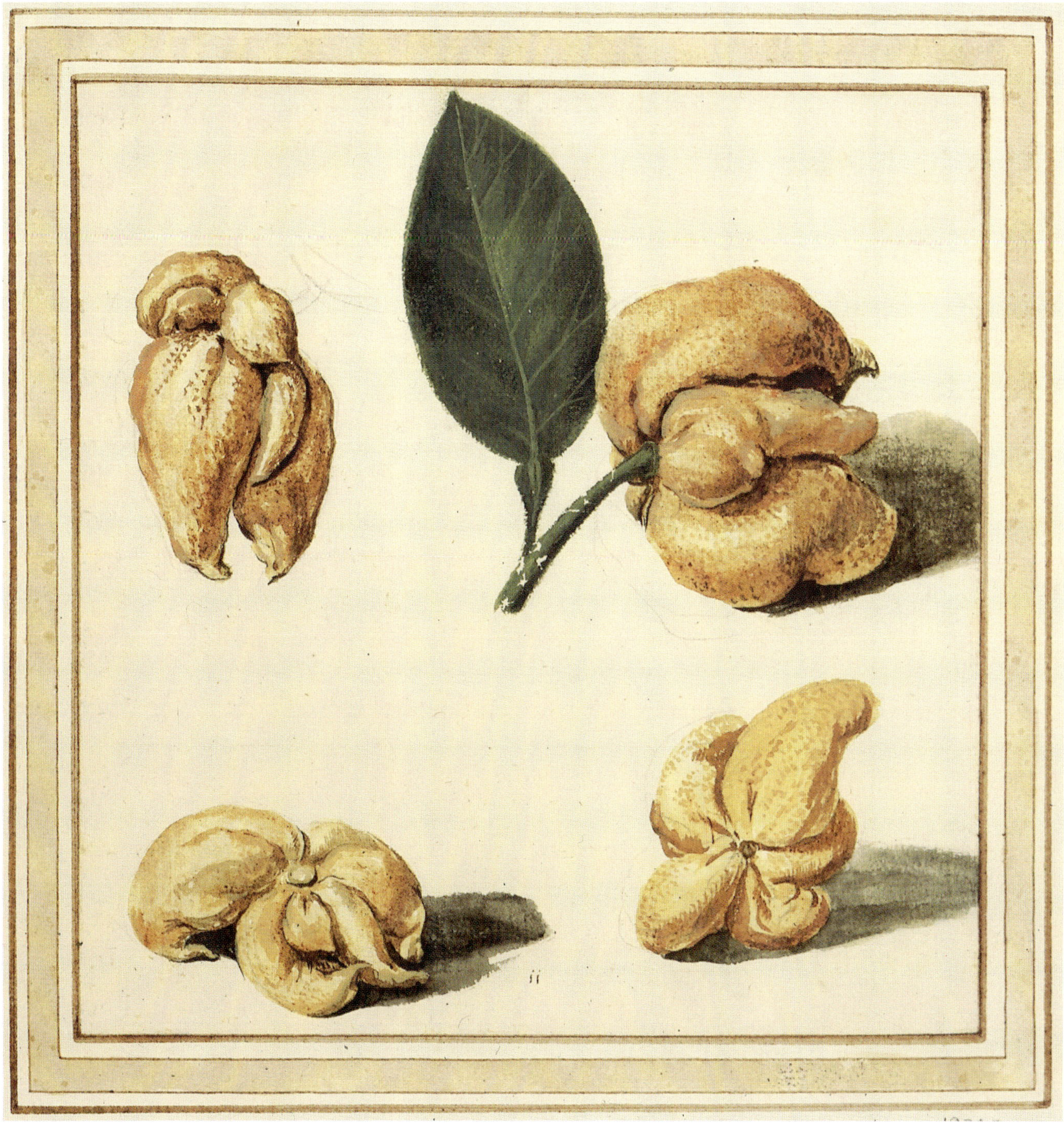

96

97. Lemon, *Citrus limon* (L.) Burm. f.: fingered whole fruit

?GIOVANNA GARZONI (1600–70)
Private collection, U.S.A.
Measurements unknown
Watercolour and bodycolour on vellum
ANNOTATION: [in a large hand] *621*

Not reproduced in Ferrari's *Hesperides*, this and the following drawing are unique in the corpus of surviving citrus drawings in that they are painted on vellum. The specimen is a misshapen lemon with lateral finger-like appendages, presumably a result of damage to the buds by *Aceria sheldoni*, or of flowers with supernumerary carpels. Both this and the following two drawings bear a certain similarity — at least in terms of their deformations — to one or two of the *Aurantia distorta* discussed by Ferrari on p. 413 and illustrated on p. 415. This and the following drawing, just as Ferrari's *Aurantium distortum* and even more the *Aurantium corniculatum* (*Hesperides*, p. 409) all resemble the specimens of the *Aurantia puteolana* in the pages inserted into one volume of Cesi's *Syntaxis Plantaria* in the Institut de France in Paris (Paris, IF, MS 976, fol. 84 bis). 'Horned' fruits of this kind are also referred to several times in Cassiano's *Notizie diverse* (Archivio Dal Pozzo MS 39, fol. 38 on *Aranzi cornuti*, for example).

Although this drawing has been located, no photograph was available at the time of going to press and the catalogue entry was compiled on the basis of an indistinct photocopy, reproduced here.

PROVENANCE: dal Pozzo; Albani; George III, by descent to George V; London art market (Mendelson); private collection, U.S.A.

REFERENCES: Baldini 1991, pp. 152–8

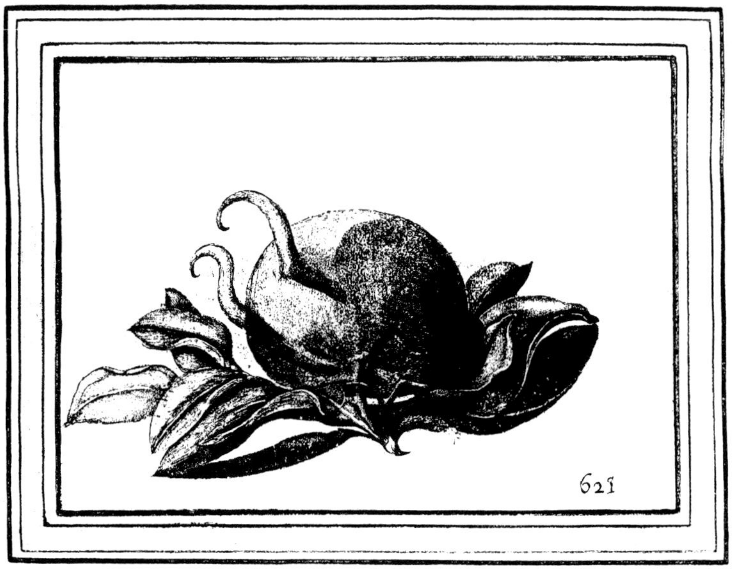

97

98. Lemon, *Citrus limon* (L.) Burm. f.: two twin-fingered fruits

?GIOVANNA GARZONI (1600–70)

Private collection, U.S.A.

Watercolour and bodycolour, with blackened lead white highlights, over black chalk, on vellum

145 × 221 mm

ANNOTATION: [right, in a large hand] *620*

In this drawing, not reproduced in the *Hesperides*, three twin lemons with lateral, finger-shaped appendages are shown, along with two transverse sections. These anomalous fruits can arise from doubly pistillate flowers, but the presence of finger-shaped appendages also suggests the more usual case of attack by *Aceria sheldoni*, or of staminal carpelloidy. Double lemons such as these are also described in Montalbani's edition of Aldrovandi's *Dendrologia* and are shown under the *Limon dulci cortice in 18 triangula divisus* and the *Limon chirodes* (Montalbani 1668, pp. 494 and 514).

To some extent these specimens, like those in the preceding and the following entries, also resemble the *Aurantium corniculatum* on p. 409 of the *Hesperides* (and in this case the *Aurantium distortum* on p. 415 too), as well as the drawing of the same type preserved in the surviving volumes of Federico Cesi's *Syntaxis Plantaria* in Paris, IF, MS 976, fol. 84 bis).

The fruits in the drawing seem to have been arranged on a step or ledge, as do those described in **71** and **107**.

PROVENANCE: dal Pozzo; Albani; George III, by descent to George V; London art market (Mendelson); acquired at a later date by N. Hobhouse

REFERENCES: Aldrovandi MS 88, c.10; Montalbani 1668, pp. 494 and 514; Duhamel du Monceau 1768, p. 305; Risso/Poiteau 1818, p. 177, pl. 85; Penzig 1887; Evreinoff 1946, p. 171; Baldini 1989, p. 163: Baldini 1991, pp. 152-8

98

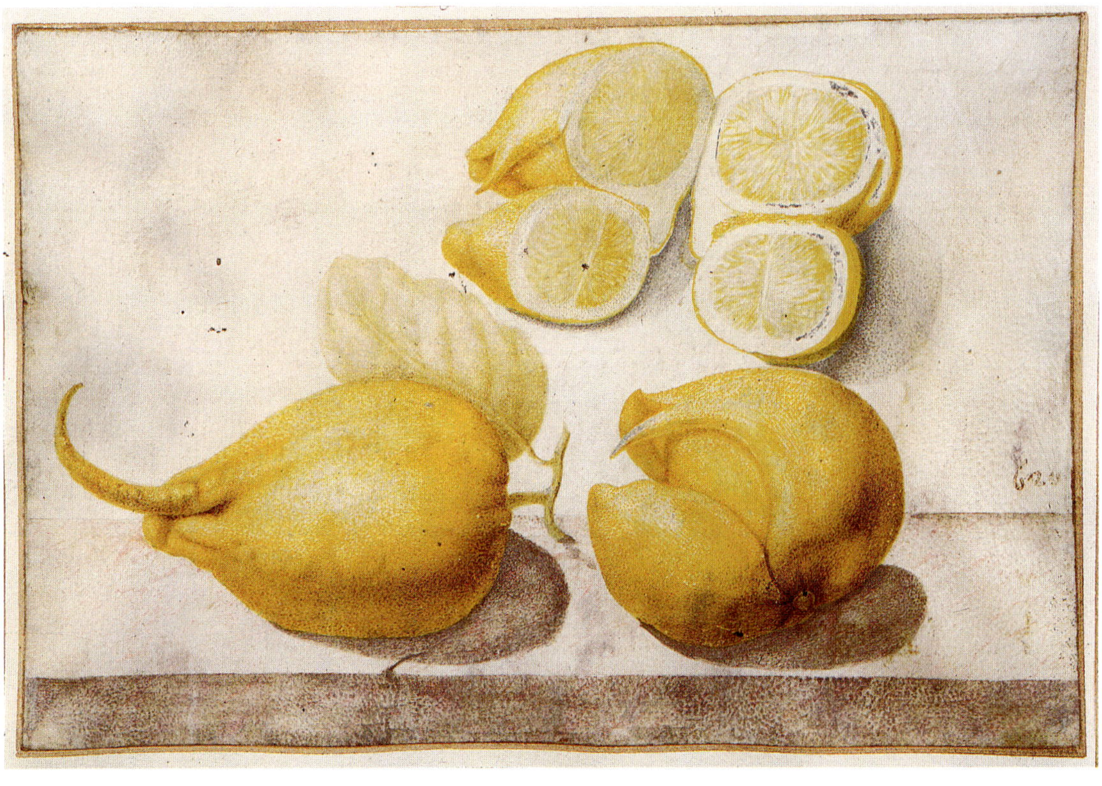

99. Lemon, *Citrus limon* (L.) Burm. f.: twin fruit

SEVENTEENTH-CENTURY ITALIAN
Private collection, England
Black chalk
175 × 143 mm. Foxed. Trimmed slightly, through shadow to right
ANNOTATION: [lower left] *848*

Not reproduced in Ferrari's *Hesperides*, this is the only surviving drawing in the corpus to have been executed solely in black chalk. It shows an ellipsoid lemon with a prominent nipple and a slender lateral appendage, in some ways similar to the etching of the *Aurantium corniculatum* on p. 409 of the *Hesperides*. As in other such cases the deformation may arise from flower-buds that have been injured by the *Aceria sheldoni* mite, from staminal concretion or other floral disorders. No drawing of the half-fruit survives among the known drawings in the Paper Museum.

PROVENANCE: dal Pozzo; Albani; George III, by descent to George V; London art market (Mendelson)

REFERENCES: Aldrovandi MS 88, fol. 8v; Montalbani 1668, pp. 492, 514; Volkamer *Nürnberg. Hesp.* II, pp. 72–4; Penzig 1887; Baldini 1989, p. 163; Baldini 1991, pp. 145–52, pl. 150

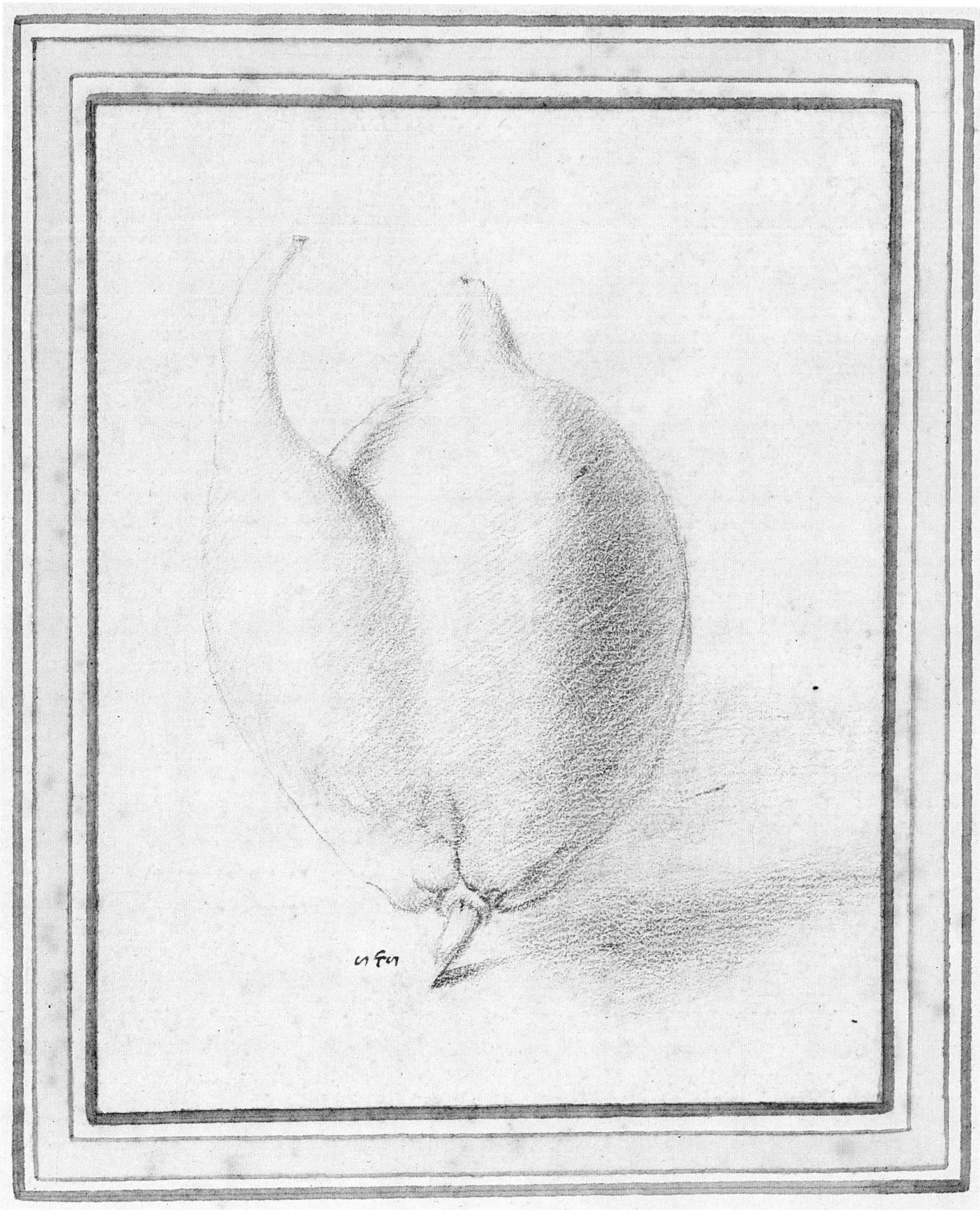

99

100. Sweet orange, *Citrus sinensis* (L.) Osb.: ribbed whole and half-fruit

VINCENZO LEONARDI (fl. 1621–46)

Windsor, RL 19345

Watercolour and bodycolour, with gum heightening, over black chalk, with incised outlines

211 × 148 mm

ANNOTATION: [below stalk] *2*

MOUNT SHEET: 587 × 434 mm. Stained, left. *Watermark*: D & C Blauw. *Annotation*: [verso: pencil, in a 20th-century hand] *64*

This drawing served as the basis for the *Aurantium striatum* etched on p. 401 of Ferrari's *Hesperides*, although in the illustration an extra leaf was engraved to the left of the cluster, presumably for compositional reasons. The addition is very clear, since the engraved technique here appears distinctly different from the etched forms of the reproduction of the present drawing. In his discussion of the striated orange on p. 397, which follows his observations on the *Aurantium virgatum* (see **103** and **105**), Ferrari notes that it is more attractive than this other, closely related form. It had been sent to Alessandro Caetani in Rome by Carlos de Cardeñas, Marquis of Laynez, whose gardens were amongst the most famous and beautiful in Naples. Cupani (1696, p. 129) called this strain in Sicilian dialect *Aranciu à vracchi di Tudiscu*, an allusion to the striped trousers of the German lansquenets. The most striking feature of this fruit are the prominent, longitudinal ribs on the rind, probably caused by a mutation involving the outer layer of the bud's apical meristems (i.e. an autogenous cytochimera). A similar event resulted in the striped fruits of *Aurantium virgatum* etched by Ferrari on p. 399 of the *Hesperides*, where greenish and orange longitudinal stripes alternate on the rind.

PROVENANCE: dal Pozzo; Albani; George III, by descent to George V; London art market (Mendelson); Rex Nan Kivell; Her Majesty The Queen

REFERENCES: Archivio Dal Pozzo MS 39, fol. 60; Ferrari 1646, p. 397; Jonston 1662, pl. XVI, p. 401; Cupani 1696, p. 129; Bimbi, Villa Med. C. 594, C. 616; Volkamer *Nürnberg. Hesp.* I, pp. 193, 197 and 199; pls. 194a,3 198, 200; Micheli *Enumeratio*, pp. 44–7, pls. 30–33; Risso/Poiteau 1818, p. 53, pl. 18; Targioni Tozzetti 1825, p. 78, pl. 77; Reuther/Batchelor/Webber II, p. 344; Tilney-Bassett 1986, p. 42; Baldini 1989, pp. 164–7; Baldini 1991, pp. 131–45

ENGRAVINGS: Ferrari 1646, p. 401

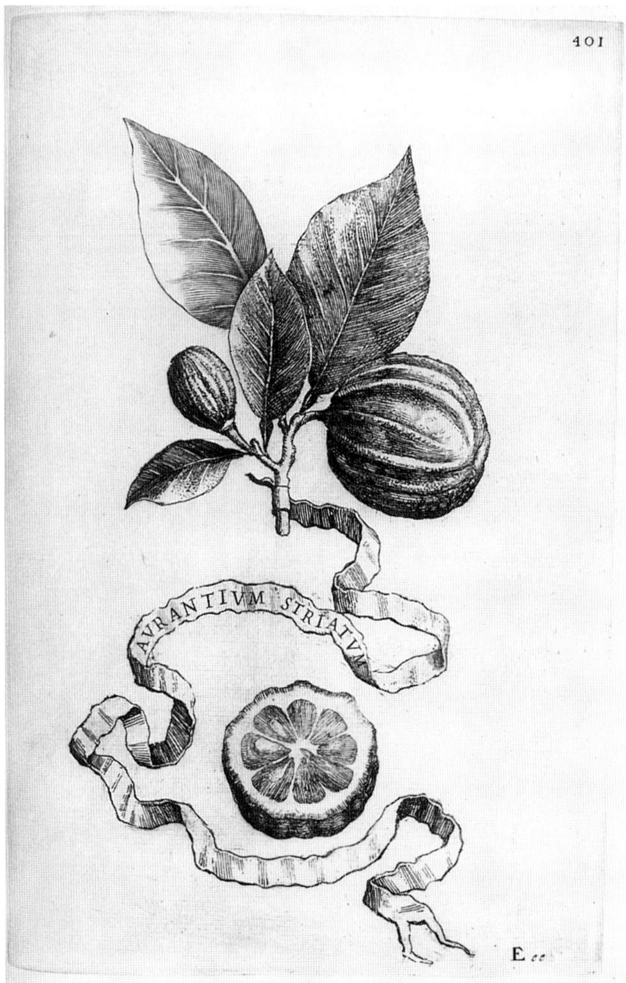

Comp. fig. 100: Ferrari 1646 p. 401

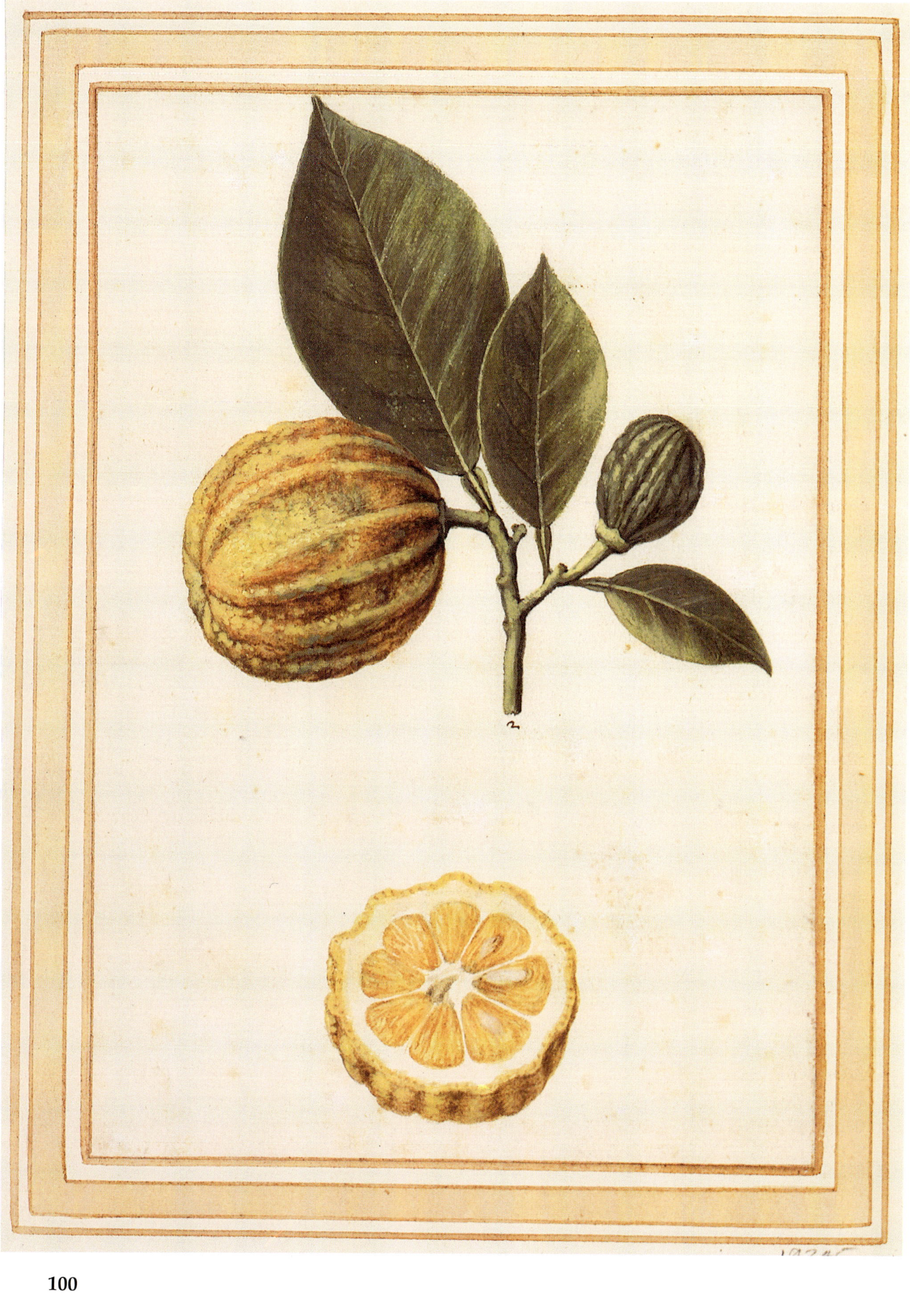

101. Citron + sour orange (?), *Citrus medica* L. + *Citrus aurantium* L. (?): chimeric whole fruit

VINCENZO LEONARDI (fl. 1621–46)

Windsor, RL 19329

Watercolour and bodycolour, with gum heightening, over black chalk and incised outlines

129 × 129 mm

MOUNT SHEET: 289 × 422 mm. Mounted with **102**, within the same wash border

In chapter X of his book on oranges, Ferrari described a particular specimen which had recently been sent to him from Naples. It was about the same size as an ordinary orange, with a most pleasant vinous taste. He called it a kind of natural miracle in miniature, with its extraordinary and beautiful 'appendix'. This excrescence, about the thickness of a finger, was very wrinkly with small veins and protuberances of green and yellow — just as in the present drawings. Here the growth seems proportionately much larger than Ferrari described it to be. He noted that these types of chimera were especially frequent in the suburb of Naples known as the Barra, where Vesuvius offers protection from cold winds.

Both the present drawings (**101** and **102**), once part of the same sheet, then separated, and once again mounted together in the 18th-century rearrangement of the drawings, served as the basis for the *Aurantium callosum multiforme* discussed in this chapter and reproduced on p. 411 of the *Hesperides*. In fact, the rind of this abnormal fruit belongs both to orange and citron, while its flesh belongs exclusively to orange. This freakish compound fruit is likely to be a typical spontaneous synthetic chimera (i.e. a graft chimera) and can be ascribed to the broad group of *Bizzarrie*, the most renowned of which is the *Malolimonia citrata-aurantia* described by Nati in 1674 (see **103** and **104**).

PROVENANCE: dal Pozzo; Albani; George III, by descent to George V; London art market (Mendelson); Rex Nan Kivell; Her Majesty The Queen

LITERATURE: Freedberg 1989a, p. 18f. and 34, pls. 10 and 11; Freedberg 1989b, p. 59, pl. 21; Freedberg 1992b, p. 301; Freedberg 1993, p. 186, pl. 112

REFERENCES: Ferrari 1646, pp. 407 and 411; Nati 1674, unnumbered plate; Risso/Poiteau 1818, p. 107, pl. 52; unpublished drawing on loose, unnumbered folio, GAG; Baldini 1989, pp. 164–7; Baldini 1991, p. 142f.; Baldini 1996b, p. 20, pl. 27

ENGRAVINGS: Ferrari 1646, p. 411

102. Citron + sour orange (?), *Citrus medica* L. + *Citrus aurantium* L. (?): chimeric half-fruit

VINCENZO LEONARDI (fl. 1621–46)

Windsor, RL 19330

Watercolour and bodycolour, with gum heightening, over black chalk and incised outlines

129 × 130 mm

ANNOTATION: *cc*

MOUNT SHEET: mounted with **101**, within the same wash border

See **101**

PROVENANCE: dal Pozzo; Albani; George III, by descent to George V; London art market (Mendelson); Rex Nan Kivell; Her Majesty The Queen

ENGRAVINGS: Ferrari 1646, p. 411

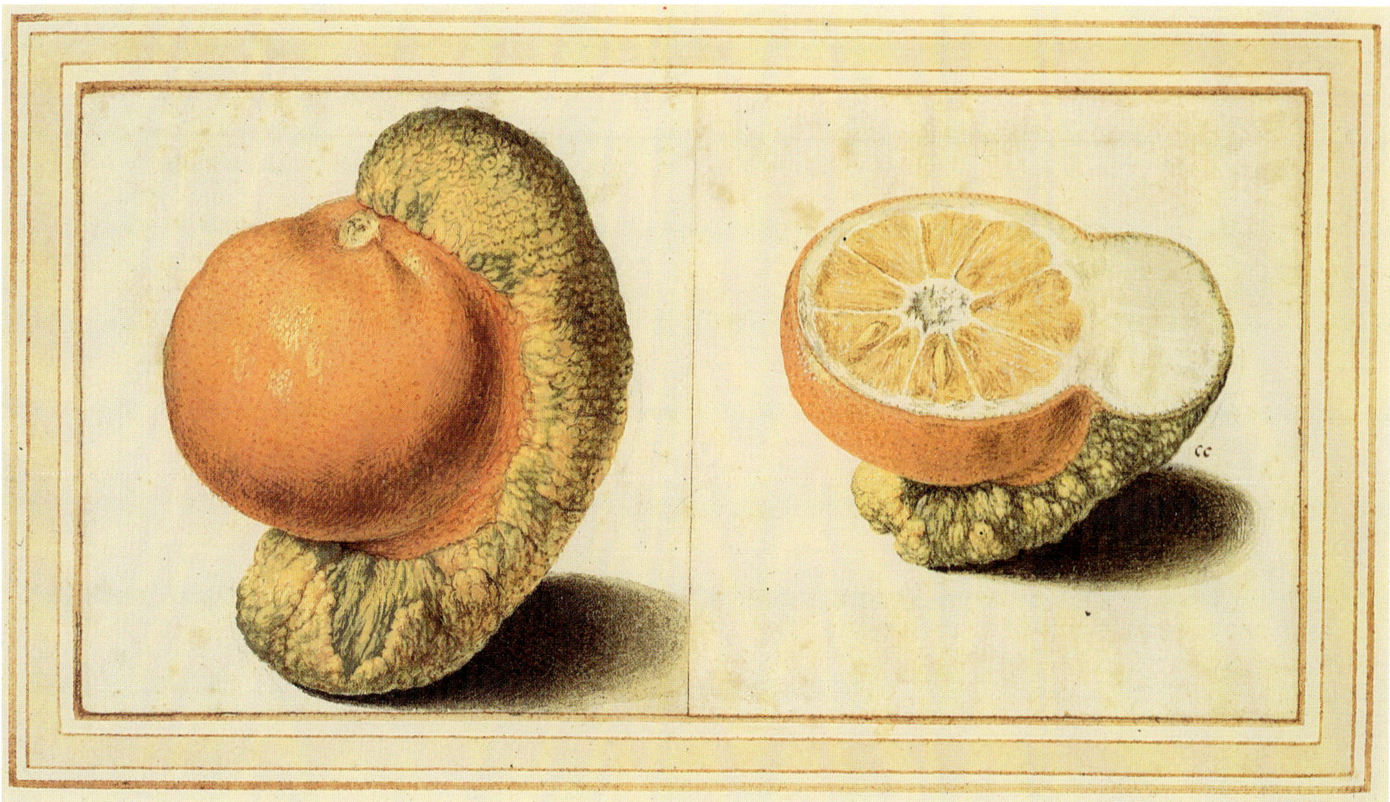

101 **102**

103. Citron + sour orange (?), *Citrus* L. + *Citrus aurantium* (L.) (?): chimeric whole and half fruits

SEVENTEENTH-CENTURY, ITALIAN

Windsor, RL 19366

Watercolour and bodycolour over graphite

362 × 232 mm. Some foxing

ANNOTATION: [bottom centre] *516*

MOUNT SHEET: 559 × 405 mm. Stained and creased. Remains of strips of tape at edges. *Watermark*: D & C Blauw.
Annotation: [verso: pencil, in a 20th-century hand] *47* [crossed out] *55* *N.F.S.*

Although not reproduced in Ferrari's *Hesperides*, both this and the following entry bear a significant resemblance to the *Aurantium callosum multiforme* (**101** and **102**). The drawing shows two views and a cross-section of a deformed fruit with an orange rind and protruding bumpy irregular ribs, similar to Gallesio's *Citrus aurantium indicum, limo citrati*. It belongs to the broad group of synthetic or graft chimeras, such as Nati's citron + sour orange *Bizzarria*, the famous *Malolimonia citrata-aurantia* observed in a garden of the Panciatichi Villa near Florence around 1630. The leaves are small and lanceolate.

For a long discussion of a citrated orange grown by Giovanni Nardi in the Medici gardens in Gualfonda in Florence, see also Archivio Dal Pozzo MS 39, fol. 106.

PROVENANCE: dal Pozzo; Albani; George III, by descent to George V; London art market (Mendelson); Rex Nan Kivell; Her Majesty The Queen

REFERENCES: Cupani 1696, p. 131; Nati 1674, unnumbered plate; Montalbani 1668, pp. 497–8; Bimbi, Villa Med. C. 594; Volkamer *Nürnberg*. *Hesp*. I, pp. 171–2a; Micheli *Enumeratio*, p. 54, pl. 41; Gallesio 1811, pp. 145–8; Risso/Poiteau 1818, p. 107, pl. 52; Targioni Tozzetti 1825, p. 28, pl. 27; unpublished drawing on loose, unnumbered folio, GAG; Gallesio 1839, p. 11; Baldini 1989, pp. 164–7; Baldini 1991, pp. 131–45; Baldini 1996b, p. 26, pl. 27

103

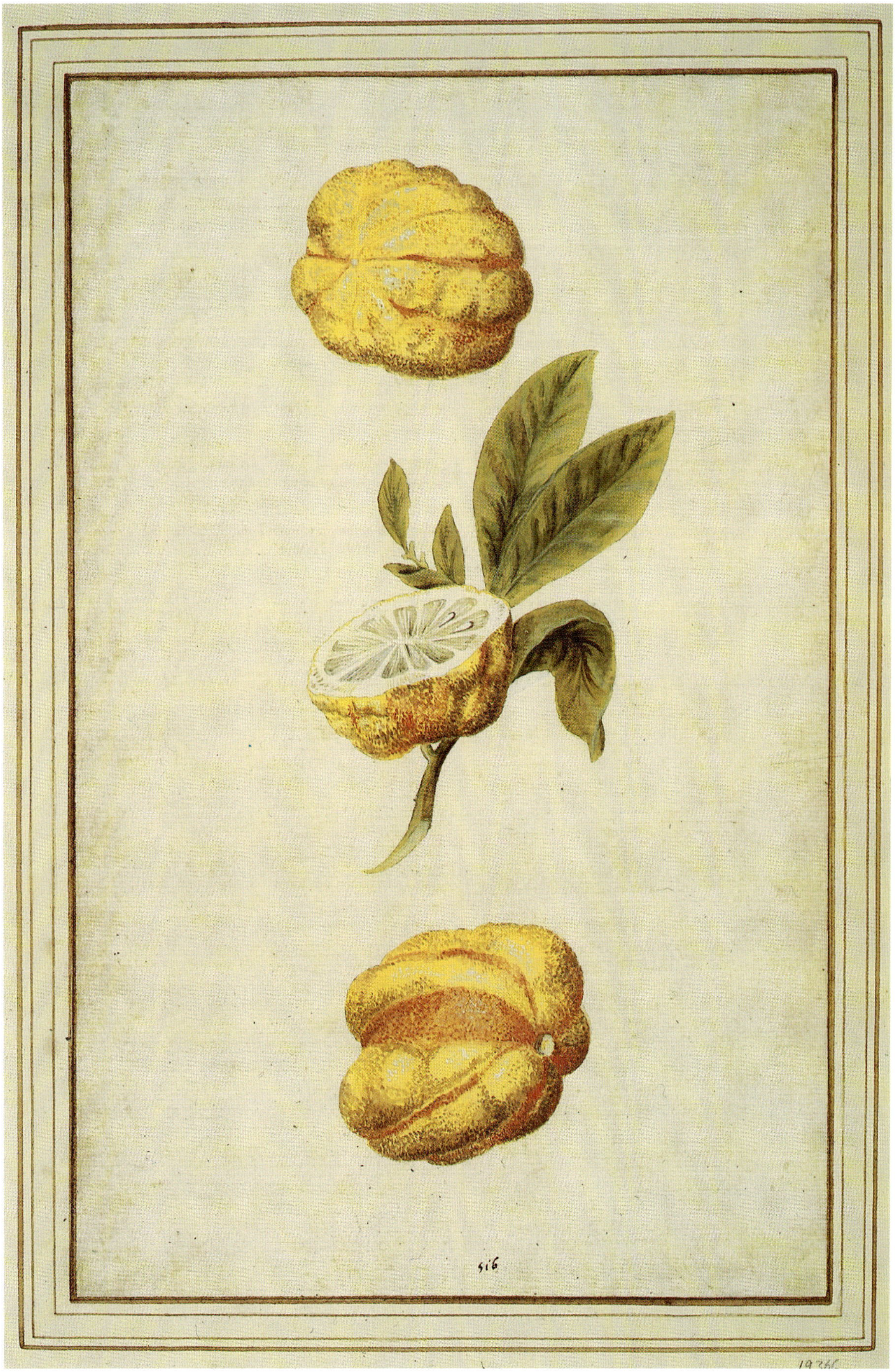

104. Citron + sour orange (?), *Citrus medica* L. + *Citrus aurantium* L. (?): chimeric whole and half fruits

SEVENTEENTH-CENTURY, ITALIAN

Windsor, RL 19367

Watercolour and bodycolour over graphite

367 × 222 mm. Stained, left. Slight foxing. *Watermark:* Figure 18B

ANNOTATION: [bottom centre] *559*

MOUNT SHEET: 586 × 435 mm. Stained and water-damaged bottom left. Stains around edges. *Watermark*: D & C Blauw

This drawing is very similar to that shown in the previous entry and, like it, was not reproduced in Ferrari's *Hesperides*. It is probable that two different specimens are shown: one represented by the top two drawings, the other by the bottom four. They are misshapen fruits, with irregular bumpy yellow-orange longitudinal ribs over a finely pitted orange peel. They can therefore be ascribed, like Nati's *Bizzarria*, to the broad group of synthetic chimeras in which Cassiano and his friends were so interested.

PROVENANCE: dal Pozzo; Albani; George III, by descent to George V; London art market (Mendelson); Rex Nan Kivell; Her Majesty The Queen

REFERENCES: Cupani 1696, p. 131; Nati 1674, unnumbered plate; Montalbani 1668, pp. 497–8; Volkamer *Nürnberg. Hesp.* I, pp. 171–2a; Micheli *Enumeratio*, p. 54, pl. 41; Gallesio 1811, pp. 145–8; Risso/Poiteau 1818, p. 107, pl. 52; Targioni Tozzetti 1825, p. 28, pl. 27; unpublished drawing on loose, unnumbered folio, GAG; Gallesio 1839, p. 11; Baldini 1989, pp. 164–7; Baldini 1991, pp. 131–45; Baldini 1996b, p. 26, pl. 27

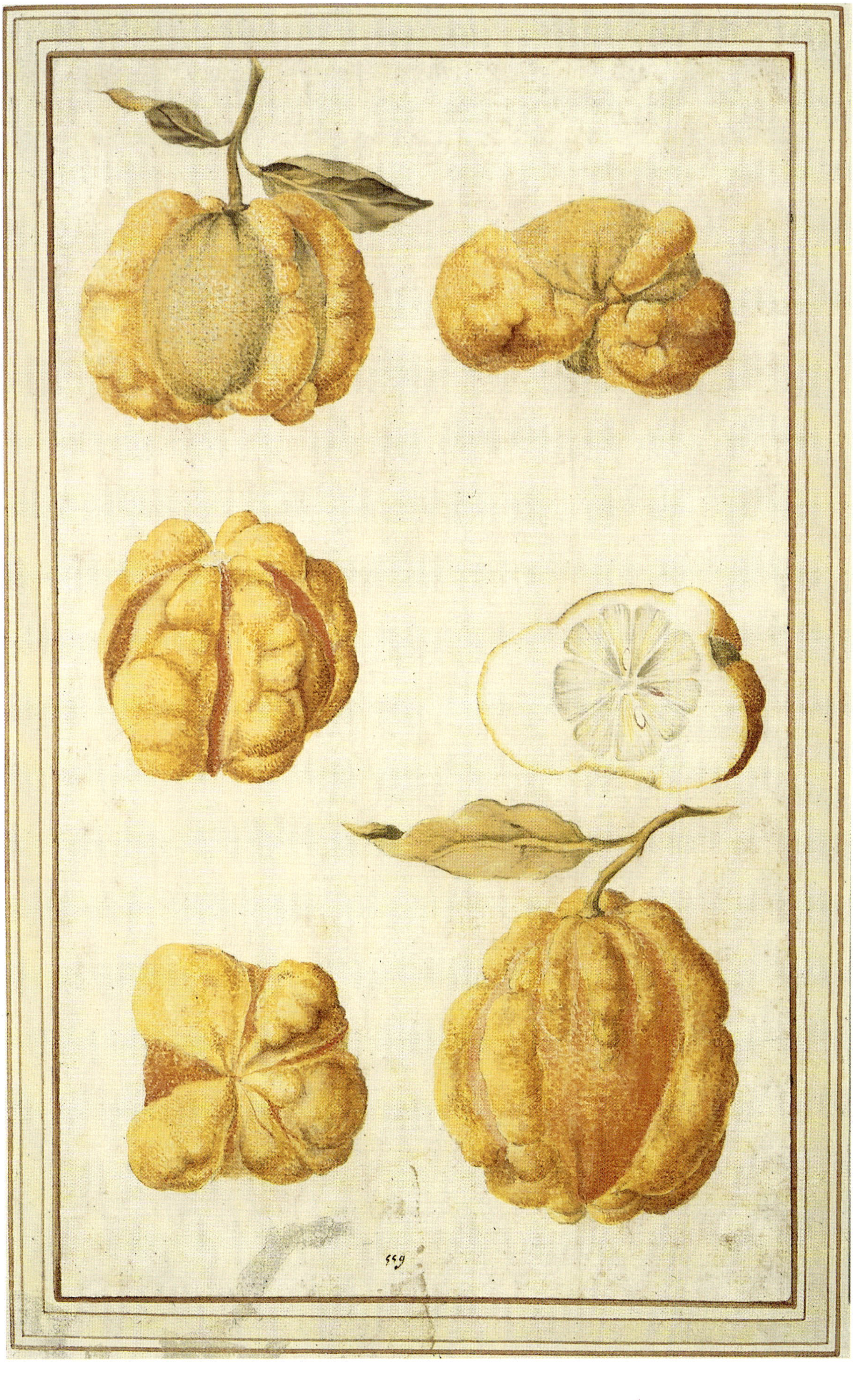
559

105. Citrus, *Citrus sp.*: chimeric, misshapen fruit

VINCENZO LEONARDI (fl. 1621–46)

Watercolour and bodycolour

Location unknown

210 × 141 mm (joint measurement with **106**. Sheet cut and rejoined).
Watermark: none recorded

ANNOTATION: *13*

MOUNT SHEET: mounted with **106**, within the same wash border

Neither this drawing nor 106 was reproduced in Ferrari's Hesperides, although the fruit does bear some resemblance to the Aurantium virgatum discussed by Ferrari on p. 397 and illustrated on p. 399 (as do the following two entries). The deformed fruit looks similar to the Citrus Aurantium Panormitanum described by Inzenga in 1875. They have elliptical leaves with broadly winged petioles. They are difficult to identify but are probably sour oranges. In both cases the surface of the rind is covered by a number of irregular, longitudinal ribs. The upper fruit is deep yellow, the lower fruit green and therefore still unripe, but with a narrow trace of yellow between the ribs. Given such features, these specimens are in all likelihood autogenous cytochimeras, although the presence of apical finger-like appendages on the upper fruit also suggests a supplementary aetiology, probably involving damage by *Aceria sheldoni*.

This drawing and **106** must originally have been on one drawing sheet, later separated, and then in the 18th century mounted together when they entered the collection of George III.

PROVENANCE: dal Pozzo; Albani; George III, by descent to George V; London art market (Mendelson); private collection, U.S.A.; Sotheby's (New York), 10 January 1995, lot 162

REFERENCES: Inzenga 1875, pp. 25–8; Micheli *Enumeratio*, p. 41f.; pls. 27–8

106. Citrus, *Citrus sp.*: chimeric, misshapen fruit

VINCENZO LEONARDI (fl. 1621–46)

Watercolour and bodycolour

Location unknown

210 × 141 mm (joint measurement with **105**. Sheet cut and rejoined).
Watermark: none recorded

ANNOTATION: *12*

MOUNT SHEET: mounted with **105**, within the same wash border

See **105**

PROVENANCE: dal Pozzo; Albani; George III, by descent to George V; London art market (Mendelson); private collection, U.S.A.; Sotheby's (New York), 10 January 1995, lot 162

REFERENCES: Inzenga 1875, pp. 25–8; Micheli *Enumeratio*, p. 41f.; pls. 27–8

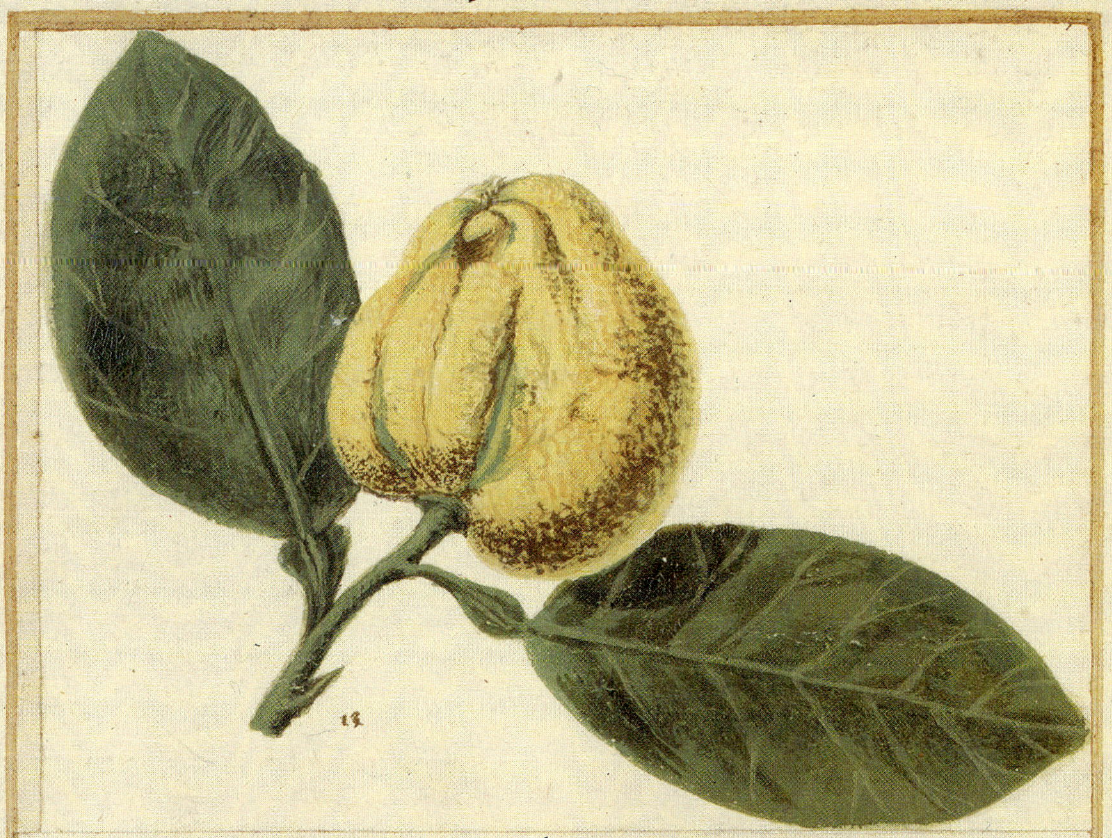

105

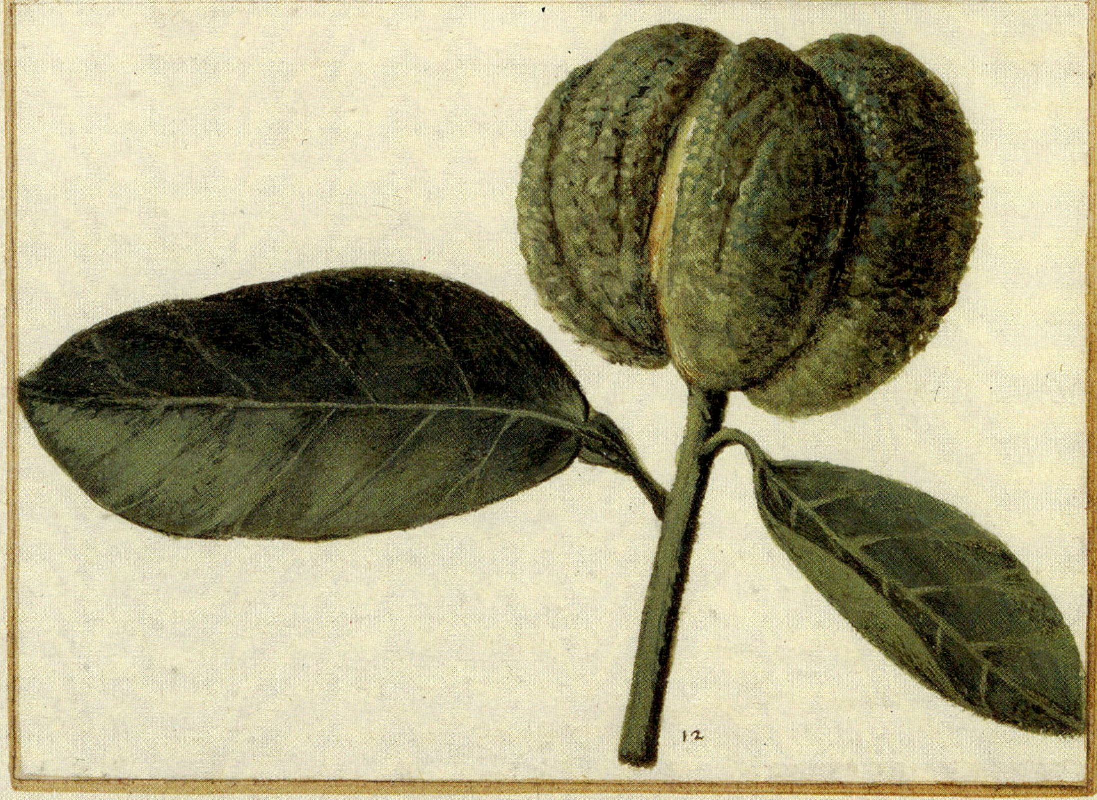

106

107. Citrus, *Citrus sp.*: ribbed fruits

?VINCENZO LEONARDI (fl. 1621–46)

Windsor, RL 19368

Watercolour with gum heightening, over black chalk; *pentimenti* in black chalk

169 × 276 mm. Prepared white paper

ANNOTATION: [bottom centre] *704* fruit

MOUNT SHEET: 435 × 586 mm. *Watermark*: D & C Blauw

Not reproduced in Ferrari's *Hesperides*, the drawing shows three yellow globose fruits, two halved ones with a thick rind, and numerous segments apparently without seeds. In all these fruits the rind is abnormally ribbed as is often the case with autogenous cyto-chimeras where the outer layer of the bud's apical meristem is affected (see **100**). Given their misshapen forms and lack of sufficiently distinctive taxonomic features, the identification of these specimens is not easy, although their yellow colour suggests that they could be citron-lemons.

A diagonal black chalk line runs along the bottom of the drawing, from left to centre, which might indicate a step or ledge, as in **71** and **98**.

PROVENANCE: dal Pozzo; Albani; George III, by descent to George V; London art market (Mendelson); Rex Nan Kivell; Her Majesty The Queen

REFERENCES: Micheli *Enumeratio*, p. 56f., pls. 43–4; Reuther/Batchelor/Webber II, p. 344; Tilney-Bassett 1986, p. 42; Baldini 1991, p. 131–45

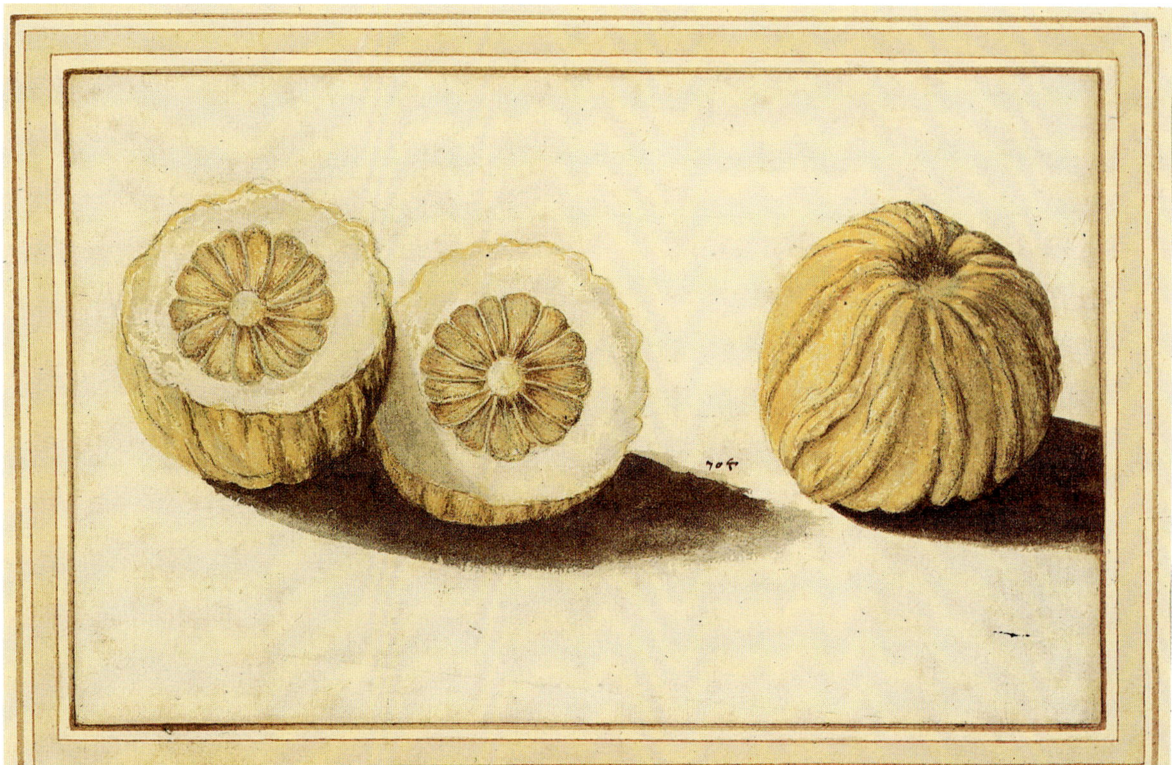

107

108. Citrus, *Citrus sp.*: whole fruit

?VINCENZO LEONARDI (fl. 1621–46)

London, Victoria and Albert Museum, E. 732-1949

Watercolour and bodycolour, with gum heightening and blackened lead white stippled highlights, over black chalk

134 × 123 mm. Trimmed through shadow. Marks of original 17th-century inlay visible (5 mm overlap)

MOUNT SHEET: 17th-century, 407 × 353 mm. Five drawing sheets inlaid. *Watermark:* Crown (not identified further). 17th-century mount sheet laid down onto 18th-century mount sheet, 584 × 435 mm. Wash borders. *Watermark*: D & C Blauw

Not reproduced in Ferrari's *Hesperides,* it is very likely that this drawing is of the same fruit as that shown as **110-112**, now mounted on the same sheet along with it. The fifth drawing mounted on this sheet (**109**) shows a more orange-coloured fruit with its leaves and with a rougher, more corrugated surface. Unfortunately from a taxonomic point of view, the present fruit is not clearly identifiable, but it does seem to have been the object of unusual interest, if indeed it is the same fruit as those shown on the drawings mounted together with it. Both the way in which the stalk joins the fruit and the arrangement of the segments within the fruit (the 'quartered' section in the centre is unique in the corpus) seem to have piqued the interest of the draughtsman or whoever it was who commissioned the drawing.

PROVENANCE: dal Pozzo; Albani; George III, by descent to George V; London art market (Mendelson); Mrs Helen de Beer

109. Citrus, *Citrus sp.*: whole rough fruit with leaves

?VINCENZO LEONARDI (fl. 1621–46)

London, Victoria and Albert Museum, E. 732-1949

Watercolour and bodycolour, with gum heightening, over black chalk

142 × 124 mm. Trimmed through shadow on right

ANNOTATION: *4*

MOUNT SHEET: 17th-century, 407 × 353 mm. Five drawing sheets inlaid. *Watermark:* Crown (not identified further). 17th-century mount sheet laid down onto 18th-century mount sheet, 584 × 435 mm. Wash borders. *Watermark*: D & C Blauw

See **108** for the problem of identifying the fruit shown here, and for its relationship with the other drawings mounted on the same sheet. It is shown with a green stalk and six small leaves, is more orange-coloured and has a rougher surface than the fruit shown on the other drawings mounted together with it.

PROVENANCE: dal Pozzo; Albani; George III, by descent to George V; London art market (Mendelson); Mrs Helen de Beer

110. Citrus, *Citrus sp.*: half-fruit with segments removed

?VINCENZO LEONARDI (fl. 1621–46)

London, Victoria and Albert Museum, E. 732-1949

Watercolour over black chalk

71 × 110 mm. Trimmed through shadows to left and right

ANNOTATION: *8*

MOUNT SHEET: 17th-century, 407 × 353 mm. Five drawing sheets inlaid. *Watermark:* Crown (not identified further). 17th-century mount sheet laid down onto 18th-century mount sheet, 584 × 435 mm. Wash borders. *Watermark*: D & C Blauw

See **108** for the problem of identifying the fruit shown here, and for its relationship with the other drawings mounted on the same sheet. Given the way in which this particular drawing seems to have been cut through the shadow both on the right and the left, it may well once have formed part of the same drawing sheet as **111**, which would thus have been shown to its left. The fact that both drawings bear the same numerical annotation (*8*) adds some (but not conclusive) weight to this hypothesis.

PROVENANCE: dal Pozzo; Albani; George III, by descent to George V; London art market (Mendelson); Mrs Helen de Beer

111. Citrus, *Citrus sp.*: half-fruit

?VINCENZO LEONARDI (fl. 1621–46)

London, Victoria and Albert Museum, E. 732-1949

Watercolour over black chalk

116 × 137 mm. Trimmed through shadow. Marks of original 17th-century inlay visible (5 mm overlap)

ANNOTATION: *8*

MOUNT SHEET: 17th-century, 407 × 353 mm. Five drawing sheets inlaid. *Watermark:* Crown (not identified further). 17th-century mount sheet laid down onto 18th-century mount sheet, 584 × 435 mm. Wash borders. *Watermark*: D & C Blauw

See **108** for the problem of identifying the half-fruit shown here, and for its relationship with the other drawings mounted on the same sheet. As noted in the previous entry, it is quite possible that the present drawing once appeared on the same drawing sheet as the previous entry (**110**), to its right. The specimen resembles the fruits represented in **108**, **110** and **112**.

PROVENANCE: dal Pozzo; Albani; George III, by descent to George V; London art market (Mendelson); Mrs Helen de Beer

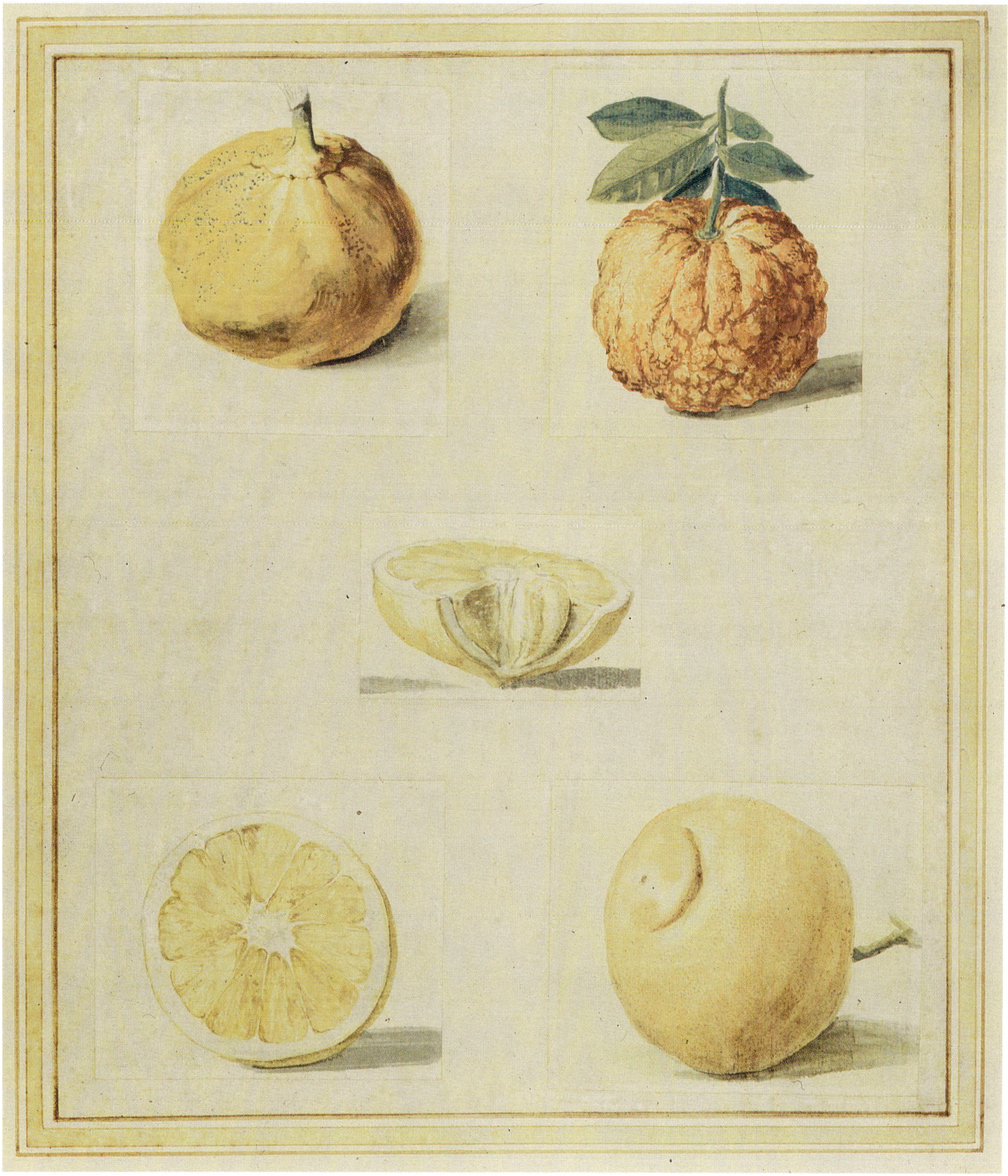

108–112

112. Citrus, *Citrus sp.*: smooth whole fruit

?VINCENZO LEONARDI (fl. 1621–46)

London, Victoria and Albert Museum, E. 732-1949

Watercolour and bodycolour

119 × 147 mm. Trimmed through shadow. Marks of original 17th-century inlay visible (5 mm overlap)

ANNOTATION: *7*

MOUNT SHEET: 17th-century, 407 × 353 mm. Five drawing sheets inlaid. *Watermark:* Crown (not identified further). 17th-century mount sheet laid down onto 18th-century mount sheet, 584 × 435 mm. Wash borders. *Watermark*: D & C Blauw

See **108** for the problem of identifying the fruit shown here, and for its relationship with the other drawings mounted on the same sheet.

PROVENANCE: dal Pozzo; Albani; George III, by descent to George V; London art market (Mendelson); Mrs Helen de Beer

The drawing shows a globose fruit (probably the same one as depicted in **108**), with a smooth rind and a flat nipple at its apex.

113. Citrus, *Citrus sp.*: ribbed misshapen fruit

SEVENTEENTH-CENTURY ITALIAN

Windsor, RL 19326

Watercolour and bodycolour, with gum heightening, over black chalk

137 × 193 mm. Survival of 5 mm overlay of original 17th-century mount at left-hand side

ANNOTATION: [right] *G*

MOUNT SHEET: 587 × 435 mm. Mounted on same sheet as **114**. Slightly stained. *Watermark*: D & C Blauw.
Annotation: [verso: pencil, in a 20th-century hand] *59*.

This drawing, not engraved in Ferrari's *Hesperides*, shows an oblate, ribbed and wrinkled fruit, with a discoloured lateral excrescence protruding on the rind, probably the result of insect infestation or bite. The fruit stalk is inserted onto a thick, leafy twig, which bears the residue of another, now detached fruit. The leaf petiole is wingless as in citrons and lemons. The species is not identifiable, although, apart from the deformation, it does bear a rough resemblance to the citrated orange or *Aurantium citratum* illustrated on p. 423 of Ferrari's *Hesperides*.

PROVENANCE: dal Pozzo; Albani; George III, by descent to George V; London art market (Mendelson); Rex Nan Kivell; Her Majesty The Queen

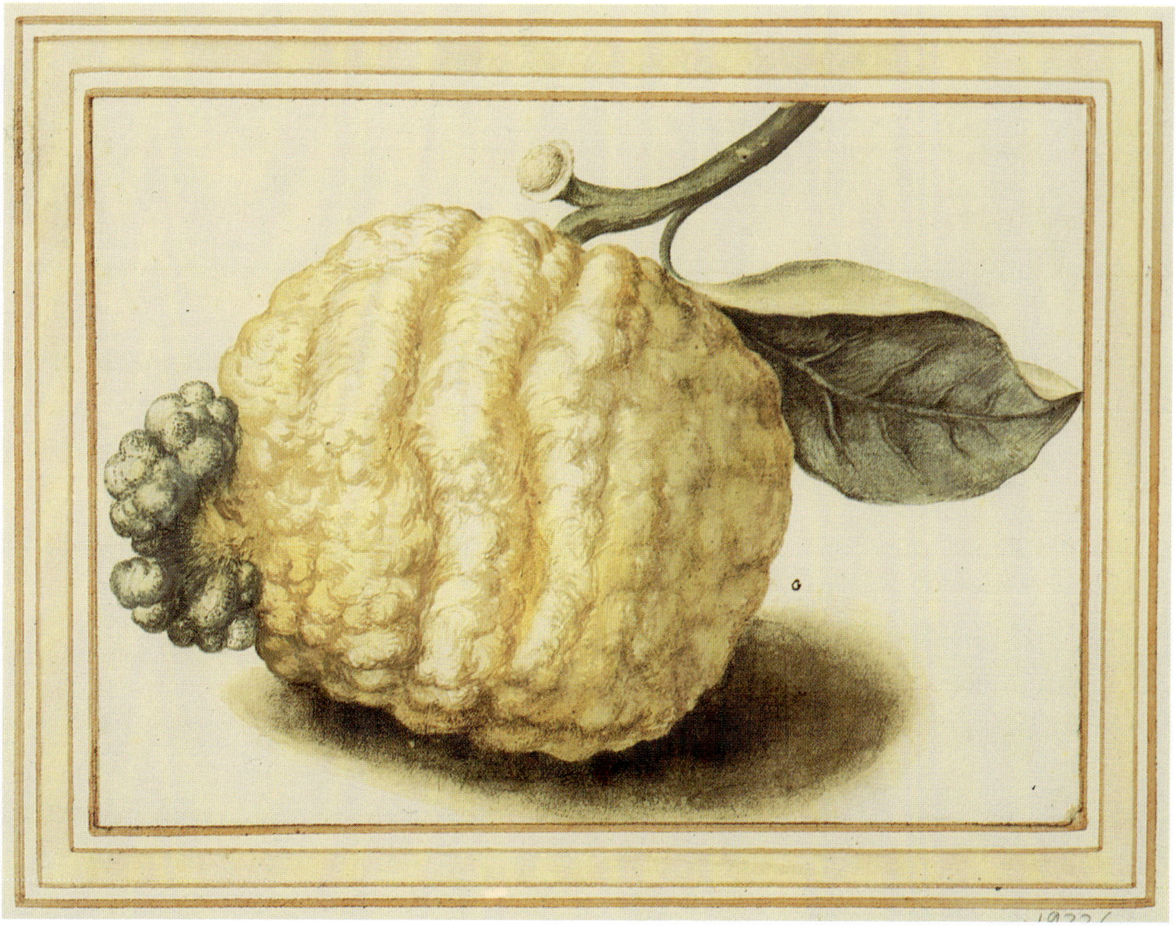

113

114. Citrus, *Citrus sp.*: misshapen half-fruit

SEVENTEENTH-CENTURY ITALIAN

Windsor, RL 19325

Watercolour and bodycolour, with gum heightening, over black chalk

121 × 134 mm

MOUNT SHEET: mounted on same sheet as **113**

This drawing, not reproduced in Ferrari's *Hesperides*, is the cross-section of the fruit shown in the preceding entry (**113**) and by the same hand; it has a relatively thick rind and fifteen seedless segments around a large central axis. Both drawings must originally have been on the same drawing sheet, later separated, and then in the 18th century mounted together when they entered the collection of George III.

PROVENANCE: dal Pozzo; Albani; George III, by descent to George V; London art market (Mendelson); Rex Nan Kivell; Her Majesty The Queen

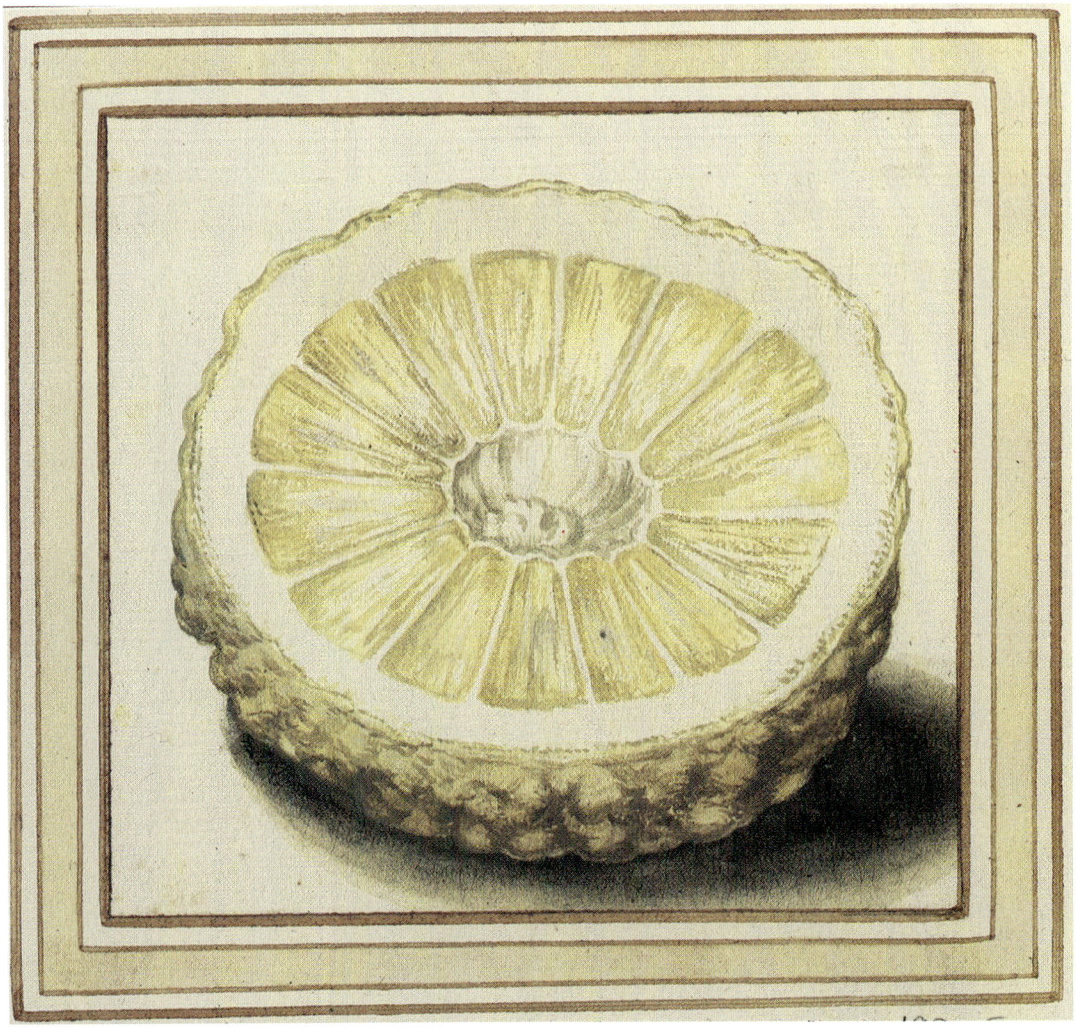

114

115. Citrus, *Citrus sp.*: whole fruit with leaves

?VINCENZO LEONARDI (fl. 1621–46)

Windsor, RL 19354

Watercolour and bodycolour, some oxidisation of white highlights over black chalk; *pentimenti* in black chalk

200 × 192 mm. Foxing

ANNOTATION: *47*

MOUNT SHEET: 391 × 337 mm. *Watermark*: D & C Blauw

Not reproduced in Ferrari's *Hesperides*, this comparatively rough drawing shows an unrecognizable, globose fruit (perhaps an orange) with a rind that is ribbed and wrinkled. It has two wide, elliptical leaves, with wingless petioles anomalously inserted at the same point on the fruit stalk.

PROVENANCE: dal Pozzo; Albani; George III, by descent to George V; London art market (Mendelson); Rex Nan Kivell; Her Majesty The Queen

115

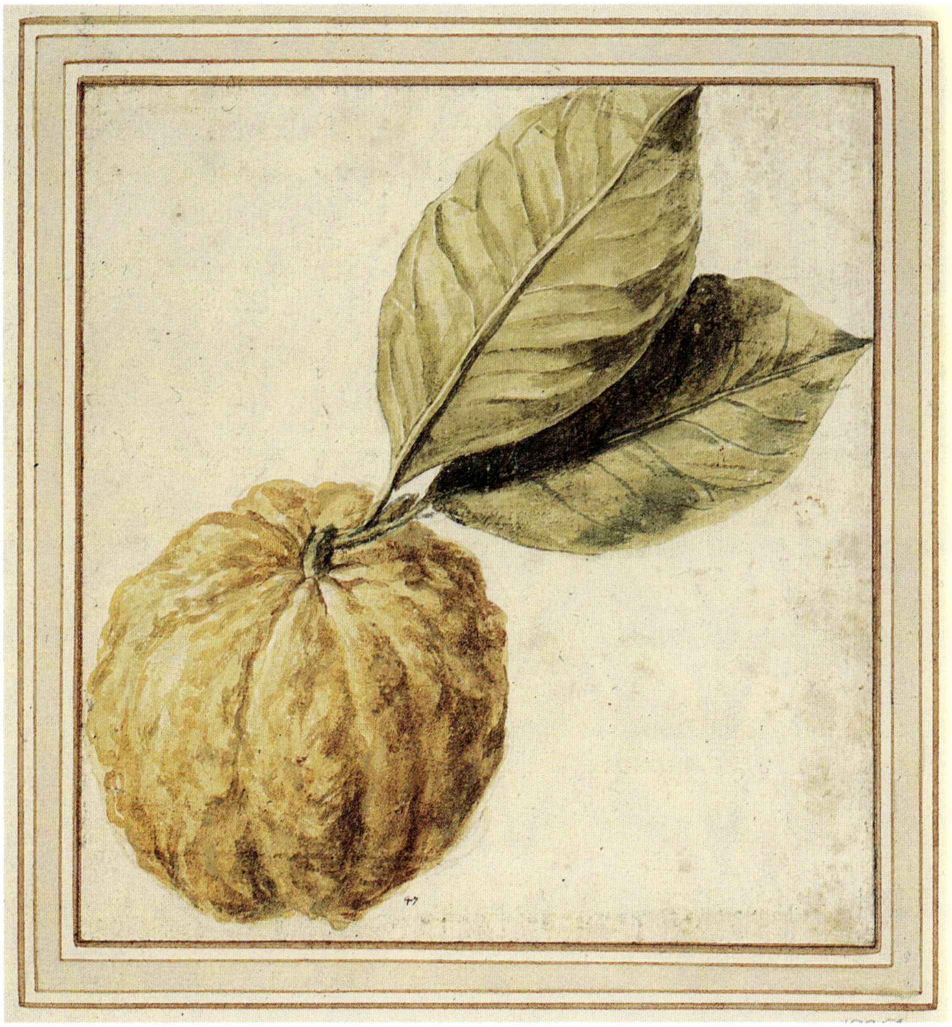

116. Citrus, *Citrus sp.*: half-fruit

?VINCENZO LEONARDI (fl. 1621–46)

Windsor, RL 19386

Watercolour over black chalk

59 × 57 mm (trimmed along right-hand edge).

Survival of 5 mm overlay of original 17th-century mount at edges. Small mark at bottom centre whitened out

MOUNT SHEET: 240 × 420 mm

Another drawing, not reproduced in the *Hesperides*, showing the cross-section of deep yellow fruit, with a thin rind and a few segments with large seeds. The drawing may be by the same hand as thc preceding entry (**115**)

PROVENANCE: dal Pozzo; Albani; George III, by descent to George V; London art market (Mendelson); Rex Nan Kivell; Her Majesty The Queen

116

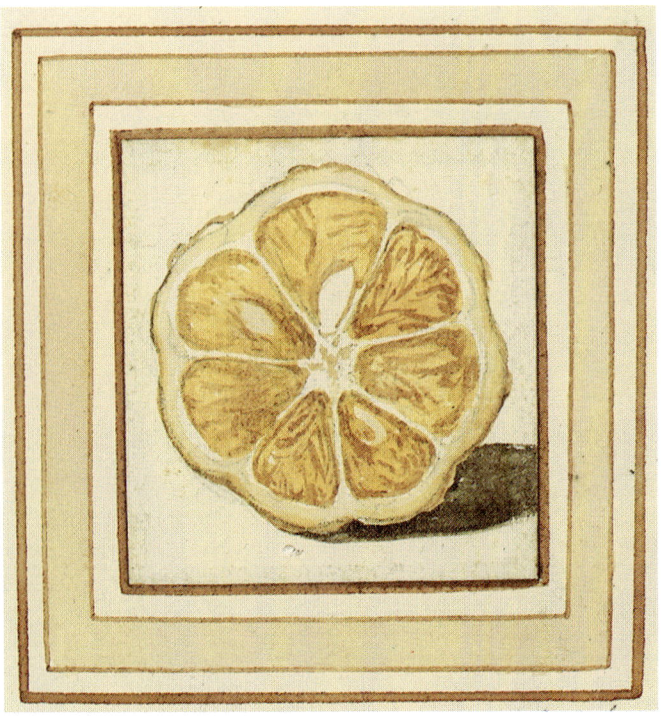

117. Citrus, *Citrus sp.*: fruiting branch

SEVENTEENTH-CENTURY ITALIAN

New York art market

Watercolour and bodycolour over graphite

312 × 205. Badly foxed. *Watermark: Figure 18* (cut)

ANNOTATION: [bottom centre] *600*

Not reproduced in Ferrari's *Hesperides*, this drawing shows a fruiting branch of an undetermined citrus species. Its leaves are elliptical and narrow, with wingless petioles. The fruits are small, globose or sub-globose, each with a dark yellow peel and a prominent nipple at the apex. This and the following drawing (**118**) are unlike all the others in the corpus of citrus illustrations that came from the dal Pozzo collection, in that they show whole branches with fruits and leaves. They may well have been intended to illustrate some kind of herbal.

PROVENANCE: dal Pozzo; Albani; George III, by descent to George V; London art market (Mendelson); James R. Herbert Boone; Sotheby's (New York), 16 September 1988, lot 154

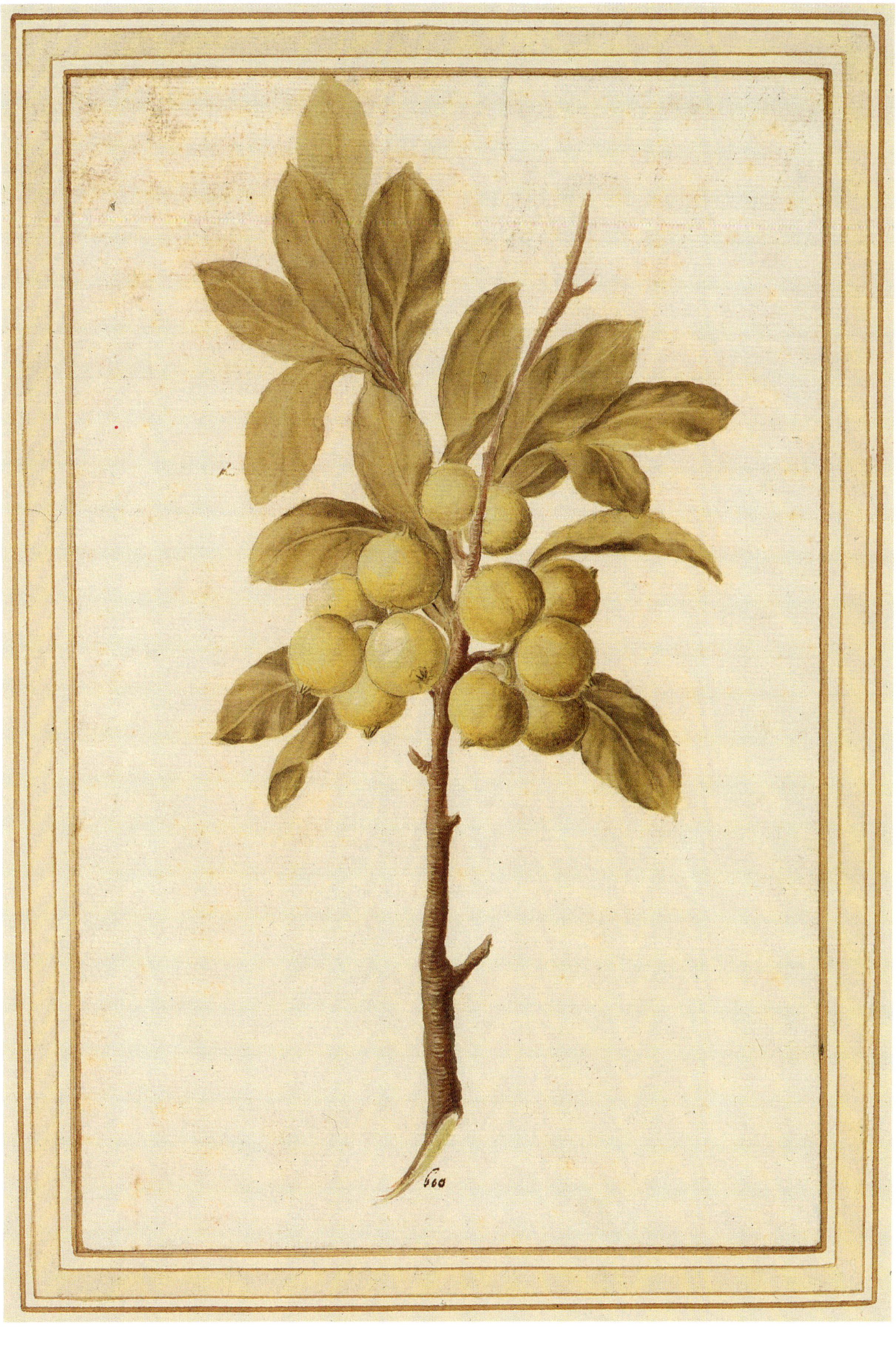

117

118. Citrus, *Citrus sp.*: anomalous fruiting branch

SEVENTEENTH-CENTURY ITALIAN

New York art market

Watercolour and bodycolour, with blackened lead white highlights, over graphite

483 × 346 mm. Foxed. *Watermark*: none recorded

ANNOTATION: [bottom centre] *517*

Like the drawing in the preceding entry, this one shows a fruiting branch. It has narrow, lanceolate leaves with wingless petioles and bears six yellowish-orange fruits. These look teratological but the variety of their shapes (globose, ovoid, prominently nippled, ribbed, furrowed) resist any attempt at taxonomic identification. This drawing was likewise not reproduced in Ferrari's *Hesperides*.

PROVENANCE: dal Pozzo; Albani; George III, by descent to George V; London art market (Mendelson); James R. Herbert Boone; Sotheby's (New York), 16 September 1988, lot 153

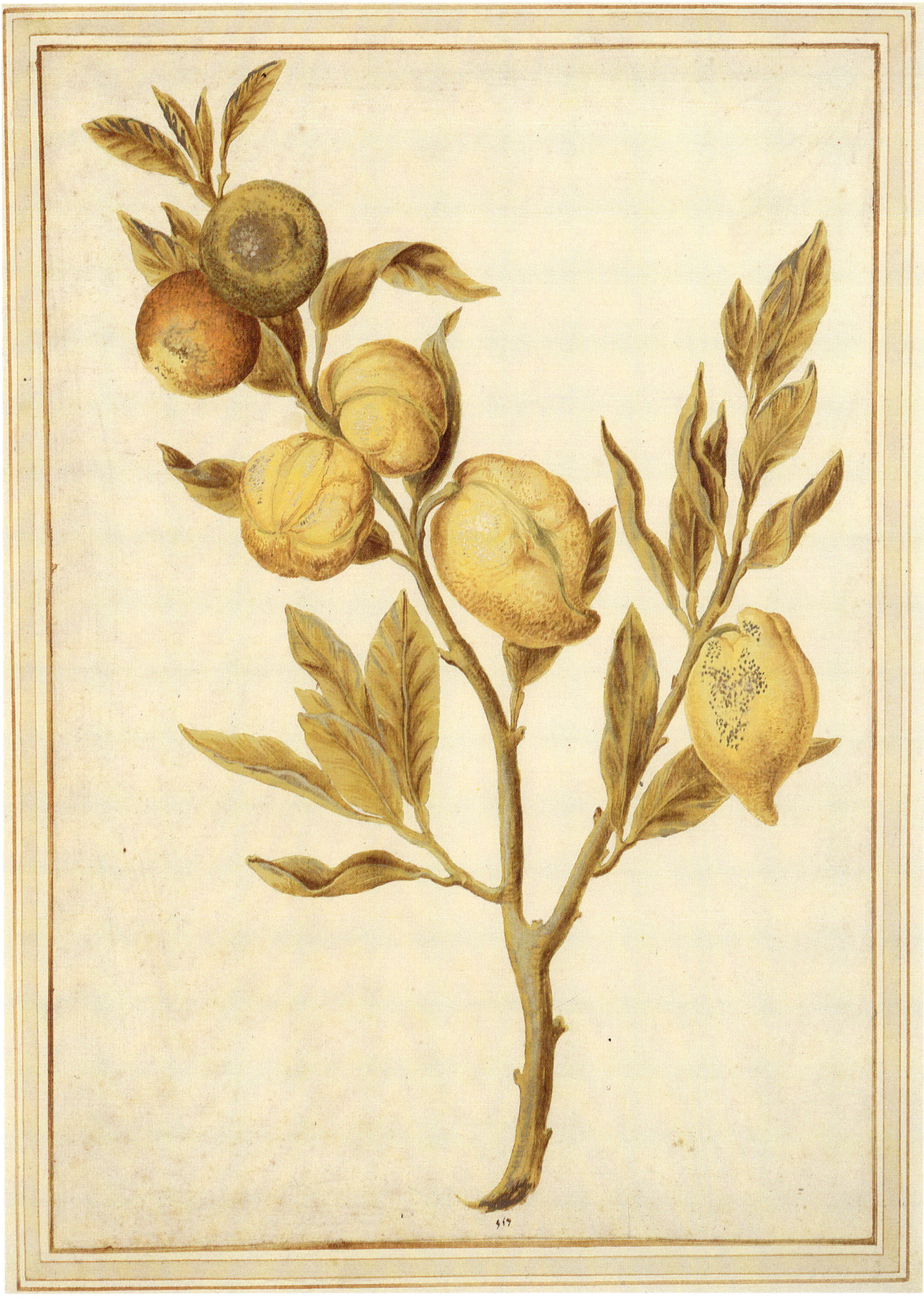

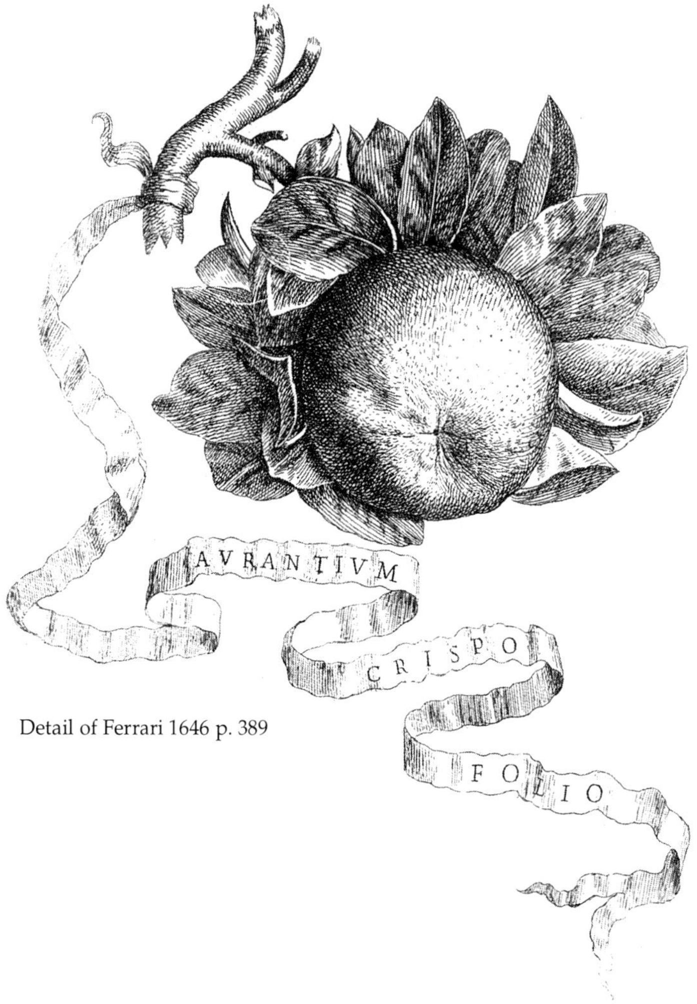

Detail of Ferrari 1646 p. 389

WATERMARK TYPES

Eileen Kinghan

A COMPREHENSIVE SURVEY of the watermarks of all the dal Pozzo drawings and their mounts was completed by Martin Clayton in 1993. This survey encompassed the drawings in the Royal Library, the British Museum, the Royal Botanical Gardens, Kew, Sir John Soane's Museum, the Stirling-Maxwell and Boone collections (prior to their respective sales), and the Paris material. Some 2,000 marks were found and photographed actual size using a luminescent light sheet, or traced where this was not possible. They were sorted into *types* (see below), of which some 32 were identified. It is planned to publish the full results of this watermark survey in due course as a 'Dictionary' in a later volume of the Catalogue Raisonné. At present, no comparable corpus of seventeenth-century Italian watermarks exists.

The typology of the watermarks identified follows the system established by Heawood, based in the first instance on the subject, known here as *type* (e.g. an anchor, bird or cross) then according to a particular *design* (e.g. anchor enclosed in circle and surmounted by six-pointed star). *Groups* of marks which are closely related by appearance were then identified; such groups probably originated from a single mould which had been subject to wear and tear (e.g. *Anchors 8* and *9*). Using accurate tracings for comparison, it was determined which marks were identical and thus made from the same mould. Pairs of marks within groups (e.g. *Hills 9* and *10*) were also identified: these may have resulted from the fact that two moulds were in use simultaneously during the papermaking process and the mouldmaker probably tried to make the marks on each pair of moulds as close in appearance as possible (see Briquet, p. 29).

Information about seventeenth-century watermarks among the citrus fruit drawings is sparse, for the following reasons. Many of the drawings could not be checked as they are now dispersed and their locations unknown (see pp. 30–1). All the drawings which were accessible had to be examined using raking light as they are laid down onto heavy eighteenth-century paper. This did not allow good tracings, and it was difficult, therefore, to make accurate comparisons of marks. A further problem, resulting from the eighteenth-century trimming of the sheets, was that a portion of almost every mark recorded had been lost.

Despite the fact that the number of marks recorded is low (there are only eleven known and identifiable watermarks) and that all are cut and some are not positively identifiable, some patterns do emerge. The marks comprise three *types*: *Figure*,

Crown and *Eagle*. Two of the *Figure* marks are of the same design; all of the *Crown* marks are from the same group. Where there is any parallel in Briquet, Heawood or Zonghi, this is stated below.

The author would like to acknowledge the help of Martin Clayton, who examined and sorted the watermarks (with assistance from Ben Hayward and Siân Smith), established their types and gave much useful advice. We are grateful to Simon Roberton who carried out the photography of all the marks.

NOTE: The photographs reproduced show clear images of each mark; they are not necessarily photographs of the drawings catalogued here.

Eagle

Water Mark	Drawing (Catalogue no)	Date	Remarks
?15	**88**	*c.* 1621–46	Same group as Woodward 63–70; cf. 64, 67 and 69

15

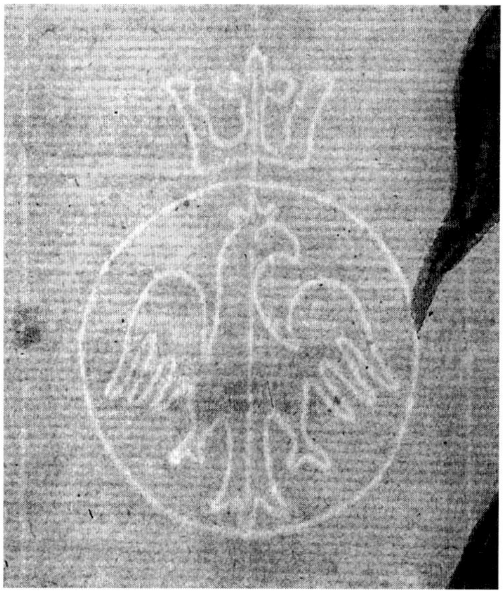

Crown

Water Mark	Drawing (Catalogue nos)	Date	Remarks
?34	**20**	*c.* 1621–46	Same group as *35* and *38*
35	**71**		Unique to this drawing
38	**65, 90**	*c.* 1621–46	Fragments of marks. Same design as Heawood 1120, 1123

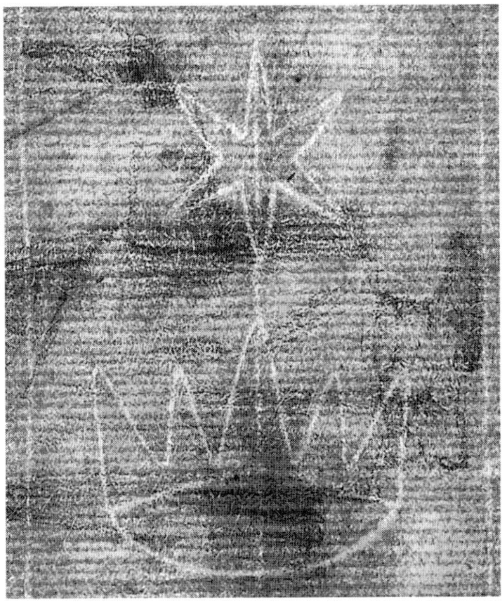

34

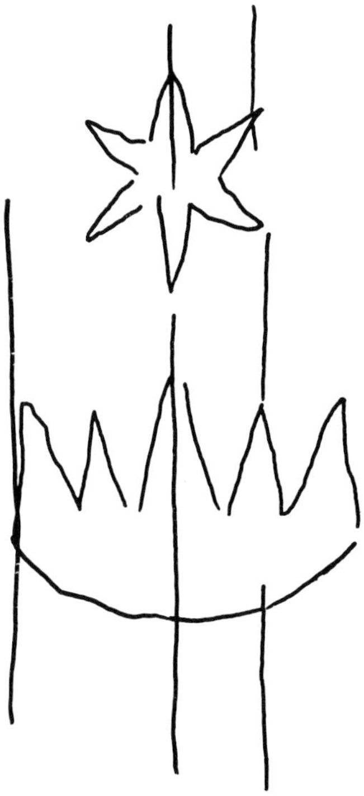

35

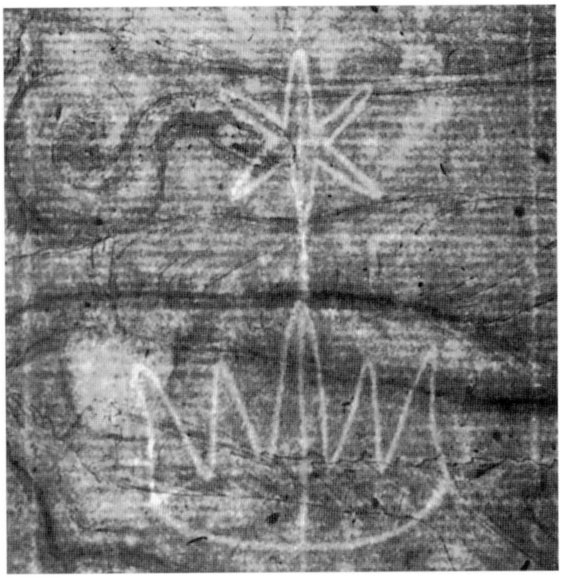

38

Figure

Water Mark	Drawing (Catalogue nos)	Date	Remarks
18	**117**		See also A.II.**177**. Possibly paired with 18B
18B	**104**		Unique to this drawing
?60	**28, 48?, 49, 55?, 61**	*c.* 1621–46	**28**, **48** and **65** show evidence of 17th-century inlay mounting

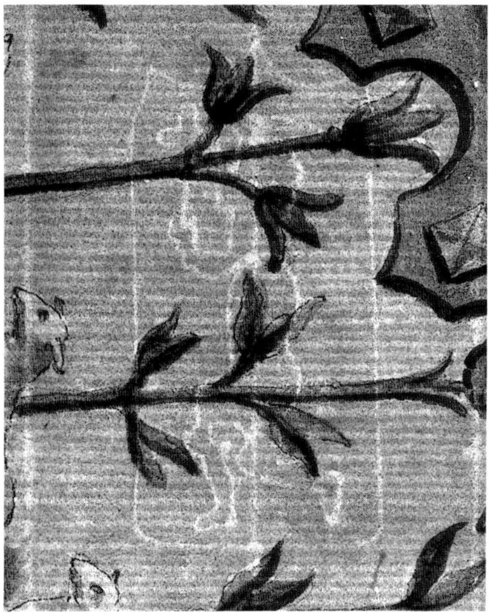

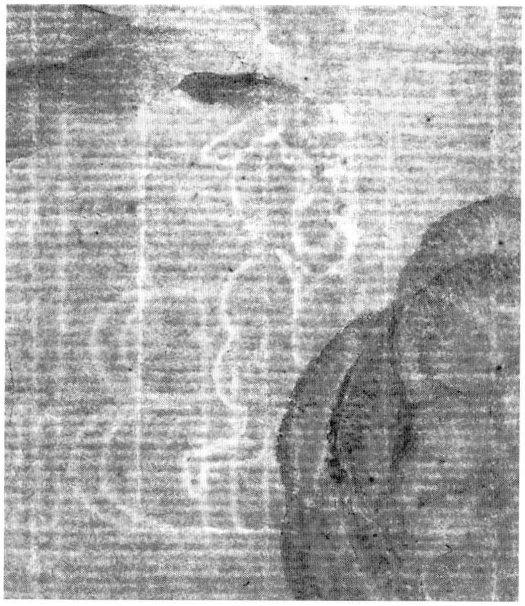

18 18B

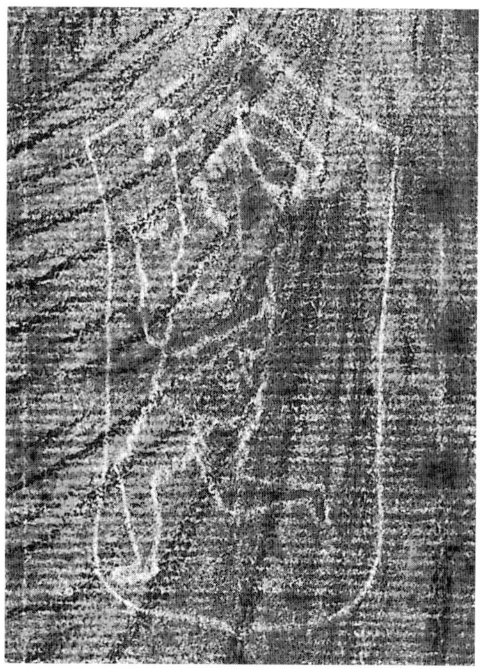

60

GLOSSARY

Enrico Baldini

Cross-references to terms in the glossary are indicated in small capitals

Abortive: (syn. defective)

Albedo: the soft, whitish and spongy MESOCARP of HESPERIDIA; it is located between the peel (EPICARP) and the segments (ENDOCARP).

Anthesis (syn. blooming, flowering): strictly speaking, it is the period of flower development when the corolla opens and pollination takes place.

Apex: the end of a shoot or a leaf or a fruit.

Apiculate: a fruit tipped with a short point.

Apomixis: the induction of asexual embryos from ovular body tissues without the fusion of sexual cells or nuclei. In *Citrus* apomixis basically corresponds to adventitious nucellar embryony.

Chimera: organ or organism composed of genetically distinct tissues in intimate association. Chimeras can arise from adventitious buds formed at the level of a graft union (synthetic chimera), or from bud MERISTEMS which become genetically heteGeneous as a consequence of naturally or artificially induced mutations (autogenous chimeras). In MERICLINAL chimeras only the outer layer of a bud, and the tissues derived from it, are genetically heterogeneous. Synthetic chimeras may be designated by a formula including a plus (+) sign between the names of the parental organisms (see Fig. 33, p. 86).

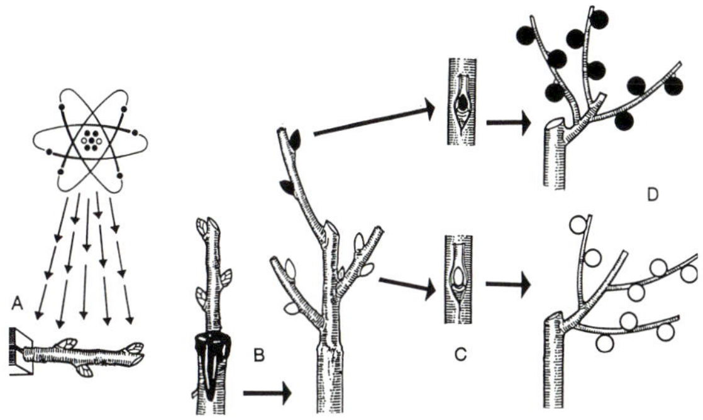

CHIMERA Diagram showing the mutagenous effect of physical treatment (radiation) on the genotype of bud meristems. A: twig exposed to X-rays. B: part of the shoot of the irradiated twig, grafted, showing buds of a different genotype (conventionally shown in black). C: vegetative propagation (grafting, budding) involving these modified buds produces a new clone (D).

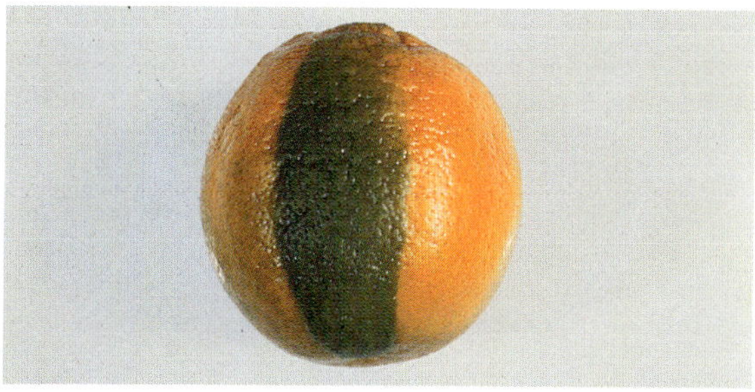

CHIMERA Sweet orange fruit with a striped rind (persistent green on orange) arising from an autogenous chimera.

Citrology: the section of POMOLOGY which deals with citrus taxonomy.

Citrus bud mite: an Acarus (*Aceria sheldoni* Ewing, fam. *Eriophydae*) which specifically affects citrus buds and flowers, markedly preferring lemons. It is 0.17-0.18 mm long; its damage consists of multiple bud formation and deformities of the twigs, leaves, blossoms and fruits.

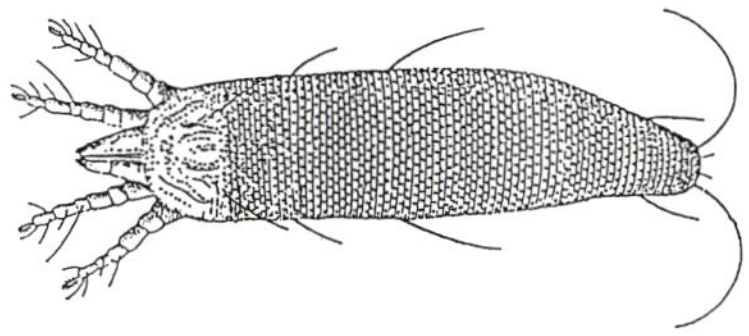

CITRUS BUD MITE Diagram showing dorsal view of the adult female of *Aceria sheldoni* Ewing (× 300). This Acarus is responsible for digitated fruits.

Clone: an assemblage of individuals which all have the same GENOTYPE. Bud mutations give rise to different clones within the same cultivar.

Cluster: an inflorescence formed by a number of flowers whose PEDICELS have a decreasing length.

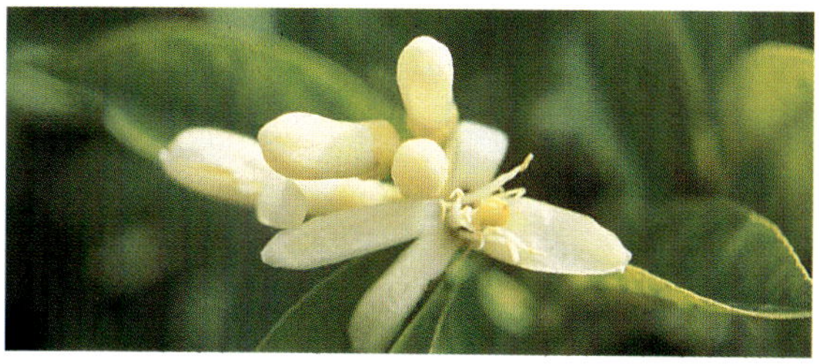

CLUSTER Sweet orange flower cluster.

Cordate: of a leaf which is heart-shaped and has two equal, rounded lobes that form a deep sinus at the base of the lamina.

Cultivar (abbreviation cv.): horticultural term, contracted from cultivated variety and defined under the International Code of Nomenclature of Cultivated Plants (1986) as an assemblage of cultivated plants which is distinguished by any character having a cultural interest (viz. fruit colour, shape and size, ripening time, tree shape and hardiness, etc.).

Cytochimera: an autogenous CHIMERA involving modification of the DNA set in a cell or a group of cells of bud MERISTEMS.

DNA: the deoxyribonucleic acid chiefly located in the nucleus of the cell and responsible for the transmission of genetic codes.

Endocarp (syn. FLESH): in *Citrus* the part of the fruit formed by the segments.

Epicarp (syn. FLAVEDO, peel): in *Citrus* the outer layer of the rind.

Epigenetic factor: an environmental or physiological factor able to modify the GENOTYPE determinism.

Everbearing plant: a plant able to bear flowers and fruits more than once a year; in *Citrus* it occurs in some lemon CULTIVARS.

Flavedo (syn. EPICARP): the outer, yellow or orange layer of the rind, rich in essential oil glands.

Flesh: idiomatic term used to indicate the ENDOCARP of citrus fruits.

Foetus (syn. embryo): foetiferous citrus plants develop abnormal fruits arising from flowers where stamens and ovaries are coalescent.

Fruitlet: the fruit in the early stages of its development, or a small, abortive fruit.

Genome: the whole set of chromosomes, hence of genes, inherited as a unit from one parent.

Genotype: the expressed or latent genetic constitution or gene makeup of an organism.

Genus (Bot.): the major subdivision of a botanical family in plant classification; it consists of one or more species.

Germplasm: a collection of GENOTYPES, that is the whole of the genetic resources of a species.

Graft: (asexual) propagation procedure generally consisting of the union of genetically different individuals such as rootstocks and scion-varieties.

Hesperidium: a citrus fruit formed by a rind (EPICARP and MESOCARP) which normally includes a number of segments containing juicy VESICLES and seeds.

Hybrid: the offspring of two plants belonging to different CULTIVARS, VARIETIES, species or GENERA. Interfertility of species is usual in *Citrus* ; the trifoliate orange (*Poncirus*) and the kumquat (*Fortunella*), two genera closely related to the genus *Citrus*, can be successfully crossed with it and some of their hybrids are somewhat fertile. Hybrids are conventionally denoted by formulas including a cross sign (X) between the names of their parent.

Hypertrophic: enhanced size of an organ due to the abnormal enlargement of its cells.

Internode: the portion of a stem or twig between nodes. Its length is related to the strength of the stem or twig.

Juvenility: the vegetative stage of plants before they begin to flower. In many woody species propagated by seed, the juvenile phase involves manifestations with respect to morphological (ex. thorny twigs) and physiological characters associated with the period before they reach their adult stage.

Lanceolate: of a leaf which is narrow and tapers toward the apex.

Mericlinal: a CHIMERA arising from a bud in which only a sector of its outer MERISTEMATIC layer has a modified GENOME.

Meristem: embryonic tissue formed by undifferentiated, actively dividing cells.

Mesocarp (syn. ALBEDO): in *Citrus* the inner, white layer of the rind.

Molecular markers: biotechnological procedures used to analyse the DNA library; they are very useful in searching ancestors or in comparing taxonomic entities which are suspected to be genetically related.

Morphogenesis: the developmental history of organisms or of their parts.

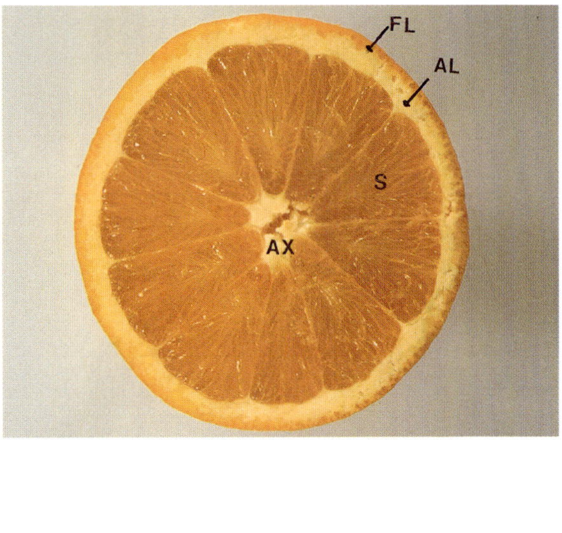

HESPERIDIUM Citrus half-fruit showing its components: FL, flavedo; AL, albedo; AX, central axis. No seeds appear in this parthenocarpic fruit.

Navel: a citrus fruit which includes an abortive fruitlet (navel) at the top of its central axis. The navel can be observed through an apical hole of the rind.

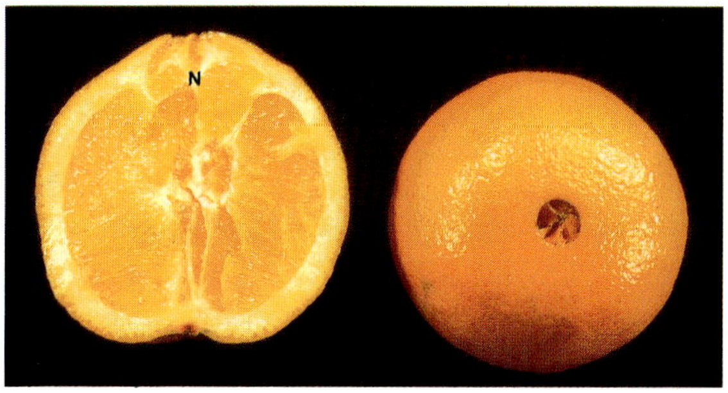

NAVEL Longitudinal section (left) of a navel orange showing the abortive fruitlet N at the top of the central axis, and (right) the apical view of the same fruit showing the stylar hole.

Nomenclature: botanical nomenclature involves Latin names of genera, species and varieties on the basis of a binomial system. Naming of CULTIVARS generally involves also specific vernacular names.

Nucellar embryo: an extra embryo arising from some of the nucellus cells.

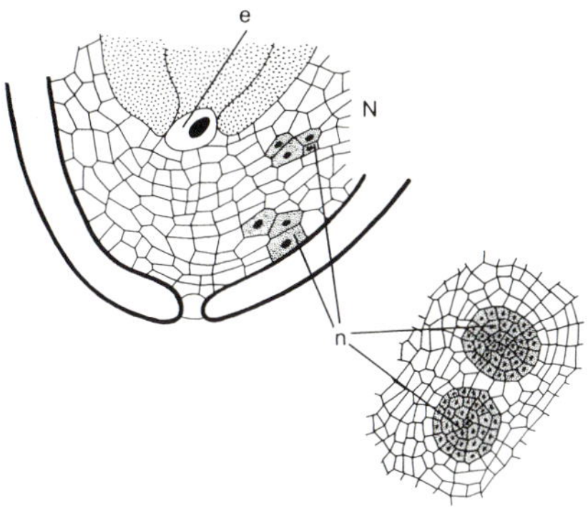

NUCELLAR EMBRYO (Apomixy, Polyembryony) Diagram showing the sexual or gametic embryo (e) inside the embryo sac and two nucellar embryos (n) arising within the nucellar sporophytic tissue of the ovule N.

Nucellus: the body (sporophytic) tissue of the ovule.

Oblate: a fruit flattened at its poles.

Obovate: a fruit inversely ovate.

Orbicular: a circular leaf or a spherical fruit.

Ovary: the enlarged, lower part of the pistil including a number of ovules. After fertilization the ovary becomes a fruit and the ovules become the seeds.

PCR: Polymeric Chain Reaction, a method of rapidly multiplying DNA *in vitro*.

Papilla (syn. nipple): a minute swelling of the fruit skin.

Parthenocarpy: the development of the fruit without fertilization; parthenocarpic fruits are seedless.

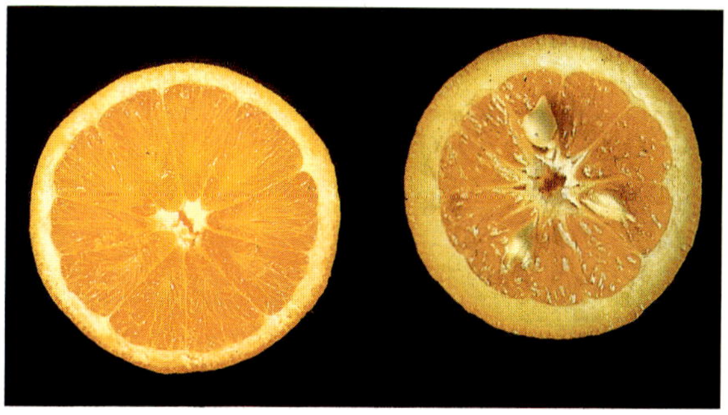

PARTHENOCARPY Parthenocarpic, seedless orange half-fruit (left) compared to a seeded orange half fruit arising from a fertilized flower (right).

Pedicel: a stalk supporting a single flower within an inflorescence.

Petal: segment of the flower corolla; in oranges, flower bud petals are white, while in lemons they are pinkish.

Petiole: the leaf stalk. It is more winged in some species of *Citrus* than in others.

Phenology: the study of plant biological processes with periodicity like bud sprouting, flowering, fruit growth and ripening.

Phenotype: the appearance of an organism resulting from the interaction between GENOTYPE and environment.

Phylogeny: the evolution of taxa like species, genera, etc.

Pistil: the female organ of a flower consisting of the ovary, the style and the stigma.

Pollen: the fertilizing grains carrying the male gametes and produced by the anthers of the flowering plants.

Polyembryony: the development of two or more embryos in one seed. In *Citrus* extra (apomictic) embryos, other than the sexual one, may be produced by nucellar tissues.

Polymorphism: the variability of characters having a taxonomic value. It can even affect the fruits of the same plant as in the everbearing lemon CULTIVARS.

Pomology: the science dealing with the description, naming and identification of fruit CULTIVARS.

Pyriform: describes a pear-shaped fruit.

Receptacle: the expanded portion of the floral axis that bears the calyx, the corolla and the reproductive organs.

RAPD: acronym for Random Amplified Polymorphic DNAs. An investigation method for DNA analysis based on *in vitro* direct amplification, via PCR, of short sequences of DNA.

Rind: idiomatic term used to indicate the outer part of the citrus fruits; it is formed by a yellow or orange EPICARP (flavedo) rich in oil glands, and by a whitish, soft MESOCARP (albedo).

Sectorial: a CHIMERA arising from a bud in which part of the outer MERISTEMATIC layer and the corpus have a modifed GENOTYPE.

Serrate: a leaf notched on its edge.

Sessile: botanical term used to indicate both a fruit or a leaf devoid of its stalk or PETIOLE and a dwarf plant.

Somatic variation (syn. bud mutation): the modification in the GENOTYPE of a bud MERISTEM; it can occur spontaneously or be promoted by physical agents.

Speciation: the process by which new species are formed.

Stamens: the male organs of a flower ending in anthers.

Taxonomy: the science dealing with identification, naming and classification of living individuals.

Teratogeny: (syn. teratogenesis): the production of monstrous forms.

Teratology: the science dealing with animal and plant monstrosities.

Thorn: a short, pointed, aborted woody shoot.

Tubercle (syn. verruca): a little wart on the skin of certain citrus fruits.

Twin fruit: concretion of a pair of fruits in the course of their development. In general double fruits are included in the same rind. They can arise from the concretion of STAMENS and OVARY or from doubly pistillate flowers.

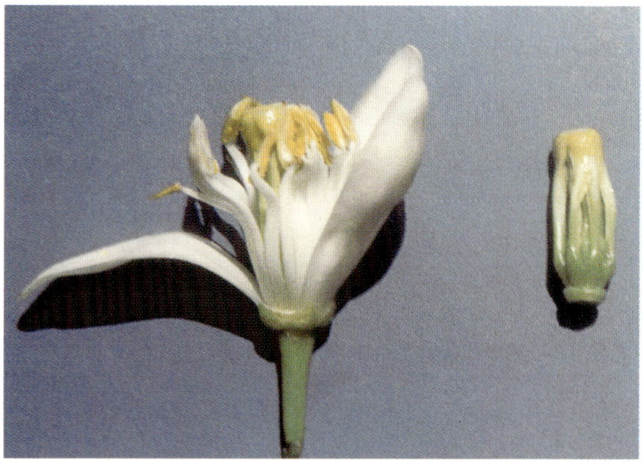

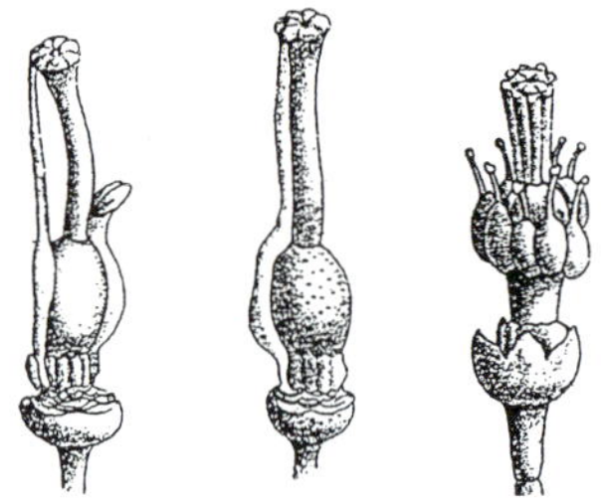

STAMINAL CARPELLOIDY These citrus flowers illustrated stamens which are variously concreted with the pistils, contributing to the development of teratological (foetiferous) fruit. In the flower on the right, the petals and outer stamens have been removed.

The diagram shows, in greater detail, the stamens concreted with the pistil.

Variety (abbreviation var.): the major subdivision of a species in plant botanical taxonomy; this term is now largely abandoned as a botanical category because of its confusion with the cultivar.

Vesicles: the small juicy sacs which form the segments (ENDOCARP) in citrus fruit.

Wings: lateral appendages of the leaf PETIOLE; in *Citrus* they are an important taxonomic character, useful to discriminate citrons and lemons, which are wingless, from other species like sour and sweet oranges, pummelo and grapefruit which are winged.

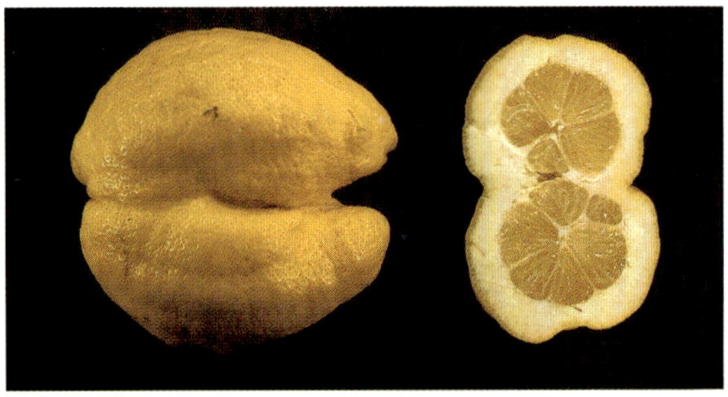

TWIN FRUIT Lateral view (left) and cross-section (right) of a twin lemon: concreted hesperides have only one rind.

GIOVANNI BATTISTA FERRARI'S 'HESPERIDES' CITRUS FRUIT AND ALLEGORICAL ILLUSTRATIONS

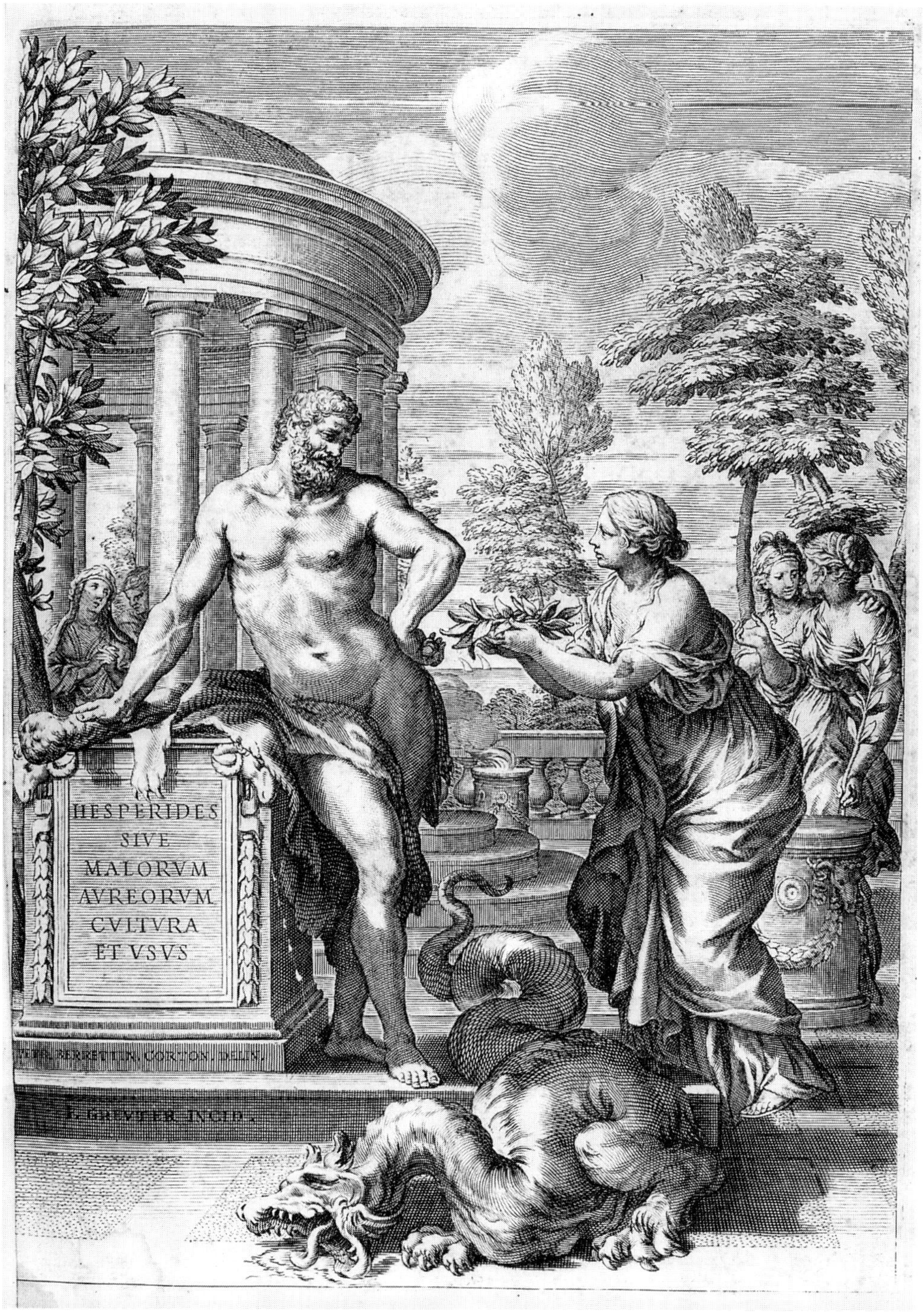

Frontispiece. Hercules as Guardian of the Garden of the Hesperides (see Fig. 13, p. 32)

The engravings and etchings illustrated in this section are identified by the page numbers in Ferrari's *Hesperides*; corresponding Catalogue numbers are given in bold. Fifty-five drawings correspond to 38 engravings; 63 were not engraved. There are no known drawings of the 42 subjects listed below:

Citri flores, p. 59
Malum citreum vulgare belluatum, p. 65
Malum citreum dulci medulla, p. 73
Limon S. Remi, p. 197
Limon vulgo pusilla pila, p. 201
Limon Caietanus, p. 205
Limon Amalphitanus, p. 207
Limon pusillus Calabriae duplex, p. 211
Limon Laurae, p. 219
Limon imperialis, p. 225
Limon dulci medulla vulgaris, p. 229
Peretta sive Spatafora striatus, p. 235
Peretta consimilis, p. 237
Limon striatus vulgatior, p. 247
Limon striatus Amalphitanus, p. 249
Limon Rosolinus, p. 255
Limon Barbadorus, p. 259
Limon scabiosus, p. 260
Limon ponzinus ligusticus, p. 291
Limon ponzinus rubens, p. 293
Limon ponzinus Chalcedonius, p. 297
Limon sponginus Calabriae rugosus, p. 301
Limon sponginus Liguriae verrucosus, p. 303
Adami pomum commune, p. 313
Lumia Valentina, p. 321
Lumia ollulae aspectu, p. 327
Lumia expers medulla, p. 329
Lima dulcis et lima agris, p. 333
Lima romana Piorum in hortis, p. 335
Alia lima citrata oblonga et mostrosa sive scabiosa, p. 337
Lima eadem rotunda, p. 339
Aurantia flores, p. 369
Aurantium acri dulcique medulla vulgare, p. 377
Aurantium sicciore medulla hibernum, p. 379
Aurantium semine carens, p. 383
Idem aurantium limonis effigie, p. 385
Aurantium crispo folio, p. 389
Aurantium stellatum et roseum, p. 395
Aurantium virgatum, p. 399
Aurantium corniculatum, p. 409
Aurantium distortum, p. 415
Aurantium citratum, p. 423

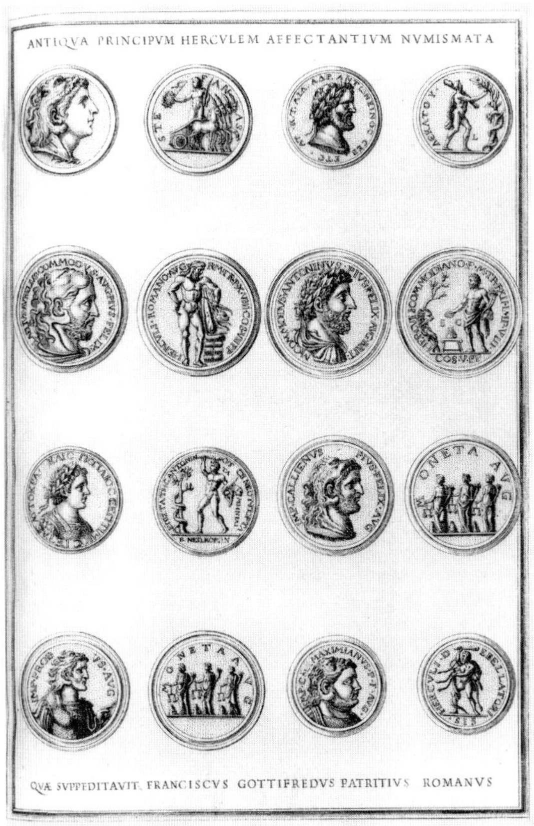

23. Ancient Coins showing Hercules, from the Collection of Francesco Gottifredi (see Fig. 31, p. 83)

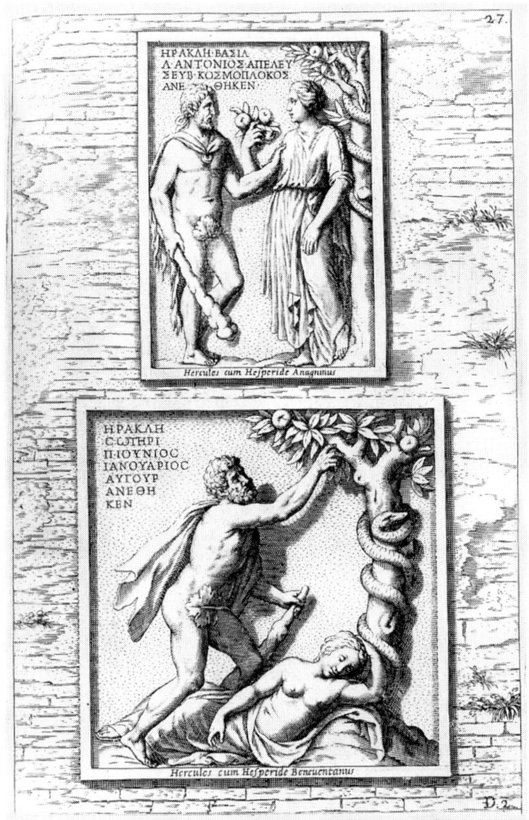

27. Antique Reliefs showing Hercules in the Garden of the Hesperides (see Fig. 14)

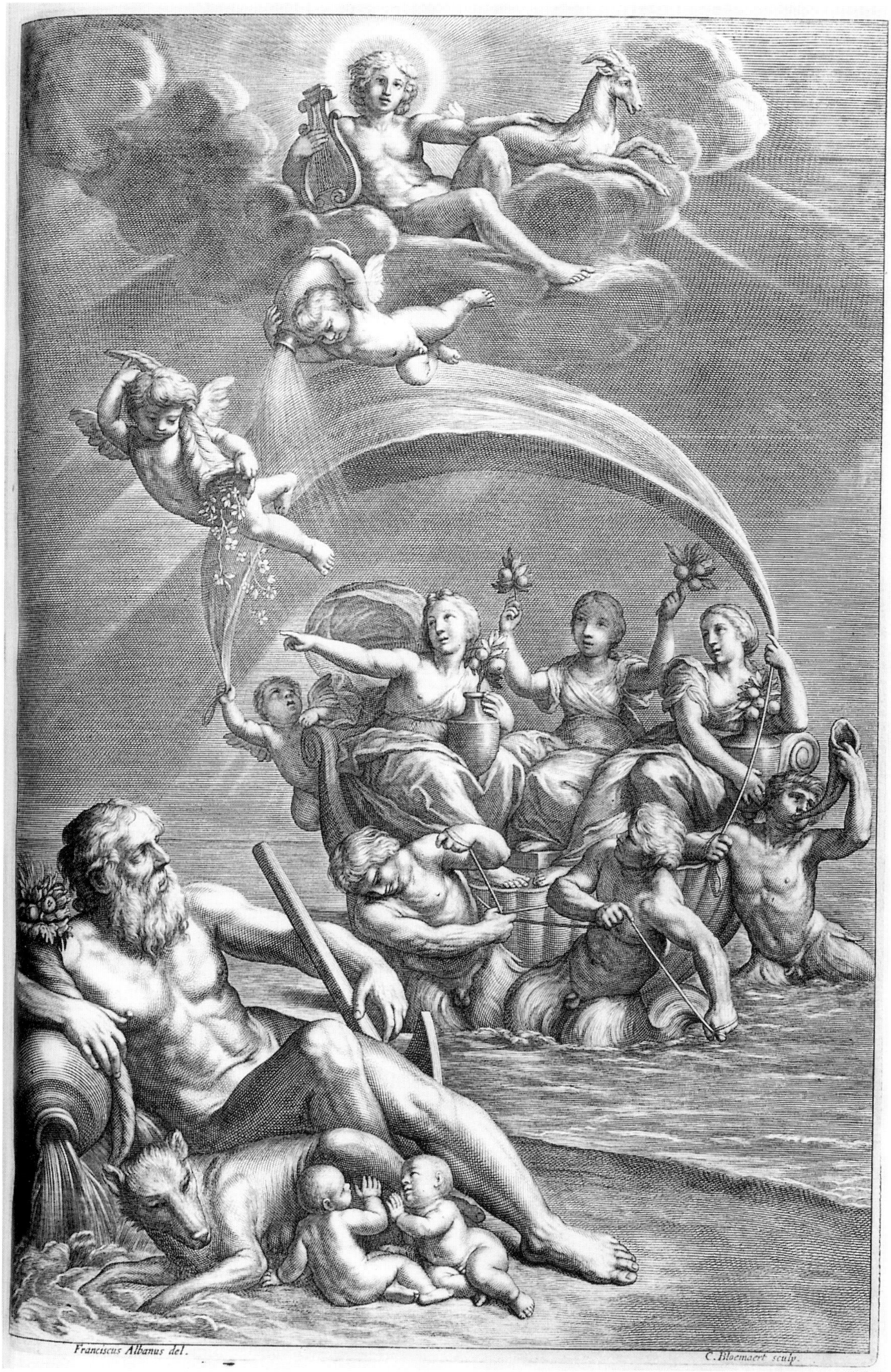

51. Nymphs of the Hesperides bringing Citrus Fruit down the Tiber to Rome (see Fig. 21. p. 65)

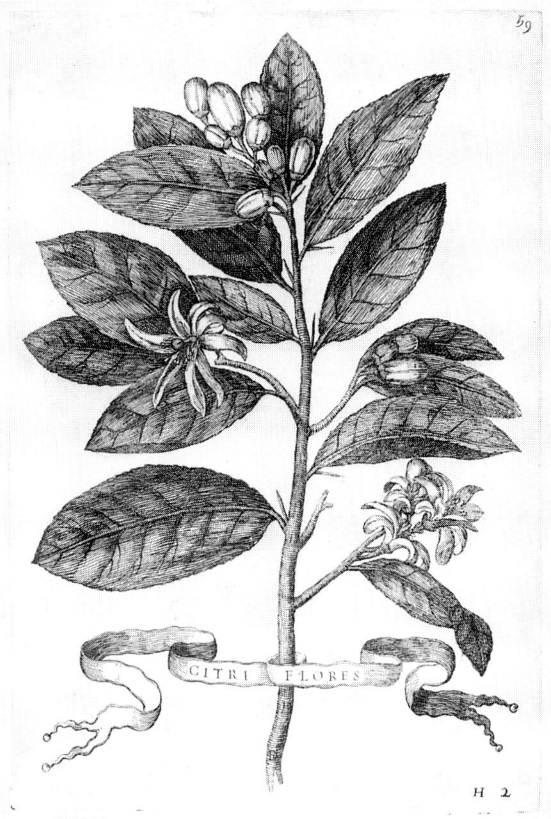

59

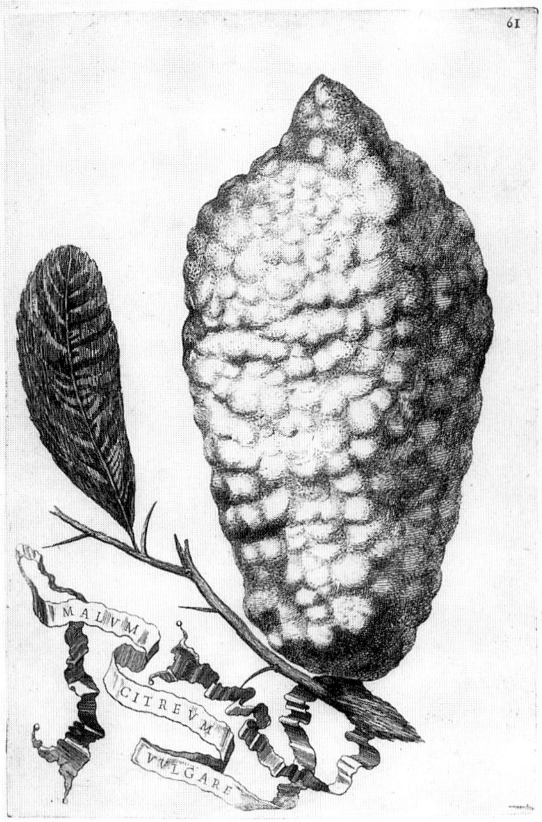

61 **1a**

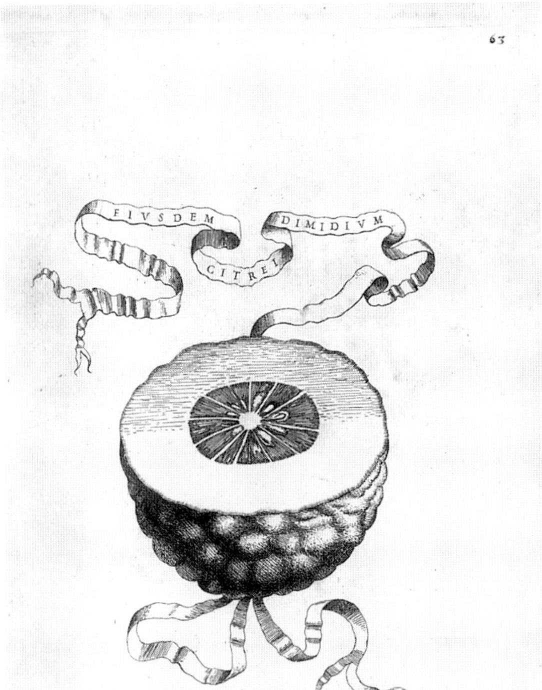

63 **1b**

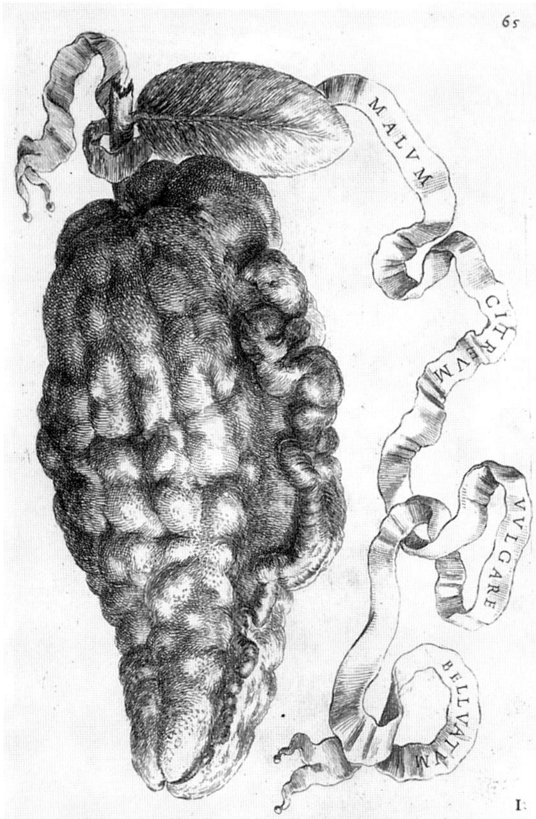

65

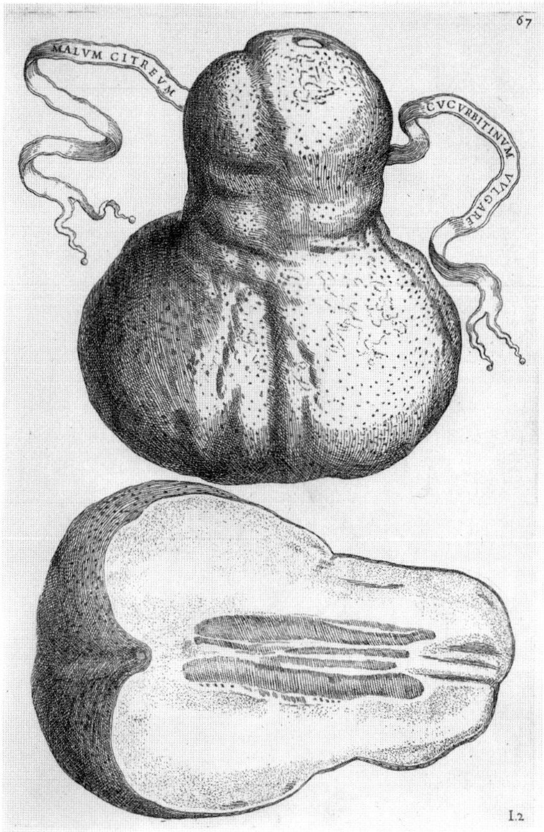

67
2, 3

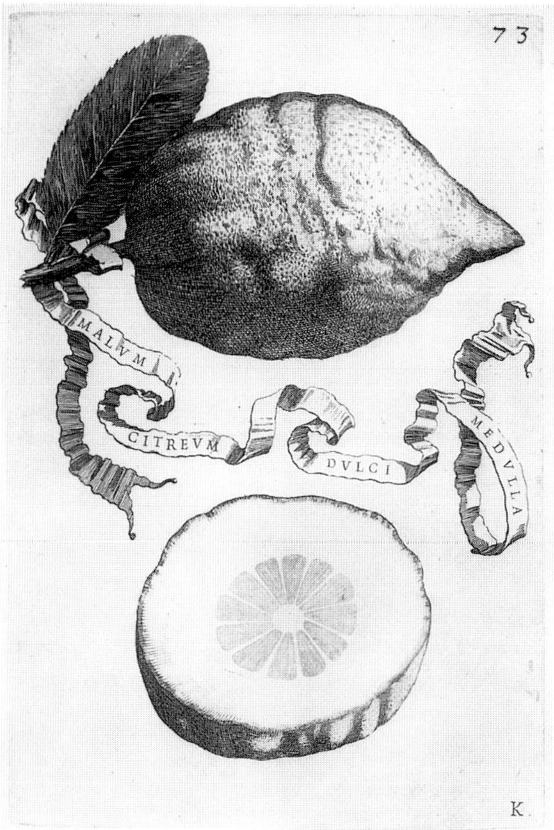

73

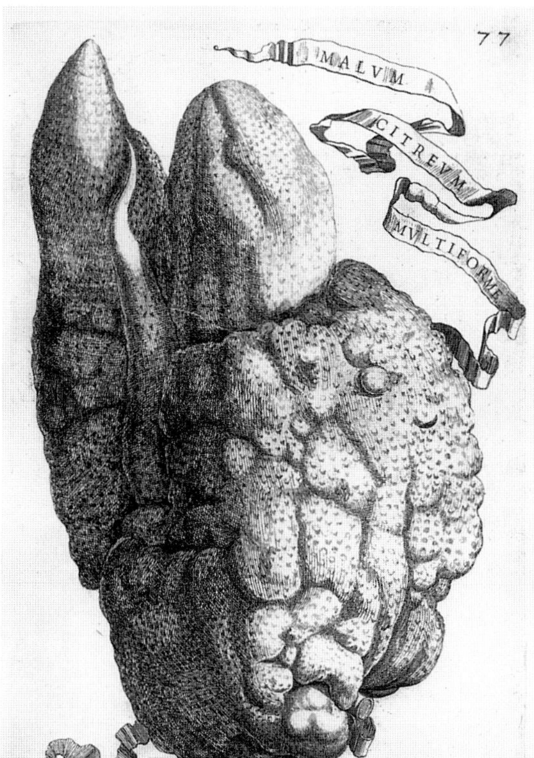

77 **72**

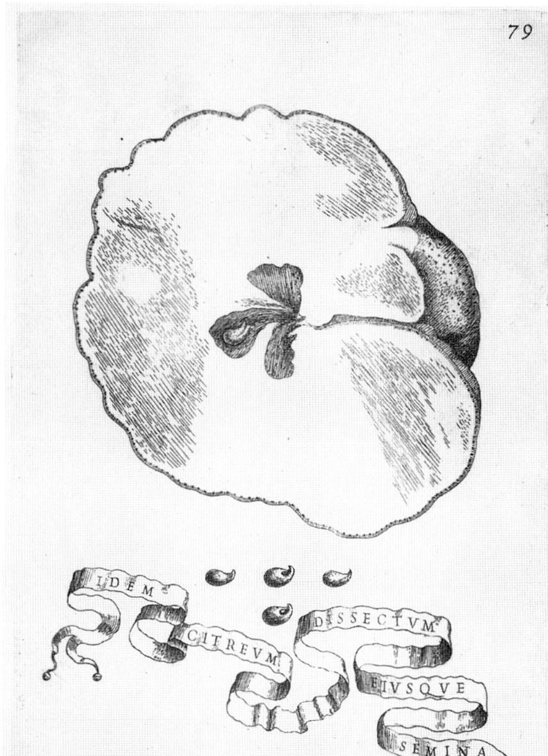

79 **73**

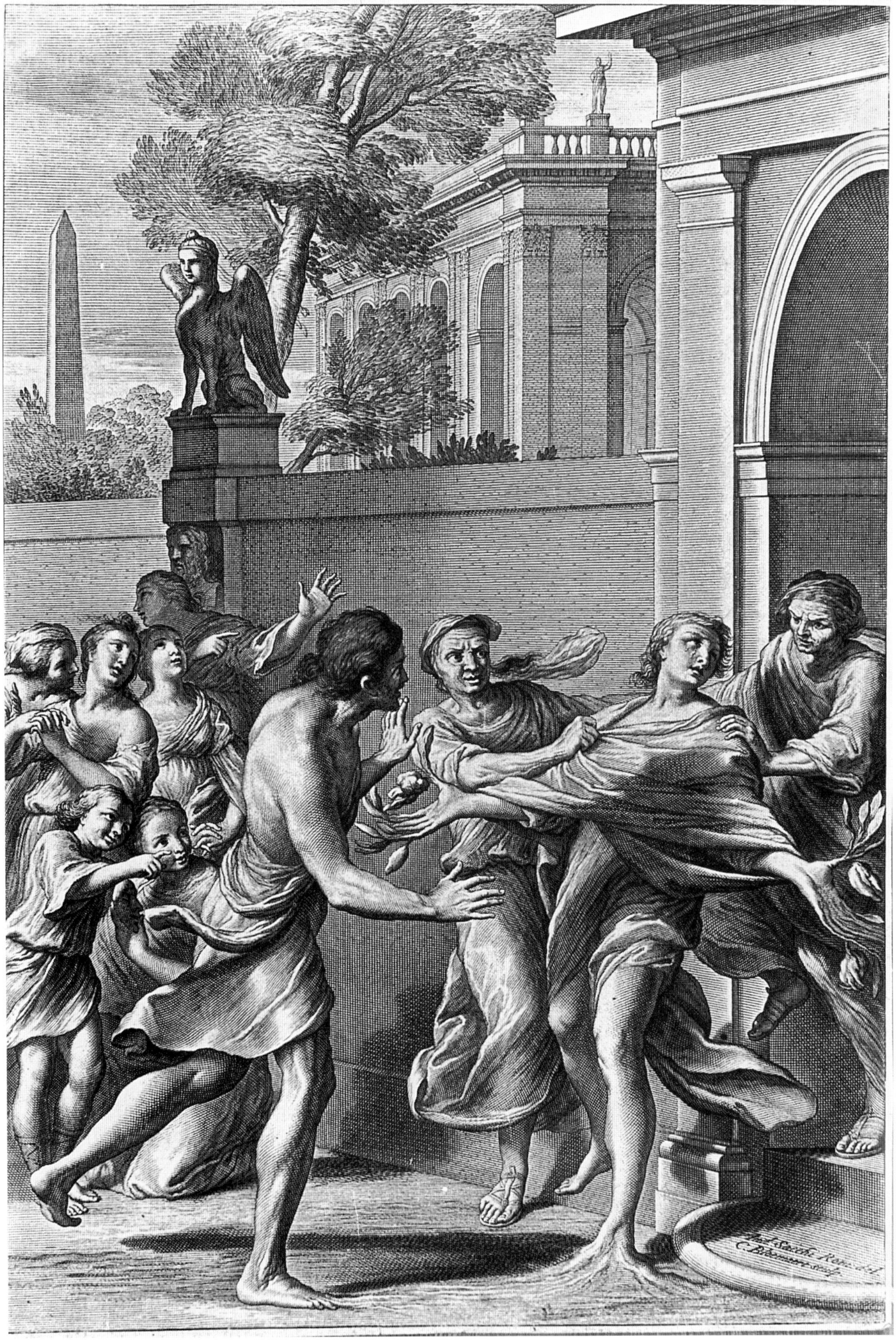

89. The Transformation of Harmonillus (see Fig. 28, p. 74)

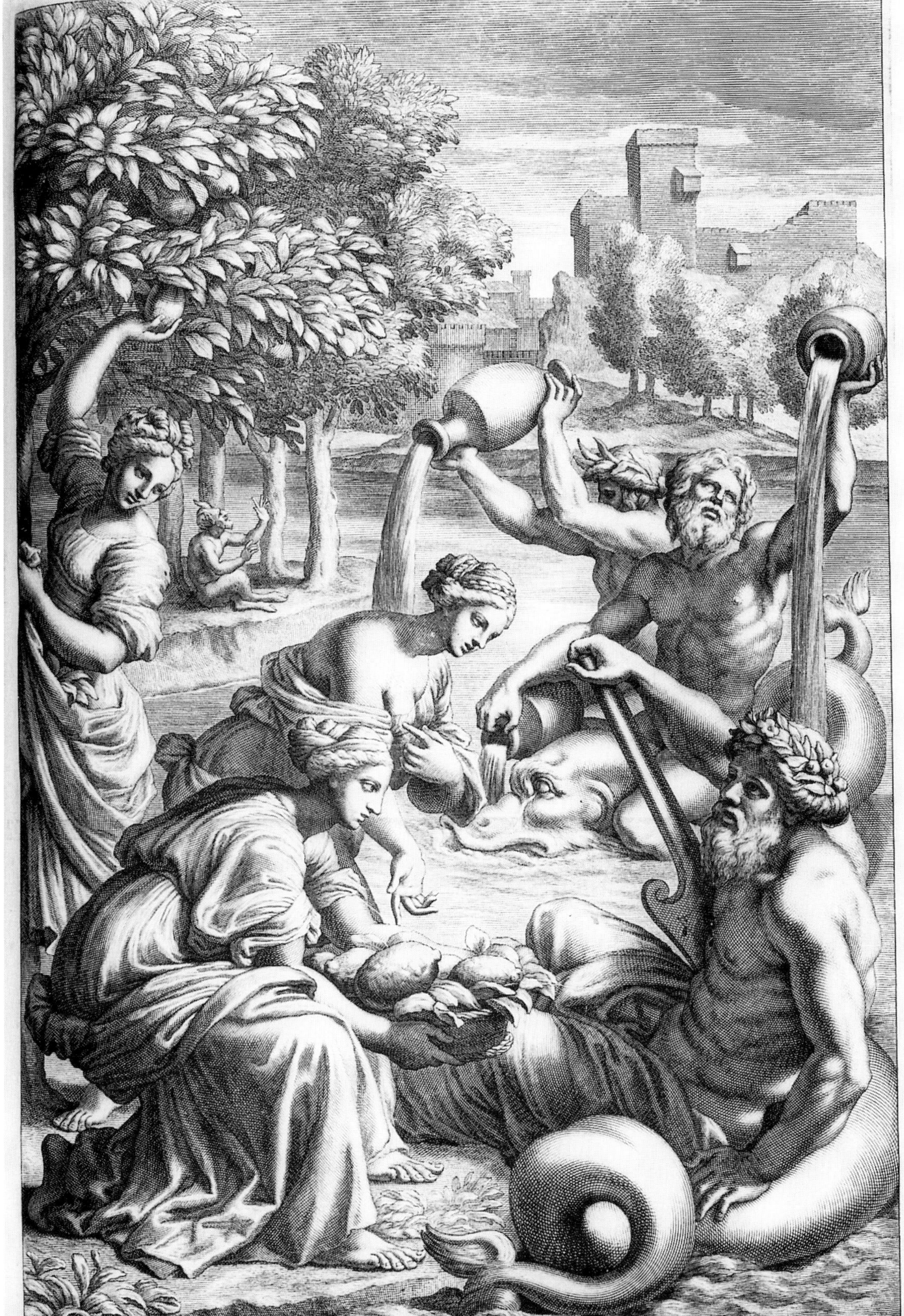

97. Nymphs of the Hesperides presenting Citrons to the Deities of Lake Garda (see Fig. 20, p. 65)

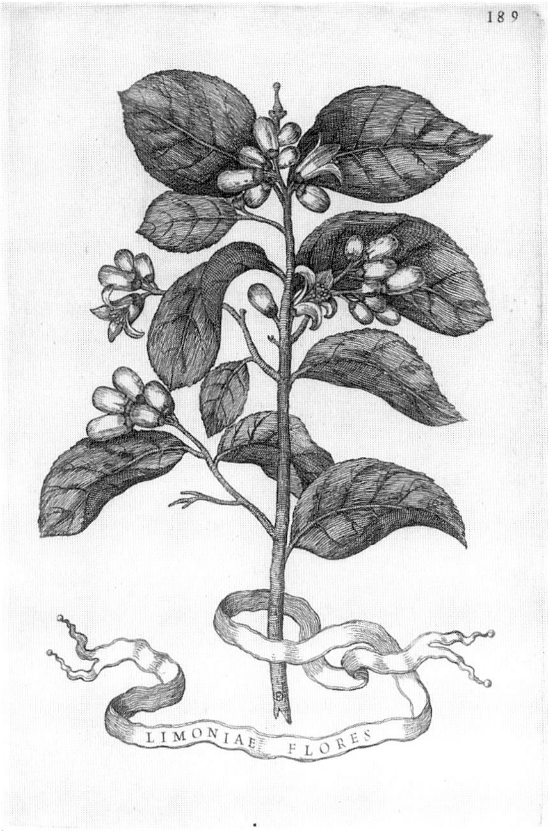

189 **18**

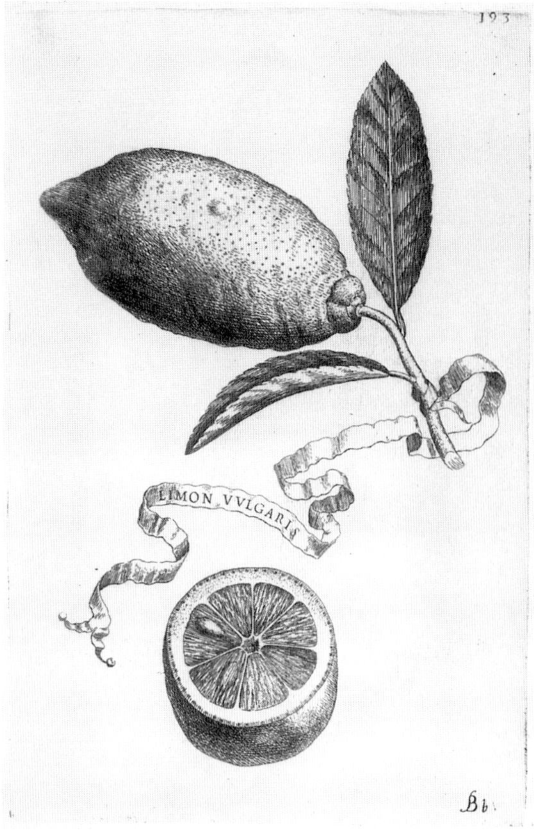

193 **19**

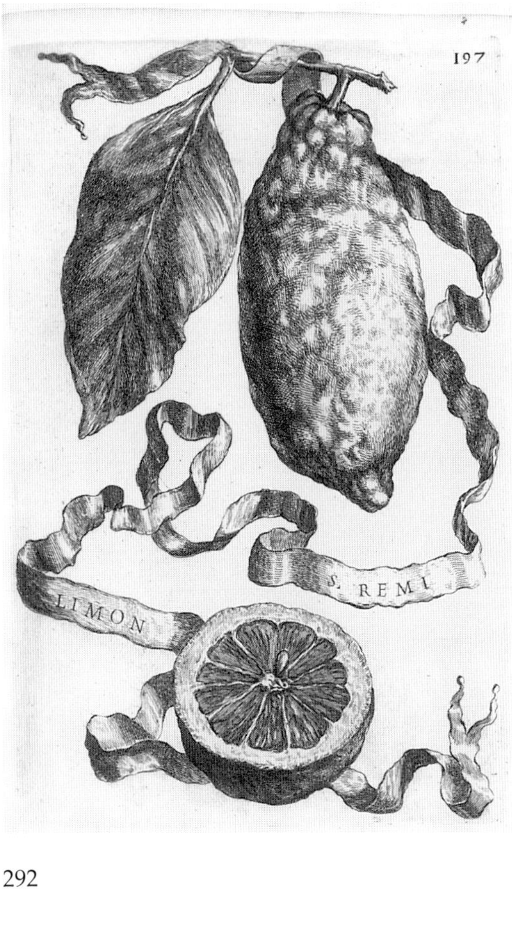

197

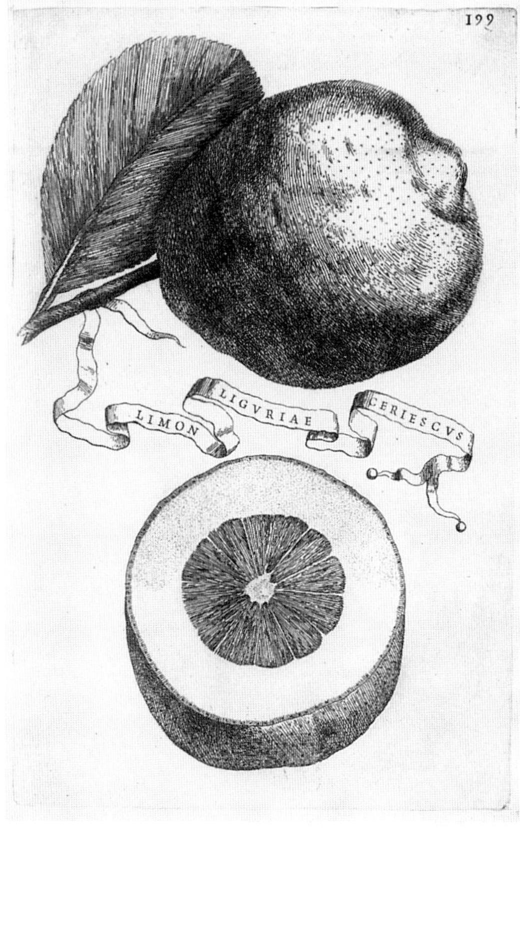

199 **43**

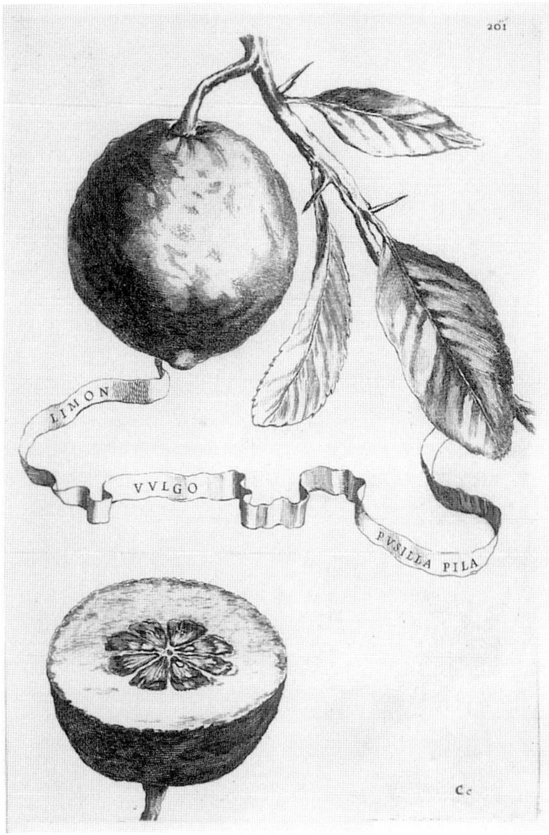

201

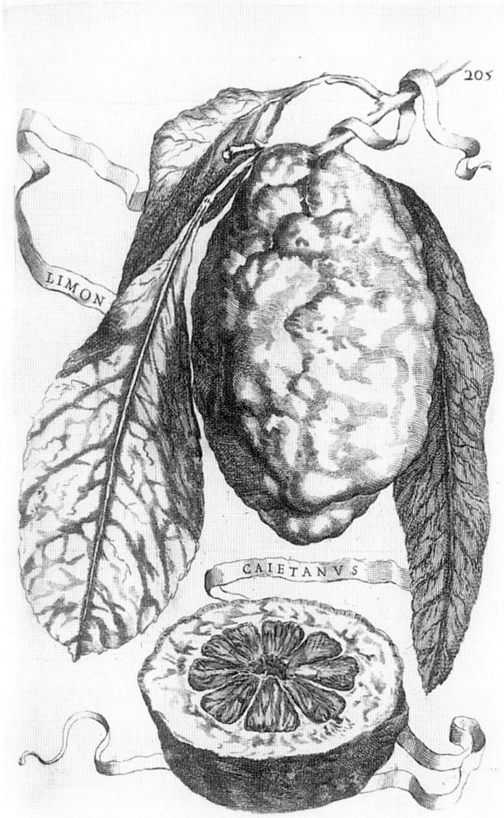

205

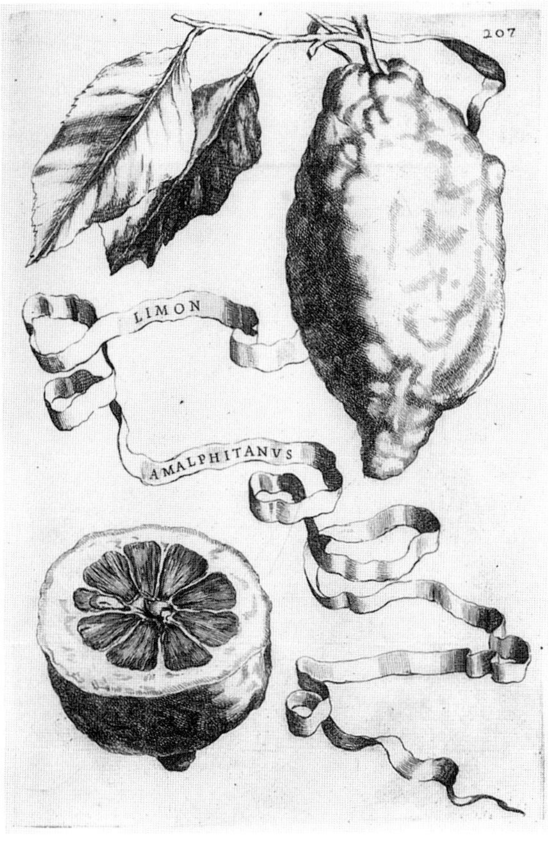

207

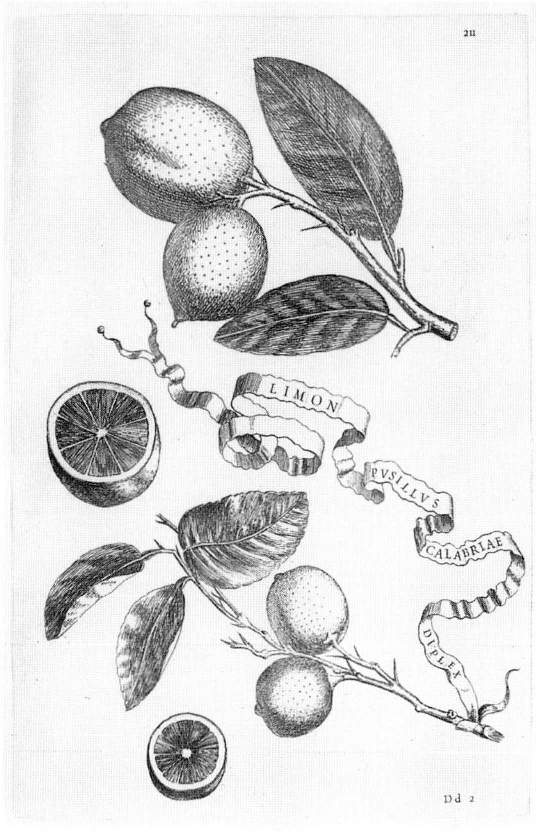

211

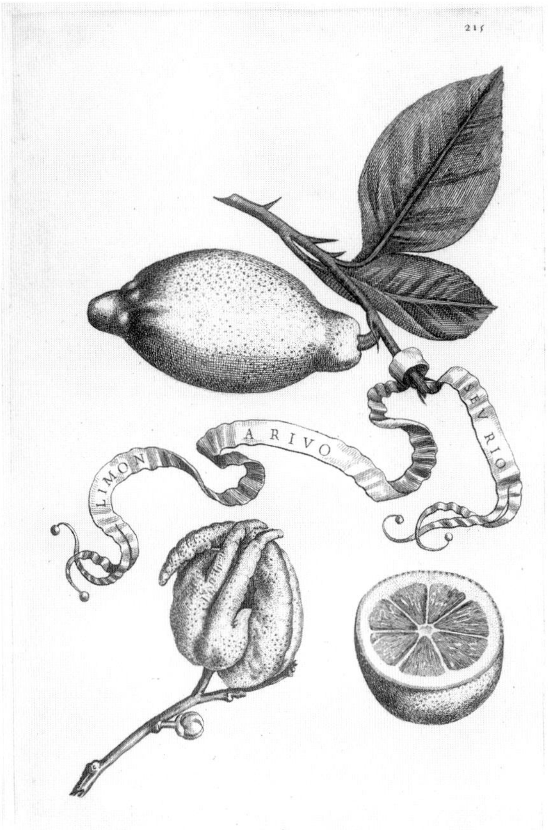

215
20, 21, 81

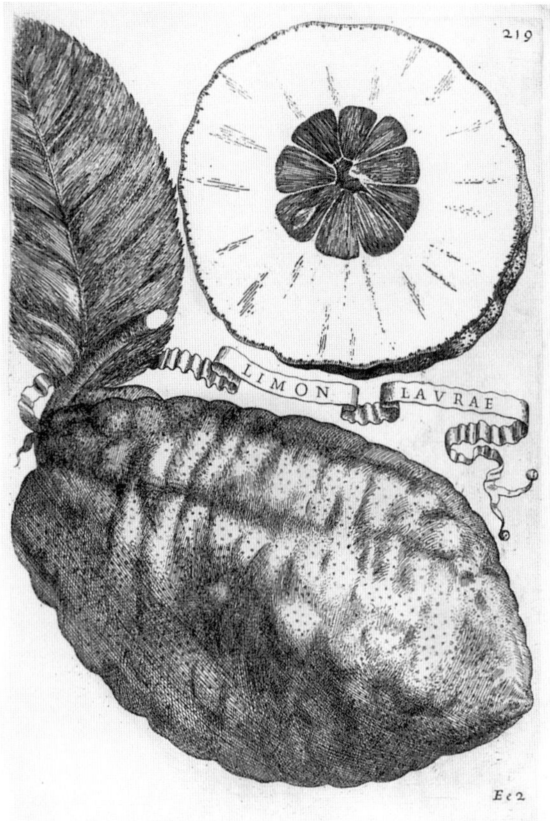

219

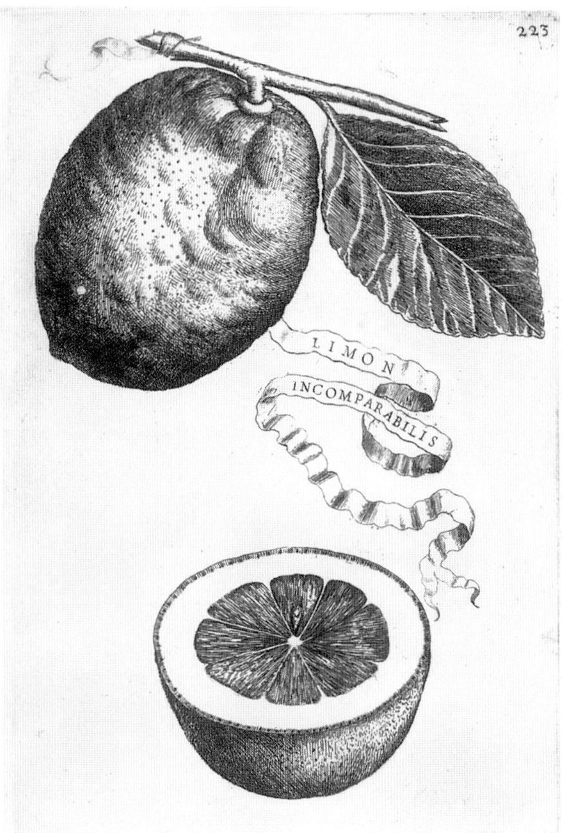

223
44

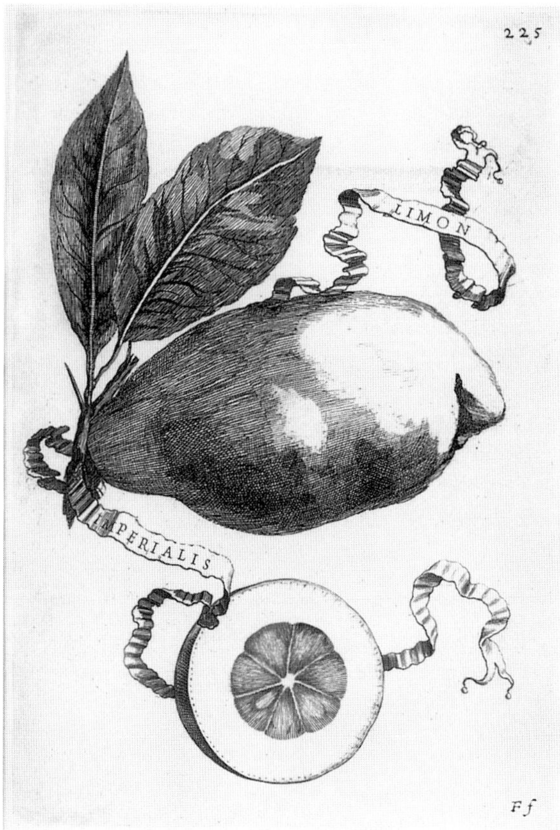

225

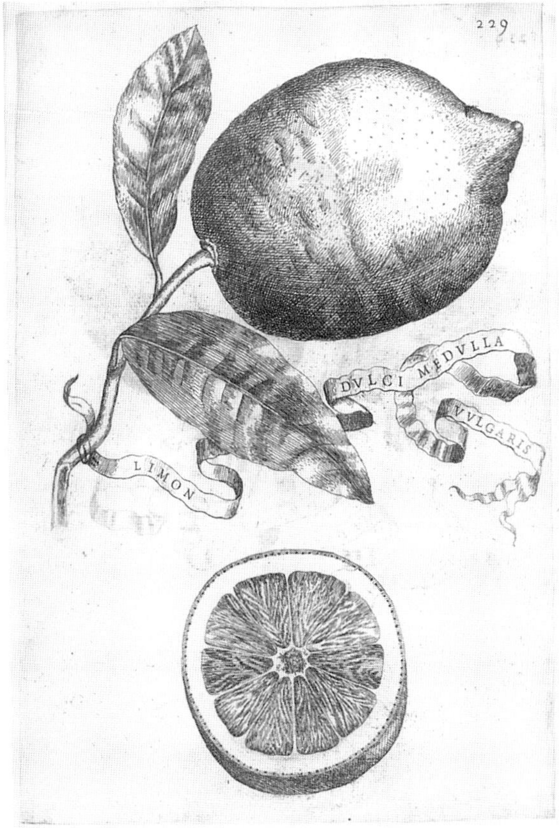

229

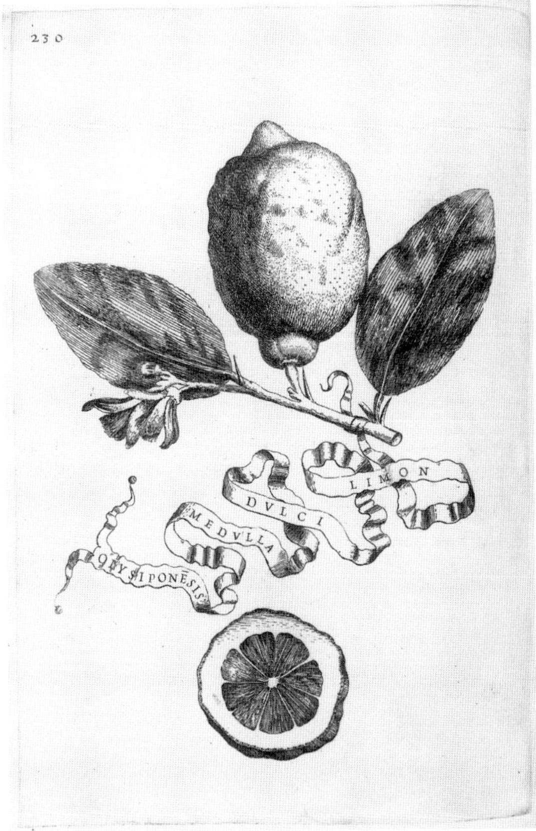

230
22, 23

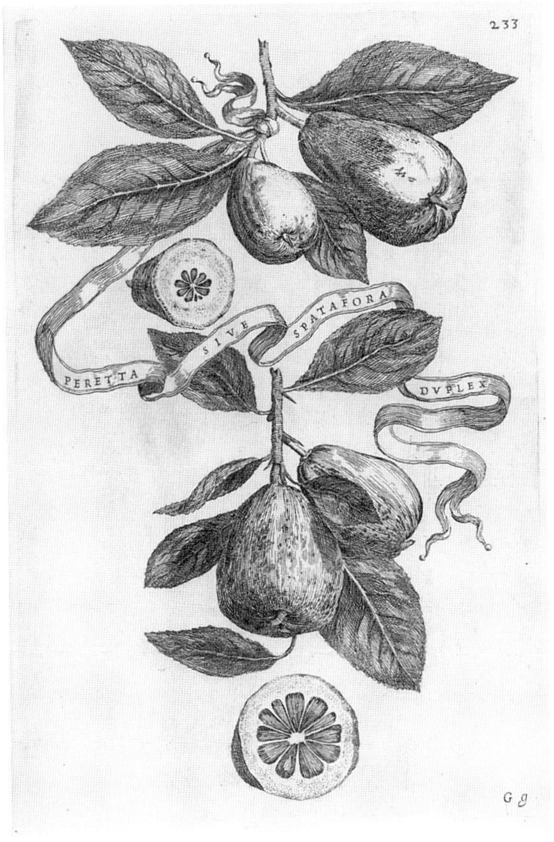

233 **45**

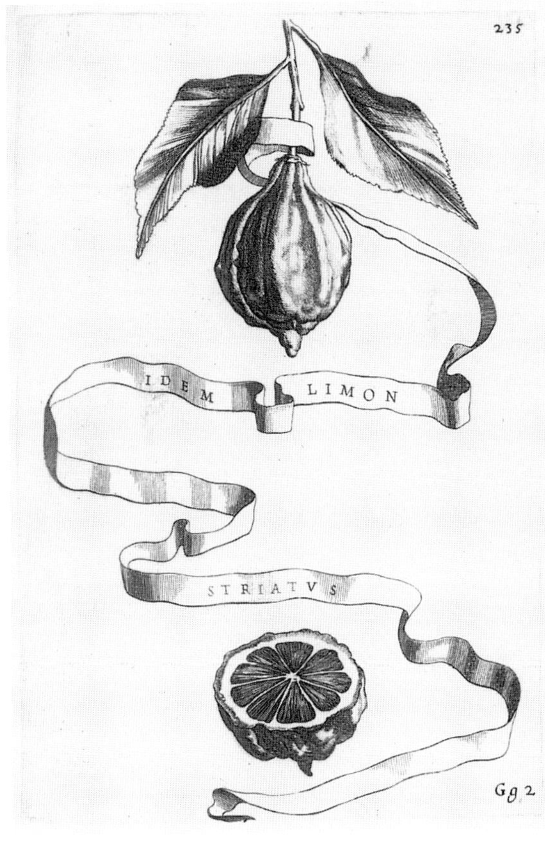

235

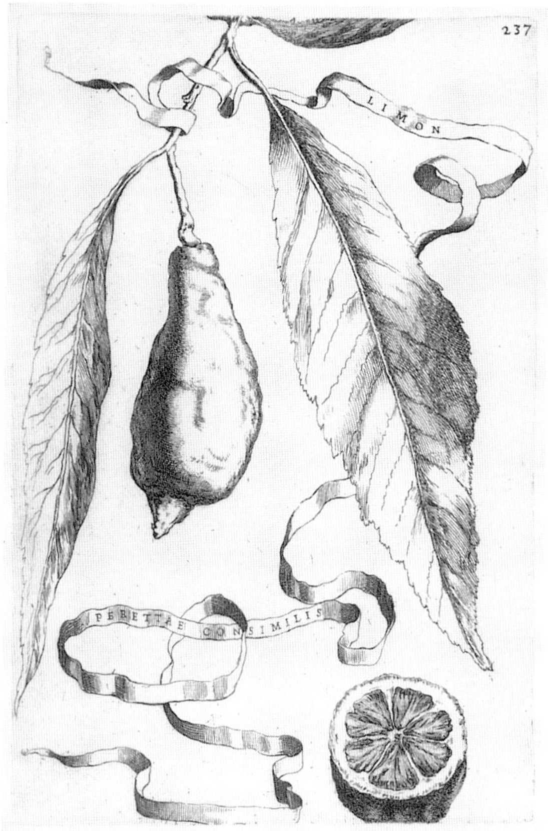

237

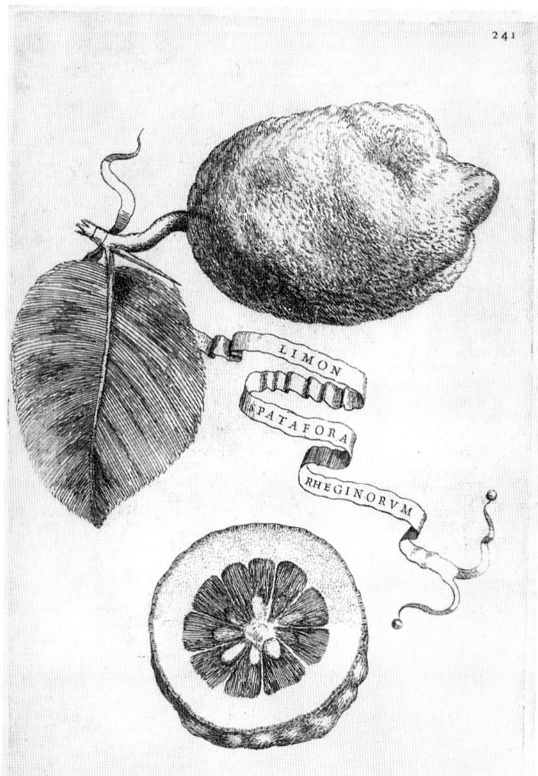

241 **46**

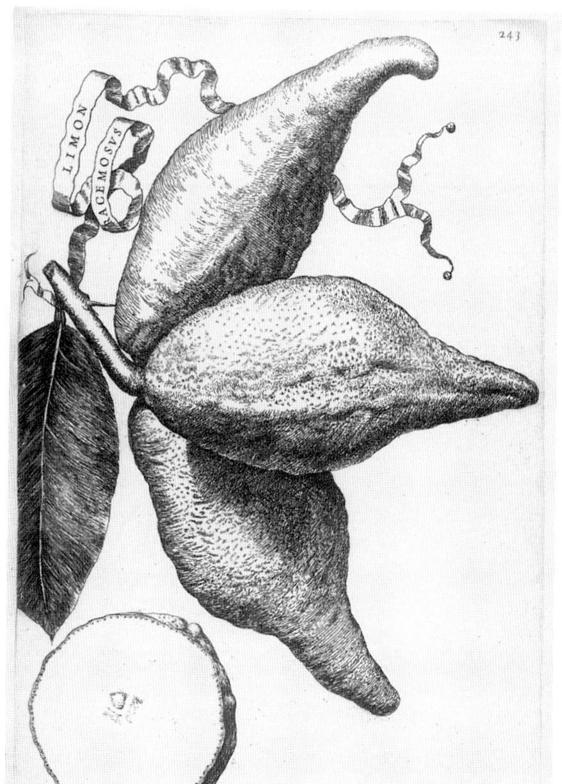

243 **47**

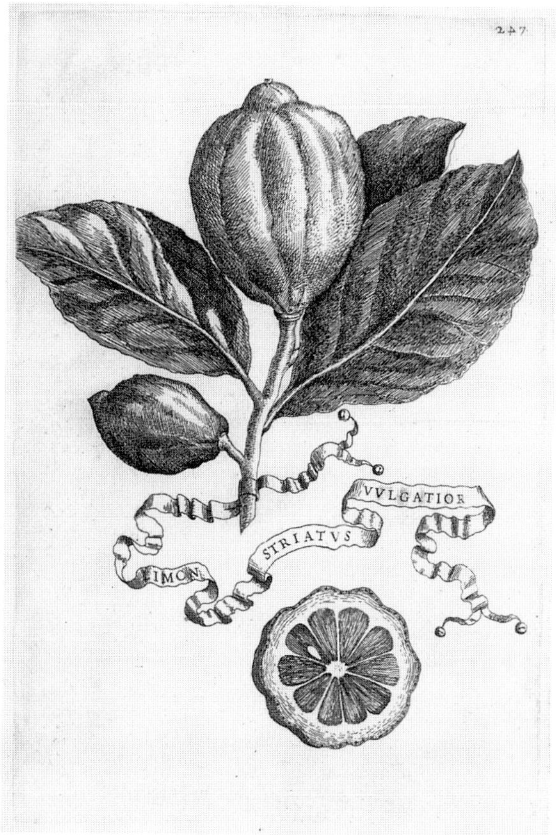

247

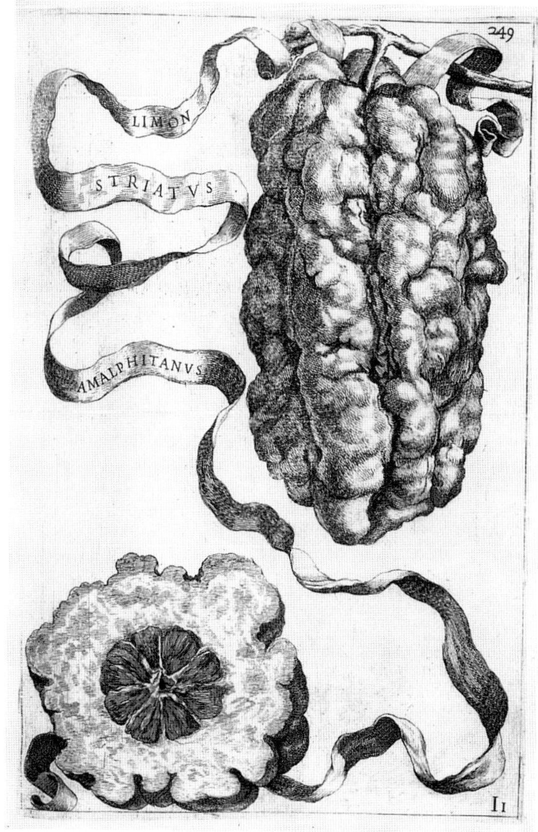

249

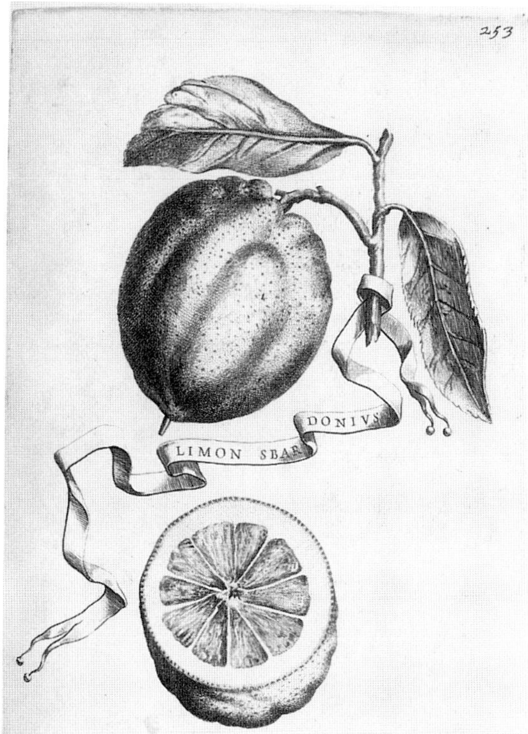

253
24, 25

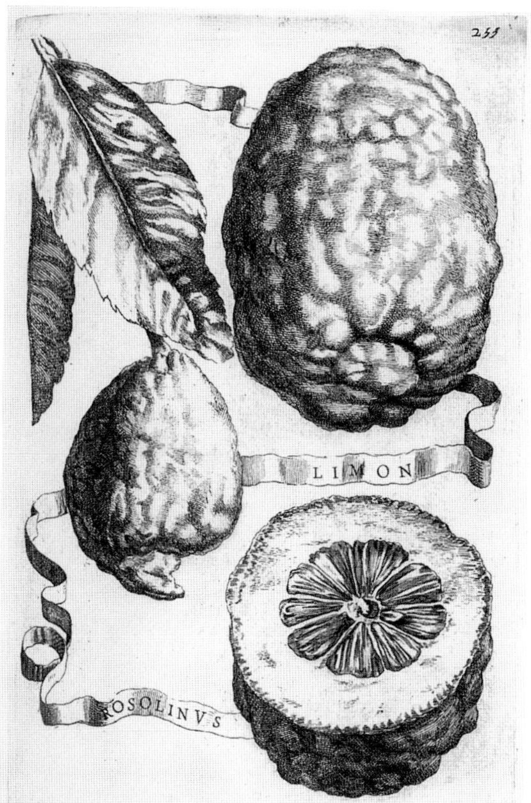

255

259

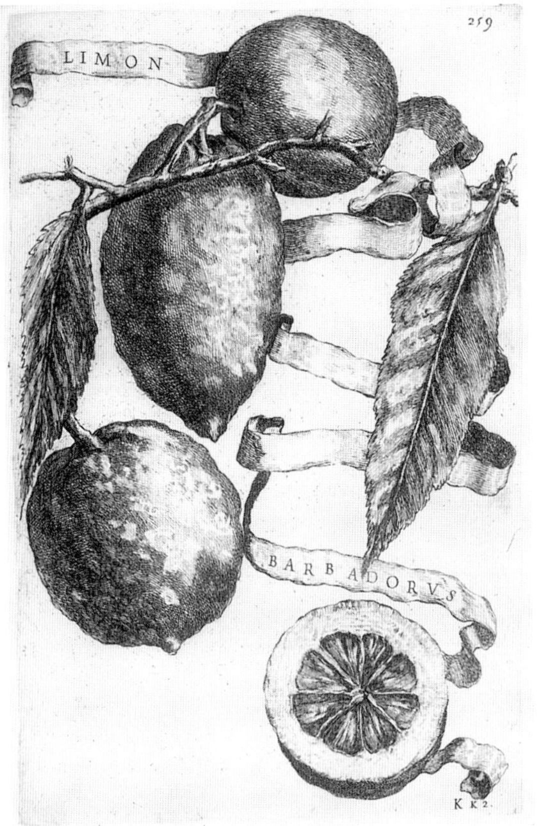

261

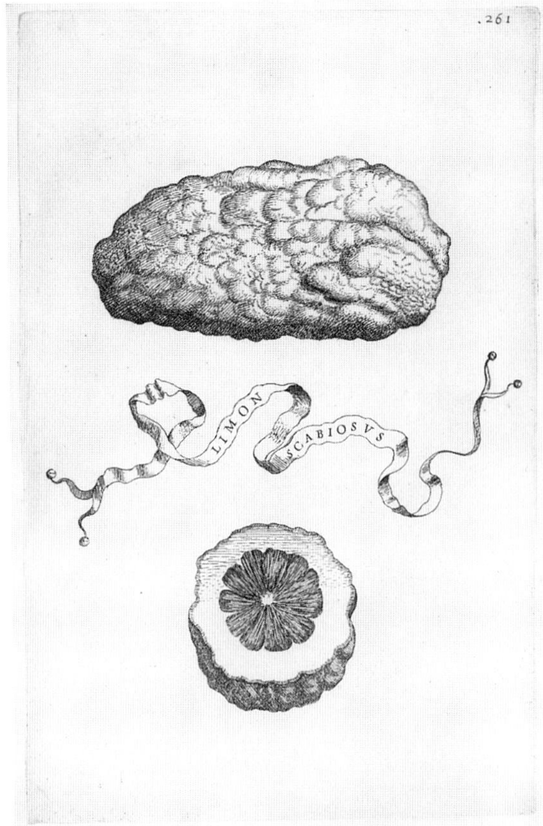

265 **48**

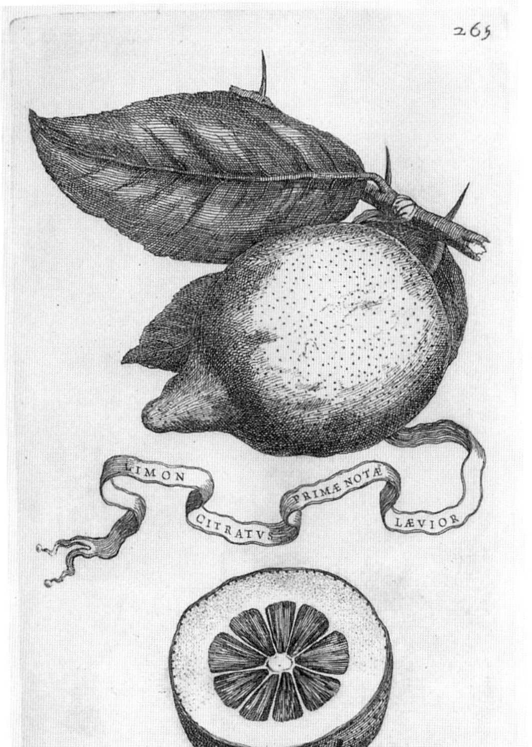

267 **49**

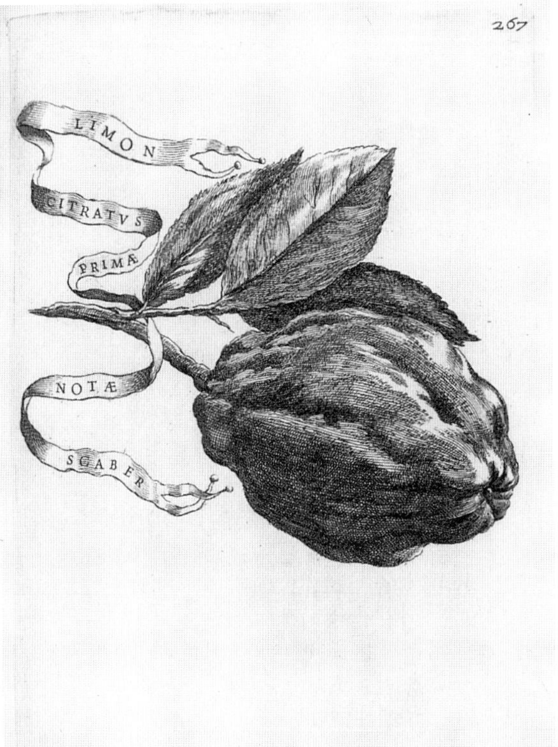

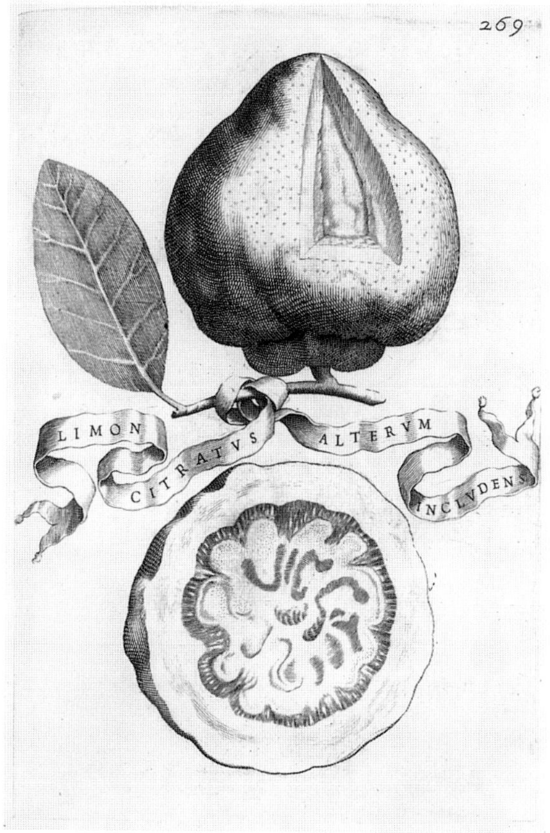

269
89, 90

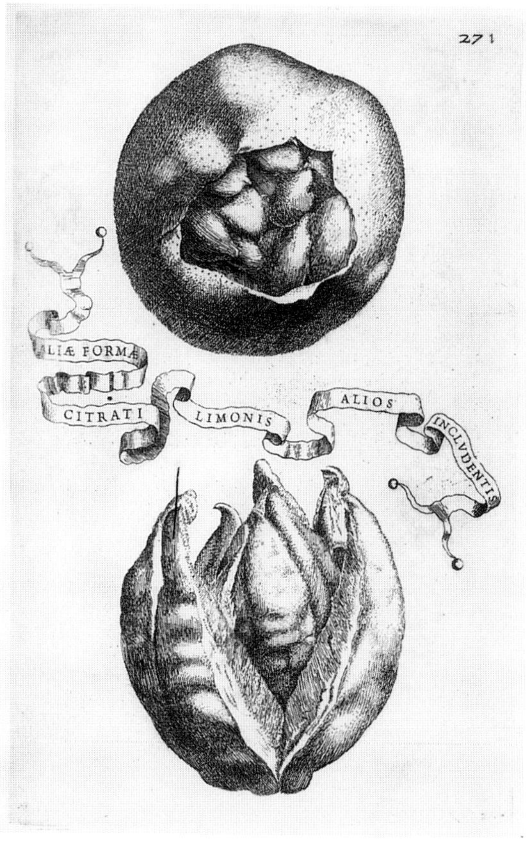

271
91, 92

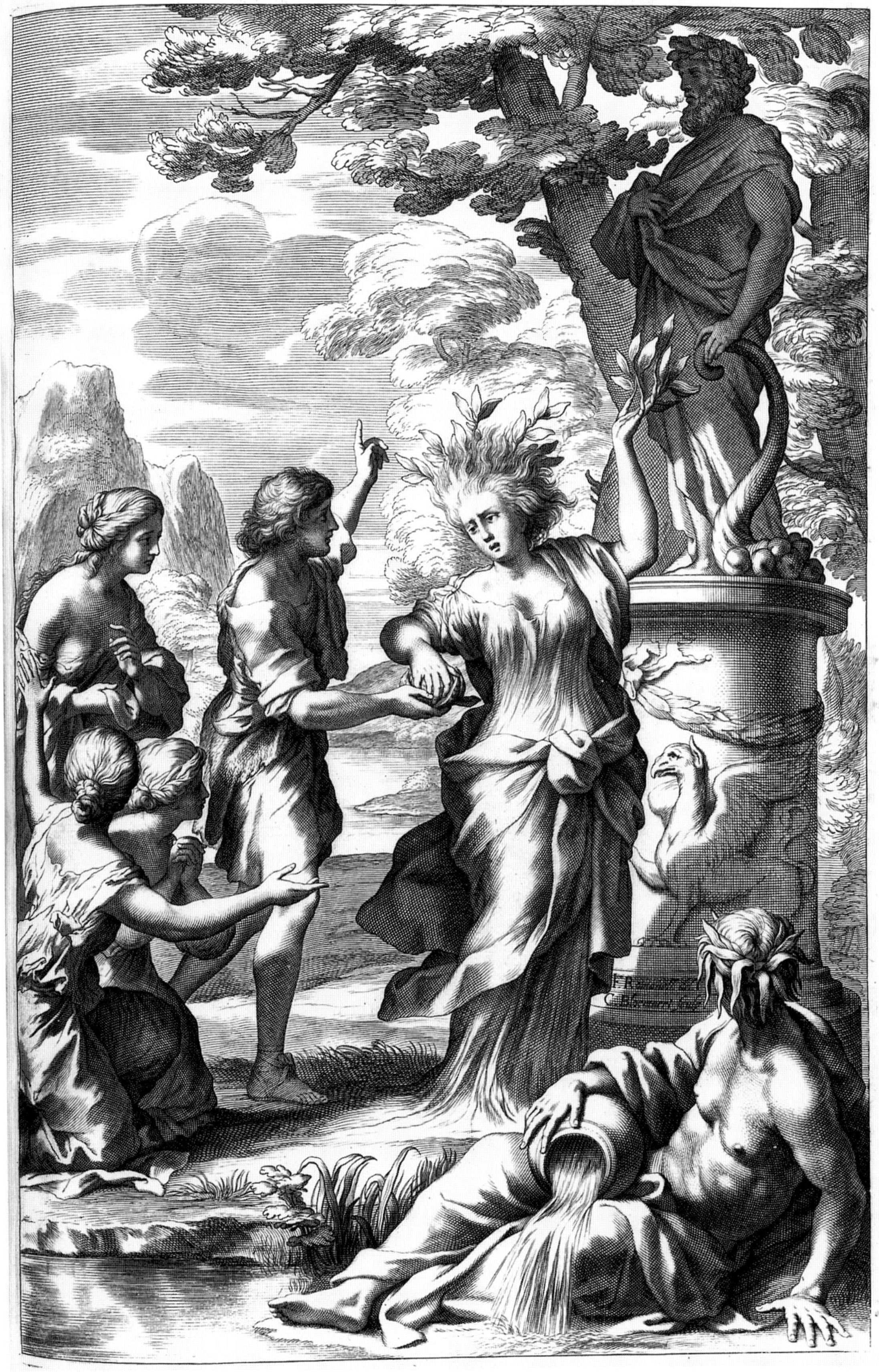

277. The Transformation of Tirsenia (see Fig. 29, p. 75)

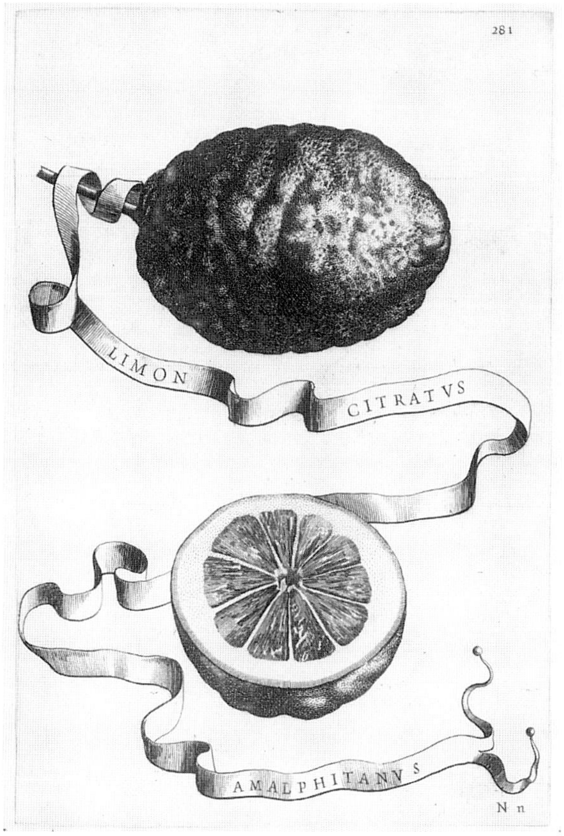

281
50, 51

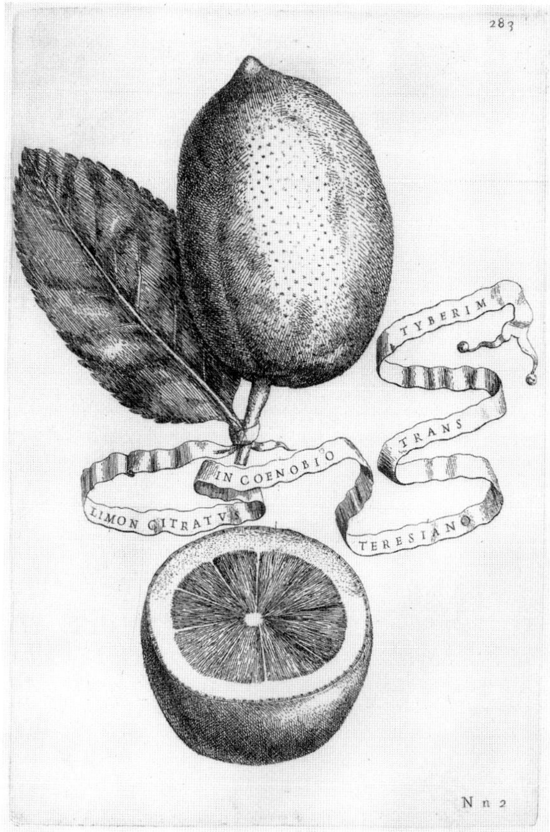

283
52,53

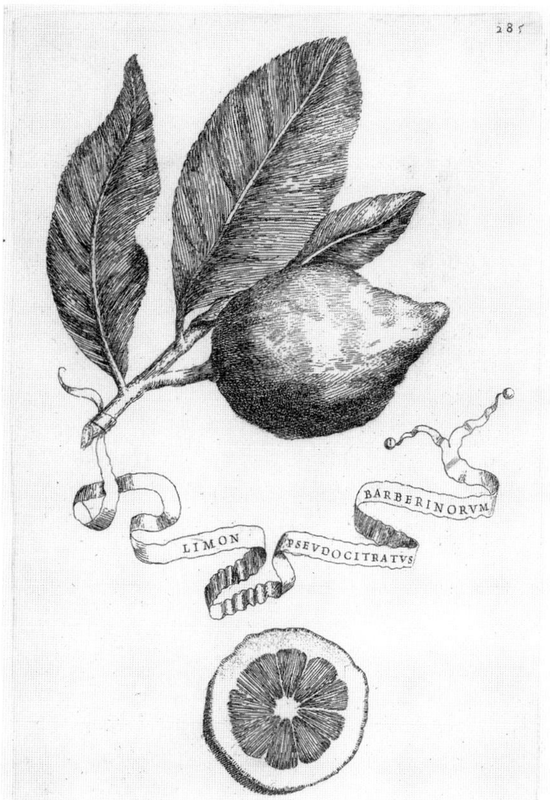

285
26, 27

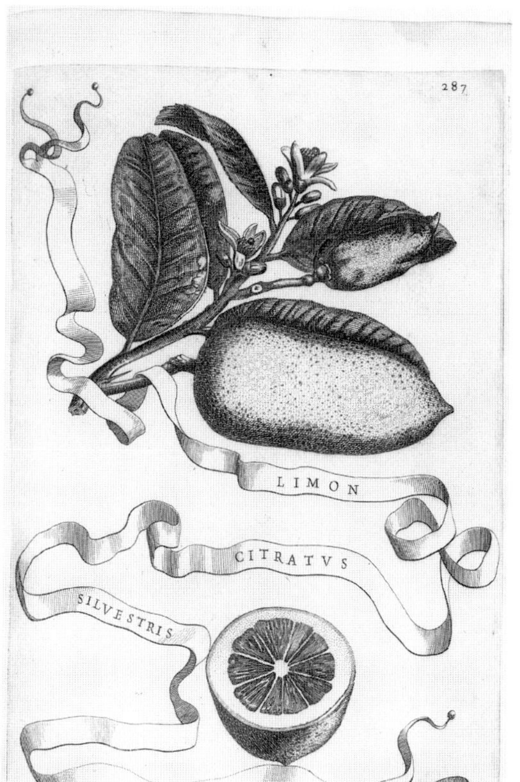

287 **54**

291

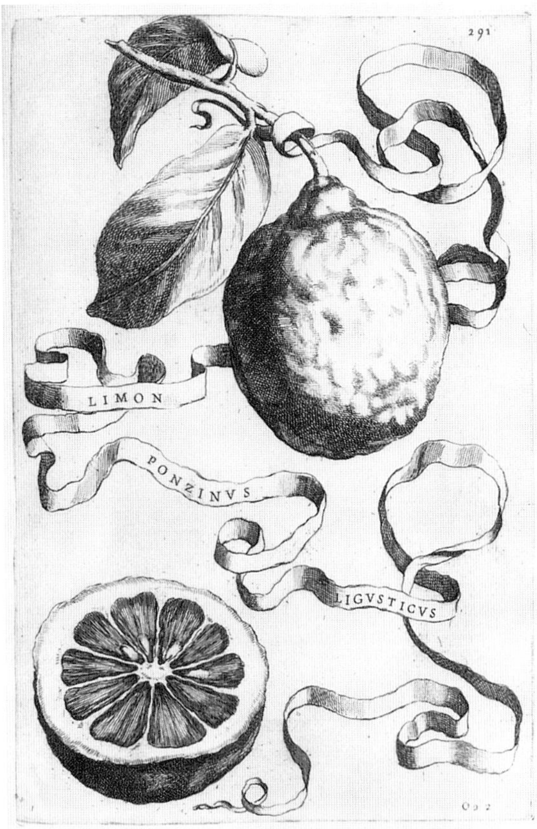

293

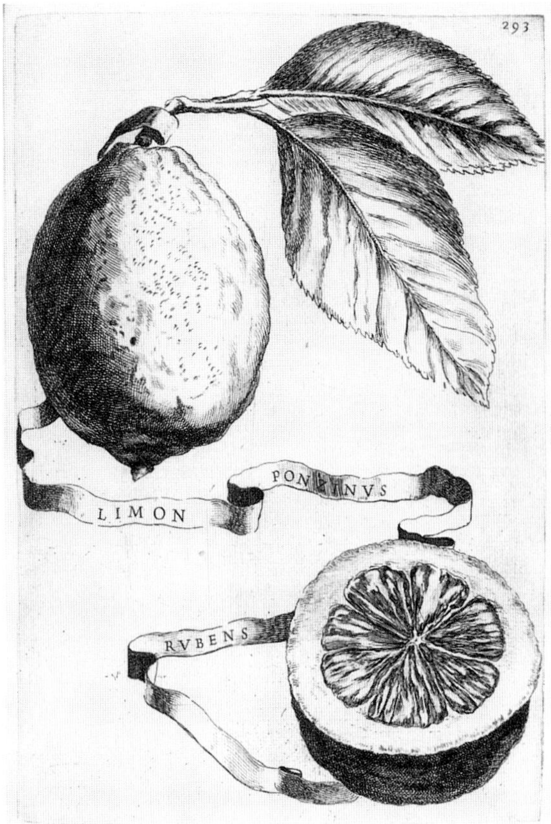

295 **55**

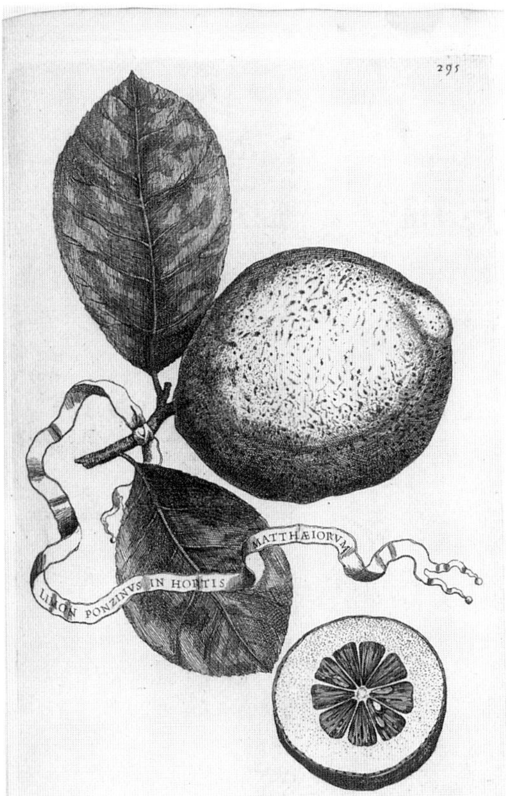

297

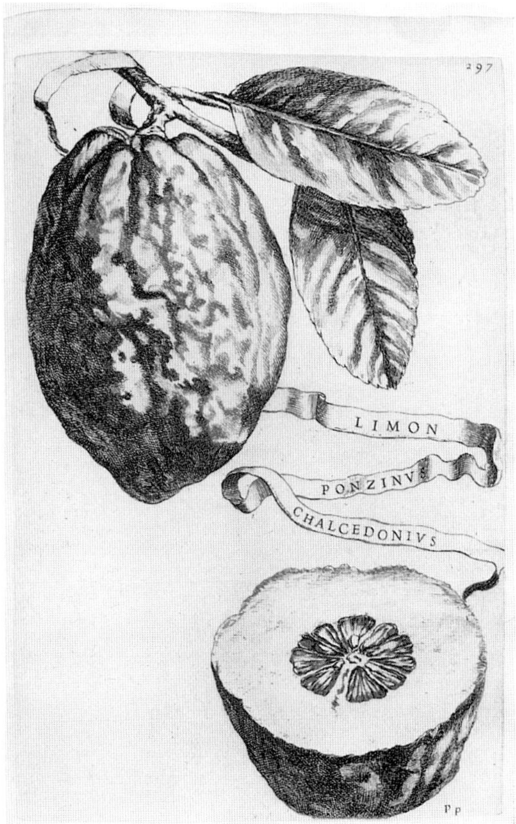

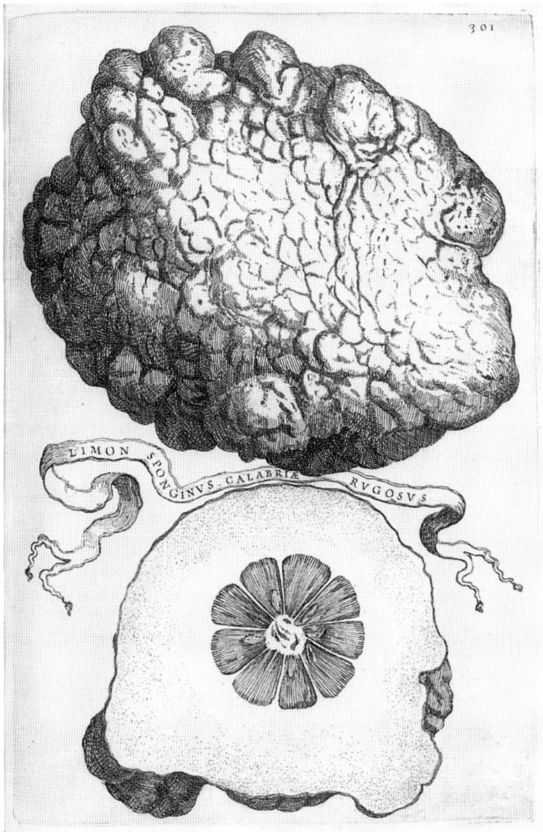

301

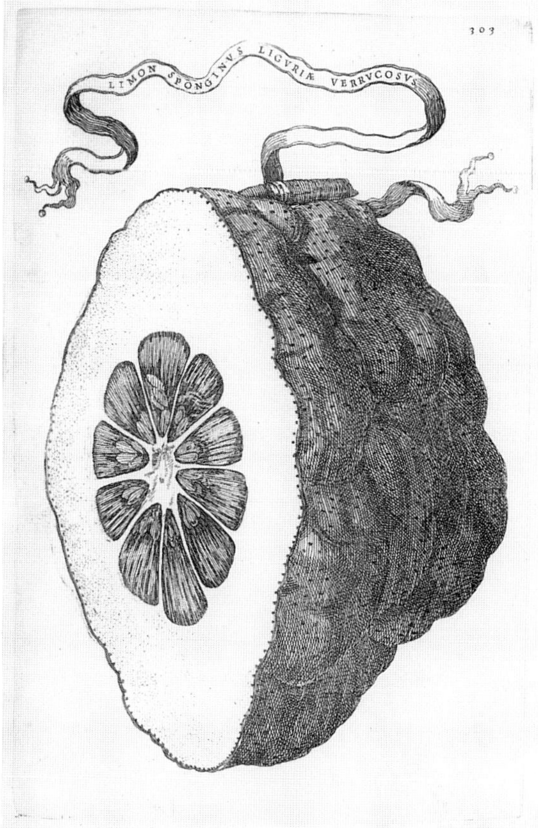

303

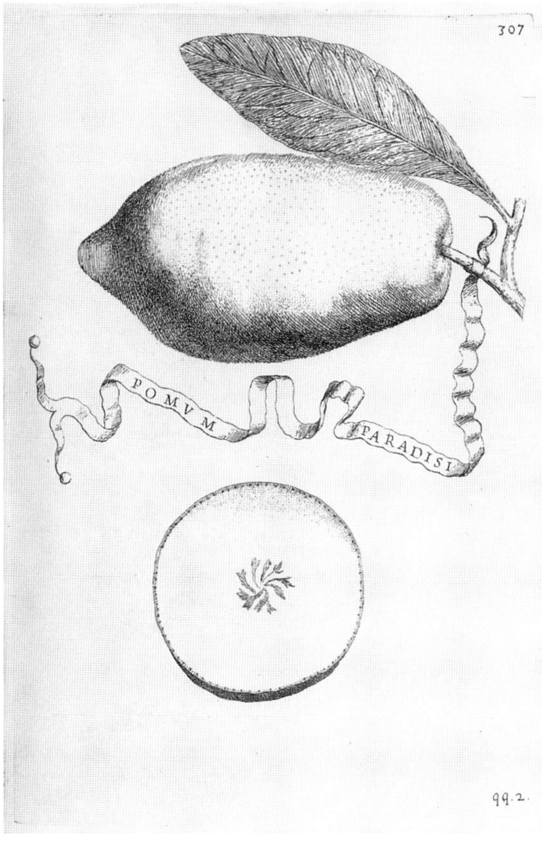

307
56, 57

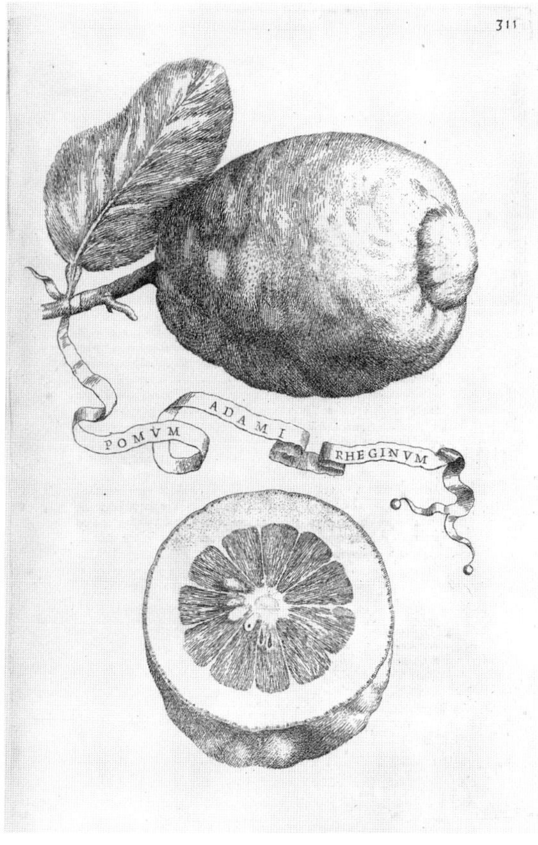

311 **58**

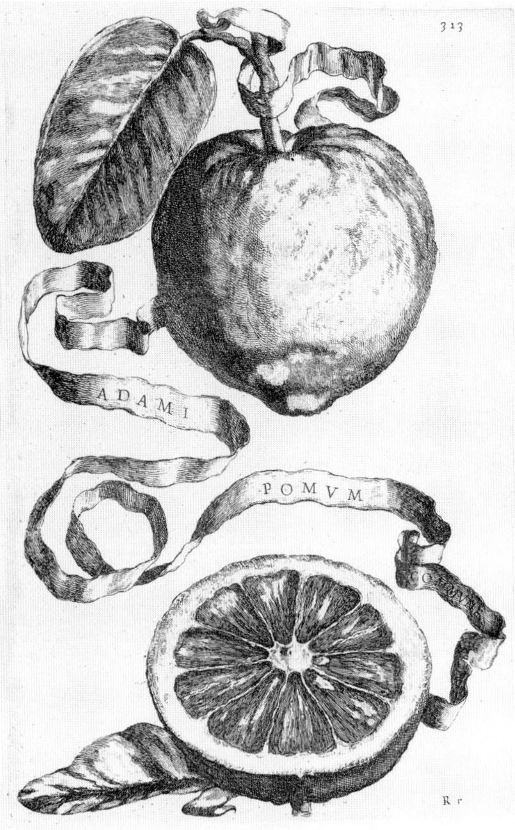

313

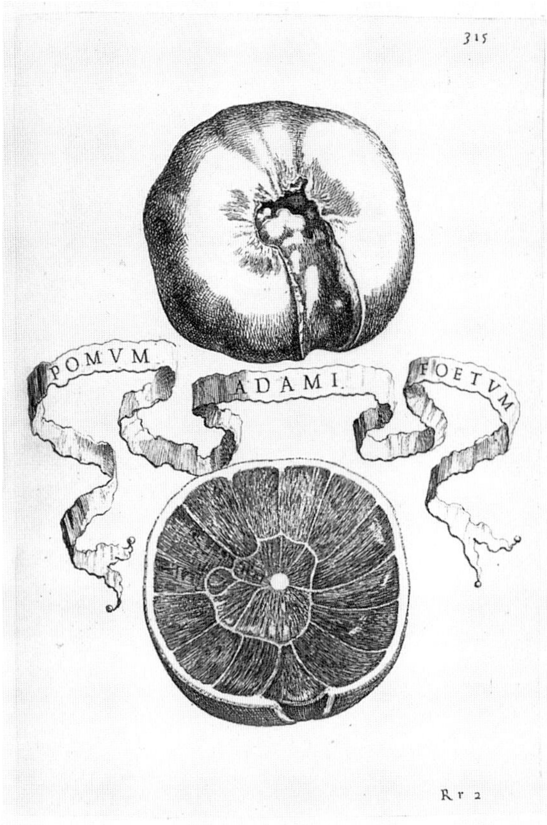

315
93, 94

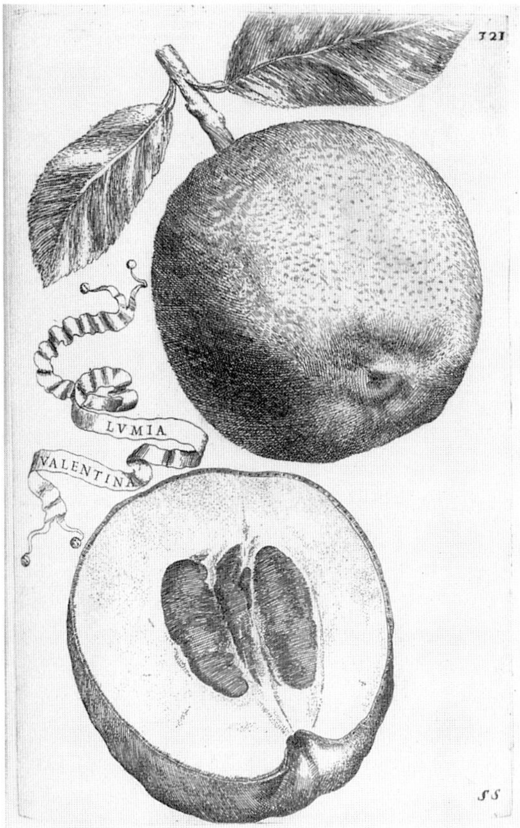

321

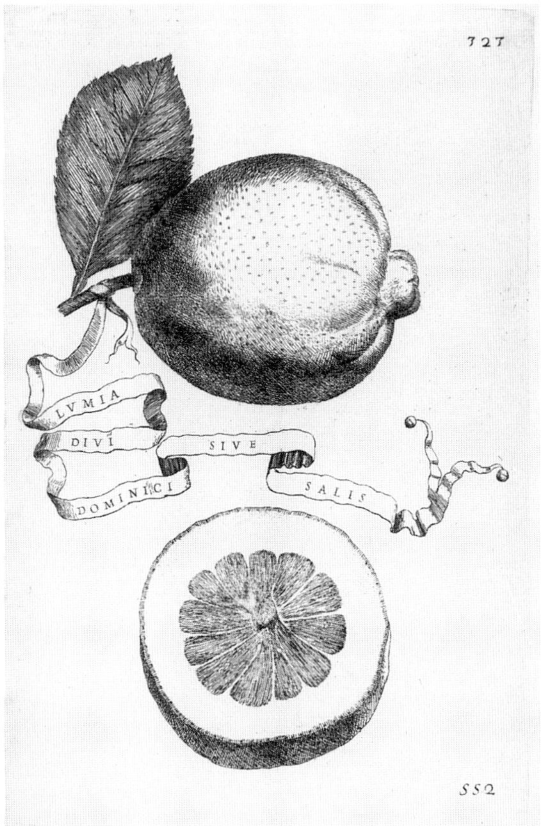

323
59, 60

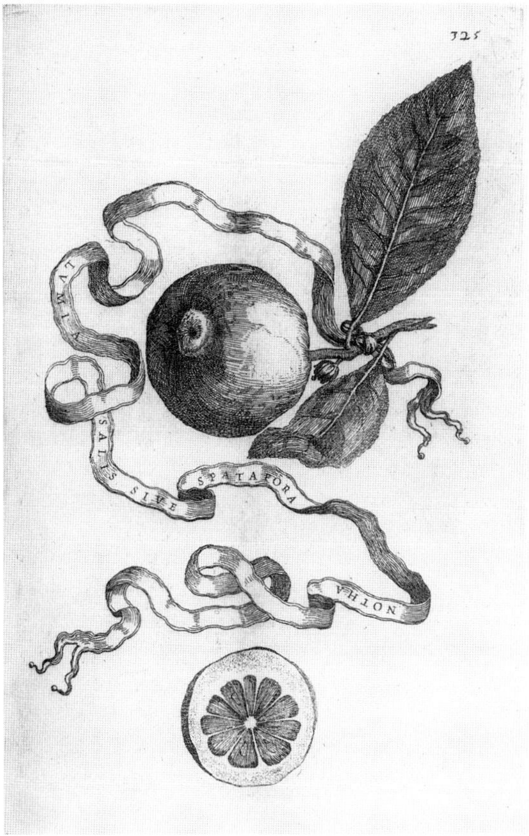

325
61, 62

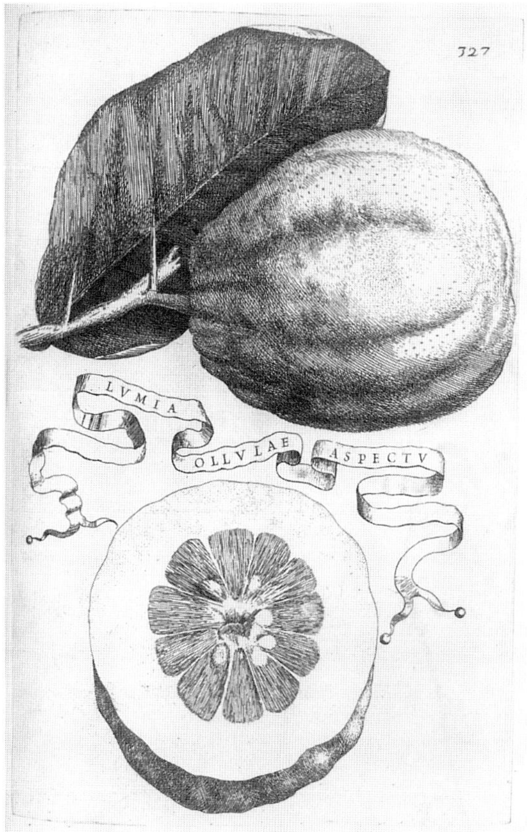

327

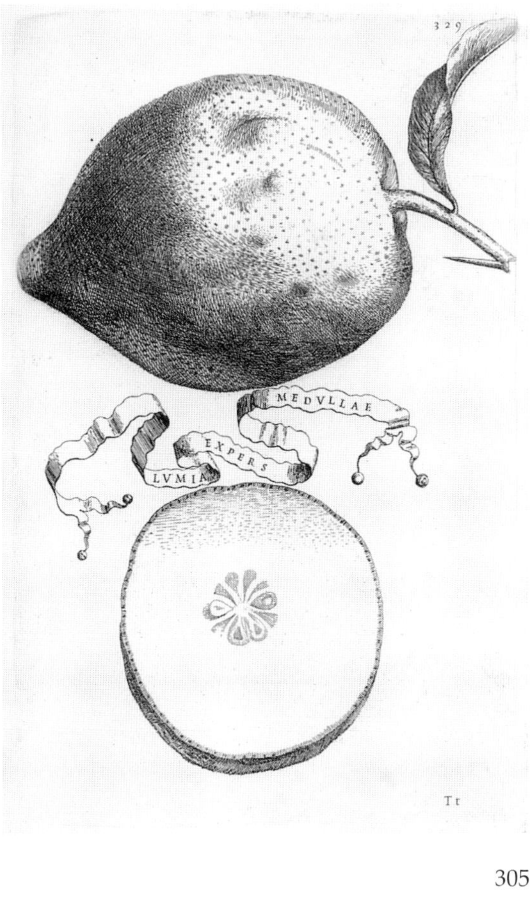

329

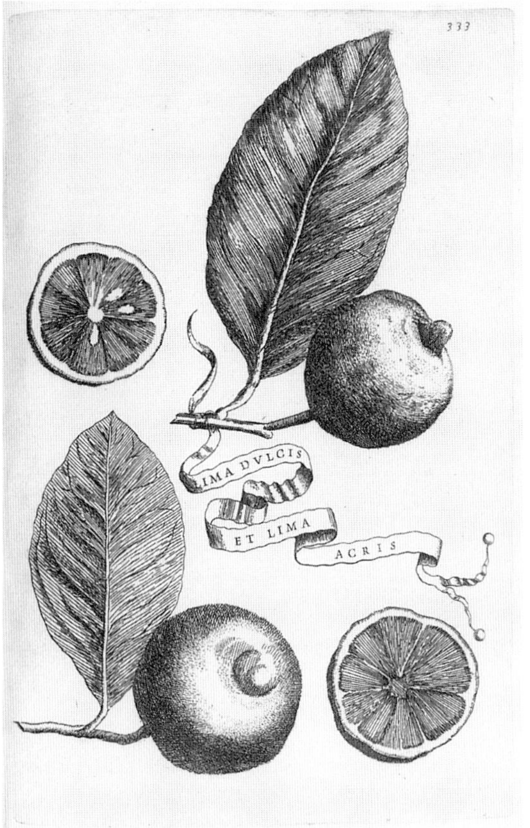

333

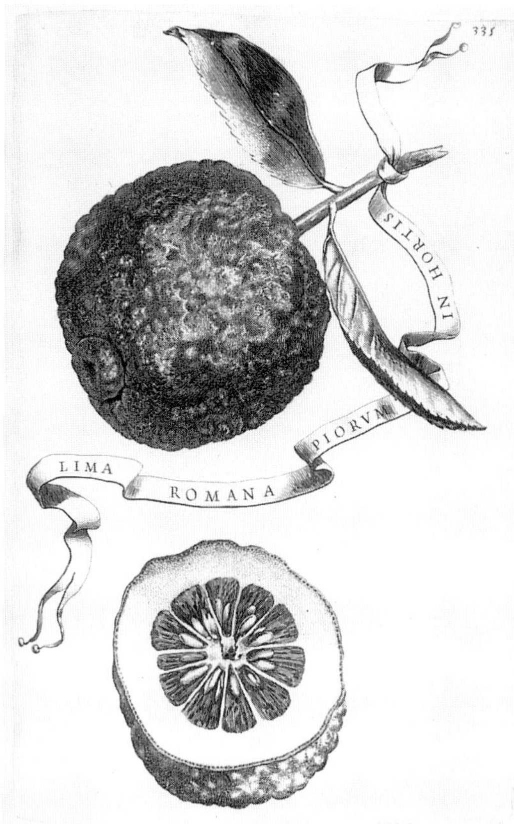

335

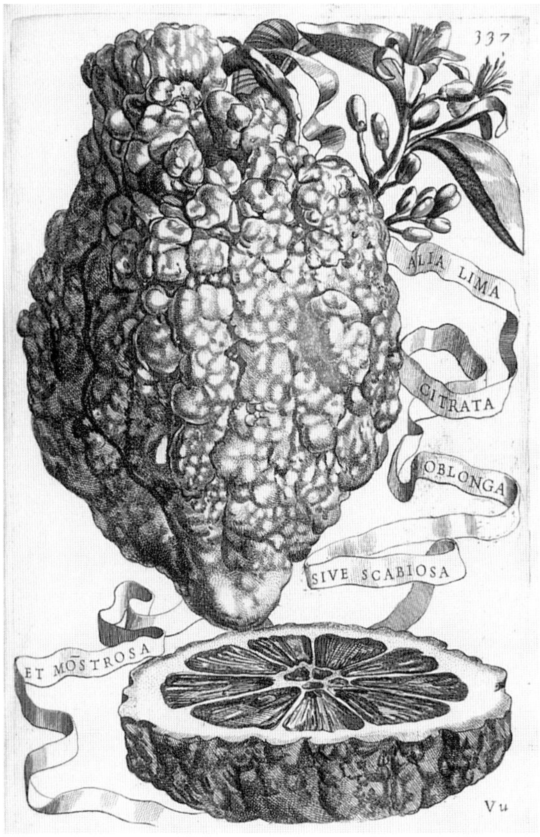

337

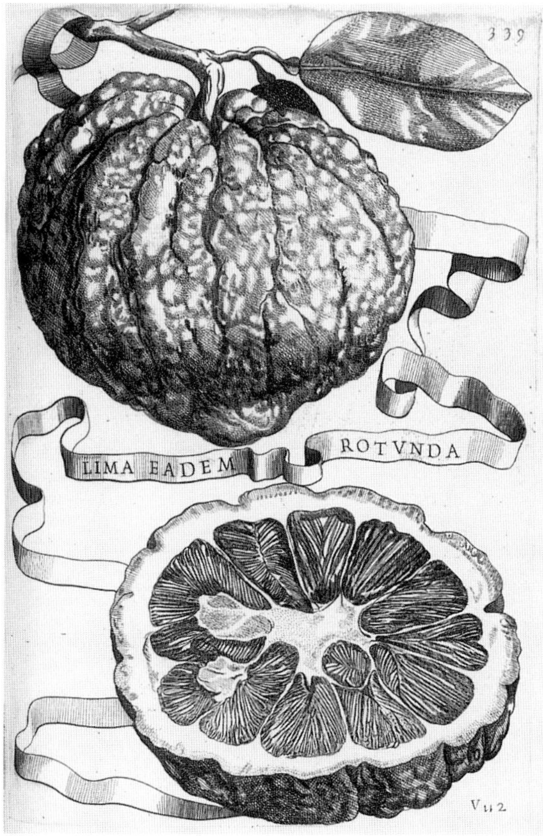

339

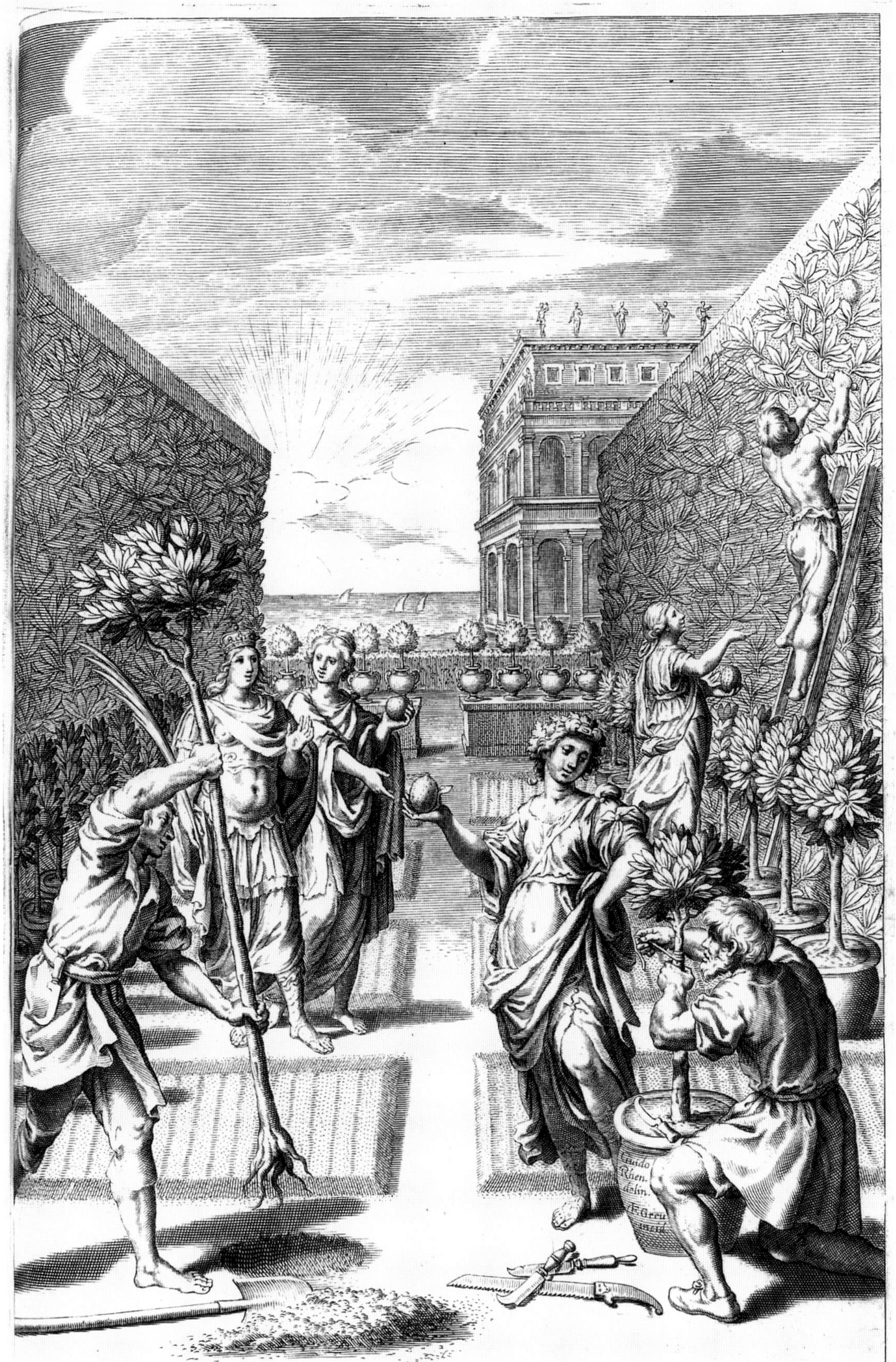

343. J. F Greuter after G. Reni, Citrus cultivation in Genoa, engraving, Ferrari 1646

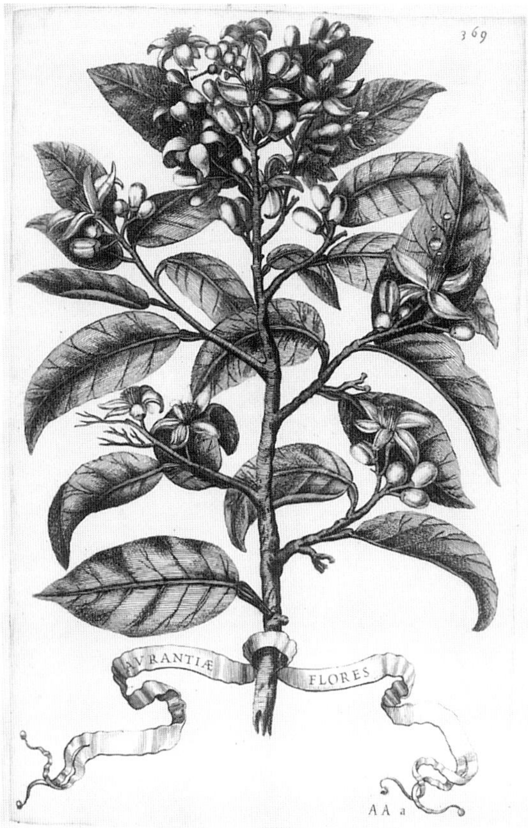
369
AVRANTIÆ
FLORES
AA a
369

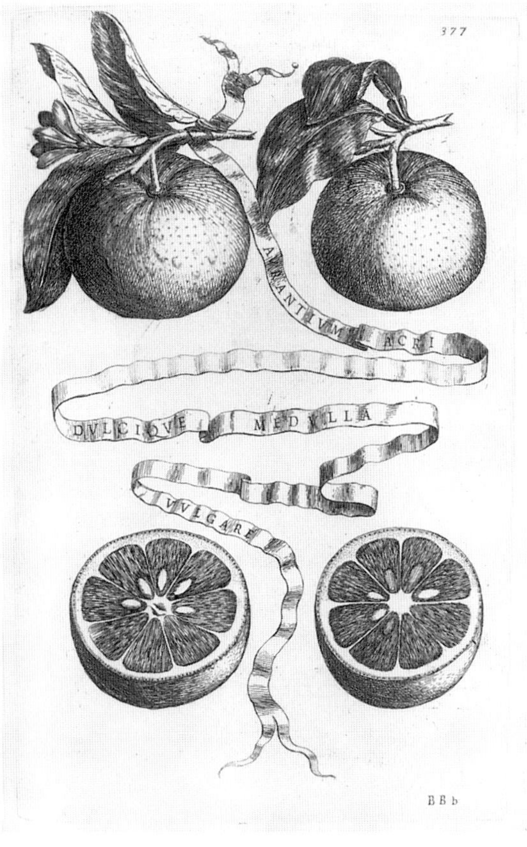
377
AVRANTIVM ACRI
DVLCIQVE MEDVLLA
VVLGARE
BB b
377

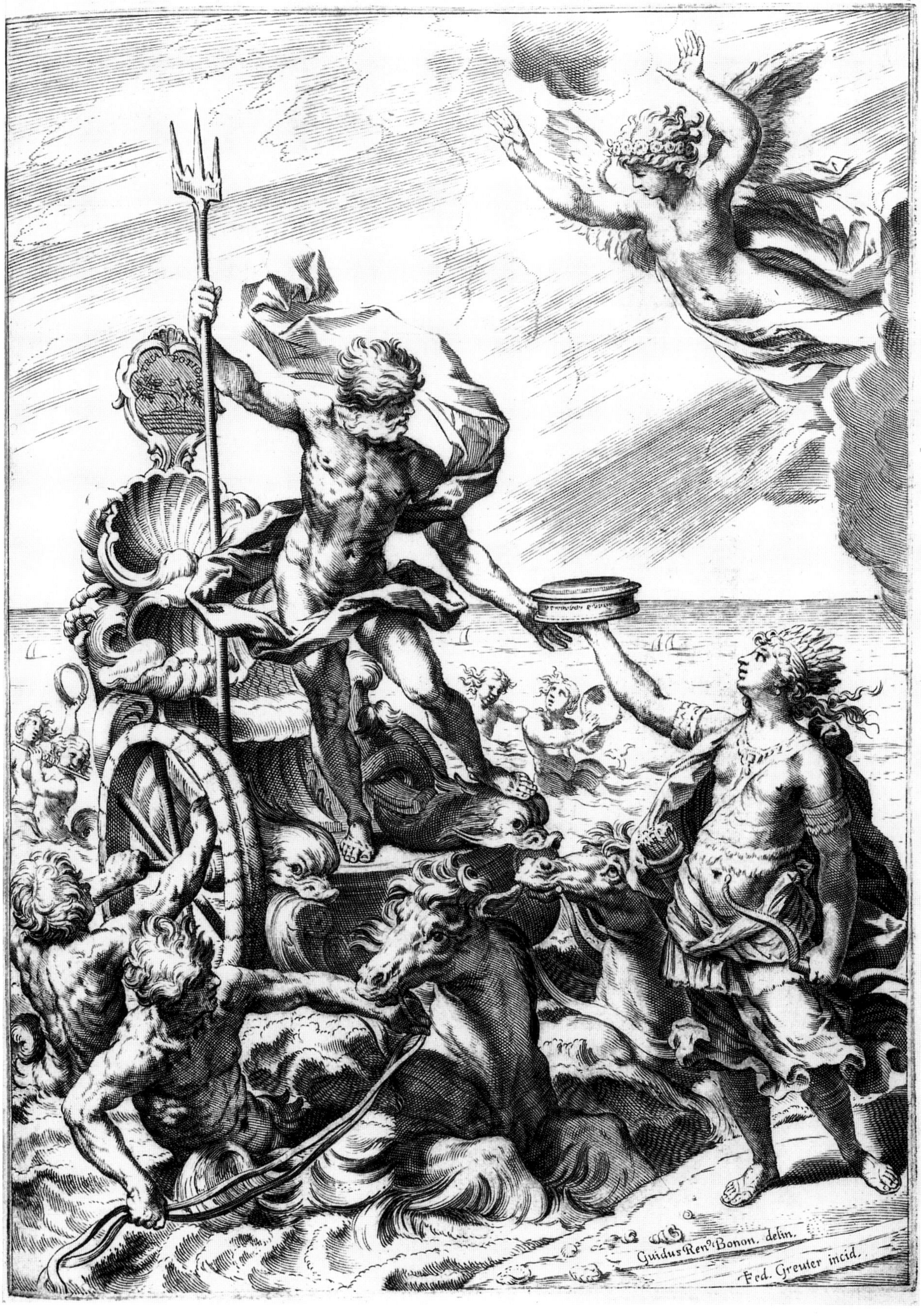

376. J.F. Greuter after G. Reni, The Indies handing seeds to Neptune to transport to the Barberini Gardens in Rome (see Fig. 17, p. 48)

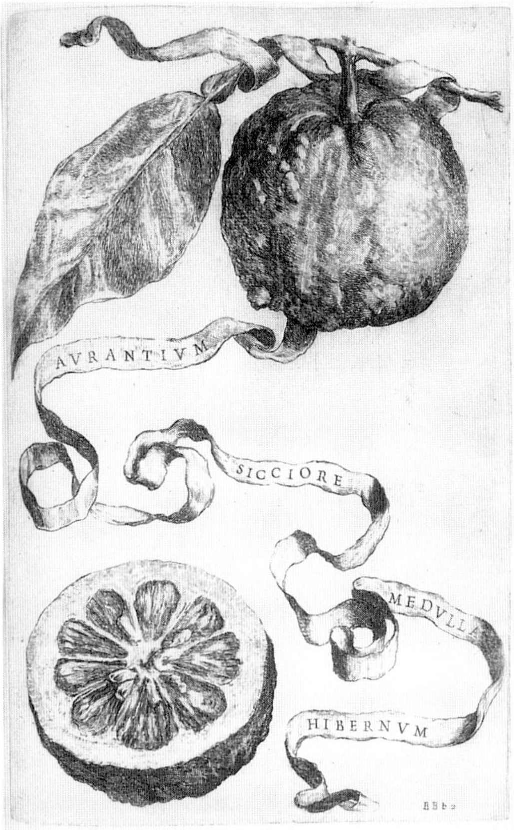

379

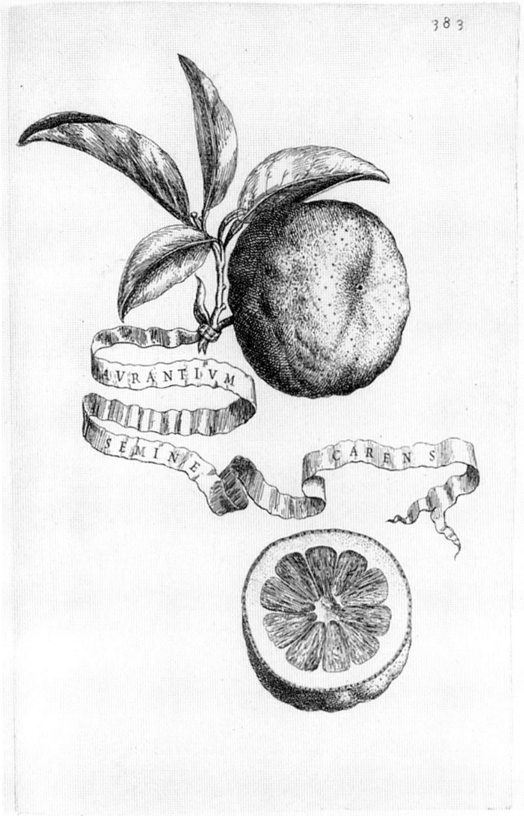

383

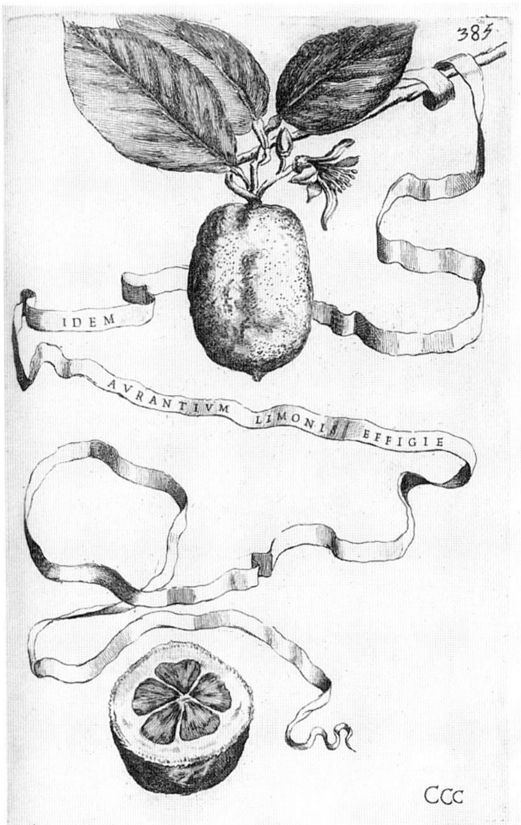

385

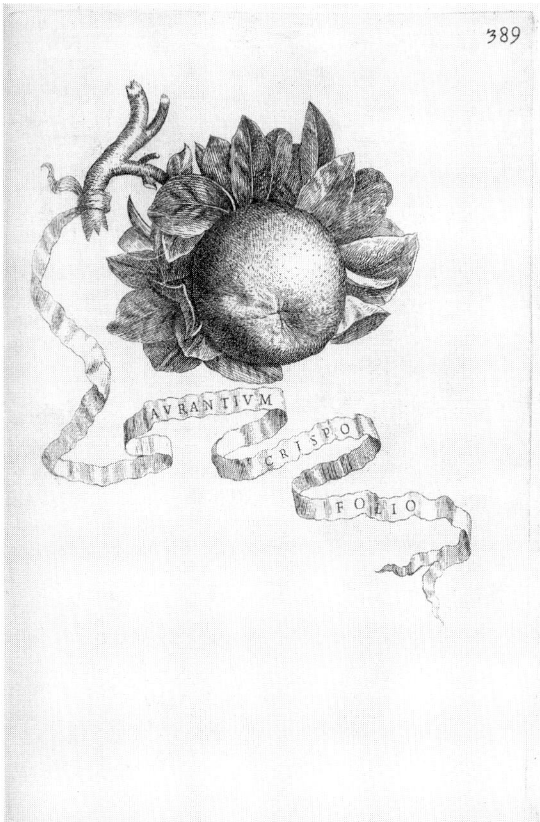

389

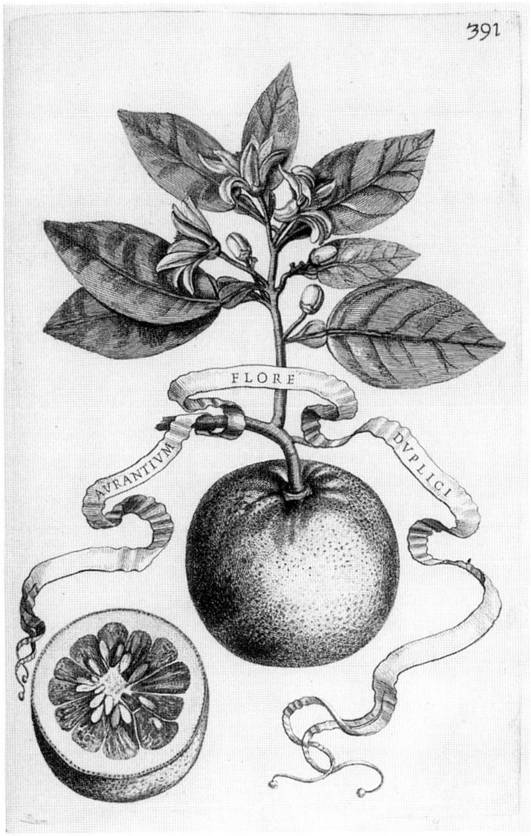

391 **31**

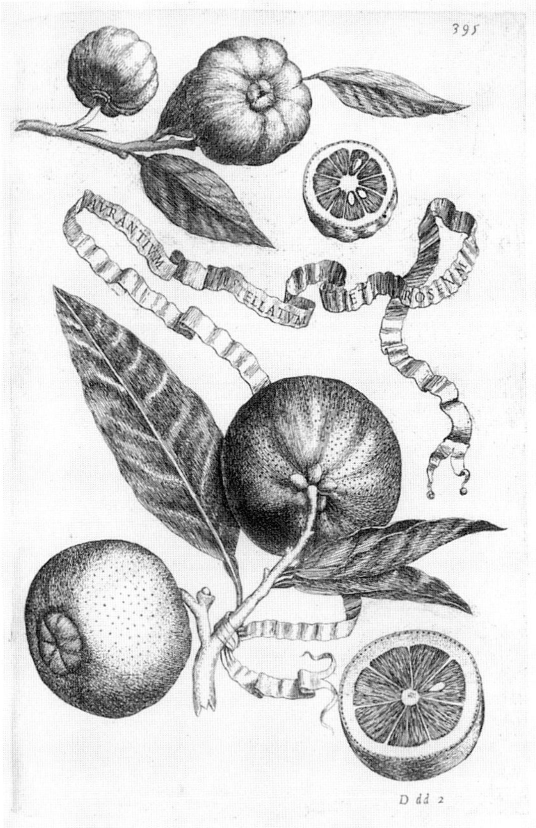

395

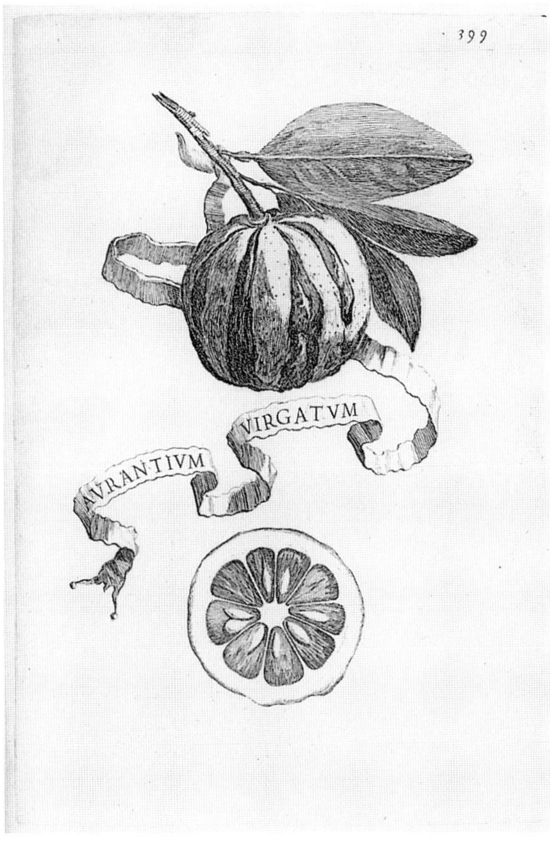

399

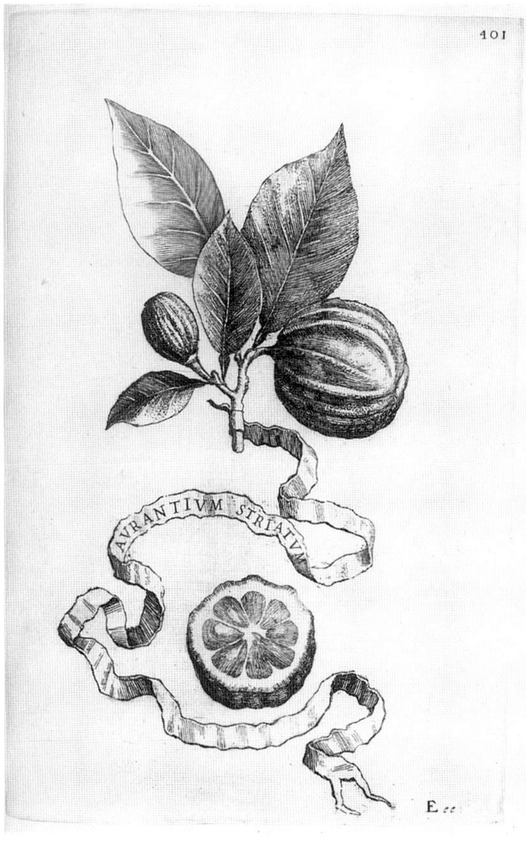

401 **100**

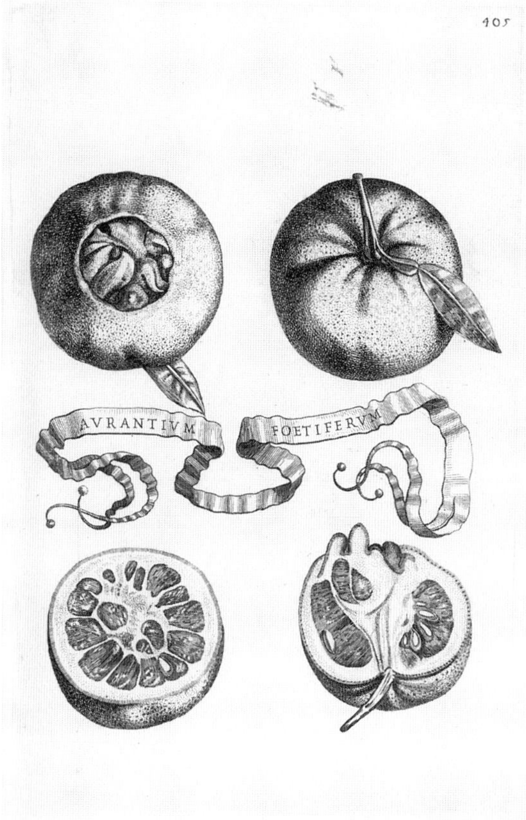

405 **95**

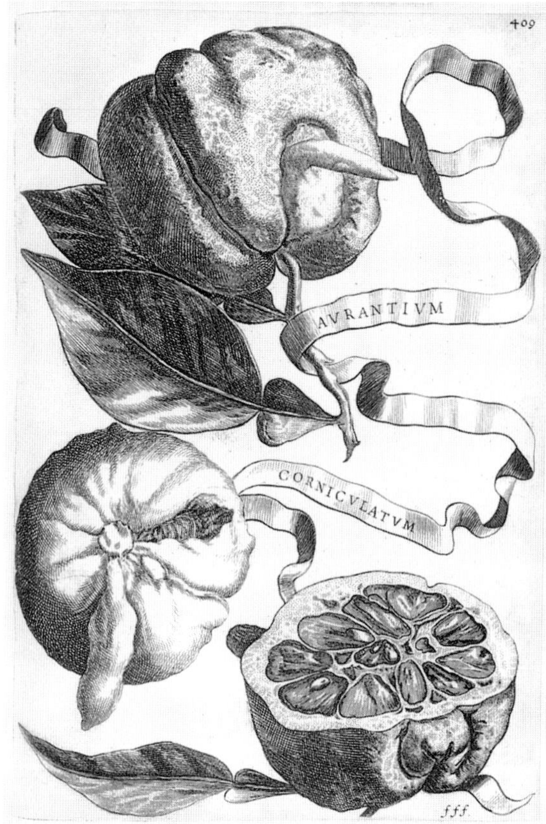

409

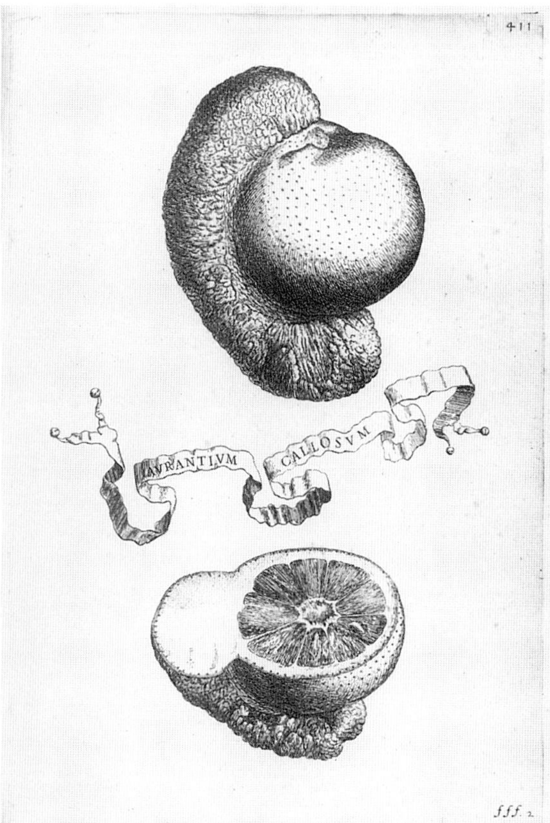

411 **101, 102**

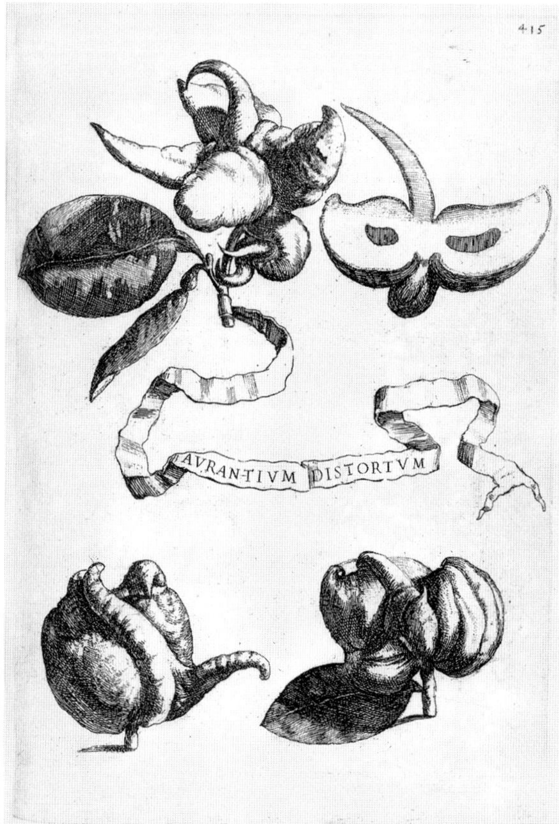

415

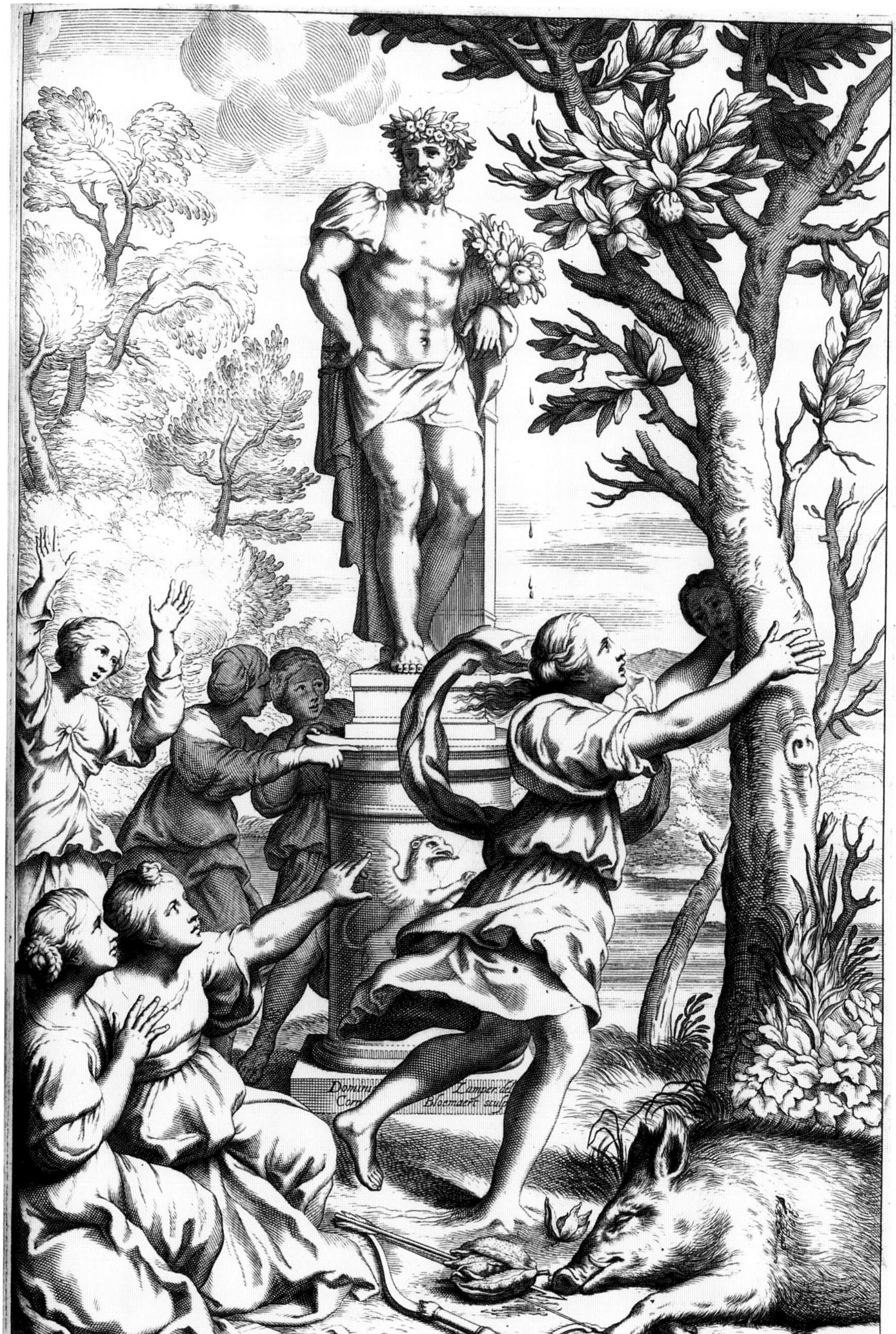

418. The Transformation of Leonilla (see Fig. 30, p. 75)

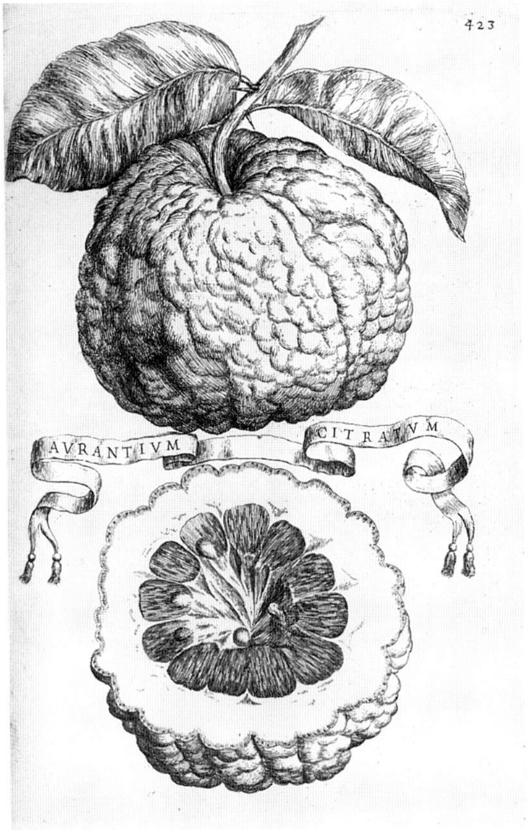

423 **40**

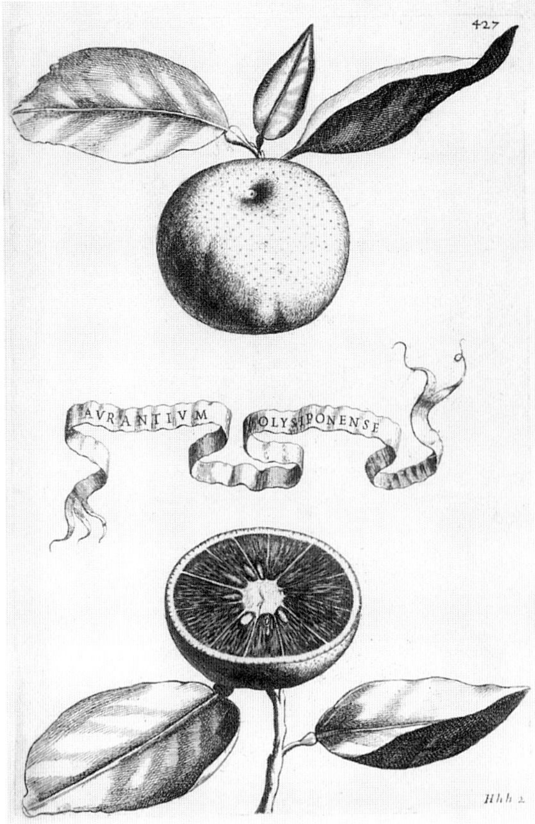

427
32, 33

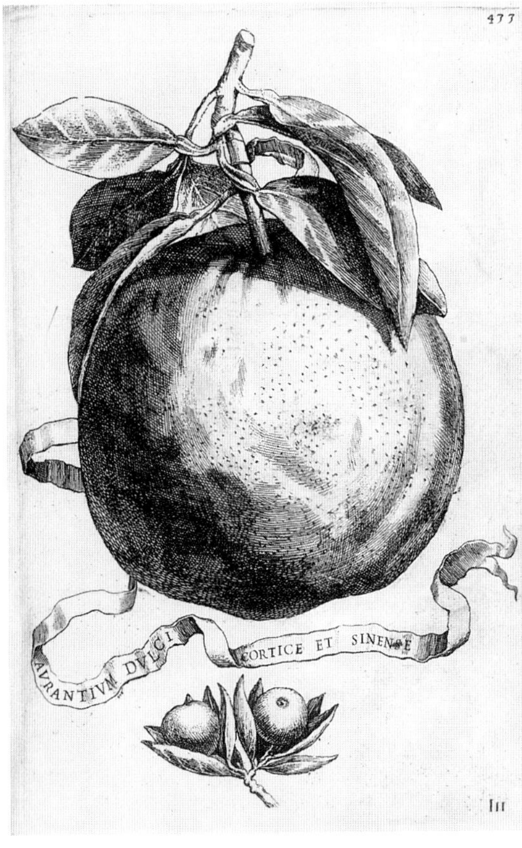

433
34, 63

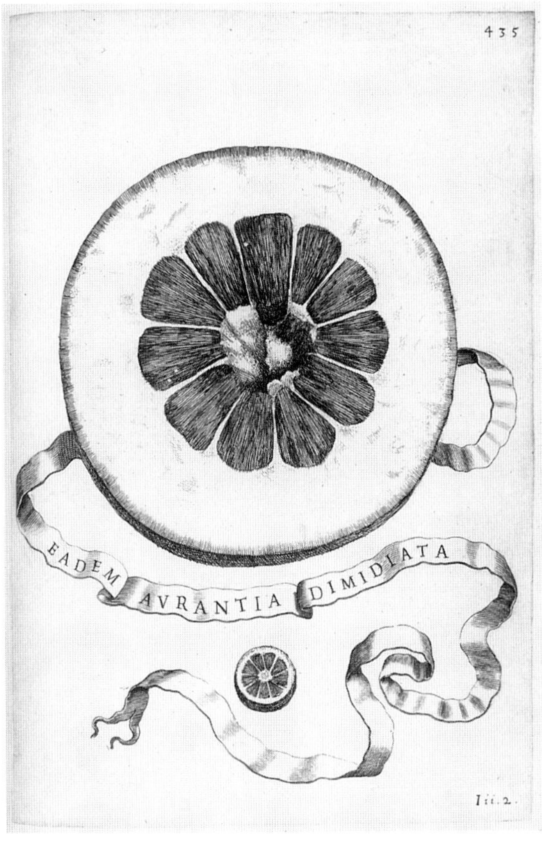

435
35, 64

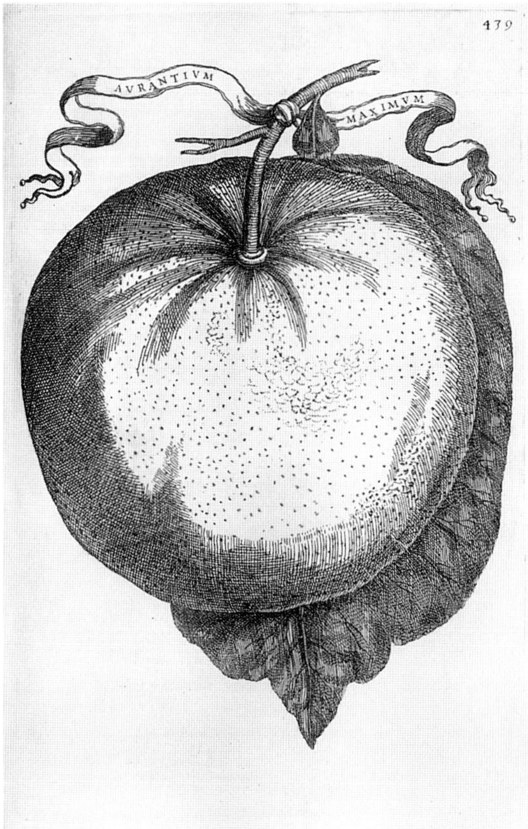

439
38

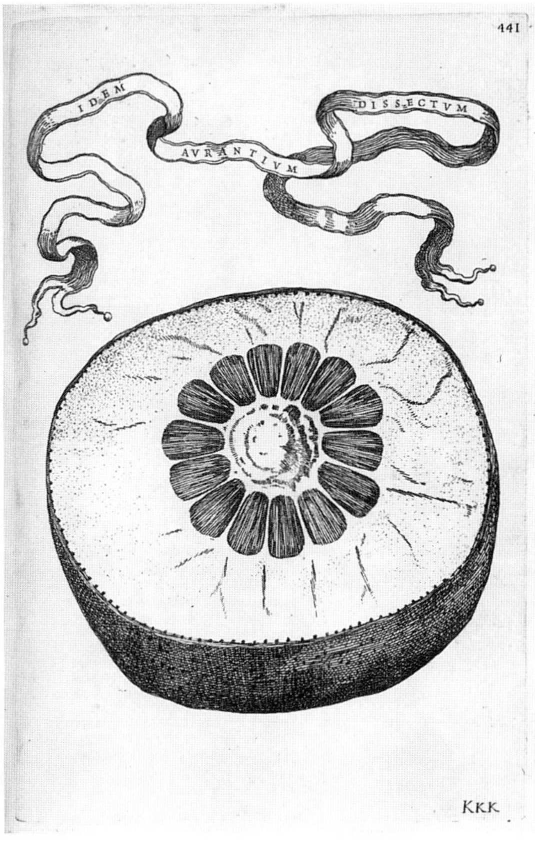

441 **39**

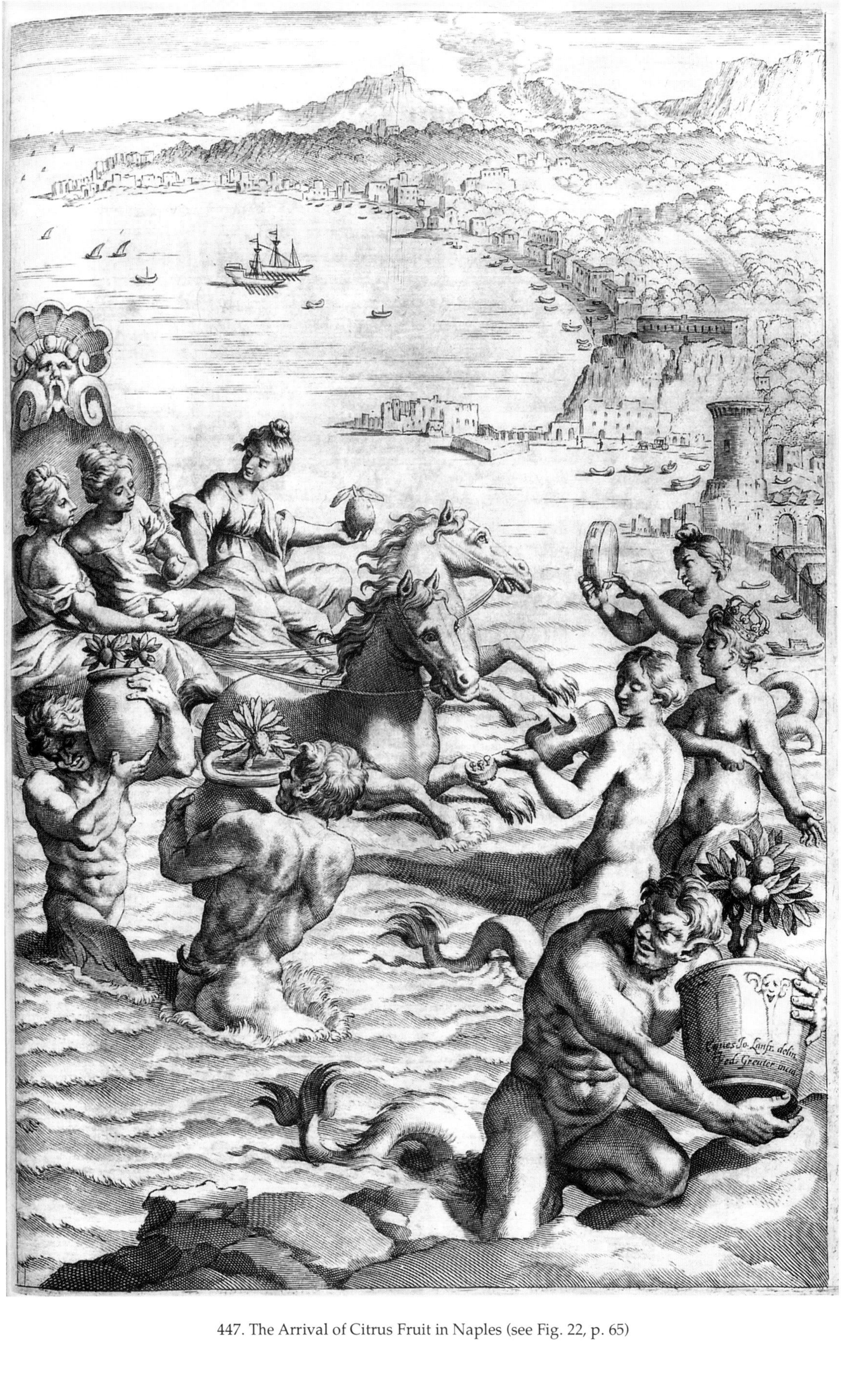

447. The Arrival of Citrus Fruit in Naples (see Fig. 22, p. 65)

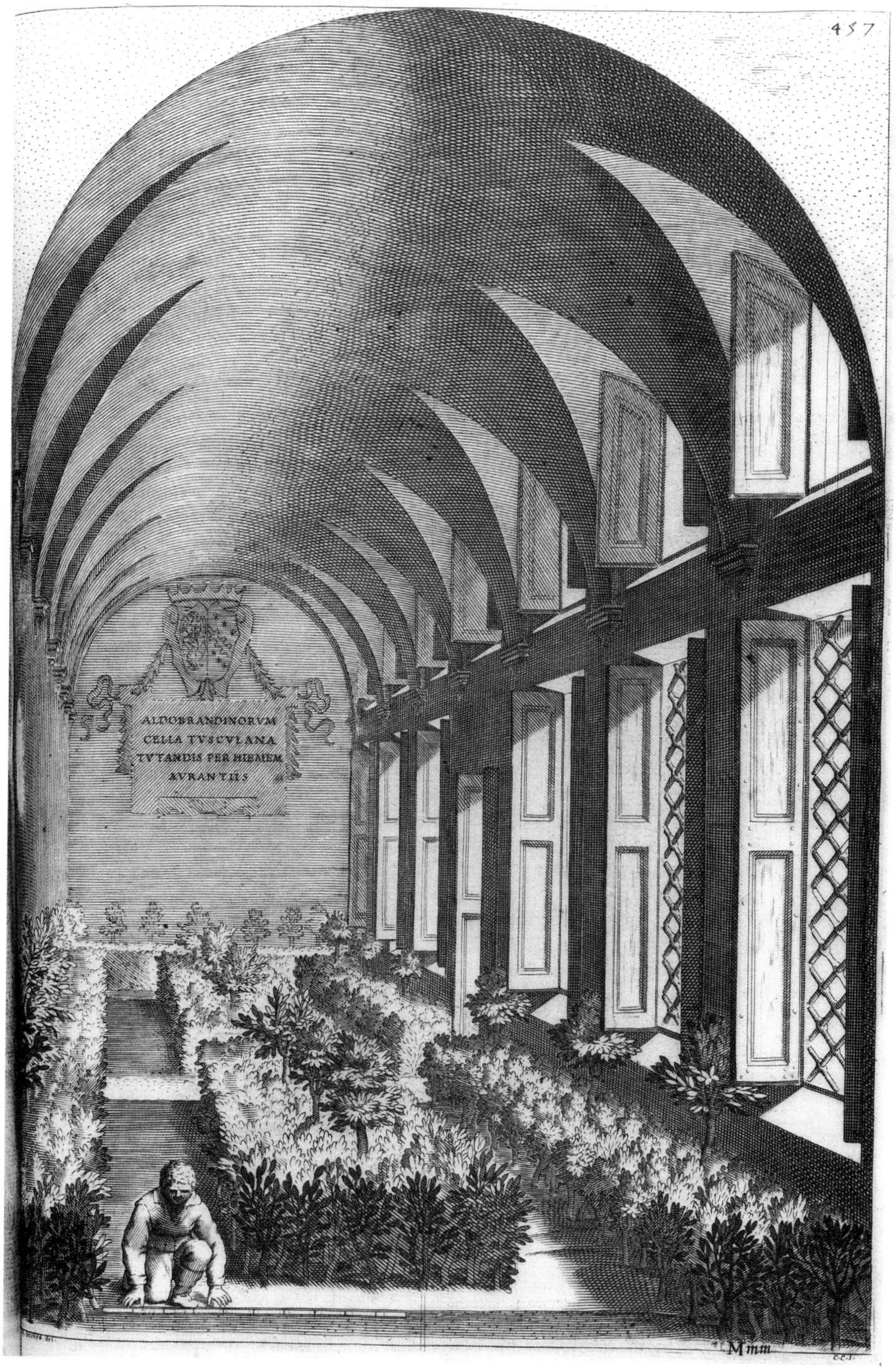

457. The Aldobrandini Orangery (see Fig. 23, p. 67).

CONCORDANCE A

Engravings in Ferrari 1646 and numbers of drawings in this volume

Page no. in Ferrari 1646	Cat. no.
61	1
63	1
67	2 & 3
77	72
79	73
189	18
193	19
199	43
215	20, 21 & 81
223	44
230	22 & 23
233	45
241	46
243	47
253	24 & 25
265	48
267	49
269	89 & 90
271	91 & 92
281	50 & 51
283	52 & 53
285	26 & 27
287	54
295	55
307	56 & 57
311	58
315	93 & 94
323	59 & 60
325	61 & 62
391	31
401	100
405	95
411	101 & 102
427	32 & 33
433	34 & 63
435	35 & 64
439	38
441	39 & 40

Cat. no.	Page no. in Ferrari 1646
1	61 & 63
2 & 3	67
18	189
19	193
20, 21 & 81	215
22 & 23	230
24 & 25	253
26 & 27	285
31	391
32 & 33	427
34 & 63	433
35 & 64	435
38	439
39	441
43	199
44	223
45	233
46	241
47	243
48	265
49	267
50 & 51	281
52 & 53	283
54	287
55	295
56 & 57	307
58	311
59 & 60	323
61 & 62	325
72	77
73	79
89 & 90	269
91 & 92	271
93 & 94	315
95	405
100	401
101 & 102	411

CONCORDANCE B

Royal Library inventory numbers and Catalogue Raisonné numbers of drawings in this volume

RL. no.	Catalogue no.
19321	76
19322	75
19323	16
19323A	17
19324	80
19324A	79
19325	114
19326	113
19327	94
19328	93
19329	101
19330	102
19331	32
19332	33
19333	41
19334	1
19335	19
19336	43
19337	25
19338	24
19339	66
19340	96
19341	51
19342	50
19345	100
19347	73
19348	65

RL. no.	Catalogue no.
19349	27
19350	26
19351	69
19352	28
19353	55
19354	115
19355	4
19356	86
19357	87
19358	88
19359	74
19360	5
19361	6
19362	90
19363	89
19364	48
19366	103
19367	104
19368	107
19372	54
19379	81
19380	82
19381	83
19382	84
19383	85
19386	116
21146	95

BIBLIOGRAPHY AND ABBREVIATIONS

Institutions

BAV	Vatican, Biblioteca Apostolica
BUB	Bologna, Biblioteca Universitaria
DBF	Dipartimento di Botanica, Università di Firenze
GAG	Genoa, Gallesio Archive (private collection)
Kew	London, Royal Botanic Gardens Library
London, BL	London, British Library
London, BM	London, British Museum
London, NHM	London, Natural History Museum
London, V&A	London, Victoria & Albert Museum
Montpellier, BEM	Montpellier, Bibliothèque, École de Médecine
Naples, BN	Naples, Biblioteca Nazionale
Paris, BN	Paris, Bibliothèque Nationale
Paris, IF	Paris, Institut de France
RL	Windsor, Royal Library
Rome, ARSI	Rome, Archivium Romanum Societatis Jesu
Rome, BAL	Rome, Biblioteca dell'Accademia Nazionale dei Lincei

Books, articles and manuscripts

Aldrovandi I.P.	U. Aldrovandi, *Iconographia Plantarum*, 10 volumes, BUB, Museo Aldrovandiano, *Tavole di piante, fiori e frutti*, not dated.
Aldrovandi MS 88	BUB, MS 124 (Fondo Aldrovandi), no. 88, not dated.
Archivio Dal Pozzo	*Il Carteggio di Cassiano dal Pozzo*, Rome BAL MS, 39 vols. For a catalogue of this see Nicolò 1991.
Ashby 1913	T.A. Ashby, 'Addenda and corrigenda to "Sixteenth-century Drawings of Roman buildings attributed to Andreas Coner"', *Papers of the British School at Rome* vi, 1913, pp. 184–210.
Ashpitel 1862	A. Ashpitel, *Catalogue of the Architectural MSS and Drawings in the Royal Library at Windsor Castle*, 1862, unpublished MS, Print Room, Royal Library.
Baldini 1982	E. Baldini, F. Rossi, V. Bassi, E. Giannini, 'Agrumi', *Agrumi, frutta e uve nella Firenze di Bartolomeo Bimbi pittore mediceo*, E. Baldini (ed.), Florence 1982, pp. 17–44.
Baldini 1986	E. Baldini, *Arboricoltura Generale*, Bologna 1986.
Baldini 1989	E. Baldini, 'Simulacri, meraviglie, prodigi e mostruosità nella Dendrologia aldrovandiana e nell'interpretazione scientifica moderna', *Rendiconti dell' Accademia delle Scienze dell'Istituto di Bologna, Classe di Scienze Fisiche* XIV, V, 1987–8, pp. 145–70.
Baldini 1990	E. Baldini 'Fruits and fruit trees in Aldrovandi's *Iconographia Plantarum*', *Advances in Horticultural Science* 4, 1990, pp. 61–73.

Baldini 1991 — E. Baldini, 'Polimorfismo e teratologia dei frutti nel genere Citrus: riscontri storici e attualità biologiche', in *Rendiconti dell'Accademia delle Scienze dell'Istituto di Bologna, Classe di Scienze Fisiche* XIV, VI, 1988–9, pp. 127–61.

Baldini 1996a — E. Baldini, 'Gli agrumi di Giorgio Gallesio: scritti e disegni inediti', *Il giardino delle Esperidi. Gli agrumi nella storia, nella letteratura e nell'arte*, A. Tagliolini and M. Azzi Visentini (eds.), Florence, 1996, pp. 59–90.

Baldini 1996b — E. Baldini, *L'Atlante citrografico di Giorgio Gallesio*, Florence 1996.

Baldini/Rossi 1983 — E. Baldini and F. Rossi, 'Gli agrumi nelle 'raccolte' medicee del XVII secolo', Proceedings of the International Congress on *Evoluzione del quadro varietale ed organizzazione della ricerca nell'agrumicoltura mediterranea*, 22–23 February 1983, pp. 157–70. Paternò 1983.

Baldini/Tagliaferri 1990 — E. Baldini and C. Tagliaferri, 'Matrici inedite dell'iconografia dendrologica di Ulisse Aldrovandi', *Memorie dell'Accademia delle Scienze dell'Istituto di Bologna, Classe di Scienze Fisiche*, 22 May 1990. Bologna 1990, pp. 1–78.

Baldini/Tosi 1994 — E. Baldini and A. Tosi, *Scienza e Arte nella Pomona Italiana di Giorgio Gallesio*, Florence 1994.

Baronio 1588 — C. Baronio, *Annales ecclesiastici*, 12 vols., Rome 1588–1607.

Bean 1959 — J. Bean, *Dessins romains du XVIIe siècle. Artistes italiens contemporains de Poussin. XXXIIIe exposition du Cabinet des Dessins*, exhib. cat., Paris 1959.

Bellori 1664 — G. P. Bellori, *Nota delli Musei, Librarie, Gallerie, & Ornamenti di Pitture ne' Palazzi, nelle Case e ne'Giardini di Roma*, Rome 1664.

Bénézit — E. Bénézit, *Dictionnaire critique et documentaire des Peintres, Sculpteurs, Dessinateurs et Graveurs...*, 8 vols., Paris 1948.

Bimbi, Villa Med. — Bartolomeo Bimbi painting in Villa Medicea, Poggio a Caiano, Florence (inventory number).

Bimbi, DBF — Bartolomeo Bimbi painting in the Museo Botanico, University of Florence (inventory number).

Biographie Universelle — *Biographie Universelle*, J.F. and L.G. Michaud (eds.), 85 vols., Paris 1811–62.

Blunt 1945 — A. Blunt, *The French Drawings ... at Windsor Castle*, Oxford/London 1945.

Blunt 1964 — *Nicolas Poussin, Lettres et propos sur l'art*, A. Blunt (ed.), Paris 1964.

Blunt 1965 — A. Blunt, 'Poussin and his Roman patrons', in *Walter Friedlaender zum 90. Geburtstag*, Berlin 1965, pp. 58–75.

Blunt 1971 — A. Blunt, 'Italian drawings: addenda and corrigenda', in E. Schilling, *The German drawings in the collection of Her Majesty the Queen at Windsor Castle*, London 1971, pp. 47–206.

Boone sale cat. — Property from the Estate of James R. Herbert Boone (sale cat.), Sotheby's, New York, 16 & 17 September 1988, lots 125–82.

Boyce/Korsmeyer 1941 — A.M. Boyce and R.B. Korsmeyer, 'The citrus mite *Eriophyes sheldoni* Ewing', *Journal of Economic Entomology* 34, 1941, pp. 745–56.

Briquet — C.M. Briquet, *Les filigranes: dictionnaire historique des marques du papier dès leur apparition vers 1282 jusqu'en 1600*, facsimile of the 1907 edition, Allan Stevenson (ed.), Amsterdam 1968.

Brummitt/Powell — R.K. Brummitt and C.E. Powell, *Authors of Plant Names*, Kew 1992.

Calabrese 1990 — F. Calabrese, 'La favolosa storia degli agrumi', *Agricoltura* 208, 1990, pp. 811–28.

Campanella 1631 — T. Campanella, *Atheismus Triumphatus Seu Reductio ad Religionem per Scientiarum Veritates*, Rome 1631.

Casella 1935 — D. Casella, 'Primo contributo sul polimorfismo del frutto in piante legnose. Il polimorfismo del frutto di alcune specie del gen. Citrus', *Annali della Stazione Sperimentale di Frutticoltura e Agrumicoltura di Acireale* 2, 1935, pp. 165–76.

Cassiano cat. 1993 — *The Paper Museum of Cassiano dal Pozzo*, exhib. cat., British Museum, London 1993. *Quaderni puteani* 4, Milan (Olivetti) 1993.

Cato 1677 — R. Cato, *L'agricoltura e cose di villa di Carlo Stefano gentil'huomo francese*, Venice 1677.

Clarici 1726 — P.B. Clarici, *Istoria e coltura delle piante che sono nel fiore più ragguardevoli e più distinte per ornare un giardino in tutto il tempo dell'anno, con un copioso trattato sugli agrumi*, Venice 1726.

Clusius 1601 — C. Clusius, *Rariorum Plantarum Historia*, Antwerp 1601.

Commelin 1676 — J. Commelin, *Nederlantze Hesperides, dat is Oeffening en gebruik van de Limoen- en oranje-boomen; Gestelt na den Aardt, en Climaat der Nederlanden*, Amsterdam 1676.

Continella/Tribulato 1992 — G. Continella and E. Tribulato, 'Germoplasma di cedro, pummelo e loro ibridi presente nell'Italia meridionale', Proceedings of the Congress on *Germoplasma frutticolo: salvaguardia e valorizzazione delle risorse genetiche*, 21–25 September 1992. Alghero 1992, pp. 437–53.

Cupani 1696 — F. Cupani, *Hortus Catholicus*, Naples 1696.

Dati 1664 — C. Dati, *Delle lodi del Commendatore Cassiano dal Pozzo*, Florence 1664.

Davies/Albrigo 1994 — F.S. Davies and L.G. Albrigo, *Citrus*, Wallingford 1994.

DBI — *Dizionario Biografico degli Italiani*, vols. 1– Rome 1960– (in progress).

De Backer/Sommervogel — A. and A. De Backer and C. Sommervogel (eds.), *Bibliothèque de la Compagnie de Jésus*, 12 vols., Paris-Brussels, 1890–1932.

Della Porta 1558 — J.B. della Porta, *Magiae naturalis libri XX*, Naples 1558.

Dempsey 1989 — C. Dempsey, 'Poussin's "Sacrament of Confirmation", the scholarship of "Roma Sotterranea", and dal Pozzo's Museum chartaceum', in Solinas 1989a, pp. 244–61.

Deng *et al.* 1995 — Z.N. Deng, A. Gentile, E. Nicolosi, G. Continella, and E. Tribulato, 'Caratterizzazione del germoplasma di agrumi: dalle descrizioni degli antichi citrologi all'analisi del DNA', *Italus Hortus* 5–6, 1995, pp. 68–79.

De Noter 1926 — R. De Noter, *Les orangers, citronniers, cédratiers et autres aurantiacées à fruit commestible*, Paris 1926.

Documentary Culture — *Documentary Culture in Florence and Rome from Grand-Duke Ferdinand I to Pope Alexander VII*, Papers from a colloquium held at the Villa Spelman, Florence 1990, E. Cropper, G. Perini, F. Solinas (eds.), Florence 1992.

Dodonaeus 1616 — R. Dodonaeus, *Stirpium Historiae Pemptades Sex sive Libri XXX*, Antwerp 1616.

Duhamel du Monceau 1768 — H.L. Duhamel du Monceau, *Traité des arbres fruitiers, contenant leur figure, déscription, leur culture, etc.*, Paris 1768.

Evelyn 1662 — John Evelyn, *Sculptura or the History and Art of Chalcography and Engraving in Copper*, London 1662.

Evreinoff 1946 — V.A. Evreinoff, 'Les fruits doubles et leur origine chez le pêcher', *Revue Horticole*, new series (XXX, 10) 2136, 1946, pp. 171–4.

Ewing 1937 — H.E. Ewing, 'A new Eriophys mite from lemon trees (Acarina: Eriophydeae)', *Proceedings of the Entomological Society* 39, Washington 1937, pp. 193–7.

Fabiani 1872 — G. Fabiani, *La Vita di Pietro Andrea Mattioli raccolta dalle sue opere, con aggiunte e annotazioni di Luciano Fabiani*, Siena 1872.

Ferrari 1622 — J.B. Ferrari, *Nomenclator Syriacus*, Rome 1622.

Ferrari 1633 — J.B. Ferrari, *De Florum Cultura*, Rome 1633.

Ferrari 1638 — J.B. Ferrari, *Flora ovvero cultura di fiori*, Rome 1638.

Ferrari 1646 — J.B. Ferrari, *Hesperides sive De Malorum Aureorum Cultura et Usu Libri Quattuor*, Rome 1646.

Fleming 1958 — J. Fleming, 'Cardinal Albani's drawings at Windsor: their purchase by James Adam for George III', *The Connoisseur* cxlii, 1958, pp. 164–9.

Foucault 1973 — M. Foucault, *The Order of Things. An Archaeology of the Human Sciences*, New York 1973. (A translation of *Les mots et les choses*, Paris 1966; originally published in the United States in 1971.)

Freedberg 1989a — D. Freedberg, 'Cassiano dal Pozzo's Drawings of Citrus Fruits', *Quaderni puteani 1*, pp. 10–36.

Freedberg 1989b — D. Freedberg, 'From Hebrew and Gardens to Oranges and Lemons: Giovanni Battista Ferrari and Cassiano dal Pozzo', Solinas 1989a, pp. 37–72.

Freedberg 1992a — D. Freedberg, 'Cassiano on the Jewish Races of Rome', *Quaderni puteani 3*, pp. 41–56.

Freedberg 1992b — D. Freedberg, 'Ferrari on the Classification of Oranges and Lemons', *Documentary Culture*, pp. 287–306.

Freedberg 1993 — D. Freedberg, 'Cassiano and the Art of Natural History', *Cassiano* cat. 1993, pp. 141–54.

Freedberg 1994 — D. Freedberg, 'The Failure of Colour', *Sight and Insight. Essays on Art and Culture in Honour of E.H. Gombrich*, ed. John Onians, London 1994, pp. 245–62.

Freedberg 1996a — D. Freedberg, 'Ferrari and the pregnant lemons of Pietrasanta', *Il giardino delle Esperidi. Gli agrumi nella storia, nella letteratura e nell'arte*, A. Tagliolini and M. Azzi Visentini (eds.), Florence, 1996, pp. 41–58.

Freedberg 1996b — D. Freedberg, 'Poussin, Ferrari, Cortona, and the *Aetas Florea*', *Poussin Actes* 1996.

Galileo 1623 — G. Galilei, *Il Saggiatore …*, Rome 1623.

Galileo 1632 — G. Galilei, *Dialogo … sopra i due massimi sistemi del Mondo tolemaico e copernicano*, Florence 1632.

Gallesio 1811 — G. Gallesio, *Traité du Citrus*, Paris 1811.

Gallesio 1839 — G. Gallesio, *Gli agrumi dei giardini botanico-agrarii di Firenze*, Florence 1839.

Gallo 1578 — A. Gallo, *Giornate dell'agricoltura e de' piaceri della villa*, Venice 1578.

Goldman 1978 — J. Goldman, *Aspects of Seicento patronage - Cassiano dal Pozzo and the amateur tradition*, unpublished PhD thesis, University of Chicago 1978.

Griffiths 1989 — A. Griffiths,'The Print Collection of Cassiano dal Pozzo', *Print Quarterly* VI, no. 1, 1989, pp. 2–10.

Haskell 1963 — F. Haskell, *Patrons and Painters: A Study in the Relations between Italian Art and Society in the Age of the Baroque*, London 1963.

Haskell 1981 — F. Haskell, 'La dispersione e conservazione del patrimonio artistico', *Storia dell'arte italiana* 10, F. Zeri (ed.), Turin 1981, pp. 5–35.

Haskell/Rinehart — F. Haskell and S. Rinehart, 'The Dal Pozzo Collection. Some New Evidence', *Burlington Magazine* CII, 1960, pp. 318–29.

Harris 1977 — A.S. Harris, *Andrea Sacchi. Complete Edition of the Paintings with a Critical Catalogue*, Oxford 1977.

Heawood — E. Heawood, *Watermarks mainly of the 17th and 18th centuries*, Hilversum 1950.

Hernandez 1651 — F. Hernandez et al, *Rerum Medicarum Novae Hispaniae Thesaurus, seu plantarum, animalium, mineralium Mexicanorum historia ...*, Rome 1651.

Hodgson 1961 — R. W. Hodgson, 'Classification and nomenclature of citrus fruits', *California citrograph* 46, 1961, pp. 256–8.

Inventory A — Manuscript inventory preserved in the Print Room, Royal Library, dating from the end of the eighteenth century; drawn up by or under the direction of George III's librarian, Richard Dalton.

Inzenga 1875 — G. Inzenga, 'Origine singolare di una nuova varietà di arancio (*Citrus Aurantium Panormitanum*)', *Atti dell'Accademia delle Scienze, Lettere ed Arti* V, Palermo 1875, pp. 25–8.

Jaffé 1988 — D. Jaffé, *Rubens Self-Portrait*, exhib. cat., National Gallery of Australia, Canberra 1988.

Jaffé 1989 — D. Jaffé, 'The Barberini circle: some exchanges between Peiresc, Rubens, and their contemporaries', *Journal of the History of Collections* 1, 1989, pp. 119–47.

Jenkins 1987 — I. Jenkins, 'Cassiano dal Pozzo's Museo Cartaceo: new discoveries in the British Museum', *Nouvelles de la République des Lettres* 1987 (II), pp. 29–41.

Jenkins 1989a — I. Jenkins, 'Newly discovered drawings from the Museo Cartaceo in the British Museum', in Solinas 1989a, pp. 131–6, 141–75.

Jenkins 1989b — I. Jenkins, 'The "Mutilated Priest" of the Capitoline Museum and a drawing from Cassiano dal Pozzo's "Museo Cartaceo"', Burlington Magazine CXXXI, 1989, pp. 543–9.

Jenkins 1992 — I. Jenkins, 'Pars pro toto: a muse from the Paper Museum', *Quaderni puteani 2*, pp. 49–65.

Jonston 1662 — J. Jonston, *Historia naturalis de arboribus et fruticibus tam nostris quam peregrini orbis libri decem*, Frankfurt 1662.

Lanzoni 1690 — G. Lanzoni, *Citrologia, seu Curiosa Citri Descriptio*, Ferrara 1690

Leroy 1953 — J.F. Leroy, 'La navelisation chez le bigaradier. Les malformations des fleurs et des fruits chez les agrumes. Aspects scientifiques de ces faits', *Revue Internationale de Botanique Appliquée* 33, 1953, pp. 414–22.

Letarouilly 1840 — P. M. Letarouilly, *Édifices de Rome moderne*, 6 vols., Liège 1840.

Lhote/Joyal 1989 — *Nicolas-Claude Fabri de Peiresc. Lettres à Cassiano dal Pozzo (1626–1637)*, J.-F. Lhote and D. Joyal (eds.), Clermont Ferrand 1989.

L'immagine e Natura — *L'immagine e Natura. L'immagine naturalistica nei codici e libri a stampa delle biblioteche Estense e Universitaria. Secoli XV–XVII* (exhib. cat.), Modena 1984.

Loiseleur-Deslongchamps 1819 — J.L.A. Loiseleur-Deslongchamps, *Nouveau Duhamel* V, Paris 1819.

Lumbroso 1874 — G. Lumbroso, 'Notizie sulla vita di Cassiano Dal Pozzo', *Miscellanea di Storia Italiana* xv, 1874, pp. 129–388.

Lumbroso 1875 — G. Lumbroso, *Notizie sulla vita di Cassiano Dal Pozzo* (offprint of Lumbroso 1874) Rome 1875.

McBurney 1989a — H. McBurney, 'History and contents of the dal Pozzo collection in the Royal Library, Windsor Castle', in Solinas 1989a, pp. 74–93.

McBurney 1989b — H. McBurney, 'A Short History of the Museo Cartaceo', in *Quaderni puteani 1*, pp. 5–9.

McBurney 1989c — H. McBurney, 'Cassiano's Drawings of Birds', *Quaderni puteani 1*, pp. 37–47.

McBurney 1989d — H. McBurney, 'The later history of Cassiano dal Pozzo's "Museo Cartaceo"', *Burlington Magazine* CXXXI, 1989, pp. 549–53.

McBurney 1992a — H. McBurney, 'Cassiano dal Pozzo as Scientific Commentator: Ornithological Texts, and Images of Birds from the *Museo Cartaceo*', *Documentary Culture*, pp. 349–62.

McBurney 1992b — H. McBurney, 'Cassiano dal Pozzo as Ornithologist', *Quaderni puteani 3*, pp. 3–22.

McBurney 1993 — H. McBurney, 'A Museum on Paper', *The Antique Collector*, April 1993.

Markovitch 1926 — B.B. Markovitch, 'Classification of the genus Citrus', *Landbouw. Buitzenzorg Giava* 2 (4), 1926, pp. 27–61.

Marsili — L.F. Marsili, *Misc. Rerum Natur.*, BUB MS 4:18, not dated.

Matthioli 1582 — P.A. Matthioli, *Dei Discorsi di M. Pietro Andrea Matthioli Sanese....Nelli sei libri di Pedacio Anazarbeo della materia medicinale*, 5 vols., Venice 1582.

Merz — Jörg Martin Merz, *Pietro da Cortona. Der Aufstieg zum führenden Maler im barocken Rom* (*Tübinger Studien zur Archäologie und Kunstgeschichte* 8), Tübingen 1991.

Michaelis 1874	A. Michaelis, 'Die Privatsammlungen antiker Bildwerke in England', *Archäologische Zeitung* xxxi, 1874, pp. 1–70.
Michaelis 1882	A. Michaelis, *Ancient Marbles in Great Britain*, trans. C. Fennell, Cambridge 1882.
Michel 1816	E. Michel, *Traité du Citronnier*, Paris 1816.
Micheli *Enumeratio*	P.A. Micheli, *Enumeratio quarundam plantarum sibi per Italiam et Germaniam observatarum iuxta Tournefortii methodum dispositarum* DBF MS 48, vol. IX, 1733–6.
Montalbani 1668	O. Montalbani, *Ulyssis Aldrovandi Dendrologiae naturalis scilicet arborum historia libri duo*, Bologna 1668.
Montalbani 1671	O. Montalbani, *Curae analiticae aliquot naturalium observationum Aldrovandicas circa historias*, Bologna 1671.
Montalenti 1960	G. Montalenti, DBI II.
Montfaucon *Ant. Exp.*	B. de Montfaucon, *L'Antiquité Expliquée et Représentée en Figures*, 5 vols. (in 10 parts), Paris 1719, and 5 supplementary vols., Paris 1724.
Montfaucon/Galliano 1987	B. de Montfaucon and A. Galliano, *Voyage en Italie — Diarium Italicum: un journal en miettes. Edizione critica*, Geneva 1987.
Mosin 1973	V. Mosin, *Anchor watermarks*, Amsterdam 1973.
Nati 1674	P. Nati, *Florentina phytologica observatio de Malolimonia citrataaurantia, vulgo Bizzarria*, Florence 1674.
Nicolò/Solinas 1987	A. Nicolò and F. Solinas, 'Cassiano dal Pozzo: appunti per una cronologia di documenti e disegni (1612–1630)', *Nouvelles de la République des Lettres* 1987 (II) pp. 59–110.
Nicolò 1991	A. Nicolò, *Il Carteggio di Cassiano dal Pozzo: Catalogo*, Florence 1991.
Olina 1622	G.P. Olina, *Uccelliera, overo discorso della natura, e proprietà di diversi uccelli ...*, Rome 1622.
Olmi 1978	G. Olmi, *Ulisse Aldrovandi. Scienza e natura nel secondo Cinquecento*, Trent 1978.
Olmi 1992	G. Olmi, 'L'inventario del mondo. Catalogazione della natura e luoghi del sapere nella prima età moderna', *Annali dell'istituto storico italo-germanico* Monografia 17, Bologna 1992.
Pena/de Lobel 1576	P. Pena and M. de Lobel, *Nova Stirpium Adversaria*, Antwerp 1576.
Penzig 1887	O. Penzig, *Studi botanici sugli agrumi e sulle piante affini*, Rome 1887.
Petau 1630	D. Petau, *Ad Iuliani opera praefatio in qua de Iuliani ipsius, et uniuerse de gentilium libris*, Paris 1630.
Petrucci 1985	F. Petrucci Nardelli, 'Il Card. Francesco Barberini Senior e la Stampa a Roma', *Archivio della Società Romana di Storia Patria*, Rome 1985.
Phillips cat. 1990	Phillips, Son and Neale, sale catalogue: Old Master Drawings, 12 December 1990, lots 219–374, London.
Popham/Wilde 1949	A.E. Popham and J. Wilde, *The Italian Drawings of the XV and XVI centuries in the Collection of His Majesty The King at Windsor Castle*, London 1949.
Poussin cat. Paris 1994	*Nicolas Poussin (1594–1665)*, P. Rosenberg and L-A. Prat (eds.), exhib. cat., Galeries Nationales du Grand Palais, Paris 1994.
Poussin cat. London 1995	*Nicolas Poussin (1594–1665)*, R. Verdi (ed.), exhib. cat., Royal Academy of Arts,London 1995.
Poussin Actes 1996	*Actes du Colloque Poussin*, A. Mérot (ed.), Paris 1996.
Quaderni puteani 1	*Il museo cartaceo di Cassiano dal Pozzo, Cassiano naturalista. Quaderni puteani 1*, Milan (Olivetti) 1989.
Quaderni puteani 2	*Cassiano dal Pozzo's Paper Museum Volume I* (Proceedings of a conference held at the British Museum and Warburg Institute, London, 14–15 December 1989). *Quaderni puteani 2*, Milan (Olivetti) 1992.

Quaderni puteani 3 — *Cassiano dal Pozzo's Paper Museum Volume II* (Proceedings of a conference held at the British Museum and Warburg Institute, London, 14–15 December 1989). Quaderni puteani 3, Milan (Olivetti) 1992.

Quaderni puteani 4 — *Cassiano* cat.1993

Reuther/Batchelor/Webber — *The Citrus industry*, W. Reuther, L.D. Batchelor and H.J. Webber (eds.), 2 vols., Berkeley 1948, 2nd edition 1967–8.

Rinehart 1960 — S. Rinehart, 'Poussin et la famille dal Pozzo', in *Nicolas Poussin. Actes de Colloques Internationaux*, Paris 1960, pp. 19–30.

Rinehart 1961 — S. Rinehart, 'Cassiano dal Pozzo (1588–1657): some unknown letters', *Italian Studies*, 1961, pp. 35–59.

Risso/Poiteau 1818 — A. Risso and A. Poiteau, *Histoire naturelle des orangers ...*, Paris 1818.

Risso/Poiteau 1872 — A. Risso and A. Poiteau, *Histoire et Culture des Orangers. Nouvelle édition entièrement revue et augmentée d'un chapitre nouveau sur la culture dans le midi et en Algérie, par A. Du Breuil*, Paris 1872.

Robert 1890 — C. Robert, *Der Pasiphae-Sarkophag* (14 Hallisches Winckelmannsprogramm), Halle 1890.

Robert *ASR* — C. Robert, *Die antiken Sarkophag-reliefs*, 3 vols. Berlin 1890–1919.

Rumpf 1750 — G. E. Rumpf, *Herbarium Amboinense*, vol. VI, Amsterdam 1750.

Russo 1986 — F. Russo, 'Tassonomia del genere *Citrus* e dei generi affini interessanti la coltivazione', *Trattato di Agrumicoltura*, Paolo Spina (co-ordinator), Bologna 1986.

Savastano 1884 — L. Savastano, 'Le forme teratologiche del fiore e del frutto degli agrumi', *Annali Regia. Scuola Superiore di Agricoltura di Portici* 1884, p. 512.

Savastano/Parrozzani 1911 — L. Savastano and A. Parrozzani, 'Di alcuni ibridi naturali degli agrumi', in *Annali della Stazione Sperimentale di Frutticoltura e Agrumicoltura di Acireale* I, Acireale 1911, pp. 37–63.

Scora 1988 — R.W. Scora, 'Biochemistry, taxonomy and evolution of modern cultivated citrus', *Proceedings of the International Society of Citriculture* 1, 1988, pp. 277–89.

Skippon — P. Skippon, 'An account of a journey made thro' part of the Low-Countries, Germany, Italy and France', published in A. Churchill, *A Collection of Voyages and Travels*, London 1732, vi, pp. 359–736 (partially reprinted in Blunt 1965, pp. 71–2).

Solinas 1989a — *Cassiano dal Pozzo. Atti del Seminario Internationale di Studi. Napoli, 18–19 decembre 1987*, F. Solinas (ed.), Rome 1989.

Solinas 1989b — F. Solinas, 'Percorsi puteani: note naturalistiche ed inediti appunti antiquari', in Solinas 1989a, pp. 95–129.

Solinas 1989c — F. Solinas, 'L'Erbario Miniato e altri fogli di iconografia botanica appartenuti a Cassiano dal Pozzo' in *Quaderni puteani 1*, pp. 52–76.

Solinas 1992 — F. Solinas, 'Sull'atelier di Cassiano dal Pozzo: metodi di recerca e documenti inediti', in *Quaderni puteani 3*, pp. 57–76.

Sparti 1990a — D. Sparti, 'Carlo Antonio dal Pozzo (1606–1689): an unknown collector', *Journal of the History of Collections* 2, 1990, pp. 7–19.

Sparti 1990b — D. Sparti, 'The Dal Pozzo Collection again: the inventories of 1689 and 1695 and the family archive', *Burlington Magazine* CXXXII, 1990, pp. 551–70.

Sparti 1992 — D. Sparti, *Le collezioni dal Pozzo: storia di una famiglia e del suo museo nella Roma seicentesca*, Modena 1992.

Spike 1983 — J.T. Spike, *Italian still life paintings from three centuries*, exhib. cat., National Academy of Design, New York 1993.

Stelluti 1638 — F. Stelluti, *Trattato del Legno Fossile Minerale Nuovamente Scoperto*, Rome 1638.

Sterbeek 1682 — F. van Sterbeek, *Citricultura*, Antwerp 1682.

Stirling-Maxwell 1891 — W. Stirling-Maxwell, *Miscellaneous Essays and Addresses*, London 1891.

Swingle/Reece 1967 W.T. Swingle and C. Reece, 'The botany of citrus and its relatives' in Reuther/Batchelor/Webber, pp. 190–430.

Tanaka 1927 T. Tanaka, 'Bizzarria, a clear case of periclinal chimera', *Journal of Genetics* 18, 1927, pp. 77–85.

Tanaka 1961 T. Tanaka, *Citrologia*, Osaka 1961.

Targioni Tozzetti 1825 A. Targioni Tozzetti, *Raccolta di fiori, frutti e agrumi disegnati dal vero da vari artisti*, Florence 1825.

Targioni Tozzetti 1853 A. Targioni Tozzetti, *Cenni storici sulla introduzione di varie piante nell'agricoltura e nell'orticoltura toscane*, Florence 1853.

Tenore 1844 M. Tenore, 'Ricerche sull'arancio fetifero', *Memorie di Matematica e Fisica della Società Italiana delle Scienze* XXII, Modena 1844, pp. 185–94.

Thieme-Becker *Allgemeines Lexikon der bildenden Künstler von der Antike bis zur Gegenwart*, U. Thieme and F. Becker (eds.), 37 vols., Leipzig 1907–1950.

Tilney-Bassett 1986 R.A.E. Tilney-Bassett, *Plant chimeras*, London 1986.

Tillson/Bamford 1938 A.M. Tillson and R. Bamford, 'The floral anatomy of the Aurantioideae', *American Journal of Botany* 25, 1938, pp. 780–3.

Tolkowsky 1938 S. Tolkowsky, *Hesperides, a History of the Culture and Use of Citrus Fruits*, London 1938.

Tongiorgi Tomasi/Ferrari L. Tongiorgi Tomasi and A. Ferrari, 'Botanica Barocca', in *Gazzetta del bibliofilo* ['semestrale fuori commercio riservati agli abbonati a *FMR*'] V, 17 (Autumn/Winter), 1986.

Turner 1992 N. Turner, 'The drawings of Pietro Testa after the Antique in Cassiano dal Pozzo's Paper Museum', in *Quaderni puteani* 3, pp. 127–44.

Turner 1993 N. Turner, 'Some of the copyists after the Antique employed by Cassiano', in *Cassiano* cat., pp. 27–37.

Vermeule 1955 C. C. Vermeule, 'Notes on a new edition of Michaelis', *American Journal of Archaeology* 59, 1955, pp. 129–150.

Vermeule 1956 C. C. Vermeule, 'The Dal Pozzo-Albani Drawings of Classical Antiquities: Notes on their content and arrangement', *Art Bulletin* XXXVIII, 1936, pp. 31–46.

Vermeule 1958 C. C. Vermeule, 'Aspects of Scientific Archaeology in the Seventeenth Century: marble reliefs, Greek vases, and minor objects in the Dal Pozzo-Albani drawings of classical antiquity', *Proceedings of the American Philosophical Society* 102, part 11, 1958.

Vermeule 1960 C. C. Vermeule, 'The Dal Pozzo-Albani Drawings of Classical Antiquities in the British Museum', *Transactions of the American Philosophical Society* 50, part 5, 1960.

Vermeule 1966 C. C. Vermeule, 'The Dal Pozzo-Albani Drawings of Classical Antiquities in the Royal Library at Windsor Castle', *Transactions of the American Philosophical Society* 56, part 2, 1966.

Vermeule 1976 C. Vermeule, 'Classical Archaeology and French Painting in the Seventeenth Century', *Bulletin of the Museum of Fine Arts Boston* lxxiv, No.370, 1976, pp. 94–109.

Volkamer *Nürnberg. Hesp.* C. Volkamer, *Nürnbergische Hesperides, oder Gründliche Beschreibung der Edlen Citronat-Citronen und Pomaranzen Früchte*, 2 vols., Nuremberg 1708 and 1714.

Woodward D. Woodward, *Catalogue of Watermarks in Italian Printed Maps ca. 1540–1600*, Chicago 1996

Zonghi Aurelio & Augusto Zonghi and A.F. Gasparinetti, *Zonghi's Watermarks*, Hilversum 1953.

LIST OF FIGURES

TEXT FIGURES

COMPARATIVE FIGURES

PHOTOGRAPHIC ACKNOWLEDGEMENTS

All catalogued items in the Royal Collection are reproduced by permission of Her Majesty The Queen. We are grateful for permission to reproduce items in the following collections (numbers in bold type are catalogue numbers): Bologna, Biblioteca Universitaria (by permission of the Ministero per i Beni Culturali e Ambientali): Figs. 32, 40; Cambridge, Fitzwilliam Museum: Fig. 4; Florence, Dipartimento di Botanica, Università di Firenze: Fig. 24; Genoa, Gallesio Archives (private collection): Fig. 25; Kew, Royal Botanic Gardens Library: Figs. 35, 36, 37, 38, 41, 44, 45, Comp. figs. 1a–b, 2–3, 18–23, 26–27, 31, 34–35, 43–62, 73, 93–94, 100, Glossary p. 284, pp. 288–9, 292–9, 301–6, 308, 310–12, 314–15; London, British Library: Figs. 13, 14, 20, 21, 22, 23, 31 and p. 286 (shelfmark: 35.g.2), 17 (shelfmark: 34.b.8), 29, 30 (shelfmark: 450.i.8), 26, 27, 42 (shelfmark: 459.b.6(1)); London, Courtauld Institute of Art: Figs. 15, 16; London, V&A Picture Library: **34, 35, 36, 70, 108, 109, 110, 111, 112**; Pisa, Archivio di Stato: Fig. 2, (S.733, n. 9, ref. 2.10.96/3113, no. 10/96); Rome, Biblioteca dell'Accademia Nazionale dei Lincei: Figs. 5, 18, 19; St Petersburg, Hermitage Museum: Fig. 3; private collectors: **13, 18, 31, 37, 44, 46, 47, 49, 63, 64, 72, 91, 92, 97, 98, 99**; W. Graham Arader III, New York: **117, 118**; Mr and Mrs Richard L. Guest, Norwalk, Connecticut: **58**; Enrico Baldini: Figs. 33, 34, 39, 43, Glossary pp. 281–4.

We are also grateful to Sotheby's, New York, for permission to reproduce photographs of drawings included in their sale on 16 September 1988 in the following catalogue entries: **10–11, 14–15, 20–23, 29–30, 39–40, 46–47, 49, 52–53, 56–57, 59–62, 67–68, 71, 77, 91–92, 117–118** (all ©1988 Sotheby's, Inc.).

BOTANICAL INDEX

Numbers in **bold** indicate catalogue numbers

GENERAL INDEX

Illustration numbers are given in *italic*, catalogue numbers in **bold**.
The names of owners indexed here include only those who acquired drawings through the London art market early in the twentieth century.